A CARIBBEAN FOREST TAPESTRY

LONG-TERM ECOLOGICAL RESEARCH NETWORK SERIES
LTER Publications Committee

Grassland Dynamics: Long-Term Ecological Research in Tallgrass Prairie
Editors: Alan K. Knapp, John M. Briggs, David C. Hartnett, and Scott L. Collins

Standard Soil Methods for Long-Term Ecological Research
Editors: G. Philip Robertson, David C. Coleman, Caroline S. Bledsoe, and Phillip Sollins

Structure and Function of an Alpine Ecosystem: Niwot Ridge, Colorado
Editors: William D. Bowman and Timothy R. Seastedt

Climate Variability and Ecosystem Response at Long-Term Ecological Sites
Editors: David Greenland, Douglas G. Goodin, and Raymond C. Smith

Biodiversity in Drylands: Toward a Unified Framework
Editors: Moshe Shachak, James R. Gosz, Steward T.A. Pickett, and Avi Perevolotsky

Long-Term Dynamics of Lakes in the Landscape: Long-Term Ecological Research on North Temperate Lakes
Editors: John J. Magnuson, Timothy K. Kratz, and Barbara J. Benson

Alaska's Changing Boreal Forest
Editors: F. Stuart Chapin III, Mark W. Oswood, Keith Van Cleve, Leslie A. Viereck, and David L. Verbyla

Structure and Function of a Chihuahuan Desert Ecosystem: The Jornada Basin Long-Term Ecological Research Site
Editors: Kris M. Havstad, Laura F. Huenneke, and William H. Schlesinger

Principles and Standards for Measuring Net Primary Production in Long-Term Ecological Studies
Editors: Timothy J. Fahey and Alan K. Knapp

Agrarian Landscapes in Transition: Comparisons of Long-Term Ecological and Cultural Change
Editors: Charles L. Redman and David R. Foster

Ecology of the Shortgrass Steppe: A Long-Term Perspective
Editors: William K. Lauenroth and Ingrid C. Burke

A Caribbean Forest Tapestry

The Multidimensional Nature of Disturbance and Response

Edited by

NICHOLAS BROKAW, TODD A. CROWL,
ARIEL E. LUGO, WILLIAM H. MCDOWELL,
FREDERICK N. SCATENA, ROBERT B. WAIDE,
AND MICHAEL R. WILLIG

LTER

OXFORD
UNIVERSITY PRESS

OXFORD
UNIVERSITY PRESS

Oxford University Press, Inc., publishes works that further
Oxford University's objective of excellence
in research, scholarship, and education.

Oxford New York
Auckland Cape Town Dar es Salaam Hong Kong Karachi
Kuala Lumpur Madrid Melbourne Mexico City Nairobi
New Delhi Shanghai Taipei Toronto

With offices in
Argentina Austria Brazil Chile Czech Republic France Greece
Guatemala Hungary Italy Japan Poland Portugal Singapore
South Korea Switzerland Thailand Turkey Ukraine Vietnam

Published by Oxford University Press, Inc.
198 Madison Avenue, New York, New York 10016

www.oup.com

Oxford is a registered trademark of Oxford University Press

Library of Congress Cataloging-in-Publication Data
A Caribbean forest tapestry : the multidimensional nature of disturbance and response / edited by
Nicholas Brokaw . . . [et al.].
p. cm.
Includes bibliographical references and index.
ISBN 978-0-19-533469-2 (hardcover : acid-free paper)
1. Forest ecology—Puerto Rico—Luquillo Mountains. 2. Forest ecology—Caribbean Area.
3. Forest ecology—Tropics. 4. Ecological disturbances—Puerto Rico—Luquillo Mountains.
5. Adaptation (Biology)—Puerto Rico—Luquillo Mountains. 6. Biotic communities—Puerto Rico—Luquillo
Mountains. 7. Forest management—Puerto Rico—Luquillo Mountains.
8. Forest conservation—Puerto Rico—Luquillo Mountains. 9. Luquillo Mountains (P.R.)—Environmental
conditions. I. Brokaw, Nicholas V. L.
QH109.P6C37 2012
577.3097295—dc23 2011035972

Dedication

This book is dedicated to two influential individuals who pioneered the fields of tropical forestry and tropical ecosystem ecology. Frank H. Wadsworth, a forester, helped to institutionalize tropical forestry in the Neotropics by organizing forest services and training foresters in several countries, establishing the regional research journal *Caribbean Forester*, and helping in the development of the Latin American Forestry Commission of the Food and Agriculture Organization some 50 years ago. Howard T. Odum, an ecologist, revolutionized the study and interpretation of tropical forests with the application of thermodynamics to ecosystem analysis and the use of large-scale studies such as the giant cylinder to study the metabolism of tropical forests. The approach and focus of this book rests on the shoulders of these two exceptional scientists who dedicated a considerable portion of their careers to understanding the functioning of the ecosystems of the Luquillo Mountains (LM).

Wadsworth was transferred to Puerto Rico by the U.S. Forest Service in 1942, and Odum first visited the LM in 1944 to learn tropical meteorology as a 2nd Lieutenant of the U.S. Army. Among the many tropical forestry issues that Wadsworth addressed throughout his career in Puerto Rico, three are immediately relevant to this book.

He collaborated with Elbert L. Little, Jr., and Roy O. Woodbury in the description of all tree species of Puerto Rico; he established what is now the oldest network of tree-growth plots in the Neotropics; and he manipulated forest basal areas in order to test ideas about tropical forest management. Wadsworth was interested in understanding tree growth in the tropics and in developing prescriptions for their management.

Odum returned to the LM in 1957 on a Rockefeller grant, and with colleagues such as Frank B. Golley, he began studying the metabolism of mangroves and wet tropical forests, including their biomass and carbon sequestration. By 1962, Odum had developed whole ecosystem models with data collected in the LM and an energy language that changed the way ecologists analyzed tropical forests. During these visits, he and Wadsworth discussed fundamental issues of tropical forest management, as is evident from the following quotation from Odum (1962:66):

> The preliminary calculations [of the organic matter flux] provide a possible solution to one question troubling Dr. Frank Wadsworth and associates, tropical foresters managing this forest. The growth rate of trees measured over 20 years has been small, 0.05 to 0.12 inches per year. The dominant trees are several hundred years old. Is this slow growth due to lack of light, lack of nutrients, or inadequate photosynthesis for other reasons? The calculation of respiration as 9 g m^{-2} day^{-1} due to leaves and 5.8 due to the soil, root, and litter indicates very little production is left for any net growth with most of it being used to sustain leaf and soil activity. The apparent reason for slow growth is thus not any inhibition of gross photosynthesis, but the full development of the ecosystem structure requiring most of the production for respiratory maintenance.

Odum's work in Sabana evolved into the Rain Forest Radiation Study, funded by the U.S. Atomic Energy Commission and hosted by the University of Puerto Rico (UPR). This study is recognized as the first example of a "big science" ecosystem project in the tropics, a harbinger of the Long Term Ecological Research (LTER) program. The outcomes of the collaboration between Wadsworth and Odum were many but are highlighted by the allocation of 180 acres of National Forest lands for exclusive research use by the UPR; the use of Forest Service facilities in support of UPR research activities, including the library and headquarters of what was then Institute of Tropical Forestry; and the sharing of information and ideas about tropical forests. These outcomes and the research infrastructure created by both Wadsworth and Odum became the backbone of the Luquillo LTER program, the results of which are summarized in this book.

However, as important as the research infrastructure has been to the LTER program, it was the ideas that emerged from the friendship between Wadsworth and Odum that constitutes their greatest legacy, a legacy that is so evident in this book. This intellectual legacy was cemented in their independent minds, their focus on experimentation at the ecosystem level, and their openness to innovation. They both understood the tropical forest as a system without ignoring the importance of its parts. It is not an accident that the Luquillo LTER has been successful in the integration of population and ecosystem ecology. Both Wadsworth and Odum successfully supported population ecology research while also maintaining a whole-system perspective and fomenting whole-system research. They both had a worldview and understood that science is a vehicle for helping resolve conservation issues and for addressing human needs. Such ideas are evident in the books in which they independently culminated their career experiences in Puerto Rico (Odum 1971; Wadsworth 1997). This book extends their points of view and celebrates the intellectual synergy that they displayed between 1963 and 1989.

Literature Cited

Odum, H. T. 1962. Man and the ecosystem. Pages 57–75 in P. E. Waggoner and J. D. Ovington, editors, *Proceedings of the Lockwood conference on the suburban forest and ecology*. New Haven, CT: Connecticut Agricultural Experiment Station.

Odum, H. T. 1971. Environment, power and society. New York: Wiley Interscience.

Wadsworth, F. H. 1997. Forest production for tropical America. USDA Forest Service Agriculture Handbook 710. Washington, DC: USDA Forest Service.

Contents

Preface

Studying Disturbance and Response to Understand Environmental Change

Tropical forests have long fascinated and intrigued scientists, arguably catalyzing the development of the major unifying theme in biology—evolution—as well as contributing the empirical observations that accelerated the maturation of ecology as a discipline and motivated legions of students to become ecologists. The chapters in this book reflect that same scientific fascination with the tropics, whether defined geographically (in which case the Luquillo Mountains is tropical) or climatologically (in which case the Luquillo Mountains is subtropical or lower montane), blending empirical and theoretical pursuits and channeling them to address some of the greatest environmental challenges of the 21st century.

Grand Challenges

Change is the main theme of the "Grand Challenges" for environmental science identified by the National Research Council (NRC) (2001). These challenges include alterations in biodiversity, alterations in biogeochemical cycles, climate change and climatic variability, and coupled human-natural ecosystems. The NRC identified these themes as grand challenges because environmental change has profound consequences for humans, including socioeconomic, ecological, aesthetic, and ethical issues. The Luquillo Long-Term Ecological Research (LTER) program anticipated the concerns formalized in the NRC (2001) report and continues to respond to these grand challenges. Consequently, our book is organized around the leitmotif of understanding environmental change as it relates to disturbance and successional responses in a tropical forest ecosystem. This is because characterizing the disturbance regime of a system, including the patterns and mechanisms of response to the suite of interacting disturbances, contributes to the predictive understanding of ecological dynamics and facilitates the development of informed management strategies (e.g., Willig and Walker 1999).

Long-term research focusing on ecological change is particularly appropriate in the Caribbean Basin, a region characterized by high cyclonic activity (chapter 2), where biotic composition and structure have been molded over evolutionary time by a disturbance regime dominated by hurricanes, and, over the past half-millennium, by increasing anthropogenic disturbances as well. Arguably, global warming will increase the number or intensity of tropical storms and hurricanes in the region (Goldenberg et al. 2001; Webster et al. 2005). At the same time, the Caribbean is experiencing a drying trend (i.e., a negative precipitation anomaly), which might be related to global warming or might represent long-term variation in rainfall (Neelin et al. 2006; chapter 3; chapter 4). Finally, the Caribbean is a global hot spot of biodiversity and an area of conservation concern, characterized by high endemism, high human population density, fragmented landscapes, and a diversity of socioecological systems. Within this context, understanding environmental change as the spatiotemporal dynamics of ecosystem structure and function is particularly germane to the future of human societies.

Changing Science and Changing Scientists

Our scientific investigation of disturbance and response emerged as a consequence of a long history of previous research in the Luquillo Mountains of Puerto Rico, and from our formal integration within the LTER Network. Our motivation to understand environmental change integrated the research efforts of a diverse group of disciplinary scientists, including population biologists, community ecologists, ecosystem scientists, foresters, landscape ecologists, and geoscientists, and gave rise to both short- and long-term empirical studies, adaptive monitoring, manipulative experimentation, and modeling efforts. Taken together, these endeavors dramatically changed our perspective of the forest as a tropical system operating at dynamic equilibrium or steady state, leading us to view it instead as a dynamic multidimensional forest that is constantly changing at a variety of spatial and temporal scales. Not only did it change the kind of science that we conduct, but our involvement in the LTER Program changed our very nature as scientists, transforming us from somewhat narrow disciplinarians to transdisciplinary scientists committed to integrating various fields of environmental science in the pursuit of understanding the dynamics of change in a complex tropical system.

In order to foresee and manage environmental change, we need to know how and why it happens. As a consequence, we focus on the drivers of change and the patterns of response, including aspects of resistance and resilience as they relate to populations, communities, and biogeochemical processes. This comprehensive approach inherently requires a long-term perspective, as these different environmental aspects of a system do not resist change in the same manner and do not share the same tempo or mode of response (Zimmerman et al. 1996; chapter 5). Often, a system's dynamics are best explored via experiments. In the Luquillo Mountains, we study the effects of natural and human disturbances as nonmanipulative "experiments" that help us to understand the complexities of environmental change. Alternatively, we employ long-term monitoring, as well as short-term and long-term manipulative experiments (e.g., chapter 8), to decouple confounded factors, such as the inputs of organic matter and the alteration of microclimate (light, temperature, humidity), that mold successional trajectories after hurricanes. In general, disturbances alter "background" patterns and processes, and studies of responses to disturbance reveal mechanisms that promote change or enhance resistance or resilience. In many cases, disturbances play a critical role in determining the structure of communities and the functioning of ecosystems, with concomitant effects on the delivery of ecosystem goods and services. As such, studies of disturbance and response must be long term and enacted at multiple spatial scales (Hobbie et al. 2003), so as to match the long-term nature of change and of environmental processes, as well as to reflect the hierarchy of spatial scales (e.g., cross-scale dynamics; Willig et al. 2007) that affect ecosystem dynamics. Consequently, in an attempt to understand the spatial and temporal dynamics of a tropical ecosystem over the long term and at multiple spatial scales, we initiated an LTER program in the Luquillo Mountains of Puerto Rico.

Coming Attractions

Since its inception, the Luquillo LTER program has pioneered research that integrates a suite of disturbances and responses at multiple scales and durations in the tropics (Waide and Lugo 1992). The Luquillo Mountains of Puerto Rico are an ideal laboratory for studying disturbance, response, and long-term environmental change. Indeed, there is a rich background of natural history and environmental research to inform contemporary studies (e.g., Odum and Pigeon 1970; Lugo and Lowe 1995; Reagan and Waide 1996). The first chapter of this book describes this breadth and depth of work and traces the evolution of environmental concepts in Puerto Rico from the island's first inhabitants through the steady-state focus of early ecologists, and on to the present dynamical view. That long evolution of concepts has contributed to an effective framework that helps us understand a changing environment (chapter 2). This framework helps us grasp how species composition and ecosystem processes vary across a landscape in relation to underlying patterns of the environment and to present and past processes that control environmental variation. The natural setting of the Luquillo Mountains and its constituent ecosystems have been well described with respect to many aspects of the abiotic and biotic environment, including studies of populations, communities, and biogeochemical processes (chapter 3). Particularly useful for understanding environmental change are the environmental gradients that are induced by elevation in the Luquillo Mountains. Along these gradients, climate change in space hints at possible consequences of climate change in time. Likewise, the land use gradient from San Juan (1.3 million inhabitants) to El Yunque National Forest (Luquillo Mountains) highlights the consequences of expanding urbanization and afforestation.

The natural and human disturbance regime in the Luquillo Mountains is multifaceted, including elements that range from single treefalls to hurricanes, or from the construction of dams to urbanization (chapter 4). Studies of these elements of the disturbance regime are particularly relevant for a changing world, in which the coupling of natural and human systems, such as in the Luquillo Mountains, is complex and characterized by multiple feedback loops. The responses to this diversity of disturbances have been well studied at the levels of populations, communities, and biogeochemical process, and from scales that range from the plot to the landscape or regional level (chapter 5). Many different response trajectories characterize environmental conditions, populations, communities, and biogeochemical processes. These responses are not the consequences of just one element of the disturbance regime; they are the dynamic outcome of multiple interacting disturbances. Many examples of resistance and resilience to disturbance illustrate stabilizing mechanisms, whereas the novel conditions imposed by human disturbance suggest the potential for divergence from well-described successional pathways, sometimes leading to the emergence of new ecosystems. The synthesis of research on disturbance and response permits an understanding within an evolutionary framework of how population dynamics account for ecosystem responses to disturbance (chapter 6). In this regard, we show how various species and taxonomic groups interact to affect dynamical responses to the disturbance regime.

Clearly, research in the Luquillo Mountains relates to a broad spectrum of environmental problems that characterize much of the globe (chapter 7). As elsewhere, climate change, land use change, and introduced species combine to alter the dimensions of biodiversity and the dynamics of biogeochemical processes in inextricably coupled natural and human ecosystems. Our long-term studies of environmental change inform the development of a comprehensive research platform, bolstered by both empirical and theoretical constructs, to understand the tempo and mode of change at multiple scales (chapter 8). Such studies will ensure that environmental science is in the vanguard of efforts to manage changing tropical environments.

We end by emphasizing the pivotal role of the Luquillo Mountains in the LTER Network of sites. Indeed, the Luquillo Mountains LTER site represents the Neotropical node in the Network, acting as the anchor for comparative research concerning a number of salient environmental gradients such as precipitation (high), temperature (high), and biodiversity (high). Moreover, our research has been, is, and will continue to be integral in addressing the environment–society interactions identified by the LTER Network's Integrative Science for Society and the Environment initiative (U.S. LTER 2007). Since the inception of the Luquillo LTER program, we have investigated the extent to which disturbances, be they short-term or long-term events, affect the structure and functioning of terrestrial and aquatic ecosystems in the Luquillo Mountains. We have increasingly extended our focus beyond tabonuco forest to include all montane ecosystems of the Luquillo Mountains, and we have more recently extended our research domain into the lowlands, including suburbanizing and urbanizing ecosystems from the base of the mountains to the highly urban coastal environs in the city of San Juan. Similarly, we have increasingly emphasized that all of the ecosystems that we study are coupled human and natural systems that provide numerous and vital services and goods to human society. We will continue to embrace multidisciplinary perspectives and collaborative interactions as a way to leverage long-term environmental research to deepen scientific understanding of complex, dynamic, and evolving living systems, and to use it to inform management and conservation action.

Acknowledgments

The partnership of the University of Puerto Rico, the USDA Forest Service, and the National Science Foundation has acted as a platform and catalyst for our integrated research and educational activities in the Luquillo Long-Term Ecological Research Program. Foremost, we acknowledge substantive financial support over the past quarter-century from the National Science Foundation (BSR-8811902, DEB-9411973, DEB-0080538, DEB-9705814, DEB-0218039, and DEB-0620910) to the Institute for Tropical Ecosystem Studies, University of Puerto Rico, and to the International Institute of Tropical Forestry, USDA Forest Service, as part of the Luquillo Long-Term Ecological Research Program. The USDA Forest Service and the University of Puerto Rico gave additional substantive support. In particular, significant financial, infrastructural, and logistical support was provided by the Institute for Tropical Ecosystem Studies (formerly the Center for Energy and Environment Research and Terrestrial Ecology Division) and the International Institute of Tropical Forestry. The administrative, technical, and support staff of both institutions facilitated our work in countless ways. Support from El Yunque National Forest (previously known as the Caribbean National Forest) and the USGS-WEBB program is noteworthy as well. The expositional clarity and scientific rigor of the entire book were improved by W. Dodds and T. Seastedt, who reviewed earlier versions of all of the chapters.

Additional support of various kinds enhanced the content of each chapter in this book. We briefly highlight such support hereinafter. We gratefully acknowledge assistance from M. Alayón, N. Fetcher, G. González, E. Helmer, R. Ostertag, B. Richardson, L. Walker, K. Vogt, and J. Zimmerman (chapter 1); B. Barker, E. Boose, B. Haines, C. Hall, M. Hall, E. Helmer, S. Presley, T. Schowalter, J. Thompson,

L. Woolbright, J. Zimmerman, and X. Zou (chapter 2); J. Bithorn, A. Estrada, E. Estrada, E. Meléndez-Colom, M. Larsen, J. Merriam, S. Moya, O. Ramos, M. Salgado, M. Sánchez, and C. Torrens (chapters 3, and 4); C. Bloch, H. Erickson, M. Gannon, R. Ostertag, S. Ward, W. Wu, and M. Yu (chapter 5); A. Covich (chapter 6); M. Alayón, N. Fetcher, and R. Myster (chapter 7); and B. Klingbeil and S. Presley (chapter 8). We thank J. Paniagua and A. McFadden for help with figures.

For additional financial support, we acknowledge the National Science Foundation (DEB-0236154 and DEB-0832652), Texas Tech University, the Center for Environmental Sciences and Engineering at the University of Connecticut, and University of New Mexico (chapter 2); the National Science Foundation (BSR-8718396, BSR-9007498, DEB-9981600, DEB-0087248, DEB-0108385, and DEB-0816727), USDA (NRICGP 9900975), and University of New Hampshire (chapter 3); and the Center for Environmental Sciences and Engineering at the University of Connecticut (chapter 8).

Our research and that of the other authors would not have been possible without the support of our home institutions, including cohorts of undergraduates, graduate students, technicians, and volunteers who have facilitated our scholarship in countless ways. Although too numerous to indentify by name, their contributions are invaluable.

Finally, we gratefully acknowledge the change in culture catalyzed by the "LTER experience." It has profoundly strengthened the spatial, temporal, and conceptual dimensions of our science and has profoundly changed our nature as scientists. Indeed, it has catalyzed our intellectual development by encouraging multidisciplinary perspectives, diverse collaborations, integration across disciplines, and synthesis from theoretical and empirical perspectives.

Literature Cited

Goldenberg, S. B., C. W. Landsea, A. M. Mestas-Nunez, and W. M. Gray. 2001. The recent increase in Atlantic hurricane activity: Causes and implications. *Science* 293:474–479.

Hobbie, J., S. R. Carpenter, N. B. Grimm, J. R. Gosz, and T. R. Seastedt. 2003. The US Long-Term Ecological Research Program. *BioScience* 53:21–32.

Lugo, A. E., and C. Lowe, editors. 1995. *Tropical forests: Management and ecology*. New York: Springer-Verlag.

National Research Council [NRC]. 2001. *Grand Challenges in Environmental Sciences*. Washington, DC: National Academies Press.

Neelin, S. J., M. Munnich, H. Su, J. E. Meyerson, and C. E. Holloway. 2006. Tropical drying trends in global warming models and observations. *Proceedings of the National Academy of Sciences of the USA* 103:6110–6115.

Odum, H. T., and R. F. Pigeon, editors. 1970. *A tropical rain forest: A study of irradiation and ecology at El Verde, Puerto Rico*. Oak Ridge, TN: U.S. Atomic Energy Commission.

Reagan, D. P., and R. B. Waide, editors. 1996. *The food web of a tropical rain forest*. Chicago: University of Chicago Press.

U.S. Long-Term Ecological Research Network [LTER]. 2007. The decadal plan for LTER: Integrative science for society and the environment. LTER Network Office Publication Series No. 24. Albuquerque, NM: LTER Network Office.

Waide, R. B., and A. E. Lugo. 1992. A research perspective on disturbance and recovery of a tropical montane forest. Pages 173–190 in J. G. Goldammer, editor, *Tropical forests in transition: Ecology of natural and anthropogenic disturbance processes*. Basel, Switzerland: Birkhäuser.

Webster, P. J., G. J. Holland, J. A. Curry, and H.-R. Chang. 2005. Changes in tropical cyclone number, duration, and intensity in a warming environment. *Science* 309:1844–1846.

Willig, M. R., C. P. Bloch, N. Brokaw, C. Higgins, J. Thompson, and C. R. Zimmermann. 2007. Cross-scale responses of biodiversity to hurricane and anthropogenic disturbance in a tropical forest. *Ecosystems* 10:824–838.

Willig, M. R., and L. R. Walker. 1999. Disturbance in terrestrial ecosystems: Salient themes, synthesis, and future directions. Pages 747–767 in L. R. Walker, editor, *Ecosystems of disturbed ground*. Amsterdam, The Netherlands: Elsevier.

Zimmerman, J. K, M. R. Willig, L. R. Walker, and W. L. Silver. 1996. Introduction: Disturbance and Caribbean ecosystems. *Biotropica* 28:414–423.

Contributors

Karen H. Beard, Utah State University
Gary E. Belovsky, Notre Dame University
Juan Felipe Blanco, Universidad de Colombia-Antioquia
Christopher P. Bloch, Bridgewater State University
Nicholas Brokaw, University of Puerto Rico-Río Piedras
Gerardo R. Camilo, Saint Louis University
Alan P. Covich, University of Georgia
Todd A. Crowl, Utah State University
Ned Fetcher, Wilkes University
Michael Gannon, Penn State University
Grizelle González, International Institute of Tropical Forestry, USDA Forest Service
Effie A. Greathouse, Oregon State University
Bruce L. Haines deceased, University of Georgia
Charles A. S. Hall, College of Environmental Science and Forestry, State University of New York
Paul Klawinski, William Jewell College
D. Jean Lodge, Center for Forest Mycology Research, Northern Research Station, USDA Forest Service
Ariel E. Lugo, International Institute of Tropical Forestry, USDA Forest Service
William H. McDowell, University of New Hampshire
Randall W. Myster, University of Central Oklahoma
Catherine M. Pringle, University of Georgia
Barbara A. Richardson, Independent, Edinburgh, U.K.

Michael J. Richardson, Independent, Edinburgh, U.K.
Frederick N. Scatena, University of Pennsylvania
Douglas A. Schaefer, Xishuangbanna Tropical Botanical Garden, Chinese Academy of Sciences
Timothy D. Schowalter, Louisiana State University
Joanne M. Sharpe, Sharplex Services, Edgecomb, Maine
Whendee L. Silver, University of California-Berkeley
Jill Thompson, The Centre for Ecology & Hydrology, U.K.
Daniel J. Vogt, University of Washington
Kristiina A. Vogt, University of Washington
Robert B. Waide, University of New Mexico
Lawrence R. Walker, University of Nevada-Las Vegas
Michael R. Willig, University of Connecticut
Lawrence L. Woolbright, Siena College
Joseph M. Wunderle, International Institute of Tropical Forestry, USDA Forest Service
Xiaoming Zou, University of Puerto Rico-Río Piedras
Jess K. Zimmerman, University of Puerto Rico-Río Piedras

A CARIBBEAN FOREST TAPESTRY

1

Ecological Paradigms for the Tropics

Old Questions and Continuing Challenges

Ariel E. Lugo, Robert B. Waide, Michael R. Willig, Todd A. Crowl, Frederick N. Scatena, Jill Thompson, Whendee L. Silver, William H. McDowell, and Nicholas Brokaw

Key Points

- The ecosystems of the Luquillo Mountains are representative of large areas of the frost-free tropical world, particularly those with high rainfall, periodic hurricane disturbances, a maritime climate, and insularity.
- The natural history of the Luquillo Mountains spans over 30 million years, whereas human presence has been an influence over the past 2,200 years.
- Indigenous peoples, Spanish conquistadors, and a steady stream of 20th and 21st century scientists have observed, studied, and experimented with the ecosystems of the Luquillo Mountains, and in the process they have left a legacy of ideas and heuristic models concerning ecosystem organization and function. The Luquillo Long-Term Ecological Research (LTER) program is rooted in this legacy.
- Important contributions to tropical science made by the Luquillo LTER program are a systematic investigation of disturbance and the identification of a number of mechanisms that contribute to the resistance and resilience of forested ecosystems.
- The LTER program has also contributed to a basic understanding of the ecology and biogeochemistry of the Luquillo Mountains and to an understanding of the long-term consequences of human activity on populations, communities, and ecosystem function.
- This book focuses on the response of the ecosystems of the Luquillo Mountains to natural and anthropogenic disturbances, with a particular focus on hurricanes and land cover change.

Introduction

The Tropics and Tropical Forests

Tropical forests cover an area of approximately 1.8 billion hectares, and they account for about 45 percent of the world's forests (Food and Agriculture Organization [FAO] 2003). Based on rainfall, ecology textbooks (e.g., Ricklefs 1997) usually represent tropical forests as belonging to one of two biomes: rain forests and dry or seasonal forests. This representation fails to appreciate the diversity of tropical forest types and perpetuates the myth that tropical forests are dichotomous in nature. The Holdridge Life Zone System (Holdridge 1967), which is based on empirical data and ecophysiological principles, provided a different picture of tropical forest types. Of the world's 112 life zones, over half (66) are tropical, and 33 include forests (out of 52 forested life zones in the world [Lugo and Brown 1991]). Thus, in climatic terms alone, tropical forests are more diverse than all other world forests combined. The diversity of tropical forest types increases even more when local factors such as geologic formation, soils, topography, and aspect are considered.

The Tropics of Cancer and Capricorn, at 23.5 degrees north and south of the equator, are usually used to define the geographical limits of the tropics. However, the distribution of the conditions amenable to the development of tropical forests does not always conform to these latitudinal criteria (figure 1-1). Tropical forest species respond to environmental factors, of which freezing temperatures is one of the most critical. Species richness decreases sharply in the presence of freezing temperatures, even within tropical latitudes, as evidenced by elevational patterns. Holdridge (1967) defined the tropics and subtropics by the absence of frost in the lowlands (figure 1-2).

Most lowland tropical species cannot tolerate frost, and this explains why tropical forests occur in frost-free areas beyond the Tropics of Cancer (India) and Capricorn (Madagascar) or contract within these geographic limits in areas such as Mexico or Australia, where frost occurs in the lowlands. Frost also occurs on tropical mountains, such as Mount Kilimanjaro, which experiences "summer every day and winter every night" (Hedberg 1997:185). In response to the dramatic diurnal temperature variation in these tropical mountain systems, plants and animals exhibit unusual adaptations such as the diurnal movement of leaves, the production of antifreeze substances, and day/night changes in behavior (Hedberg 1964). The distribution of tropical forests in relation to frost-free conditions is an example of how ecological space, defined by the distribution of environmental factors, differs from a distribution based on geographic space (i.e., the Tropics of Cancer and Capricorn) (see chapter 2).

The diversity of tropical forests is a challenge to ecologists. The task of describing the diversity of forest types is daunting, and it becomes even more complicated when considering forest function and responses to natural and anthropogenic disturbances. The tradition in tropical ecology was to compare forests with little consideration of differences in their climate or disturbance regimes (e.g., Gentry 1990). However, our research in Puerto Rico, and that of colleagues in other parts

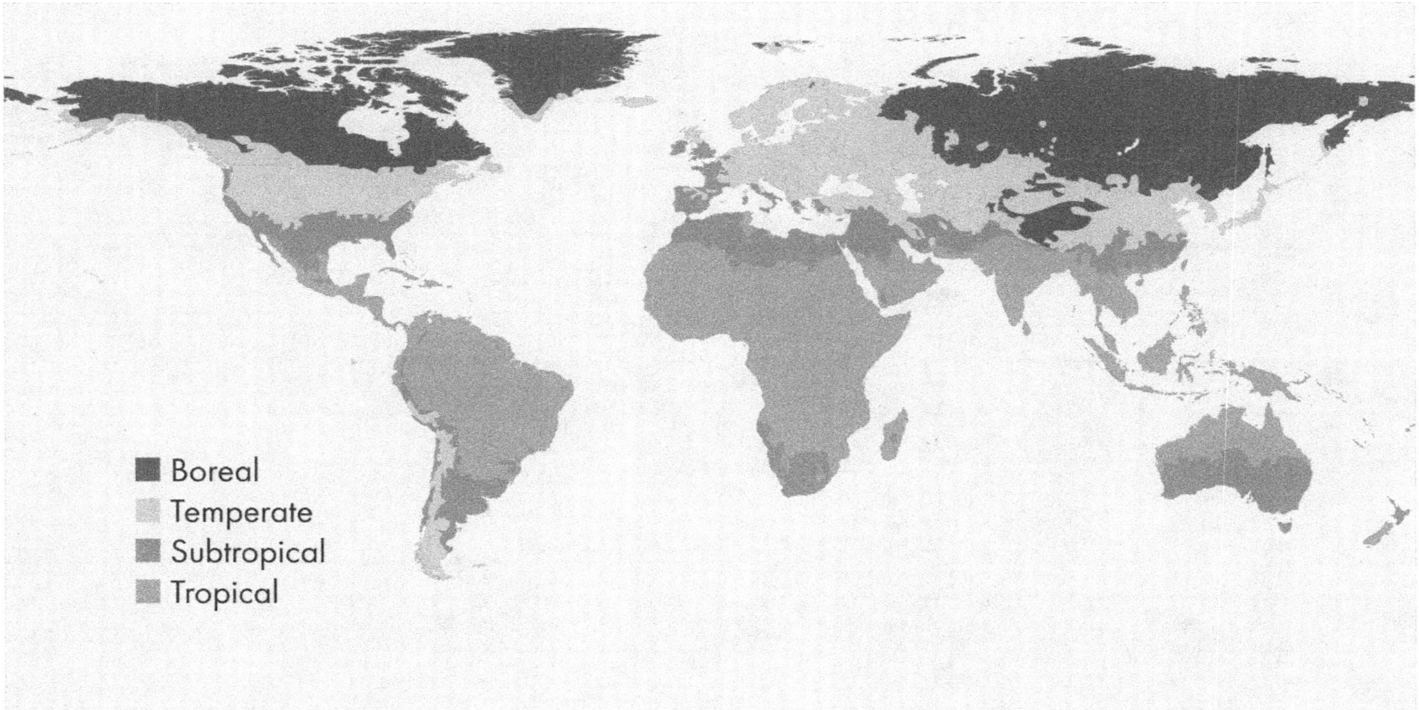

Figure 1.1 The global frost line defines the ecological space in which tropical forests occur. This map, prepared by R. P. Neilson, illustrates four levels of frost throughout the world. The tropics correspond to the area with a 12-month growing season and no frost. The long-term average monthly temperatures for all 12 months exceed the average monthly temperature associated with spring green-up (Neilson 1995). The other thermal zones are as follows: Boreal = supercooled freezing point (–40°C) reached annually; long-term minimum average monthly temperature < –16°C. Temperate = hard frosts annually (24 h < 0°C); long-term minimum average monthly temperature < –1.25°C. Subtropical = frequency of hard frosts ranging from less than annual to relatively rare; nearly zero days annually when the maximum temperature is < 0°C; long-term minimum average monthly temperature < 13°C. The definitions of "tropical" and "subtropical" in this system are different from the designations used by Holdridge (1967), who considers subtropical zones as also frost-free. Thus, "tropical" in this map coincides with "tropical" and "subtropical" in the Holdridge Life Zone System (figure 1-2).

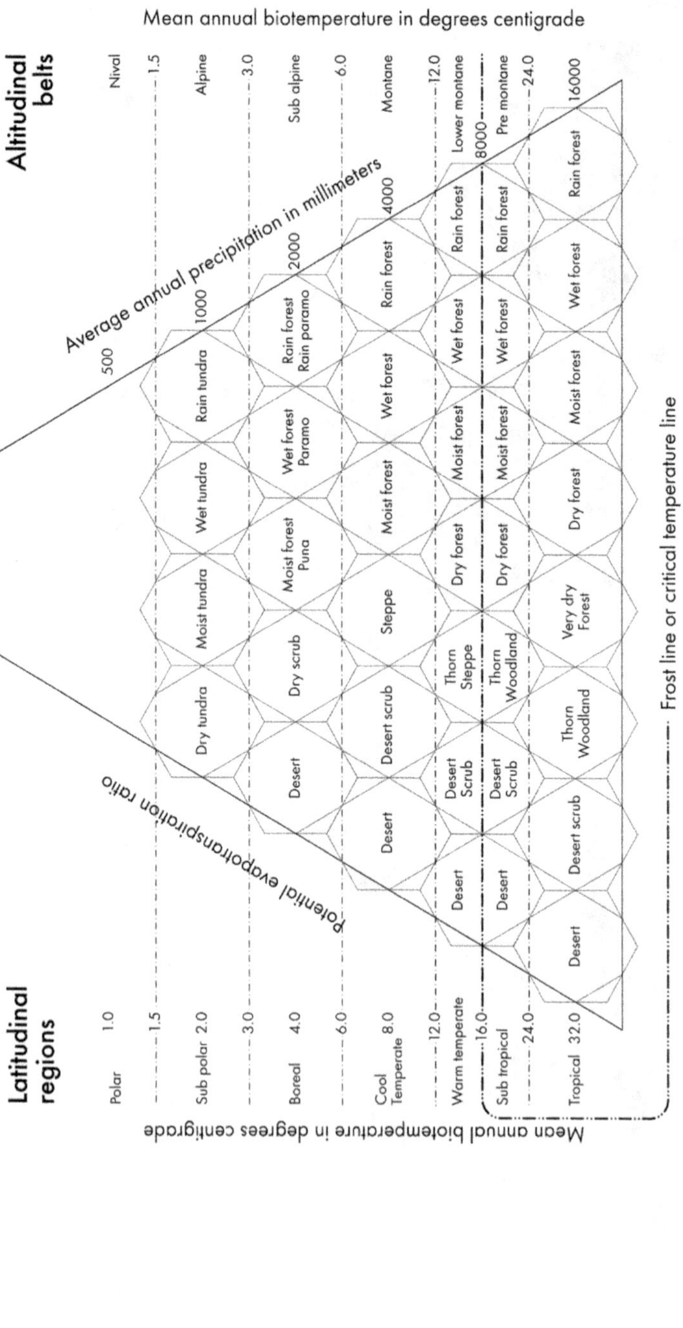
Latitudinal regions
Altitudinal belts
Mean annual biotemperature in degrees centigrade
Average annual precipitation in millimeters
Potential evapotranspiration ratio
Frost line or critical temperature line
Polar 1.0
1.5
Sub polar 2.0
3.0
Boreal 4.0
6.0
Cool Temperate 8.0
12.0
Warm temperate
16.0
Sub tropical
24.0
Tropical 32.0
Nival
Alpine
Sub alpine
Montane
Lower montane
Pre montane
500
1000
2000
4000
8000
16000
Dry tundra
Moist tundra
Wet tundra
Rain tundra
Desert
Dry scrub
Moist forest Puna
Wet forest Paramo
Rain forest Rain paramo
Desert
Desert scrub
Steppe
Moist forest
Wet forest
Rain forest
Desert
Desert Scrub
Thorn Steppe
Dry forest
Moist forest
Wet forest
Rain forest
Desert
Desert Scrub
Thorn Woodland
Dry forest
Moist forest
Wet forest
Rain forest
Desert
Desert scrub
Thorn Woodland
Very dry Forest
Dry forest
Moist forest
Wet forest
Rain forest

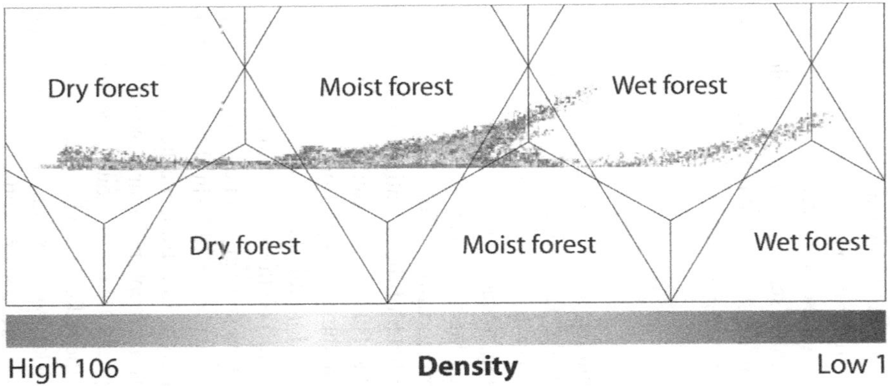

Figure 1.2 The Holdridge ternary classification system (opposite page; Holdridge 1967) defining life zones with latitudinal regions, altitudinal belts, and potential eva potranspiration ratios based on biotemperature and precipitation. A Holdridge Life Zone map program generated the scatter plot above indicating the continuous distributions of classifications in Holdridge ternary space. This scatter plot represents the biotemperature x precipitation setting in Puerto Rico, that is, the climatic ecological space of the island. Both panels, the program that generates life zones for particular locations, and the scatter plot are from Helmer and Plume (personal communication, 2005) and Plume and Helmer (2005).

of the world, shows that comparisons among tropical forests require knowledge and a consideration of environmental conditions, the age of forest stands, and the disturbance regime under which forests function (Lugo et al. 2002). Different features of the forest ecosystem, such as species composition, canopy structure, and rates of primary productivity, respond differently to various driving forces. These features affect comparisons among forests and generalizations about their processes. Dry and wet forests, for example, might respond similarly to wind in terms of their canopy structure but differently in terms of their phenology, as a result of water availability and species composition (Lugo et al. 2002).

Using the life zone approach, the guiding principle underlying the definition of "tropics" and the diversity of tropical forests is that environmental conditions—or ecological space, as discussed in chapter 2—dictate the organization, composition, and functioning of ecosystems from local to global scales. Therefore, ecological comparisons among ecosystems require a clear understanding of the environmental conditions that are relevant at the various spatial, temporal, and biological scales.

Puerto Rico and the Luquillo Mountains

Puerto Rico is within the geographic tropics and the global frost-free zone (figure 1-1), but it falls within the subtropical belt of the Holdridge Life Zone System because of its temperature regime (figure 1-2). The location of Puerto Rico within the Caribbean basin results in the island's being subjected to frequent hurricanes (chapter 4). Ocean and trade winds moderate the island's climate. One of the deepest spots in the Atlantic Ocean is several kilometers northwest of the Luquillo Mountains, a factor that, coupled with the long wind fetch of the Atlantic, contributes to high-energy conditions on the north coast of the island. A mountain chain in the middle of the island creates a rain shadow so that the annual precipitation in Puerto Rico spans a gradient of almost 5,000 mm from the Luquillo Mountains on the windward north coast to the Guánica dry forest (800 mm) on the leeward south coast.

The Luquillo Mountains loom large to observers from any vantage in the northeastern corner of Puerto Rico (figure 1-3). They rise to over 1,000 m above sea level, and the El Yunque peak is only 8 km in a straight line from the nearest beach. Because of their height, the Luquillo Mountains intercept moist air blown from the Atlantic Ocean by the steady trade winds; the peaks are under cloud cover most of the time. In comparison with tropical forests in the Atlantic lowlands of Costa Rica and the lowlands of central Panama—other well-known sites of long-term research activity (Gentry 1990)—the Luquillo Mountains are cooler, wetter, and less seasonal (Scatena 1998). Dry periods in these mountains last days and weeks rather than months and are only moderately seasonal in occurrence. Rainfall in the Luquillo Mountains has a nutrient-rich oceanic chemical signature (figure 1-4) with a high frequency of low-intensity showers punctuated by periodic high-intensity storms.

Discussions about the ecology of Puerto Rico raise the issue of insularity. Insularity has well-documented effects on the rate of species migrations and turnover (MacArthur and Wilson 1967), but the implications of insularity for the functional aspects of forests or the density of species remain poorly understood (Whittaker 1998). The effects of hurricanes and human disturbance on

Figure 1.3 The Luquillo Mountains. (Photo by A. E. Lugo.)

ecosystems in the Luquillo Mountains are difficult to disentangle from the effects of insularity.

This Book

This book is a synthesis of ecological knowledge about the Luquillo Mountains and its application to the conservation of biodiversity and the improvement of paradigms in the biological, ecological, and earth sciences. In this first chapter we review and synthesize ecological studies from eight decades of research, beginning with Gleason and Cook's (1926) vegetation survey and Wadsworth's (1947) examination of long-term forest growth 15 to 20 years after Hurricanes San Felipe (1928) and San Ciriaco (1932) struck the forest. An examination of the history of ecological research in the Luquillo Mountains reveals the gradual development of more refined and complex conceptual models, as well as the punctuated development of ideas across decades of research by different groups of scientists. All of these investigations have their conceptual roots in long-term assessments of the biotic and abiotic characteristics of the Luquillo Mountains. As part of our synthesis, we demonstrate in this chapter how our conceptualization of the ecosystems of the Luquillo Mountains contributes to a general understanding of the dynamics of forested ecosystems. In other chapters, the focus is principally on research conducted by the Luquillo Long-Term Ecological Research (LTER) program in response to Hurricane Hugo, after which we launched a series of studies in order to understand the effects of disturbances on forest dynamics, structure, and composition.

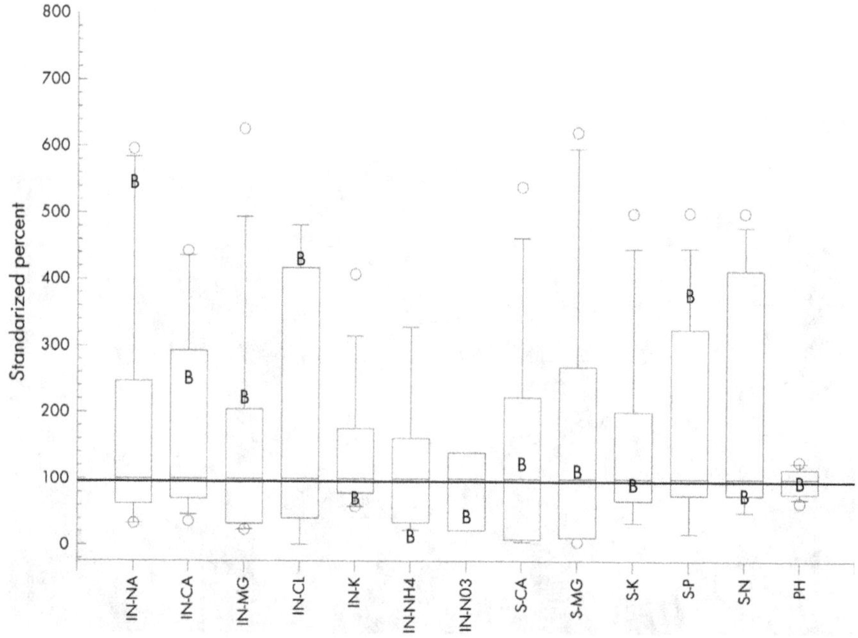

Figure 1.4 Box plots of standardized values of the bulk precipitation and soil pool size for various humid tropical forests (Scatena 1998). Values for Bisley (B), Luquillo Experimental Forest, are highlighted for comparison with other sites. Each box encompasses the 25th through 75th percentiles and has horizontal lines at the 10th and 90th percentiles. Circles represent data outside the range of the 10th and 90th percentiles. Dividing the value from a particular site by the meridian value of all the sites and then multiplying by 100 gives standardized values. The abbreviations IN-NA, IN-CA, IN-MG, IN-CL, IN-K, IN-NH_4, and IN-NO_3 denote the annual average inputs by bulk precipitation for sodium, Ca, Mg, Cl, K, NH_4-N, and NO_3, respectively. The abbreviations S-CA, S-MG, S-K, S-P, S-N, and pH denote the concentrations of extractable soil nutrients in surface soils. The coefficients of variation and sample size are as follows: IN-NA = 0.96, 11; IN-CA = 0.77, 14; IN-MG = 1.04, 14; IN-CL = 0.90, 9; IN-K = 0.65, 14; IN-NH_4 = 0.82, 7; IN-NO_3 = 0.73, 8; S-CA = 1.42, 23; S-MG = 1.18, 23; S-K = 1.36, 23; S-P = 0.88, 17; S-N = 0.76, 17; pH = 0.20, 23.

Models of Forest Structure and Functioning

Humans have visited and modified the Luquillo Mountains since prehistoric time. Each wave of visitors, including modern scientists, has no doubt marveled at the beauty and contemplated the mysteries of these mountains. Each group has also formulated questions and sought answers in an effort to understand the sights and sounds and to derive benefits from the ecosystems of these mountains.

A number of conceptual models of the ecosystems of the Luquillo Mountains have emerged. The Taíno Indians were among the first inhabitants of Puerto Rico and most likely generated the first conceptual models of the Luquillo Mountains (Domínguez Cristóbal 1989, 2000). Several scientists have summarized the scientific understanding of the Luquillo Mountains during the 20th century (Gleason and

Cook 1926; Holdridge 1947; Beard 1949; Wadsworth 1949, 1950; Odum 1970a; Lugo and Scatena 1995; Reagan and Waide 1996). Robinson (1997) recently published a popular version of the natural history and historic events of the Luquillo Mountains. Each of the works mentioned above represents a particular concept of the world that was descriptive of the state of knowledge at the time of its formulation. These works mentioned the role of large and infrequent disturbances but gave little attention to them. In this book, we offer a new synthesis of information that will certainly be modified in the future as our understanding of ecological phenomena increases through additional research.

Indigenous Peoples—The Forest as a Sacred Place

Humans arrived in Puerto Rico some 2,200 years ago by island hopping from South America (Domínguez Cristóbal 2000). These indigenous peoples included three successive groups or cultures: the Saladoides, the Taínos, and the Caribs. The Saladoides were the first to arrive via the Orinoco River and from Saladero, across the sea in Venezuela. They were hunters and gatherers who were replaced in Puerto Rico by the Taínos, who had mastered agriculture. By 1490, indigenous peoples had spread throughout the island. Their activities modified the flora and fauna by introducing new species to Puerto Rico (Francis and Liogier 1991) and caused the extinction of numerous native animal species (Brash 1987). The Carib Indians, known for their superior navigational skills, were becoming prominent in the Caribbean region when the Europeans interrupted their expansion after 1493.

The Taíno Indians are the best known among the three indigenous groups. They left rock carvings within the Luquillo Mountains that depict creatures, both alive and dead (dead people were represented with the soul leaving the body above the deceased's head). The writings of early European observers and subsequent inquiries suggested that the Taíno's view of the Luquillo Mountains was both religious and pragmatic (Domínguez Cristóbal 2000). To the Taíno people, the Luquillo Mountains were a sacred place where the good god *yucahu* or *yucayú* resided; this god protected them from the bad god *mabuya* or *juracán*. The modern term *hurricane* originates from the Taíno word *juracán*. The existence of this term and its connection to the Taíno religion suggests some knowledge of the most severe natural disturbance of the Luquillo Mountains. Clearly, questions about the nature and origin of hurricane disturbances and forest recovery from them had to be of concern to these early inhabitants of the Luquillo Mountains. Long-term records of hurricane tracks show two lanes to the north and south of Puerto Rico with a high number of tracks, and a lower number of hurricane tracks over the island (Neuman et al. 1978). This pattern, locally known as the "Puerto Rico split," correlates with the Taíno belief that the Luquillo Mountains somehow influenced the passage of hurricanes and protected their island.

Taínos also believed in totems and, possibly inspired by the Luquillo Mountains and the Central Cordillera, visualized the whole island as being carried by a large animal, which evolved into a figure with a human face and feet (Domínguez Cristóbal 2000). This animal figure is a *cemí* (figure 1-5), representations of which are sold today as decorations and as a tourist curiosity. The movement of the cemí was thought to contribute to the periodic earthquakes that affected the island.

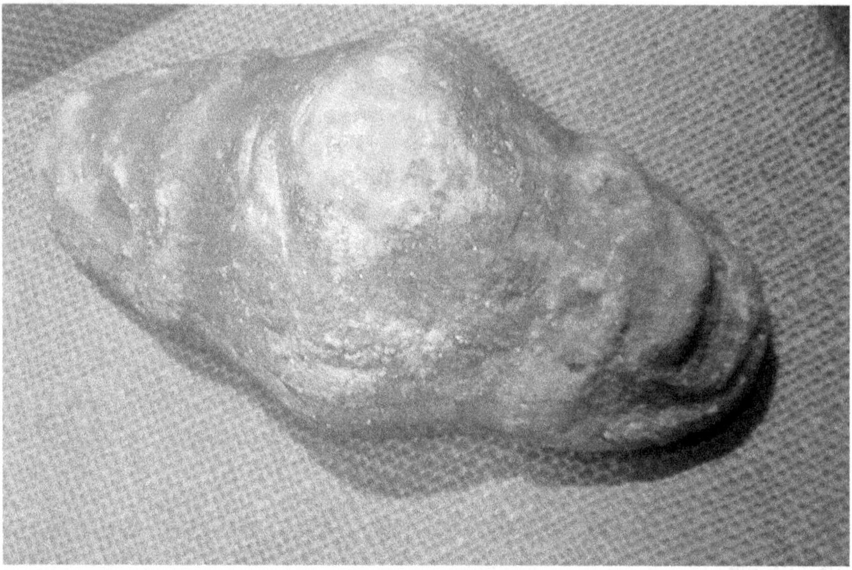

Figure 1.5 A Taíno cemí illustrates the idea that the island of Puerto Rico is steadied by an anthropogenic being. (Photo by Jerry Bauer.)

The sacred nature of the Luquillo Mountains, however, did not stop the Taínos from using natural products from its forests. They used the resin of the tabonuco tree (*Dacryodes excelsa*) to caulk their canoes, a custom that the Spaniards adopted after arriving on the island. Although the Taínos used tabonuco and other plants and animals for food, construction materials, and medicinal purposes, there is no evidence that suggests a sophisticated understanding of the relationship between ecosystem disturbance and response. Taínos also used the Luquillo Mountains as a haven during their conflict with the Spanish conquistadors (Scatena 1989).

Spanish Conquistadors—The Forest as a Resource

Europeans first saw Puerto Rico during the second voyage of Columbus in 1493 (Morison 1974). As Columbus's ships approached from the east, it is likely that the Luquillo Mountains were the first part of Puerto Rico that the sailors saw, which made them believe they were the tallest mountains on the island. The conquistadors subsequently used the Luquillo Mountains as a beacon to guide their ships as they sailed between the Atlantic and the Caribbean. The predominant paradigm of Spanish colonization involved economics, focusing on the exploitation of people and resources. The Taínos disappeared as a people under the 400 years of Puerto Rico's Spanish rule. The main focus of the Spaniards in Puerto Rico was on products, and what was available on the island for export to Spain or to support local Spanish activities. The conquistadors began an inventory of the island's wood and minerals in order to exploit them. These inventories represent the first description of the biodiversity and ecosystem services provided by the Luquillo Mountains and Puerto Rico (Domínguez Cristóbal 1992).

The Spanish government established a forest service in Puerto Rico (*Inspección de Montes*) between 1876 and 1889 (Domínguez Cristóbal 1992). This agency focused on timber production and land management. Land management and harvesting plans were developed, and timber harvesting had begun in the Luquillo Mountains by 1880. Tabonuco and ausubo (*Manilkara bidentata*) were species targeted for extraction. A tabonuco tree was valued at 1.50 pesos, whereas an ausubo tree was worth 2.25 pesos, assuming a minimum height and circumference for extraction of 8.5 and 1.58 m (0.5 m in diameter at breast height), respectively (a Spanish peso in the 19th century was equivalent to 60 U.S. cents in modern currency). Trees were harvested by the end of January and extracted during the somewhat drier months of January to March. A 9,000 ha area in the Luquillo Mountains yielded 19,630 m^3 of wood, or 15,857 pesos y^{-1}. Enforcement activities involved arrests, as was reported in 1889 when two people were arrested for cutting dozens of laurel sabino (*Magnolia splendens*), an endemic timber tree species.

The Spanish government's approach to forestry included the planned use of the forests and the protection of their watershed value. The government passed laws and proclamations to protect the forest timber for the crown, and they also designated buffer areas along rivers and streams in order to protect the water quality (Wadsworth 1949). A large area of the Luquillo Mountains and other forest locations in Puerto Rico were designated as public forests in 1876, making it one of the earliest such designations in the Western hemisphere. These actions anticipated the modern understanding of sustainable management practices and the effects of anthropogenic disturbance. However, no evidence suggests that the Spanish government actually estimated the watershed values of the Luquillo Mountains. This would not occur until the 1990s, when scientists in the LTER program developed a technical justification for the protection of these resources.

Early Foresters—Focus on Forest Management

North American foresters started writing about the Luquillo Mountains immediately after the Spanish–American War of 1898 (Hill 1899; Gifford 1905). The foresters surveyed the forest resources of the Luquillo Mountains from a utilitarian viewpoint. However, their approach touched on modern issues of functional diversity, ecosystem resilience, and species introductions. For example, they noted the abundance of "useless palms" (*Prestoea montana*) and asked how to control them (Gifford 1905). One suggestion was to import pigs to eat the palm fruit and thereby control the palm populations. Murphy (1916) published a comprehensive analysis of forestry in Puerto Rico and predicted that if timber exploitation continued at the rate observed, all forest cover would be lost from the island in the next 11 years. Murphy also considered the best approaches for the reforestation of slopes degraded by subsistence agriculture, but the early foresters did not know which species to use on particular sites or how to plant them. Early foresters spent a short time in Puerto Rico, and their contribution was observational rather than experimental.

In 1903, 2 years before the establishment of the U.S. Forest Service and the National Forest System, the U.S. government created the Luquillo Forest Reserve. In 1907, the Luquillo Forest Reserve was proclaimed the Luquillo National Forest, and in 1935 it was named the Caribbean National Forest. The use of the forest for research purposes was recognized in 1956, when the Caribbean National Forest was also designated as the Luquillo Experimental Forest; this is the only example in the National Forest System of a National Forest that also is designated as an Experimental Forest. In 1976, the Luquillo Experimental Forest was designated as a United Nations Educational, Scientific, and Cultural Organization (UNESCO) Biosphere Reserve, and in 2007 the Caribbean National Forest was renamed the El Yunque National Forest.

Natural Historians—Focus on Biodiversity

In the beginning of the 20th century, over a period of some 30 years, Nathaniel Lord Britton led an impressive number of scientists from the New York Academy of Sciences and the University of Puerto Rico on a scientific survey to describe the natural history of Puerto Rico and the Virgin Islands (Britton 1919). Figueroa Colón (1996) updated many aspects of this survey. The natural historians answered many taxonomic and botanical questions and created the taxonomic foundation for most of the research that would follow on the Luquillo Mountains. Expeditions from the New York Academy of Sciences made fundamental contributions to many subjects, including geology (Meyerhoff 1933), botany (Britton and Wilson [1923, 1924, 1925, 1926] 1930), ecology (Gleason and Cook 1926), paleobotany (Hollick 1928), Pteridophyta (Maxon 1926), bryophytes (Britton 1924; Crum and Steere 1957), fungi (Seaver and Chardón 1926; Seaver et al. 1932; Hagelstein 1932), and mammals (Anthony 1925). On the 80th anniversary of the beginning of the scientific survey, the state of knowledge on birds (Wiley 1996) and insects (Maldonado Capriles 1996) was updated, as were other topics (Figueroa Colón 1996).

Gleason and Cook (1926) were the first to propose models on the successional relations of vegetation in Puerto Rico. Their studies initiated investigations into the community ecology on the island and specifically addressed the relationship between community composition and disturbance. They were interested in the effects of agricultural activities on the species composition of plant associations. The work of Gleason and Cook (1926) provided the basis for subsequent long-term studies and experiments.

Modern Foresters—Focus on Control of Production

A series of hurricanes struck Puerto Rico between 1928 and 1932 and had severe effects that changed the land uses and economy of the island. In the 1940s, research turned once again toward methods to stimulate tree growth and the harvesting of forest timber. The first mechanistic studies and long-term experiments began during this period. J. S. Beard (1942, 1945) and Frank Wadsworth (1949), for example, both addressed questions about the "useless palms" that grew on steep slopes with saturated soils and on the sites of large landslides (Lugo et al. 1995). Studies of the tree growth of both palms and dicotyledonous trees led to the conclusion that although palm brakes had no potential for wood production, they were of significant watershed value because they grew on the wettest slopes of the Luquillo Mountains and protected significant catchment areas for lowland water supplies.

The failure of reforestation efforts within and outside of the Luquillo Mountains led to the establishment in 1939 of the Tropical Forest Experiment Station, later to become the International Institute of Tropical Forestry. The mission of this institution was to develop a scientific basis for effective reforestation and ecosystem restoration (Wadsworth 1995).

Leslie Holdridge, the first scientist of the Tropical Forest Experiment Station, addressed the relationship between vegetation and climate and developed the concept of the life zone based on observations about the Luquillo Mountains and the mountains of Haiti (Holdridge 1947, 1967). Ewel and Whitmore (1973) published a map of the life zones of Puerto Rico; however, they performed no validation of the correspondence of life zones with vegetation parameters such as species composition or the physiological limits of plant growth. Nevertheless, life zone studies highlighted the importance of environmental gradients in the distribution of communities and ecological processes. They provided the groundwork for later studies on the interrelationships among disturbance, vegetation, and climate, and they established the baseline information for the depiction of ecological space in Puerto Rico (see chapter 2).

Frank Wadsworth, a U.S. Department of Agriculture (USDA) Forest Service scientist, asked whether tree growth could be accelerated to make all trees in a stand grow as fast as the fastest-growing ones. He addressed the relationship between disturbance and productivity by establishing forest inventory plots under a variety of conditions in which he cut down trees in order to manipulate the basal area and measured the growth of the remaining trees. Wadsworth also measured the natural rates of tree growth and their variation over time (Wadsworth 1947). Wadsworth and other USDA Forest Service scientists continued these long-term studies (Crow and Weaver 1977; Weaver 1979, 1983; Wadsworth et al. 1989). However, with one exception (Crow 1980), the temporal changes in the structural and functional characteristics of forest stands received little attention until the 1980s.

José Marrero (1947, 1950) conducted tree-planting experiments in collaboration with Charles Briscoe and Frank Wadsworth, who worked in the tree plantation program of the USDA Forest Service. These foresters provided information on the correspondence between tree species and site conditions. Additional autecological research resulted in detailed life history observations for a number of tree species (McCormick 1995) and a summary of the silviculture of tropical tree species in Puerto Rico and the Caribbean (Francis and Lowe 2000). Plantation experiments led to greater success in reforestation efforts (Francis 1995; Wadsworth 1995), but no consideration was given to the effects of the planted species on the soils, site productivity, and successional outcome of planted sites. More recently, research on nutrient cycling, carbon dynamics, plant succession, and the relationship between diversity and function in plantations (Lugo et al. 1990b; Cuevas et al. 1991; Lugo 1992; Cuevas and Lugo 1998; Silver et al. 2000, 2004) has been carried out and continues today in plantations that now exceed 70 years of age (figure 1-6). Two syntheses of forestry research and experience in Puerto Rico and their applications in tropical forest production were recently published (Wadsworth 1997; Francis and Lowe 2000). Lugo et al. (2003) summarized the experience with mahogany plantations.

Monitoring Soils, Climate, and Hydrology

Scientists in the Agriculture Experiment Station of the University of Puerto Rico studied the soils of Puerto Rico and developed a detailed map of the island's 165 soil series (Roberts 1942) that is still in use today. According to Beinroth et al. (1996), the state of soil characterization in Puerto Rico is unmatched anywhere else in the tropics. In one data set, they report an analysis of a soil pedon per approximately 4,500 ha, or one data point for every plot in a grid of 6.5 by 6.5 km. Soils in Puerto Rico are extremely diverse and include 10 of the 12 soil orders of the USDA Soil Classification System. The monitoring of climate in Puerto Rico is also comprehensive. The National Weather Service operates over 90 weather stations, of which 12 have been keeping continuous records for over 100 years (Larsen 2000). Some stations contain records from the time of the Spanish government. These weather stations are complemented with an island-wide U.S. Geological Survey network of stream-gauging and well-monitoring stations.

Figure 1.6 A mature artificial forest in the *Dacryodes excelsa* zone of the Luquillo Mountains. Dr. F. H. Wadsworth (right) was responsible for the management of this site. (Photo by A. E. Lugo.)

Modern Ecologists

H. T. Odum's Rain Forest Project

Howard T. Odum and dozens of scientists and technicians, supported by the Atomic Energy Commission and the University of Puerto Rico, conducted the first large-scale ecosystem study on the effects of disturbance on the tabonuco forest. Their experiments using radiation and tree harvesting established the foundation for long-term ecological research in Puerto Rico and the tropics in general (Odum and Pigeon 1970). The Odum research legacy in the Luquillo Mountains transcends the radiation experiment in which he found that the tabonuco ecosystem's structure and function had a high resistance to radiation (Odum 1970a). Odum also described in detail the climate of the Luquillo Mountains (Odum et al. 1970b) and demonstrated how to measure forest metabolism on a grand scale, by isolating a section of forest within a giant plastic cylinder (Odum and Jordan 1970). Scientists involved in the Rain Forest Project also made comparisons with other tropical forests, both insular and continental (Odum 1970a, 1970c), and raised many questions for future studies. Many of Odum's questions concerned key methodological and monitoring approaches of the time, and others emphasized fundamental issues for research in tropical forests (table 1-1).

Studies of nutrient cycling in the tabonuco forest addressed plant biomass and nutrient content (Ovington and Olson 1970), soil nutrients (Edmisten 1970b), nitrogen (N) (Edmisten 1970a) and phosphorus (P) (Luse 1970) cycles, nutrient input in litterfall (Wiegert 1970a), litter decomposition (Wiegert and Murphy 1970), and nutrient losses in leachate (Sollins and Drewry 1970; Tukey 1970). These and other subsequent studies identified key aspects of biogeochemical cycling in tropical forests that have been applied to other tropical forest environments. These include the following:

- Tukey (1970) measured phosphorus leaching and foliar phosphorus absorption in bromeliads, thus demonstrating a mechanism by which these epiphytes receive nutrients from the atmosphere.
- Odum et al. (1970a), Witkamp (1970), Kline (1970), and others measured the radioactive fallout retention of epiphytes, epiphylls, and other forest surfaces,

Table 1.1 Ecosystem functioning questions raised by Odum (1970a: I-273–I-274) with annotations regarding any progress made

- What controls forest functioning? This continues to be a research priority.
- How much diversity is needed for stability and control? This hotly debated question remains unanswered.
- What is the weight of nervous tissue in the forest? Nervous tissue per unit area was proposed as an index of the animal contribution to the control system of the forest, but its value has not been determined beyond the early estimates of Canoy (1970).
- What is the significance of regenerative specialists for the planning of systems of man and nature where complex chemicals are used? There is considerable interest in designing new ecosystems (Lugo 1997), but we have very little knowledge of the functions of individual species that would compose these ecosystems.

thus showing the global connection of tropical forests to atmospheric systems originating in temperate latitudes.

- The global role of tropical forests in accumulating carbon in biomass and soils was a subject of study and synthesis in the Luquillo Mountains, beginning with a study by Odum and Pigeon (1970), which was followed by the work of Brown and Lugo (1982), Lugo and Brown (1982), Brown et al. (1984), Aide et al. (1995), and Silver et al. (2000). These studies have documented that tropical forests in Puerto Rico are carbon sinks under both natural and human-altered conditions.

Jordan et al. (1972) summarized the results of mineral cycling research in the Luquillo Mountains and proposed several hypotheses regarding the ways in which elements cycle in temperate ecosystems compared to cycling in tropical ecosystems. They also hypothesized that the size of a given ecosystem compartment would have a proportional effect on variations in the mineral cycling. For example, in the tabonuco forest of the Luquillo Mountains the relatively small litter pool would be more sensitive to disturbances than would the stemwood or root biomass pools, owing to their larger size and greater potential buffering capacity.

Before the LTER program, the animal species that received the greatest research attention included termites (McMahan 1970; Wiegert 1970b), earthworms (Lyford 1969), birds (Kepler and Kepler 1970; Recher 1970; Snyder et al. 1987), lizards, frogs (Turner and Gist 1970, Pough et al. 1983; Stewart and Pough 1983; Stewart 1985; Narins and Smith 1986), and snails (Heatwole et al. 1970; Stiven 1970). Collectively, these studies represent an effort to understand the population and community dynamics of animals in the forest, explore the relationship between their activity and vegetation dynamics, and hypothesize their importance to the overall functioning of the forest.

For Odum, the Luquillo Mountains functioned as an integrated ecosystem connected to the rest of the globe via regional flows of energy and cycling of materials. He recognized the connection between the tabonuco forest and latitudinal wind patterns through inputs of water and nutrients. Odum also recognized the role of wind and hurricanes in shaping the canopy of the forest (Odum 1970b), demonstrated the hierarchical nature of forest function, and integrated the functions of organisms from microbes to humans (Odum 1970b). Through research on the fundamental ecosystem structure and function, Odum developed models of sustainable land use for the tropics, including the design of ecosystems for human uses such as waste recycling or wood production (Odum 1995).

Government and Academic Scientists

Even today, we lack answers for many of the questions posed by Odum and other scientists. Many institutions are currently working together to answer these questions, including government (U.S. Department of the Interior [USDI] Geological Survey, USDA Forest Service, U.S. Department of Energy, USDI Fish and Wildlife Service, U.S. National Aeronautic and Space Administration) and academic (University of Puerto Rico and other universities in Puerto Rico and mainland United

States) institutions. Together we are addressing many lines of research in order to advance our knowledge of the Luquillo Mountains from the 1960s to the present. The overarching contribution of these efforts is to diversify the scope of science in the Luquillo Mountains, introduce the latest research technologies, and consolidate the reputation of the Luquillo Mountains as a tropical site with intense monitoring and experimental investigations of ecological phenomena. In the following sections we present summaries of the major conclusions of these studies, so as to lay the foundation for new material presented later in this book.

Managed Systems in Relation to Soils and Succession Research regarding tree plantations has received considerable attention in the Luquillo Mountains. These studies have attempted to increase site productivity by matching selected tree species to particular site conditions and have thus addressed issues relevant to our present ideas about ecological space. Some of the oldest tree plantations in the tropics (approximately 70 years old) grow in the Luquillo Mountains. These plantation ecosystems were established by the USDA Forest Service in the 1930s, and the subsequent monitoring of these ecosystems has provided an unprecedented opportunity to conduct comparative research in managed and unmanaged forests. Research on tree plantations has focused on key ecosystem storages and fluxes, including the following:

- the determination of standing stocks, flow rates, and nutrient-use efficiencies in pine (*Pinus caribaea*) and mahogany (*Swietenia macrophylla*) plantations in comparison with those in nearby secondary forests of similar age (Cuevas et al. 1991; Lugo 1992);
- the documentation of species differences in rates of nutrient retranslocation and nutrient use efficiency (Cuevas and Lugo 1998);
- the quantification of the effects of tree plantations on soil carbon and nutrient dynamics (Lugo et al. 1990b; Silver et al. 2004) and on organismal diversity (Cruz 1987, 1988; González et al. 1996); and
- the assessment of the responses of tree plantations to hurricanes (Fu et al. 1996; Wang and Scatena 2003; Ostertag et al. 2005).

Trees in plantations grow faster at low elevations (<500 m), where soils are better aerated and the rainfall is lower than at high elevations. Plantations attempted at higher elevations and rainfall levels have generally failed. On degraded sites, plantations contributed to the recovery of nutrient and organic matter pools in the soil (Lugo et al. 1990b; Cuevas et al. 1991; Cuevas and Lugo 1998). This recovery took decades (Silver et al. 2004). In addition, the plantation understories were invaded by a number of native tree species, although their richness was not as high as that of natural forest stands (Lugo 1992). Native species reinvade plantations, eventually contributing to the tempo and direction of succession. The long-term consequences of this reinvasion are still under investigation (Parrotta and Turnbull 1997).

Topographic Control of Vegetation and Soils Tree species are not evenly distributed across ridges, slopes, and valleys (Wadsworth 1949; Weaver 1987). Some species, such as the tabonuco, prefer ridge habitats with well-aerated soils, whereas others—for example, *Pterocarpus officinalis*—grow in valleys where the soils are saturated. Still other species, such as *Sloanea berteriana*, occur on slopes. The pattern of species distributions on these catenas occurs throughout the Luquillo Mountains, but the relationships among the topography, disturbance regimes, and vegetation dynamics were not understood in a comprehensive fashion until the LTER program explored them (Silver et al. 1994; Scatena and Lugo 1995; see also chapter 3).

Nutrient Cycles and Soil Organic Matter In addition to the studies in the tabonuco forest discussed above, nutrient cycles have been studied in the Luquillo Mountains in mature (Lugo 1992; Silver 1992, 1994; Silver and Vogt 1993; Silver et al. 1994; McDowell 1998), successional (Lugo 1992; Silver 1992; Scatena et al. 1996; Silver et al. 1996), and plantation forests (Lugo et al. 1990b; Cuevas et al. 1991; Lugo 1992; Fu et al. 1996; Cuevas and Lugo 1998; Silver et al. 2004). These studies provided estimates of the storages and the main fluxes of nutrients and organic matter, the relative distribution of organic matter and nutrients between above- and belowground compartments, and the efficiency of nutrient cycling. The results indicate the following:

- Nitrogen and calcium (Ca) do not limit the productivity of tabonuco forest, but phosphorus (P) and potassium (K) might be limiting.
- The distribution of nutrients and biomass in a forest is a function of the forest's age, topographic position, and climate.
- The efficiencies of cycling differ among nutrients (for example, high for P and low for N).
- The large quantity of belowground nutrients and organic matter contributes to the resilience of the forest.
- Nutrients in natural forest stands cycle at faster rates with relatively less storage in biomass as compared to that in plantations.

In the late 1960s, the Arnold Arboretum of Harvard University conducted a set of integrated studies of the biology, ecology, and ecophysiology of elfin forests in the Luquillo Mountains. These studies contributed basic information about, and some of the first observations of, the biogeochemistry in upper montane elfin forest plants and soils (Howard 1968, 1969, 1970; Lyford 1969; Wagner et al. 1969). Studies included the description of the canopy soil (complete with earthworms), the saturated surface soils, and the foliar chemistry of upper montane forest species. This interdisciplinary study gave us the first comprehensive overview of the short stature elfin forest as a saturated wetland on a mountaintop, and it described the many biotic adaptations of the flora and fauna to the extreme conditions of wind and wetness of this forest.

Soil Oxygen and Greenhouse Gases Wet tropical forests, such as those in the Luquillo Mountains, are commonly characterized by low or fluctuating soil oxygen availability, a factor that has a significant effect on the structure and functioning of the ecosystem. High-clay soils, warm temperatures, and abundant water lead to conditions in which the oxygen consumption by roots and soil organisms exceeds the rate of replacement from the atmosphere. These conditions, together with pockets of saturated and waterlogged soils (Wadsworth and Bonnet 1951), led to the description of the forests of the Luquillo Mountains as "slope wetlands" (Frangi 1983; Lugo et al. 1990a). The most extreme case of poorly oxygenated upland soils occurs on the peaks and ridges of the upper elevation elfin forests, where Lyford (1969) found organic soils at all levels of the forest from the ground to the canopy. He suggested that the canopy roots and arboreal earthworms were escaping the strongly reducing conditions of the terrestrial soil environment. Odum (1970c) noted mottling and gleying in the soil profile, a tendency for roots to concentrate near the surface, and the low redox potential of these soils. Silver et al. (1999) quantified the soil oxygen concentrations in soils along elevation and topographic gradients (figure 1-7). They found that the soil oxygen decreased with increasing annual rainfall, as well as from ridgetops to valley bottoms, and they interpreted the importance of these observations for patterns in tree species richness, nutrient cycling, primary productivity, and greenhouse gas emissions.

Tropical forests are involved in the circulation of greenhouse gases, and the dynamic redox of the Luquillo soils contributes to strong patterns in the production and emissions of carbon dioxide, nitrous oxide, and methane (Keller et al. 1986; Steudler et al. 1991; Silver et al. 1999; McGroddy and Silver 2000; Silver et al. 2001; Teh et al. 2005). Patterns in carbon dioxide emissions are complex and are affected by a combination of the rates of net primary production, the soil redox status, and past disturbance (Keller et al. 1986; McGroddy and Silver 2000). Tropical forests are the largest natural source of nitrous oxide, a potent greenhouse gas. Flood plains and upper elevation soils are large natural sources of this gas in the Luquillo Mountains (Keller et al. 1986; Erickson et al. 2001; McSwiney et al. 2001;

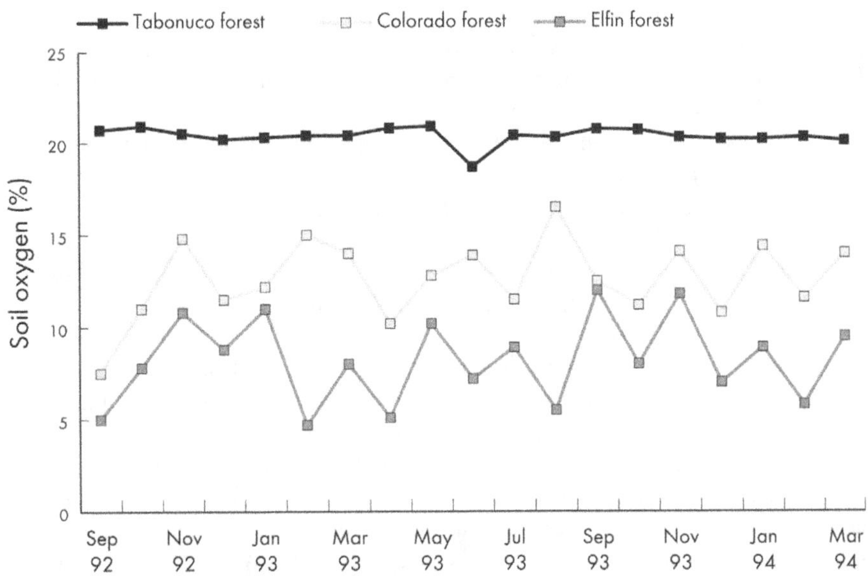

Figure 1.7 The soil oxygen gradient in the Luquillo Mountains (Silver et al. 1999).

Silver et al. 2001). Tropical wetlands are a natural source of methane, which is also a potent greenhouse gas, but upland tropical forests were thought to be a net sink. Research in the Luquillo Mountains suggests that slope wetlands and valley bottoms are actually a net source, owing to the abundance of anaerobic soil microsites, even in well-drained soils (Keller et al. 1986; Silver et al. 1999; Teh et al. 2005).

Nutrient cycles are fundamentally affected by soil oxygen as redox states change with the onset of anaerobic conditions. This is particularly true for phosphorus, an element generally thought to limit the net primary production on highly weathered tropical soils. Phosphorus cycling is tightly coupled with the redox state of iron, and it can be released through the reduction of abundant iron oxides (Silver et al. 1999). Soil redox dynamics, both spatial and temporal, create sharp interfaces in the soil (aerobic versus anaerobic microbial physiologies) and between plants and soil (anoxia tolerant versus intolerant species) that help shape the structure and function of the forest.

Root Grafting and Tree Unions Wadsworth and Englerth (1959) observed the resistance of trees on ridges to high winds, and later Odum (1970a, 1970c) reported the presence of root grafting in the tabonuco forest. These two independent observations have profound importance for the understanding of several phenomena in the tabonuco forest. They might help explain the success of monospecific stands of tabonuco on ridges, which are the most oxygen-rich sites in the forest (Silver et al. 1999). They might also help explain the high respiration rates of tabonuco shade leaves (Odum et al. 1970c), which could benefit from the translocation of sugars among tabonuco trees connected through root grafts. Chapter 3 reviews the long-term ecological research that placed these early observations in context with regard to the dominance and functioning of tabonuco forests on ridges.

Food Webs and Functional Diversity The animal species richness of the Luquillo Mountains is less than that found in similar-sized mainland tropical forests (Waide 1987), in part because of biogeographical and historical conditions. Communities comprise a small number of abundant or functionally important animal species, and this provides an excellent opportunity to examine the influence that animals have on ecosystem structures and processes.

Vertebrates are abundant in the Luquillo Mountains. Frogs and lizards each average more than two individuals per square meter; this density is among the highest recorded for these types of animals (Reagan 1996; Stewart and Woolbright 1996). On average, the body size of vertebrates is smaller in the Luquillo Mountains than in mainland tropical forests. However, because of their high density in Puerto Rico, vertebrates have a significant effect on the movement of mass, nutrients, and materials in forest stands. For example, lizard and frog populations consume about a million insects per hectare per day (Reagan 1996; Stewart and Woolbright 1996). Birds, bats, and insects pollinate and disperse seed for most tree species (Garrison and Willig 1996; Waide 1996; Willig and Gannon 1996). Among invertebrates, termites accelerate the decomposition of woody materials (Wiegert 1970b; Wiegert and Murphy 1970), and earthworms aerate low-oxygen-saturated soils (Lyford

1969) and accelerate litter decay (González and Seastedt 2001). These studies have been expanded greatly in the LTER program (see chapter 3).

The functional relationships among taxa emerged only as a result of a research program at the El Verde Field Station in the 1980s. Reagan and Waide (1996) synthesized over 30 years' worth of research to develop a comprehensive picture of the food web structure in the tabonuco forest (see chapter 3). Several important observations about food webs in the tabonuco forest have influenced contemporary theory about the trophic organization of ecosystems. The following five examples from Reagan et al. (1996) illustrate the insights gained from this long-term and intensive study of the biota in the tabonuco forest at El Verde (see also chapter 3).

- Although feeding loops or cycles (e.g., species A consumes species B, which in turn consumes species A) within a community should be rare for theoretical reasons associated with the destabilization of population dynamics, they are quite common in the tabonuco forest. Approximately one-third of all (ca. 20,000) food chains in the forest involve at least one species that is part of a food loop. The more dominant vertebrate taxa, such as frogs and lizards, participate in food loops through ontogenic dietary shifts (e.g., adult vertebrates consume some invertebrates that in turn consume immature vertebrates).
- Connectance in a food web, or the proportion of possible feeding relations realized in a community, is hypothesized to be an invariant characteristic of communities such that, on average, each species should interact trophically with approximately 14 percent of the other species. This pattern is present in the tabonuco forest at the lowest level of trophic species resolution (100 to 300 species). However, as the trophic species resolution increases (>1500 species), the connectance decays to a value of approximately 2 percent.
- Although trophic ratios are hypothesized to be scale-invariant, the ratios of basal to intermediate to top species and links among top to intermediate to basal species in the tabonuco forest vary significantly with the number of trophic species. In particular, top predators, even at the lowest level of taxonomic resolution, were significantly less common (by an order of magnitude) than theoretically predicted.
- Food web theory and thermodynamic constraints indicate that omnivory (species feeding on multiple trophic levels) should be rare and that food chains should be short (three to five links). Nonetheless, in the tabonuco forest, omnivores are pervasive and include about one-quarter of the bird species (Waide 1996), many of the bat species (Willig and Gannon 1996), and keystone species of frogs (Stewart and Woolbright 1996) and anoline lizards (Reagan 1996). Similarly, the range of food chain lengths in tabonuco forest is from 2 to 19 links, with a mean of 8.6 and a mode of 8.
- The reticulate hypothesis of food web organization suggests that the compartmentalization of trophic interactions should not occur within well-defined habitats. In contrast, in the tabonuco forest, a strong dichotomy in the food web organization exists, with nocturnal and diurnal compartments dominated by frogs and lizards, respectively.

The structure and dynamics of foods webs in the Luquillo Mountains are closely related to the biogeography, habitat heterogeneity, and disturbance, as discussed in chapters 2 through 5.

Land–Water Interactions A review of the available literature on land–water interactions in the Luquillo Mountains prior to the LTER program suggested that the connectivity within and among the ecosystems of the Luquillo Mountains is enhanced by interfaces involving water (Lugo 1986). During heavy rains, a continuous film of water covers the land from all surfaces of high-elevation forests (palm, colorado, and elfin forests) to streams and via the two-way movement of biota between the tops of the mountains and the ocean. The ecological importance of this connectivity rests in the coupling of the ecosystems of the Luquillo Mountains through a variety of alternative avenues for the exchange of materials and organisms. The support for this proposal includes the following:

- The cloud condensation level is around 600 m, which means that the whole aboveground structure of the forests above this elevation is immersed frequently within clouds. This increases humidity, decreases radiation input, saturates all plant and soil surfaces, and supports epiphytic growth and aquatic systems in tank bromeliads and other crevices.
- High rainfall coupled with clay-rich soils results in low redox soils and saturated decaying logs. The high clay content also limits the infiltration of rain, and this contributes to overland runoff. This also results in a high proportion of fine roots located near or on the soil surface and in the canopy of plants, instead of deep in the soil. Therefore, the typical role of roots in water uptake from deep in the soil is reduced compared to that in lowland ecosystems. Waterlogged decaying logs on the forest floor also become sites of water storage, with reduced rates of wood decay by aerobic organisms. The waterlogged soils and logs form a continuous film of water that aquatic organisms can use for mobility or the transport of larvae and, in the case of algae, spores.
- The forest canopy and tank bromeliads harbor 126 species of aquatic algae (Foerster 1971). These tank bromeliads are aquatic microcosms within the terrestrial community that support aquatic food chains connected to terrestrial food webs. Maguire (1970) found that these communities had as many as 76 kinds of aquatic organisms. He described a minimum of eight stable associations of fauna with two consistent community characteristics: they were highly resistant to ionizing radiation, and they showed rapid and effective dispersal mechanisms. In 12 days, 40 experimental microcosms with distilled water accumulated 180 different types of organisms.
- Sedges and aquatic plants with abundant aerenchyma and specialized gas exchange structures occur in bogs at high elevations. The presence of lowland wetland species (e.g., the sawgrass *Cladium jamaicensis*) in bogs within elfin forests is associated with long hydroperiods, saturated soils, and the possibility of groundwater movement along catenas.

- Shrimp, fish, mollusks, and crustaceans in the streams of the Luquillo Mountains migrate to the ocean to reproduce. These migrations establish another form of contact among montane forests, estuaries, and coastal systems. The food webs of streams and rivers are also connected to terrestrial food webs in the Luquillo Mountains (Covich and McDowell 1996).

Long-Term Ecological Research Program: An Integration of Approaches

Most of the research conducted in the Luquillo Mountains until 1988 was of relatively short duration (from less than a year to a decade). Even Odum's Rain Forest Project, which in its time was the most comprehensive study ever conducted of a tropical forest, lasted only 5 years (from 1963 to 1968). Notable longer and ongoing studies include (from 1942) the monitoring of tree growth and survival under natural and managed conditions (Brown et al. 1983), the recovery of vegetation after ionizing radiation (Taylor et al. 1995), and the recovery project for the endangered Puerto Rican Parrot (Snyder et al. 1987). The establishment of an LTER site in the Luquillo Mountains in 1988 initiated a new research focus on ecosystem-forcing functions of long duration, infrequent occurrence, or incremental effect. The passage of Hurricane Hugo in 1989 and Hurricane Georges 9 years later directed attention to the key role that repeated disturbances play in tabonuco forests. These hurricanes also brought into focus the qualitative and quantitative differences in the types of disturbances common in tabonuco forests, such as hurricanes, floods, droughts, landslides, treefalls, and a wide range of human activities. Finally, LTER studies have identified the interactions among different kinds of disturbances as an important factor when interpreting the existing distributions of organisms, biomass, and nutrients. Each of these conceptual advances has contributed to the understanding of the Luquillo Mountains that we put forward in this book.

The LTER program has encouraged coordination among a variety of scientific disciplines, as well as broadened our understanding and stimulated comparisons of the fundamental characteristics of the different ecosystems of the Luquillo Mountains. Pre-LTER observations of forests become increasingly relevant and important for current research, as they provide a context for the long-term study of natural phenomena. Since 1988, we have studied the Luquillo Mountains using a coordinated research program involving the population, the community, and the ecosystem, as well as landscape ecologists, hydrologists, soil scientists, geologists, foresters, climatologists, atmospheric scientists, and modelers. Results from the LTER program have been compared to those from other sites in the LTER Network (e.g., decomposition, N cycling, productivity, landscape diversity, watershed hydrology, and disturbance effects) and from other national (Land Margin Ecosystem Research, Long-Term Intersite Decomposition Team Project) and international (Flow Regimes from International Experimental and Network Data, Soil Biology and Fertility Program of UNESCO—Man and the Biosphere Program, Taiwan Ecological Research Network, Chinese Ecological Research Network, Center for Tropical Forest Science) programs studying soil organisms, landslide revegetation, hydrological characteristics, hurricane/typhoon effects, and tree communities, as well as to data from the Smithsonian's Center for Tropical Forest Science forest dynamics plot network. The breadth of the current research program is proving to be an asset in support of comparative ecological studies and synthesis across scales that transcend the Luquillo Experimental Forest.

Paradigm Shifts and the Improvement of Understanding

The analysis of the ecological effects of the passage of Hurricane Hugo over the Luquillo Mountains provided first-hand evidence in support of a paradigm shift that had started decades earlier concerning forest ecosystems. Until 1989, the focus of ecological research in the Luquillo Mountains had been the functioning of forest stands from the perspective of microbes, plants, and animals, and the priority for ecologists was an understanding of the structure and function of complex tropical forests, without an emphasis on natural disturbances as an integrating force (Odum and Pigeon 1970). In part, this priority arose because Puerto Rico had not been impacted by a major hurricane since 1932 or a tropical storm since 1956. Therefore, scientists had not had the opportunity to study windstorm events and their effect on forests. Research in Luquillo, as in temperate and other tropical forests, had mainly explored the long-term effects of small, discrete disturbances such as branch falls, treefalls, and clearing (Whitmore 1978, 1984, 1989; Denslow 1980, 1984; Frangi and Lugo 1985; Pickett and White 1985). Notable exceptions are the studies of Whitmore (1974), Garwood et al. (1979), and Foster (1980). At the same time as when these short-term studies were being published, evidence of the crucial effects of hurricanes in the Caribbean was slowly building. For example, Odum (1970a, 1970c) explained the canopy structure of Caribbean forests as being the result of trade winds and hurricanes. Doyle (1981) created a model that suggested that hurricanes maintained the species richness of the tabonuco forest. Crow (1980) interpreted structural changes in tabonuco forests as being caused by the 1932 hurricane. Willig et al. (1986) suggested that key consumer species might play a critical role in nutrient dynamics and succession after disturbance. Lugo et al. (1983) documented the effects of Hurricane David on the tabonuco forests of Dominica. Together, these observations formed the core basis for the first proposal for an LTER site in the Luquillo Mountains.

The recognition of the importance of disturbances to the species composition, structure, and functioning of ecosystems has its roots in early 20th century science, and the development of this disturbance paradigm has been the subject of several literature reviews (Walker 1999; White and Jentsch 2001). In fact, the rain forest experiment carried out by Odum and many colleagues (Odum and Pigeon 1970) was an experiment in human-produced disturbance. Nevertheless, the passage of Hurricane Hugo over our research sites was a significant turning point in our LTER research, as it unequivocally demonstrated the critical nature of such events with regard to forests (Walker et al. 1991, 1996). Our research benefited from the presence of a research infrastructure that allowed scientists to take full advantage of the event and from a strong partnership between the University of Puerto Rico, the U.S. Forest Service, and the National Science Foundation LTER program.

After Hurricane Hugo, we were surprised by the rapid rate of forest recovery in the Luquillo Mountains. It became evident that hurricanes and other disturbances had continually disrupted the forest and that this state of constant change had led to the development of adaptations to disturbance by forest organisms. In particular, the large-scale effects of hurricanes meant that most areas in the Luquillo Mountains were always responding to previous disturbances. Large and infrequent disturbances introduce pulses of high primary productivity with lagging respiration, followed by long periods of gradual forest changes and readjustment toward maturity (figure 1-8). The LTER program that we review and synthesize in this book signaled the completion of a shift toward an integrated disturbance paradigm and an effort to quantify the processes associated with the dynamic responses to disturbances in the Luquillo Mountains.

The disturbance paradigm has had fundamental effects on many other ecological ideas used to guide research and conservation in the Luquillo Mountains. For example, under a steady-state paradigm, forests were believed to be fragile, as any disturbance would shift them away from maturity and balance into states that were deemed incompatible with long-term stability. However, as hurricanes are recurrent, it was immediately obvious that the species in the forests of the Luquillo Mountains possessed adaptations that allowed them to resist or recover when confronted with these disturbance events. This insight led to a search for mechanisms that provided resilience in these forests (Lugo et al. 2002). Similarly, the changes in plant and animal population distributions and abundance observed after the passage of Hurricanes Hugo and Georges (Walker et al. 1991, 1996; Gannon and Willig 1994; Secrest et al. 1996) contributed to the recognition of environmental gradients (Hall et al. 1992) that shift in time and space. We recognized that geographic space couldn't always be equated to ecological space. The shift of environmental gradients results in the dynamic redistribution of the biota (plant and animal) and in the demonstration by individual species of characteristic types and rates of response to changing environmental conditions. The discovery of the importance of these dynamic responses provided an impetus for the integration of ideas about environmental gradients, disturbance, and response into a conceptual model relating geographic and ecological space (see chapter 2).

Our improved understanding of how species and ecosystems respond to environmental change provides a means with which to approach the restoration of degraded tropical forests. An example is the experience with plantations on degraded sites and the rapid establishment of species-rich understories within these plantations in the Luquillo Mountains (Lugo 1992, 1997). The rapid recovery of species richness and biomass in abandoned pastures further illustrated the possibilities for the restoration of tropical forests (Aide et al. 1995, 1996). From studies of severely degraded sites, we see that some alterations to ecosystems take them beyond the ecological conditions that they have experienced in their evolutionary history. This leads to the dominance of nonforest or nonnative plant species in some systems in the early stages of their recovery. Although there are many examples of the negative effects of nonnative species, in some cases their presence accelerates the recovery of ecosystems, and therefore the dominance of introduced species in the early

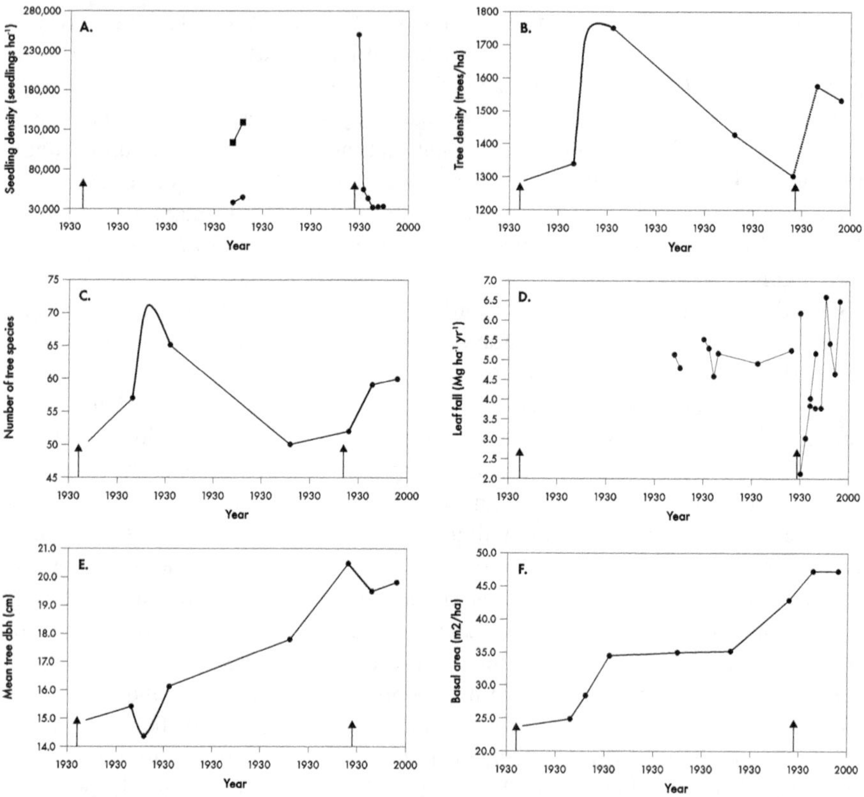

Figure 1.8 Changes in the structure and functioning of tabonuco stands (*Dacryodes excelsa*) for a period of over 60 y. This diagram is an updated version of the one in Lugo et al. (2000) and incorporates unpublished data available from the USDA Forest Service International Institute of Tropical Forestry and information from Smith (1970), Scatena et al. (1996), Lugo and Zimmerman (2002), and Lugo and Fu (2003). The data correspond to the El Verde-3 long-term plot of the USDA Forest Service (B., C., E., F) and other studies at El Verde in the immediate vicinity of the long-term plot (A., D.). This sector of the forest was on the leeward side of Hurricane Hugo and was not as affected as windward sectors of the Luquillo Mountains. The increase in the seedling density in the 1960s corresponds to a canopy opening experiment by Smith (1970). The diameter of the tree (dbh, in cm) with the mean basal area (BA, in m^2 ha^{-1}) was estimated by the following formula: dbh in cm = √ (BA in m^2 ha^{-1}) (12732.30)/tree density in trees ha^{-1}. The vertical arrows in 1932 and 1989 correspond to Hurricanes San Ciriaco and Hugo, respectively.

stages of recovery need not always cause alarm (Lugo and Helmer 2004; Lugo and Brandeis 2005), as their dominance at any location appears to be temporary (Wadsworth and Birdsey 1983; Lugo 2004). However, some introduced species have naturalized (Liogier 1990; Francis and Liogier 1991), and their dominance in certain sites is changing the trajectory of succession so that future forests will be different from what originally occupied the site (Lugo 1997, 2004; Lugo and Helmer 2004).

Metaphors for Forest Complexity

The search for a holistic overview of the Luquillo Mountains has led us to perceive the Luquillo Mountains in terms of a series of heuristic metaphors and conceptual models (see chapter 2). These metaphors and models summarize our understanding of the Luquillo Mountains, which we pass to others much as the Taínos did millennia before us. One metaphor considers the Luquillo Mountains as a *tapestry* interwoven with elements of topography, water, geology, soil oxygen, and living organisms that is exposed continually to light, wind, rain, and periodic violent combinations of these elements. Such exposure usually nourishes and maintains the components of the tapestry, but occasionally it reorganizes them through large and infrequent disturbances, keeping the mountain in a continuous state of flux. Each event defines a new pattern in the tapestry, and the cumulative effects of these events leave an imprint on the genotypes, abundances, and distributions of the organisms that constitute the tapestry. In short, the tapestry metaphor implies that at any moment visible patterns are the result of a complex history of disruption and repair occurring through processes of either self-organization or management for conservation goals.

The metaphor of a tapestry evolved from field observations before and after Hurricane Hugo. Before the hurricane, studies documented the high diversity of algae (126 epiphytic algal species as reported by Foerster [1971]) and other aquatic organisms in specialized aquatic habitats (such tank bromeliads, as reviewed by Lugo [1986]) in the forests of the Luquillo Mountains. We surmised that saturated air and a continuous film of water (a tapestry of moisture) that periodically covers the Luquillo Mountains permit connections between bromeliads and the organisms they contain and other aquatic habitats, including communities in streams, rivers, and estuaries. After Hurricane Hugo, a dense mat of vines and climbers resulted in a continuous leaf cover that appeared like a green tapestry over the affected forests (figure 1-9).

The metaphor of a tapestry, however, is superficial with regard to the literal sense of the word. The tapestry that we see reveals only the surface details of the Luquillo Mountains while hiding the underlying dynamics of the forest. A more apt metaphor is that of a *palimpsest*, a manuscript page that has been written on more than once, with earlier messages only partially erased and still visible (Hubbell 1979). In ecological usage, a palimpsest is an area that reflects its history and highlights the notion of the changing organizational patterns that reflect the effects of contemporary, recent, and ancient disturbances. The geologic and topographic structure underlying the Luquillo Mountains results from a series of tectonic events that occurred eons ago and continue at a very slow pace. The biotic composition of the Luquillo Mountains arises from a series of relatively recent (in geological time scales) immigration and extinction events, each of which has left an evolutionary and paleontological record. Repeated contemporary disturbance events such as hurricanes or human land use are preserved as changes in the ecosystem characteristics that are visible through the examination of the forest's composition and structure, including its soils. Each of these tectonic, biogeographic, or disturbance events is recorded in the biotic and abiotic structure of the Luquillo Mountains, and taken together they determine the composition of the palimpsest we see today.

Figure 1.9 A tapestry of leaves over a *Dacryodes excelsa* forest in the Luquillo Mountains. (Photo by A. E. Lugo.)

Our goal in this book is to elucidate the effects of recent disturbances on the Luquillo Mountains and to interpret them in the geologic, geographic, and climatic context of the mountains and Puerto Rico (chapter 3). In this context, we examine how disturbances restructure environmental gradients and affect ecosystem heterogeneity (chapter 4), how these changes interact with the biota (chapter 5), and how the biota affect environmental gradients. Each cycle of disturbance and response adds another layer to the historical record, partially obscuring previous patterns. The patterns that remain visible through the more recent layers (i.e., legacies) signal the importance of historical events and provide a better understanding of the ecosystems of the Luquillo Mountains. Our synthesis is another step in the development of the relationship between generations of people that care about and depend on nature, the Luquillo Mountains, and all tropical forests. Our hope is that our integrated and synthetic understanding of ecological dynamics in time and space will allow us to better appreciate and conserve the beauty of these mountains.

Summary

The forests of the Luquillo Mountains are representative of insular wet tropical forests subjected to frequent hurricane disturbances. Over 2,000 years of human activity have also affected the species composition and structure of forests in parts of these mountains. We identify and discuss seven models of forest structure and functioning in these mountains as reflected in the views of indigenous people, conquistadors, early foresters, natural historians, modern foresters, government agencies, and modern ecologists. These models embody the outcomes of the history of human experience, including research, in the Luquillo Mountains and provide the foundation for a new long-term and disturbance-based research paradigm reported in this book. The paradigm shift from short-term studies leads to an improved understanding of the ecosystems of the Luquillo Mountains by focusing on the forest's response to natural and anthropogenic disturbances, with particular attention paid to hurricanes and land cover change. The information is the product of long-term ecological research involving collaboration between the National Science Foundation, the University of Puerto Rico, and the U.S. Forest Service.

Literature Cited

Aide, T. M., J. K. Zimmerman, L. Herrera, M. Rosario, and M. Serrano. 1995. Forest recovery in abandoned tropical pastures in Puerto Rico. *Forest Ecology and Management* 77:77–86.

Aide, T. M., J. K. Zimmerman, M. Rosario, and H. Marcano. 1996. Forest recovery in abandoned cattle pastures along an elevational gradient in northeastern Puerto Rico. *Biotropica* 28:537–548.

Anthony, H. E. 1925. Mammals of Porto Rico, living and extinct—Chiroptera and Insectivora. In *Scientific survey of Porto Rico and the Virgin Islands* 9(2):97–141 + plates. New York: New York Academy of Sciences.

Beard, J. S. 1942. Montane vegetation in the Antilles. *Caribbean Forester* 3:61–74.

Beard, J. S. 1945. The progress of plant succession on the Soufriere of St Vincent. *Journal of Ecology* 33:1–9.

Beard, J. S. 1949. Natural vegetation of the windward and leeward islands. *Oxford Forestry Memoirs* 21:1–192.

Beinroth, F. H., M. A. Vázquez, V. A. Snyder, P. F. Reich, and L. R. Pérez-Alegría. 1996. *Factors controlling carbon sequestration in tropical soils: A case study of Puerto Rico.* Mayagüez, Puerto Rico: University of Puerto Rico at Mayagüez and USDA Natural Resources Conservation Service.

Brash, A. R. 1987. The history of avian extinction and forest conversion on Puerto Rico. *Biological Conservation* 39:97–111.

Britton, E. G. 1924. The mosses of El Yunque. *The Bryologist* 27:64–65.

Britton, N. L. 1919. History of the survey. In *Scientific survey of Porto Rico and the Virgin Islands* 1:1–10. New York: New York Academy of Sciences.

Britton, N. L., and P. Wilson. 1923, 1924, 1925, 1926, 1930. Botany of Porto Rico and the Virgin Islands. Descriptive flora—Spermatophyta. In *Scientific survey of Porto Rico and the Virgin Islands* 5(1–4):6. New York: New York Academy of Sciences.

Brown, S., A. Glubczynski, and A. E. Lugo. 1984. Effects of land use and climate on the organic carbon content of tropical forest soils in Puerto Rico. Pages 204–209 in *New forests for a changing world: Proceedings of the Convention of the Society of American Foresters, Portland, Oregon.* Washington, DC: Society of American Foresters.

Brown, S., and A. E. Lugo. 1982. The storage and production of organic matter in tropical forests and their role in the global carbon cycle. *Biotropica* 14:161–187.

Brown, S., A. E. Lugo, S. Silander, and L. H. Liegel. 1983. Research history and opportunities in the Luquillo Experimental Forest. General Technical Report SO-44. New Orleans, LA: USDA Forest Service.

Canoy, M. J. 1970. Deoxyribonucleic acid in rain forest leaves. Chapter G-6 in H. T. Odum and R. F. Pigeon, editors, *A tropical rain forest: A study of irradiation and ecology at El Verde, Puerto Rico*. Oak Ridge, TN: U.S. Atomic Energy Commission.

Covich, A. P., and W. H. McDowell. 1996. The stream community. Pages 433–459 in D. P. Reagan and R. B. Waide, editors, *The food web of a tropical forest*. Chicago: University of Chicago Press.

Crow, T. R. 1980. A rainforest chronicle: A 30-year record of change in structure and composition at El Verde, Puerto Rico. *Biotropica* 12:42–55.

Crow, T. R., and P. L. Weaver. 1977. Tree growth in a moist tropical forest of Puerto Rico. Forest Service Research Paper ITF-22. Río Piedras, PR: Institute of Tropical Forestry, USDA Forest Service.

Crum, H. A., and W. C. Steere. 1957. The mosses of Porto Rico and the Virgin Islands. In *Scientific survey of Porto Rico and the Virgin Islands* 7:395–599. New York: New York Academy of Sciences.

Cruz, A. 1987. Avian community organization in a mahogany plantation in a neotropical island. *Caribbean Journal of Science* 23:286–296.

Cruz, A. 1988. Avian resource in a Caribbean pine plantation. *Journal of Wildlife Management* 52:274–279.

Cuevas, E., S. Brown, and A. E. Lugo. 1991. Above and belowground organic matter storage and production in a tropical pine plantation and a paired broadleaf secondary forest. *Plant and Soil* 135:257–268.

Cuevas, E., and A. E. Lugo. 1998. Dynamics of organic matter and nutrient return from litterfall in stands of ten tropical tree plantation species. *Forest Ecology and Management* 112:263–279.

Denslow, J. S. 1980. Gap partitioning among tropical rain forest trees. *Biotropica* 12(supplement):47–55.

Denslow, J. S. 1984. Influence of disturbance on species diversity: Reply to T. C. Whitmore. *Biotropica* 16:240.

Domínguez Cristóbal, C. M. 1989. La situación forestal pre-hispánica de Puerto Rico [Puerto Rico's pre-Hispanic forestry situation]. *Acta Científica* 3:63–66.

Domínguez Cristóbal, C. M. 1992. Los planes de aprovechamiento forestal de la Inspección de Montes de Puerto Rico en la Sierra de Luquillo durante los años forestales de 1880–81 al 1888–89 [Forest harvesting plans of the Puerto Rico *Inspección de Montes* for the Luquillo Mountains during forest years 1880–81 to 1888–89]. *Acta Científica* 6:71–75.

Domínguez Cristóbal, C. M. 2000. *Panorama histórico forestal de Puerto Rico* [*Historical overview of Puerto Rico's forestry activity*]. Río Piedras, PR: Editorial Universidad.

Doyle, T. W. 1981. The role of disturbance in the gap dynamics of a montane rain forest: An application of a tropical successional model. Pages 55–73 in D. C. West, H. H. Shugart, and D. B. Botkin, editors, *Forest succession: Concepts and application*. New York: Springer-Verlag.

Edmisten, J. 1970a. Preliminary studies of the nitrogen budget of a tropical rain forest. Chapter H20 in H. T. Odum and R. F. Pigeon, editors, *A tropical rain forest: A study of irradiation and ecology at El Verde, Puerto Rico*. Oak Ridge, TN: U.S. Atomic Energy Commission.

Edmisten, J. 1970b. Soil studies in the El Verde rain forest. Chapter H3 in H. T. Odum and R. F. Pigeon, editors, *A tropical rain forest: A study of irradiation and ecology at El Verde, Puerto Rico*. Oak Ridge, TN: U.S. Atomic Energy Commission.

Erickson, H., M. Keller, and E. A. Davidson. 2001. Nitrogen oxide fluxes and nitrogen cycling during postagricultural succession and forest fertilization in the humid tropics. *Ecosystems* 4:67–84.

Ewel, J. J., and J. L. Whitmore. 1973. The ecological life zones of Puerto Rico and the U.S. Virgin Islands. USDA Forest Service Research Paper ITF-18. Río Piedras, PR: Institute of Tropical Forestry.

FAO [Food and Agriculture Organization]. 2003. *La situación de los bosques del mundo 2003* [*State of the world's forests 2003*]. Rome: Food and Agriculture Organization.

Figueroa Colón, J. C., editor. 1996. The scientific survey of Puerto Rico and the Virgin Islands. An eighty-year reassessment of the island's natural history. *Annals of the New York Academy of Sciences* 776:1–273.

Foerster, J. W. 1971. The ecology of an elfin forest in Puerto Rico, 14. The algae of Pico del Oeste. *Journal of the Arnold Arboretum* 52:86–109.

Foster, R. B. 1980. Heterogeneity and disturbance in tropical vegetation. Pages 75–92 in M. E. Soulé and B. A. Wilcox, editors, *Conservation biology: An evolutionary-ecological perspective*. Sunderland, MA: Sinauer.

Francis, J. K. 1995. Forest plantations in Puerto Rico. Pages 210–223 in A. E. Lugo and C. A. Lowe, editors, *Tropical forests: Management and ecology*. New York: Springer-Verlag.

Francis, J. K., and H. A. Liogier. 1991. Naturalized exotic tree species in Puerto Rico. General Technical Report SO-82. New Orleans, LA: USDA Forest Service.

Francis, J. K., and C. A. Lowe, editors. 2000. Silvics of native and exotic trees of Puerto Rico and the Caribbean islands. General Technical Report IITF-15. Rio Piedras, PR: USDA Forest Service.

Frangi, J. L. 1983. Las tierras pantanosas de la montaña Puertorriqueña [Montane wetlands of Puerto Rico]. Pages 233–247 in A. E. Lugo, editor, *Los bosques de Puerto Rico* [*The forests of Puerto Rico*]. San Juan, PR: USDA Forest Service Institute of Tropical Forestry and Puerto Rico Department of Natural Resources.

Frangi, J. L., and A. E. Lugo. 1985. Ecosystem dynamics of a subtropical floodplain forest. *Ecological Monographs* 55:351–369.

Fu, S., C. Rodríguez Pedraza, and A. E. Lugo. 1996. A twelve-year comparison of stand changes in a mahogany plantation and a paired natural forest of similar age. *Biotropica* 28:515–524.

Gannon, M. R., and M. R. Willig. 1994. The effects of Hurricane Hugo on the bats of the Luquillo Experimental Forest of Puerto Rico. *Biotropica* 26:320–331.

Garrison, R. W., and M. R. Willig. 1996. Arboreal invertebrates. Pages 183–245 in D. P. Reagan and R. B. Waide, editors, *The food web of a tropical rain forest*. Chicago: University of Chicago Press.

Garwood, N. C., D. P. Janos, and N. Brokaw. 1979. Earthquake-caused landslides: A major disturbance to tropical forests. *Science* 140:997–999.

Gentry, A. H. 1990. *Four neotropical rainforests*. New Haven, CT: Yale University Press.

Gifford, J. C. 1905. The Luquillo Forest Reserve, Porto Rico. Division of Forestry Bulletin No. 54. Washington, DC: U.S. Department of Agriculture.

Gleason, H. A., and M. T. Cook. 1926. Plant ecology of Porto Rico. In *Scientific survey of Porto Rico and the Virgin Islands* 7:1–173 + plates. New York: New York Academy of Sciences.

González, G., and T. R. Seastedt. 2001. Soil fauna and plant litter decomposition in tropical and subalpine forests. *Ecology* 82:955–964.

González, G., X. Zou, and S. Borges. 1996. Earthworm abundance and species composition in abandoned tropical croplands: Comparison of tree plantations and secondary forests. *Pedobiologia* 40:385–391.

Hagelstein, R. 1932. Revision of Myxomycetes. In *Scientific survey of Porto Rico and the Virgin Islands* 8:241–248. New York: New York Academy of Sciences.

Hall, C. A. S., J. A. Stanford, and F. R. Hauer. 1992. The distribution and abundance of organisms as a consequence of energy balances along multiple environmental gradients. *Oikos* 65:377–390.

Heatwole, H., A. Rossy, I. Colorado, and R. Amadeo. 1970. Effects of radiation on a population of the Puerto Rican tree snail, *Caracolus caracolla*. Chapter E-1 in H. T. Odum and R. F. Pigeon, editors, *A tropical rain forest: A study of irradiation and ecology at El Verde, Puerto Rico*. Oak Ridge, TN: U.S. Atomic Energy Commission.

Hedberg, O. 1964. Features of afroalpine plant ecology. *Acta Physiographica Suecia* 49: 1–144.

Hedberg, O. 1997. High mountain areas of tropical Africa. Pages 185–197 in F. E. Wielgolaski, editor, *Polar and alpine tundra*. Amsterdam: Elsevier.

Hill, R. T. 1899. Notes on the forest conditions of Porto Rico. Bulletin 25. Washington, DC: U.S. Department of Agriculture, Division of Forestry.

Holdridge, L. R. 1947. Determination of world plant formations from simple climatic data. *Science* 105:367–368.

Holdridge, L. R. 1967. *Life zone ecology*. San José, Costa Rica: Tropical Science Center.

Hollick, A. 1928. Paleobotany of Porto Rico. In *Scientific survey of Porto Rico and the Virgin Islands* 7:177–393. New York: New York Academy of Sciences.

Howard, R. A. 1968. The ecology of an elfin forest in Puerto Rico, 1. Introduction and composition studies. *Journal of the Arnold Arboretum* 49:381–418.

Howard, R. A. 1969. The ecology of an elfin forest in Puerto Rico, 8. Studies of stem growth and form and of leaf structure. *Journal of the Arnold Arboretum* 50:225–261.

Howard, R. A. 1970. The summit forest of Pico del Oeste, Puerto Rico. Chapter B-20 in H. T. Odum and R. F. Pigeon, editors, *A tropical rain forest: A study of irradiation and ecology at El Verde, Puerto Rico*. Oak Ridge, TN: U.S. Atomic Energy Commission.

Hubbell, S. P. 1979. Tree dispersion, abundance, and diversity in a tropical dry forest. *Science* 202:1299–1309.

Jordan, C. F., J. R. Kline, and D. S. Sasscer. 1972. Relative stability of mineral cycles in forest ecosystems. *The American Naturalist* 106:237–253.

Keller, M., W. A. Kaplan, and S. C. Wofsy. 1986. Emission of N_2O, CH_4, and CO_2 from tropical forest soils. *Journal of Geophysical Research* 91:11791–11802.

Kepler, C. B. and A. K. Kepler. 1970. Preliminary comparison of bird species diversity and density in Luquillo and Guánica forests. Chapter E-14 in H. T. Odum and R. F. Pigeon, editors, *A tropical rain forest: A study of irradiation and ecology at El Verde, Puerto Rico*. Oak Ridge, TN: U.S. Atomic Energy Commission.

Kline, J. R. 1970. Retention of fallout radionuclides by tropical forest vegetation. Chapter H-17 in H. T. Odum and R. F. Pigeon, editors, *A tropical rain forest: A study of irradiation and ecology at El Verde, Puerto Rico*. Oak Ridge, TN: U.S. Atomic Energy Commission.

Larsen, M. C. 2000. Analysis of 20th century rainfall and streamflow to characterize drought and water resources in Puerto Rico. *Physical Geography* 21:494–521.

Liogier, H. A. 1990. The plants introduced in the West Indies after the discovery of America and their impact on the ecology. San Juan, PR: Comisión Puertorriqueña para la Celebración del Quinto Centenario del Descubrimiento de America y Puerto Rico.

Lugo, A. E. 1986. Water and the ecosystems of the Luquillo Experimental Forest. General Technical Report SO-63. New Orleans, LA: USDA Forest Service.

Lugo, A. E. 1992. Comparison of tropical tree plantations with secondary forests of similar age. *Ecological Monographs* 62:1–37.

Lugo, A. E. 1997. The apparent paradox of re-establishing species richness on degraded lands with tree monocultures. *Forest Ecology and Management* 99(100):9–19.

Lugo, A. E. 2004. The outcome of alien tree invasions in Puerto Rico. *Frontiers in Ecology and the Environment* 2:265–273.

Lugo, A. E., M. Applefield, D. J. Pool, and R. B. McDonald. 1983. The impact of Hurricane David on the forests of Dominica. *Canadian Journal of Forest Research* 13:201–211.

Lugo, A. E., A. Bokkestijn, and F. N. Scatena. 1995. Structure, succession, and soil chemistry of palm forests in the Luquillo Experimental Forest. Pages 142–177 in A. E. Lugo and C. Lowe, editors, *Tropical forests: Management and ecology*. New York: Springer-Verlag.

Lugo, A. E., and T. J. Brandeis. 2005. A new mix of alien and native species coexist in Puerto Rico's landscapes. Pages 484–509 in D. F. R. P. Burslem, M. A. Pinard, and S. E. Hartley, editors, *Biotic interactions in the tropics: Their role in the maintenance of species diversity*. Cambridge, England: Cambridge University Press.

Lugo, A. E., and S. Brown. 1982. The role of tropical forests in the global carbon cycle: State of understanding. Pages 115–130 in S. Brown, editor, *Global dynamics of biospheric carbon*. CONF-8108131-UC-11. Washington, DC: U.S. Department of Energy.

Lugo, A. E., and S. Brown. 1991. Comparing tropical and temperate forests. Pages 319–330 in J. Cole, G. Lovett, and S. Findlay, editors, *Comparative analysis of ecosystems: Patterns, mechanisms, and theories*. New York: Springer-Verlag.

Lugo, A. E., S. Brown, and M. M. Brinson. 1990a. Concepts in wetland ecology. Pages 53–85 in A. E. Lugo, M. M. Brinson, and S. Brown, editors, *Forested wetlands*. New York: Elsevier.

Lugo, A. E., E. Cuevas, and M. J. Sánchez. 1990b. Nutrients and mass in litter and top soil of ten tropical tree plantations. *Plant and Soil* 125:263–286.

Lugo, A. E., J. C. Figueroa Colón, and M. Alayón, editors. 2003. *Big-leaf mahogany: genetics, ecology, and management*. New York: Springer-Verlag.

Lugo, A. E., J. Figueroa Colón, and F. N. Scatena. 2000. The Caribbean. Pages 593–622 in M. G. Barbour and W. D. Billings, editors, *North American terrestrial vegetation*. Cambridge, England: Cambridge University Press.

Lugo, A. E., and S. Fu. 2003. Structure and dynamics of mahogany plantations in Puerto Rico. Pages 288–328 in A. E. Lugo, J. C. F. Colón, and M. Alayón, editors, *Big-leaf mahogany: Genetics, ecology, and management*. New York: Springer-Verlag.

Lugo, A. E., and E. Helmer. 2004. Emerging forests on abandoned land: Puerto Rico's new forests. *Forest Ecology and Management* 190:145–161.

Lugo, A. E., and F. N. Scatena. 1995. Ecosystem-level properties of the Luquillo Experimental Forest with emphasis on the tabonuco forest. Pages 59–108 in A. E. Lugo and C. Lowe, editors, *Tropical forests: Management and ecology*. New York: Springer-Verlag.

Lugo, A. E., F. N. Scatena, W. Silver, S. Molina Colón, and P. G. Murphy. 2002. Resilience of tropical wet and dry forests in Puerto Rico. Pages 195–225 in L. H. Gunderson and L. Pritchard, Jr., editors, *Resilience and the behavior of large-scale systems*. Washington, DC: Island Press.

Lugo, A. E., and J. K. Zimmerman. 2002. Ecological life histories. Pages 191–213 in J. A. Vozzo, editor, *Tropical tree seed manual*. Agriculture Handbook 721. Washington, DC: USDA Forest Service.

Luse, R. A. 1970. The phosphorus cycle in a tropical rain forest. Chapter H-11 in H. T. Odum and R. F. Pigeon, editors, *A tropical rain forest: A study of irradiation and ecology at El Verde, Puerto Rico*. Oak Ridge, TN: U.S. Atomic Energy Commission.

Lyford, W. H. 1969. The ecology of an elfin forest in Puerto Rico, 7. Soil, root, and earthworm relationships. *Journal of the Arnold Arboretum* 50:210–224.

MacArthur, R. H., and E. O. Wilson. 1967. *The theory of island biogeography*. Princeton, NJ: Princeton University Press.

Maguire, B., Jr. 1970. Aquatic communities in bromeliad leaf axils and influence of radiation. Chapter E-6 in H. T. Odum and R. F. Pigeon, editors, *A tropical rain forest: A study of irradiation and ecology at El Verde, Puerto Rico*. Oak Ridge, TN: U.S. Atomic Energy Commission.

Maldonado Capriles, J. 1996. The status of insect alpha taxonomy in Puerto Rico after the scientific survey. *Annals of the New York Academy of Sciences* 776:201–216.

Marrero, J. 1947. A survey of forest plantations in the Caribbean National Forest. Master's thesis. School of Forestry and Conservation, University of Michigan, Ann Arbor.

Marrero, J. 1950. La reforestación de tierras degradadas de Puerto Rico [The reforestation of degraded lands in Puerto Rico]. *Caribbean Forester* 11:16–24.

Maxon, W. R. 1926. Pteridophyta. In *Scientific survey of Porto Rico and the Virgin Islands* 6:273–521. New York: New York Academy of Sciences.

McCormick, J. F. 1995. A review of the population dynamics of selected tree species in the Luquillo Experimental Forest, Puerto Rico. Pages 224–257 in A. E. Lugo and C. Lowe, editors, *Tropical forest: Management and ecology*. New York: Springer-Verlag.

McDowell, W. H. 1998. Internal nutrient fluxes in a tropical rain forest. *Journal of Tropical Ecology* 14:521–536.

McGroddy, M., and W. L. Silver. 2000. Variations in belowground carbon storage and soil CO_2 flux rates along a wet tropical climate gradient. *Biotropica* 32:614–624.

McMahan, E. A. 1970. Radiation and the termites at El Verde. Chapter E-8 in H. T. Odum and R. F. Pigeon, editors, *A tropical rain forest: A study of irradiation and ecology at El Verde, Puerto Rico*. Oak Ridge, TN: U.S. Atomic Energy Commission.

McSwiney, C. P., W. H. McDowell, and M. Keller. 2001. Distribution of nitrous oxide and regulators of its production across a tropical rainforest catena in the Luquillo Experimental Forest, Puerto Rico. *Biogeochemistry* 56:265–286.

Meyerhoff, H. A. 1933. *Geology of Puerto Rico*. Monograph Series B, No. 1. Río Piedras, PR: University of Puerto Rico.

Morison, S. E. 1974. *The European discovery of America. The southern voyages*. New York: Oxford University Press.

Murphy, L. S. 1916. Forests of Puerto Rico: Past, present, and future and their physical and economic environment. Bulletin 354. Washington, DC: U.S. Department of Agriculture.

Narins, P. M., and S. L. Smith. 1986. Clinal variation in anuran advertisement calls: Basis for acoustic isolation? *Behavioral Ecology and Sociobiology* 19:135–141.

Neilson, R. P. 1995. A model for predicting continental-scale vegetation distribution and water balance. *Ecological Applications* 5:362–385.

Neuman, C. J., G. W. Caso, and B. R. Jaruinen. 1978. *Tropical cyclones of the north Atlantic Ocean, 1871–1978*. Asheville, NC: U.S. Department of Commerce, National Oceanic and Atmospheric Administration, National Climatic Center.

Odum, H. T. 1970a. An emerging view of the ecological system at El Verde. Pages I-191 to I-289 in H. T. Odum and R. F. Pigeon, editors, *A tropical rain forest: A study of irradiation and ecology at El Verde, Puerto Rico*. Oak Ridge, TN: U.S. Atomic Energy Commission.

Odum, H. T. 1970b. Introduction to section F. Pages F-3–F-7 in H. T. Odum and R. F. Pigeon, editors, *A tropical rain forest: A study of irradiation and ecology at El Verde, Puerto Rico*. Oak Ridge, TN: U.S. Atomic Energy Commission.

Odum, H. T. 1970c. Rain forest structure and mineral cycling homeostasis. Chapter H-3 in H. T. Odum and R. F. Pigeon, editors, *A tropical rain forest: A study of irradiation and ecology at El Verde, Puerto Rico*. Oak Ridge, TN: U.S. Atomic Energy Commission.

Odum, H. T. 1995. Tropical forest systems and the human economy. Pages 343–393 in A. E. Lugo and C. Lowe, editors, *Tropical forests: Management and ecology*. New York: Springer-Verlag.

Odum, H. T., G. A. Briscoe, and C. B. Briscoe. 1970a. Fallout radioactivity and epiphytes. Chapter H-13 in H. T. Odum and R. F. Pigeon, editors, *A tropical rain forest: A study of irradiation and ecology at El Verde, Puerto Rico*. Oak Ridge, TN: U.S. Atomic Energy Commission.

Odum, H. T., G. Drewry, and J. R. Kline. 1970b. Climate at El Verde, 1963–1966. Chapter B-22 in H. T. Odum and R. F. Pigeon, editors, *A tropical rain forest: A study of irradiation and ecology at El Verde, Puerto Rico*. Oak Ridge, TN: U.S. Atomic Energy Commission

Odum, H. T., and C. Jordan. 1970. Metabolism and evapotranspiration of the lower forest in a giant plastic cylinder. Chapter I-9 in H. T. Odum and R. F. Pigeon, editors, *A tropical rain forest: A study of irradiation and ecology at El Verde, Puerto Rico*. Oak Ridge, TN: U.S. Atomic Energy Commission.

Odum, H. T., A. Lugo, G. Cintrón, and C. F. Jordan. 1970c. Metabolism and evapotranspiration of some rain forest plants and soil. Pages I103–I164 in H. T. Odum and R. F. Pigeon, editors, *A tropical rain forest: A study of irradiation and ecology at El Verde, Puerto Rico*. Oak Ridge, TN: U.S. Atomic Energy Commission.

Odum, H. T., and R. F. Pigeon, editors. 1970. *A tropical rain forest: A study of irradiation and ecology at El Verde, Puerto Rico*. Oak Ridge, TN: U.S. Atomic Energy Commission.

Ostertag, R., W. L. Silver, and A. E. Lugo. 2005. Factors affecting mortality and resistance to damage following hurricanes in a rehabilitated subtropical moist forest. *Biotropica* 37:16–24.

Ovington, J. D., and J. S. Olson. 1970. Biomass and chemical content of El Verde lower montane rain forest plants. Chapter H-2 in H. T. Odum and R. F. Pigeon, editors, *A tropical rain forest: A study of irradiation and ecology at El Verde, Puerto Rico*. Oak Ridge, TN: U.S. Atomic Energy Commission.

Parrotta, J. A., and J. W. Turnbull, editors. 1997. Catalyzing native forest regeneration on degraded tropical lands. *Forest Ecology and Management* 99:1–290.

Pickett, S. T. A., and P. S. White, editors. 1985. *The ecology of natural disturbance and patch dynamics*. Orlando, FL: Academic Press.

Plume, D. A., and E. H. Helmer. 2005. ArcMap/Visual Basic Holdridge life zones classification and mapping program. USDA Forest Service Technical Report. Río Piedras, PR: International Institute of Tropical Forestry and Rocky Mountain Research Station.

Pough, F. H., T. L. Taigen, M. M. Stewart, and P. F. Brussard. 1983. Behavioral modification of evaporative water loss by a Puerto Rican frog. *Ecology* 64:244–252.

Reagan, D. P. 1996. Anoline lizards. Pages 321–345 in D. P. Reagan and R. B. Waide, editors, *The food web of a tropical rain forest*. Chicago: University of Chicago Press.

Reagan, D. P., G. R. Camilo, and R. B. Waide. 1996. The community food web: Major properties and patterns of organization. Pages 461–510 in D. P. Reagan and R. B. Waide, editors, *The food web of a tropical rain forest*. Chicago: University of Chicago Press.

Reagan, D. P., and R. B. Waide, editors, 1996. *The food web of a tropical rain forest*. Chicago: University of Chicago Press.

Recher, H. F. 1970. Population density and seasonal changes of the avifauna in a tropical forest before and after gamma irradiation. Chapter E-5 in H. T. Odum and R. F. Pigeon, editors, *A tropical rain forest: A study of irradiation and ecology at El Verde, Puerto Rico*. Oak Ridge, TN: U.S. Atomic Energy Commission.

Ricklefs, R. E. 1997. *The economy of nature: A textbook in basic ecology*, 4th ed. New York: W.H. Freeman and Company.

Roberts, R. C. 1942. *Soil survey of Puerto Rico*. Series 1936, No. 8. Washington, DC: U.S. Department of Agriculture.

Robinson, K. 1997. *Where dwarfs reign: A tropical rain forest in Puerto Rico*. Río Piedras: University of Puerto Rico Press.

Scatena, F. N. 1989. An introduction to the physiography and history of the Bisley Experimental Watersheds in the Luquillo Mountains of Puerto Rico. General Technical Report SO-72. New Orleans, LA: USDA Forest Service.

Scatena, F. N. 1998. A comparative ecology of the Bisley biodiversity plot and experimental watersheds, Luquillo Experimental Forest, Puerto Rico. Pages 213–230 in F. Dallmeier and J. A. Comiskey, editors, *Forest biodiversity in North, Central and South America and the Caribbean*. New York: Parthenon.

Scatena, F. N., and A. E. Lugo. 1995. Geomorphology, disturbance, and the soil and vegetation of two subtropical wet steepland watersheds of Puerto Rico. *Geomorphology* 13:199–213.

Scatena, F. N., S. Moya, C. Estrada, and J. D. Chinea. 1996. The first five years in the reorganization of aboveground biomass and nutrient use following Hurricane Hugo in the Bisley Experimental Watersheds, Luquillo Experimental Forest, Puerto Rico. *Biotropica* 28:424–440.

Seaver, F. J., and C. E. Chardón. 1926. Mycology. In *Scientific survey of Porto Rico and the Virgin Islands* 8:1–208. New York: New York Academy of Sciences.

Seaver, F. L., C. E. Chardón, R. A. Toro, and F. D. Kern. 1932. Botany of Porto Rico and the Virgin Islands, supplement to mycology. In *Scientific survey of Porto Rico and the Virgin Islands* 8(2):209–311. New York: New York Academy of Sciences.

Secrest, M. F., M. R. Willig, and L. L. Lind. 1996. The legacy of disturbance on habitat associations of terrestrial snails in the Luquillo Experimental Forest, Puerto Rico. *Biotropica* 28:502–514.

Silver, W. 1992. Effects of small-scale and catastrophic disturbance on carbon and nutrient cycling in a lower montane subtropical wet forest in Puerto Rico. Ph.D. dissertation. Yale School of Forestry and the Environment, New Haven, CT.

Silver, W. L. 1994. Is nutrient availability related to plant nutrient use in humid tropical forests? *Oecologia* 98:336–343.

Silver, W. L., S. Brown, and A. E. Lugo. 1996. Effects of changes in biodiversity on ecosystem function in tropical forests. *Conservation Biology* 10:17–24.

Silver, W. L., D. J. Herman, and M. K. Firestone. 2001. Dissimilatory nitrate reduction to ammonium in upland tropical forest soils. *Ecology* 82:2410–2416.

Silver, W. L., L. M. Kueppers, A. E. Lugo, R. Ostertag, and V. Matzek. 2004. Carbon sequestration and plant community dynamics following reforestation of tropical pasture. *Ecological Applications* 14:1115–1127.

Silver, W. L., A. E. Lugo, and M. Keller. 1999. Soil oxygen availability and biogeochemistry along rainfall and topographic gradients in upland wet tropical forest soils. *Biogeochemistry* 44:301–328.

Silver, W. L., R. Ostertag, and A. E. Lugo. 2000. The potential for carbon sequestration through reforestation of abandoned tropical agricultural and pasture lands. *Restoration Ecology* 8:394–407.

Silver, W. L., F. N. Scatena, A. H. Johnson, T. G. Siccama, and M. J. Sánchez. 1994. Nutrient availability in a montane wet tropical forest in Puerto Rico: Spatial patterns and methodological considerations. *Plant and Soil* 164:129–145.

Silver, W. L., and K. A. Vogt. 1993. Fine root dynamics following single and multiple disturbances in a subtropical wet forest ecosystem. *Journal of Ecology* 81:729–738.

Smith, R. F. 1970. The vegetation structure of a Puerto Rican rain forest before and after short-term gamma irradiation. Pages D103–D140 in H. T. Odum and R. F. Pigeon, editors, *A tropical rain forest: A study of irradiation and ecology at El Verde, Puerto Rico.* Oak Ridge, TN: U.S. Atomic Energy Commission.

Snyder, N. F. R., J. W. Wiley, and C. B. Kepler. 1987. *The parrots of Luquillo: Natural history and conservation of the Puerto Rican Parrot.* Los Angeles: Western Foundation of Vertebrate Zoology.

Sollins, P., and G. Drewry. 1970. Electrical conductivity and flow rate of water through the forest canopy. Chapter H-10 in H. T. Odum and R. F. Pigeon, editors, *A tropical rain forest: A study of irradiation and ecology at El Verde, Puerto Rico.* Oak Ridge, TN: U.S. Atomic Energy Commission.

Steudler, P. A., J. M. Melillo, R. D. Bowden, M. S. Castro, and A. E. Lugo. 1991. The effects of natural and human disturbances on soil nitrogen dynamics and trace gas fluxes in a Puerto Rican wet forest. *Biotropica* 23:356–363.

Stewart, M. M. 1985. Arboreal habitat use and parachuting in a subtropical forest frog. *Journal of Herpetology* 19:391–401.

Stewart, M. M., and F. H. Pough. 1983. Population density of tropical forest frogs: Relation to retreat sites. *Science* 221:570–572.

Stewart, M. M., and L. L. Woolbright. 1996. Amphibians. Pages 273–320 in D. P. Reagan and R. B. Waide, editors, *The food web of a tropical rain forest.* Chicago: University of Chicago Press.

Stiven, A. E. 1970. Respiration in the snail *Caracolus caracolla* and an estimate of the relative density and biomass of litter snails. Chapter I-5 in H. T. Odum and R. F. Pigeon, editors, *A tropical rain forest: A study of irradiation and ecology at El Verde, Puerto Rico.* Oak Ridge, TN: U.S. Atomic Energy Commission.

Taylor, C. M., S. Silander, R. B. Waide, and W. Pfeiffer. 1995. Recovery of a tropical forest after gamma irradiation: A 23-year chronicle. Pages 258–285 in A. E. Lugo and C. Lowe, editors, *Tropical forests: Management and ecology.* New York: Springer-Verlag.

Teh, Y. A., W. L. Silver, and M. E. Conrad. 2005. Oxygen effects on methane production and oxidation in humid tropical forest soils. *Global Change Biology* 11:1283–1297.

Tukey, H. B., Jr. 1970. Leaching of metabolites from foliage and its implication in the tropical rain forest. Chapter H-11 in H. T. Odum and R. F. Pigeon, editors, *A tropical rain forest: A study of irradiation and ecology at El Verde, Puerto Rico.* Oak Ridge, TN: U.S. Atomic Energy Commission.

Turner, F. B., and C. S. Gist. 1970. Observations of lizards and tree frogs in an irradiated Puerto Rican forest. Chapter E-2 in H. T. Odum and R. F. Pigeon, editors, *A tropical rain forest: A study of irradiation and ecology at El Verde, Puerto Rico.* Oak Ridge, TN: U.S. Atomic Energy Commission.

Wadsworth, F. H. 1947. Growth in the lower montane rain forest of Puerto Rico. *Caribbean Forester* 8:27–44.

Wadsworth, F. H. 1949. The development of the forest and land resources of the Luquillo Mountains, Puerto Rico. Ph.D. dissertation. University of Michigan, Ann Arbor.

Wadsworth, F. H. 1950. Notes on the climax forests of Puerto Rico and their destruction and conservation prior to 1900. *Caribbean Forester* 11:38–47.

Wadsworth, F. H. 1995. A forest research institution in the West Indies: The first 50 years. Pages 33–56 in A. E. Lugo and C. Lowe, editors, *Tropical forests: Management and ecology.* New York: Springer-Verlag.

Wadsworth, F. H. 1997. *Forest production for tropical America.* Agriculture Handbook 710. Washington, DC: USDA Forest Service.

Wadsworth, F. H., and R. A. Birdsey. 1983. Un nuevo enfoque de los bosques de Puerto Rico [A new focus to the forests of Puerto Rico]. Pages 12–27 in *Puerto Rico Department of Natural Resources Ninth Symposium on Natural Resources.* San Juan: Puerto Rico Department of Natural Resources.

Wadsworth, F. H., and J. A. Bonnet. 1951. Soil as a factor in the occurrence of two types of montane forests in Puerto Rico. *Caribbean Forester* 12:67–70.

Wadsworth, F. H., and G. H. Englerth. 1959. Effects of the 1956 hurricane on forests in Puerto Rico. *Caribbean Forester* 20:38–51.

Wadsworth, F. H., B. R. Parresol, and J. C. Figueroa Colón. 1989. Tree increment indicators in a subtropical wet forest. Pages 205–212 in W. M. Wan Razali, H. T. Chan, and S. Appanah, editors, *Proceedings of seminar on growth and yield mixed/moist forests.* Kuala Lumpur, Malaysia: Forest Research Institute.

Wagner, R. A., A. B. Wagner, and R. J. Howard. 1969. The ecology of an elfin forest in Puerto Rico, 9. Chemical studies of colored leaves. *Journal of the Arnold Arboretum* 50:556–565.

Waide, R. B. 1987. The fauna of Caribbean ecosystems: Community structure and conservation. *Acta Científica* 1:64–71.

Waide, R. B. 1996. Birds. Pages 363–398 in D. P. Reagan and R. B. Waide, editors, *The food web of a tropical rain forest.* Chicago: University of Chicago Press.

Walker, L. R., editor. 1999. *Ecosystems of disturbed ground.* Amsterdam: Elsevier.

Walker, L. R., D. J. Lodge, N. V. L. Brokaw, and R. B. Waide, editors. 1991. Plant, animal, and ecosystem responses of hurricanes in the Caribbean. *Biotropica* 23:313–521.

Walker, L. R., J. K. Zimmerman, M. R. Willig, and W. L. Silver, editors. 1996. Long-term responses of Caribbean ecosystems to disturbance. *Biotropica* 28:414–614.

Wang, H. H., and F. N. Scatena. 2003. Regeneration after hurricane disturbance of big-leaf and hybrid mahogany plantations in Puerto Rico. Pages 237–257 in A. E. Lugo, J. C. Figueroa Colón, and M. Alayón, editors, *Big-leaf mahogany: Genetics, ecology, and management.* New York: Springer-Verlag.

Weaver, P. L. 1979. Tree growth in several tropical forests of Puerto Rico. Research Paper SO-152. New Orleans, LA: USDA Forest Service.

Weaver, P. L. 1983. Tree growth and stand changes in the subtropical life zones of the Luquillo Mountains of Puerto Rico. Research Paper SO-190. New Orleans, LA: USDA Forest Service.

Weaver, P. L. 1987. Structure and dynamics in the colorado forests of the Luquillo Mountains of Puerto Rico. Ph.D. dissertation. Michigan State University, East Lansing, MI.

White, P. S., and A. Jentsch. 2001. The search for generality in studies of disturbance and ecosystem dynamics. *Progress in Botany* 62:399–449.

Whitmore, T. C. 1974. Change with time and the role of cyclones in tropical rain forest on Kolombangara, Solomon Islands. Institute Paper No. 46. Oxford, England: Commonwealth Forestry Institute, University of Oxford.

Whitmore, T. C. 1978. Gaps in the forest canopy. Pages 639–655 in P. B. Tomlinson and M. H. Zimmerman, editors, *Tropical trees as living systems.* Cambridge, England: Cambridge University Press.

Whitmore, T. C. 1984. Gap size and species richness in tropical rain forests. *Biotropica* 16:239.

Whitmore, T. C. 1989. Canopy gaps and the two major groups of forest trees. *Ecology* 70:536–538.

Whittaker, R. J. 1998. *Island biogeography: Ecology, evolution, and conservation.* Oxford, England: Oxford University Press.

Wiegert, R. G. 1970a. Effects of ionizing radiation on leaf fall, decomposition, and litter microarthropods of a montane rain forest. Chapter H-4 in H. T. Odum and R. F. Pigeon, editors, *A tropical rain forest: A study of irradiation and ecology at El Verde, Puerto Rico.* Oak Ridge, TN: U.S. Atomic Energy Commission.

Wiegert, R. G. 1970b. Energetics of nest-building termite *Nasutitermes costalis* (Holmgren) in a Puerto Rican forest. Chapter I-4 in H. T. Odum and R. F. Pigeon, editors, *A tropical rain forest: A study of irradiation and ecology at El Verde, Puerto Rico.* Oak Ridge, TN: U.S. Atomic Energy Commission.

Wiegert, R. G., and P. G. Murphy. 1970. Effect of season, species, and location on the disappearance rate of leaf litter in a Puerto Rican rain forest. Chapter H-5 in H. T. Odum and R. F. Pigeon, editors, *A tropical rain forest: A study of irradiation and ecology at El Verde, Puerto Rico.* Oak Ridge, TN: U.S. Atomic Energy Commission.

Wiley, J. W. 1996. Ornithology in Puerto Rico and the Virgin Islands. *Annals of the New York Academy of Sciences* 776:149–179.

Willig, M. R., and M. R. Gannon. 1996. Mammals. Pages 399–431 in D. P. Reagan and R. B. Waide, editors, *The food web of a tropical rain forest.* Chicago: University of Chicago Press.

Willig, M. R., R. W. Garrison, and A. Bauman. 1986. Population dynamics and natural history of a neotropical walking-stick, *Lamponius portoricensis* Rehn (Phasmatodea: Phasmatidae). *Texas Journal of Science* 38:121–138.

Witkamp, M. 1970. Mineral retention by epiphyllic organisms. Chapter H-14 in H. T. Odum and R. F. Pigeon, editors, *A tropical rain forest: A study of irradiation and ecology at El Verde, Puerto Rico.* Oak Ridge, TN: U.S. Atomic Energy Commission.

2

Conceptual Overview

Disturbance, Gradients, and Ecological Response

Robert B. Waide and Michael R. Willig

Key Points

- The abundance and distribution of organisms and the attendant ecosystem processes vary across the landscape of the Luquillo Mountains in relation to underlying patterns of spatial heterogeneity and gradients of environmental factors.
- The ecosystems of the Luquillo Mountains are affected by frequent climate-induced disturbances such as treefalls, landslides, tropical storms, and droughts, as well as by human-induced disturbances associated with land use (i.e., agriculture and forest harvest).
- The term "ecological space" refers to multivariate dimensions defined by a suite of environmental characteristics. Disturbances can disrupt or create gradients by altering the mapping of ecological characteristics onto geographic space.
- Because the relationship between geographic space and ecological space is dynamic, the relationship between the physical template and the distribution and abundance of animal, plant, and microbial species cannot be understood without reference to the disturbance regime.
- The resilience of an ecosystem to anthropogenic disturbances might be low because such disturbances often produce severe modifications to the environment, creating novel combinations of environmental characteristics that are outside of the ecological space that was characteristic of the site or which are characterized by the absence of biological residuals.

- Historical factors, as well as contemporary geology, topography, and abiotic or biotic conditions, interact to create spatial variability in ecological characteristics. This variability ultimately determines the abundance and distribution of species in the Luquillo Mountains.

Introduction

The importance of environmental conditions and resources in determining the distribution and abundance of organisms is a fundamental tenet of ecology (Shelford 1951; Andrewartha and Birch 1954; Maguire 1976; Krebs 1985; Tilman 1988; Smith and Huston 1989). As Lugo et al. point out in chapter 1, spatial gradients in environmental conditions underlie the geographic distribution of ecosystems at global scales and affect the variation within ecosystems at smaller scales. The number of studies of physical and climatic gradients in the ecological literature demonstrates the importance attached to environmental conditions and resources as controls of ecosystem structure and function. Thus, knowledge of the long-term spatial and temporal patterns of environmental factors is critical if one is to understand the dynamics of ecosystems.

The ecosystems of the Luquillo Mountains are affected by frequent disturbances, as defined below and as described in chapter 4. The Luquillo Long Term Ecological Research (LTER) program has focused much effort over the past 20 years on understanding the impacts of two hurricanes, Hugo and Georges, in the context of a disturbance regime that also includes treefalls, landslides, tropical storms, and droughts and which has included human-dominated land uses such as agriculture and forest harvest in the past. This chapter provides an overview of an integrated research framework that incorporates theoretical elements from studies of disturbance and environmental variation.

Field observations supported by experiments and modeling during the past 45 years have led to the formation of an overarching conceptual model for integrating the spatial and temporal dynamics of pattern and process that define the contemporary tapestry of the Luquillo Mountains. In this model, the geological template and the geographic context change slowly over long time scales (figure 2-1) and are driven by processes such as tectonic activity, changes in physiography, and sea level changes. In contrast, regional climate, driven by topography, geography, and global climate, is potentially more dynamic and might change at the scale of centuries or less. Finally, frequent local disturbances provoke dynamism in the system at the annual or decadal scale. Together, geology, topography, regional climate, and disturbance produce heterogeneity or variation in the abiotic environment. The abiotic environment, the disturbance regime, and the regional species pool determine the composition of the biota, which then feeds back to modify the disturbance regime, the structural environment, and the abiotic environment. This last set of relationships—those among the abiotic environment, the structural environment, the biotic environment, and the disturbance regime—provides the focus for the remainder of this chapter. Definitions of key concepts that relate to these relationships supply critical background for further exploration of these concepts in later chapters.

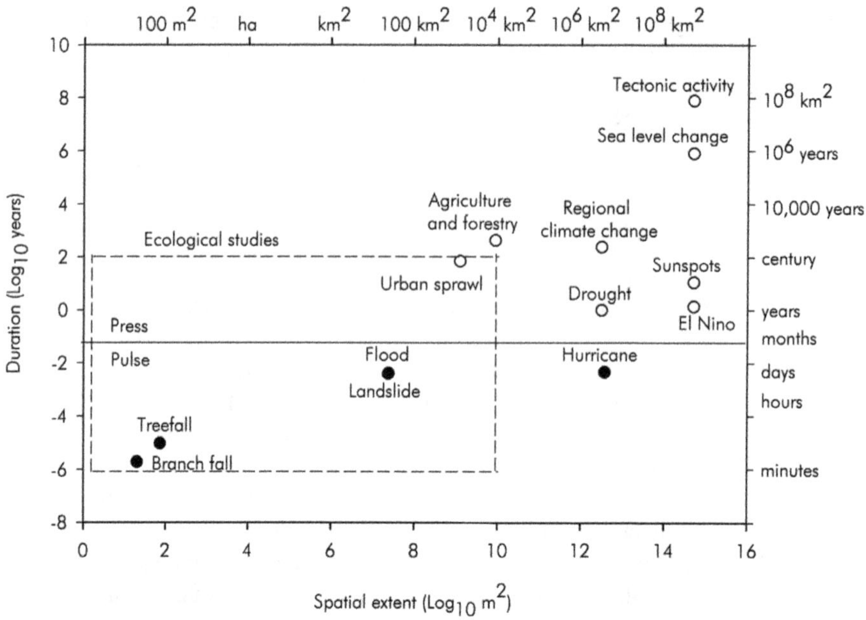

Figure 2.1 Diagrammatic representation of the temporal and spatial relationships of processes that interact to generate heterogeneity in the ecosystems of the Luquillo Mountains. "Duration" refers to the extent of time over which a process acts. Circles represent the approximate median values for each process in time and space. The horizontal solid line separates press disturbances (open circles), which act over long time periods, from pulse disturbances (solid circles), which act over short periods of time. The dashed box bounds the spatial and temporal extent of most ecological studies in Puerto Rico. The study of the full spatial or temporal extent of some processes requires collaboration with other disciplines (e.g., geology, paleoecology, climatology) or comparative studies using syntopic networks. For example, a full understanding of hurricanes requires information about storms with a wide range of physical characteristics, as well as information about storm impacts under different socioecological conditions. See the text for further explanation.

Components of the Environment of the Luquillo Mountains

The abundance and distribution of organisms, as well as the attendant ecosystem processes, vary across the landscape of the Luquillo Mountains in relation to underlying patterns of spatial heterogeneity and gradients of environmental factors. This variation reflects contemporary, past, and ancient processes operating at multiple spatial and temporal scales (figure 2-1) and results in the abiotic and biotic layers that imbue the current ecological tapestry of the Luquillo Mountains with structure. The abiotic and biotic layers interact within the contexts of geography (surrounding continents and oceans), climate, and regional species pools to determine the spatial and temporal patterns of the ecosystems of the Luquillo Mountains. A complex disturbance regime that includes hurricanes, tropical storms, landslides, and treefalls, as well as anthropogenic disturbances associated with forest management, urbanization, and other land uses, makes these patterns dynamic and increases the environmental variability. The following overview of environmental patterns and disturbance provides the necessary background for a discussion of ecological space and presages the more detailed treatments of these subjects in chapters 3, 4, and 5.

The Abiotic Environment

Geographic Context and Geologic Template: Changes over Long Time Scales

Puerto Rico lies as a fulcrum between the Greater and Lesser Antilles as a consequence of tectonic movements that have persisted for tens of millions of years and which continue to the present. More specifically, Puerto Rico was formed by volcanic activity and tectonic uplift during the Albian to the upper Eocene (120 to 140 million years ago), followed by erosion and later sedimentary deposition. The same geological processes that created Puerto Rico molded the topographic irregularities of the island, and the local variation in the topography adds heterogeneity to the larger-scale pattern. These ancient and ongoing processes placed Puerto Rico in the midst of oceanic currents, airsheds, and atmospheric fronts, thereby determining to a great extent the prevailing climatic conditions and disturbance regime of the island. Processes such as glacial fluctuations also determined the geographic relationship of Puerto Rico to other islands, continents, and bodies of water. Such historical factors shaped the biogeographic affinities of the island and influence the current biotic composition and richness of Puerto Rico.

Climate and Topography: Changes over Moderate Time Scales

The steep topographic gradient of the Luquillo Mountains (sea level to 1000 m in 8 km) produces landscape-scale variability in the local climate, abiotic characteristics (e.g., soil nutrient levels), and disturbance regime. Across this short spatial extent, the temperature decreases and precipitation increases from sea level to the summit. The direction of the prevailing winds from the northeast creates windward-leeward variation in temperature, rainfall, and windspeeds. Periodic shifts in ocean currents modify the near-shore environment and influence meteorological patterns over the land. Across the smaller scale of land forms, local relief contributes to spatial variation in the abiotic conditions (e.g., pH, oxygen content, soil moisture, temperature, insolation) because of differential exposure to sunlight and flows of water and air from ridges downslope to valleys. The interaction of topography and regional climate creates clouds that frequently shade the upper elevations in the Luquillo Mountains and consequently modify abiotic gradients. Many of the geophysical and geochemical characteristics of soils are determined by the nature of the rock underlying them (e.g., volcaniclastic versus igneous soils). Biogeochemical fluxes of nutrients vary with topography and soil characteristics. The topographic characteristics of the land influence both large-scale disturbances (the effects of hurricanes are more severe on windward sides of mountains) and small-scale disturbances (landslides are most likely to occur on steep slopes; see chapter 4).

Disturbance: Changes over Short Time Scales

Disturbance is a pervasive feature of ecological systems (Pickett and White 1985; Walker and Willig 1999; Willig and Walker 1999) and is a primary driver that produces temporal dynamics in the abiotic and biotic characteristics associated with geographic space. Although disturbance has been defined in a variety of ways (e.g., Sousa 1984), we follow the modified definition of Pickett and White (1985) that was offered by Walker and Willig (1999:3): a disturbance is a relatively discrete event in time and space that alters the structure of populations, communities, and ecosystems, including their attendant processes. Pulse (acute) disturbances are those that transpire over short periods relative to the dynamics of the focal system (e.g., hours, as in hurricanes in a tropical forest), whereas press (chronic) disturbances are those that transpire over longer periods (e.g., months, as in drought in tropical forests). Although the pulse/press dichotomy is a simplification of a continuum of disturbance characteristics, the principal disturbances in the Luquillo Mountains do separate into two classes according to their temporal attributes (figure 2-1). The effects of disturbance are detected as changes in the density, biomass, or spatial distributions of the biota; as alterations in the availability and distribution of resources and substrate; or as alterations in the physical environment. Consequently, disturbance creates patches, affects spatial heterogeneity, and modifies the spatial gradients of environmental factors.

The degree to which ecosystem characteristics remain unaffected by disturbance is referred to as *resistance* (figures 2-2[A] and 2-2[B]). The time required for an ecosystem to return to conditions that are indistinguishable from those prior to a disturbance represents the system's *resilience*. Systems that return more quickly to predisturbance conditions are more resilient than those that return more slowly (figures 2-2[C] and 2-2[D]). Nonetheless, even when two ecosystems are equally resilient, one can undergo more dramatic changes in its ecological characteristics than the other (figures 2-2[D] and 2-2[E]).

The combination of resistance and resilience to disturbance produces ecological patterns over time (figure 2-3). In some cases, disturbances can be sufficiently severe as to arrest ecosystem development for extended periods or to prevent the system from returning to predisturbance conditions (Carpenter 2001). Importantly, assessments of stability, resistance, and resilience are each scale dependent. For example, at a small focal scale such as a plot (e.g., square meters), sites might not return to predisturbance conditions with respect to the species composition for long periods, if ever. But at a larger focal scale, such as a watershed (e.g., square kilometers), sites might more quickly return to predisturbance conditions, because the variation among plots within watersheds is amalgamated into the larger spatial unit. Thus, at large scales, systems can be quite stable even if they are markedly unstable or even hypervariable at smaller constituent focal scales. In such situations, the larger system can act as a metacommunity that exhibits metastability (see Wu and Loucks 1995; Ingegnoli 2004).

The Biotic Environment

The composition of the biota of the Luquillo Mountains is determined by a complex combination of factors that act over a wide range of spatial and temporal scales. The insular species pool is regulated largely by the biogeographic factors of the island's

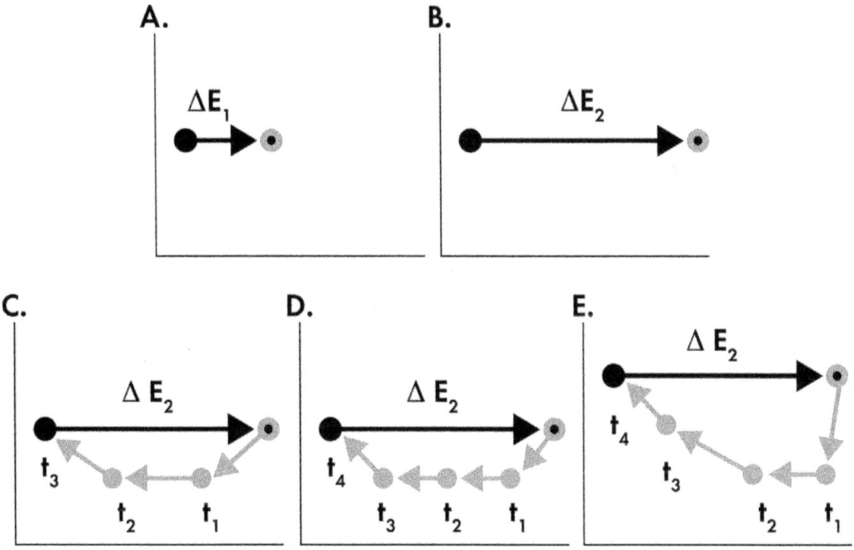

Figure 2.2 Representation of resistance and resilience in ecological space (E) defined by two axes. Each axis can represent aspects of the abiotic (e.g., soil moisture and soil temperature), structural (e.g., foliage height diversity and litter depth), or biotic (e.g., abundances of *Piper glabrescens* and *Piper hispidum*) environment. Change in these ecological attributes as a result of a disturbance (ΔE, illustrated by a solid arrow) quantifies resistance. Panel (A) illustrates a more resistant system (i.e., ΔE_1 is small) than does panel (B) (i.e., ΔE_2 is large). The time needed for a system to return to predisturbance conditions quantifies resilience (number of gray arrows). Panel (C) illustrates a more resilient system (i.e., the time to recovery is short [three time steps]) than does panel (D) (i.e., the time to recovery is long [four time steps]). Although two systems can be equally resilient (i.e., panels [D] and [E] both represent a return to predisturbance conditions in four time steps), secondary succession might evince different trajectories of recovery, with some moving the system to states quite distinct from those of the pre- or even postdisturbance (immediate) environmental conditions (i.e., compare panel [E] to panel [D]).

size and location, the distance from source biotas, and the colonizing ability of different species. Thus, the biota of Puerto Rico (both present and fossil) lacks many groups of large mammals (Willig and Gannon 1996) and birds (Waide 1996) that are characteristic of mainland tropical forests. Invasions and extinctions continue at a slow-to-moderate pace (e.g., five bird species lost within the past 100 years [Raffaele 1989]) and contribute to species turnover and spatial heterogeneity. On the island, a north-south rainfall gradient and an east-west disturbance gradient (figure 2-4), as well as biotic interactions such as competition, predation, and mutualism, affect the distribution of species. Local variation in the topography, edaphic characteristics, and a multitude of abiotic factors adds heterogeneity to the larger-scale pattern.

Legacies of human intervention complicate the biogeographic patterns. Habitat modification is severe and widespread in Puerto Rico; at one time, forests occupied less than 5 percent of the island (Birdsey and Weaver 1982). Socioeconomic changes since 1950 have led to gradual reforestation, but new forest

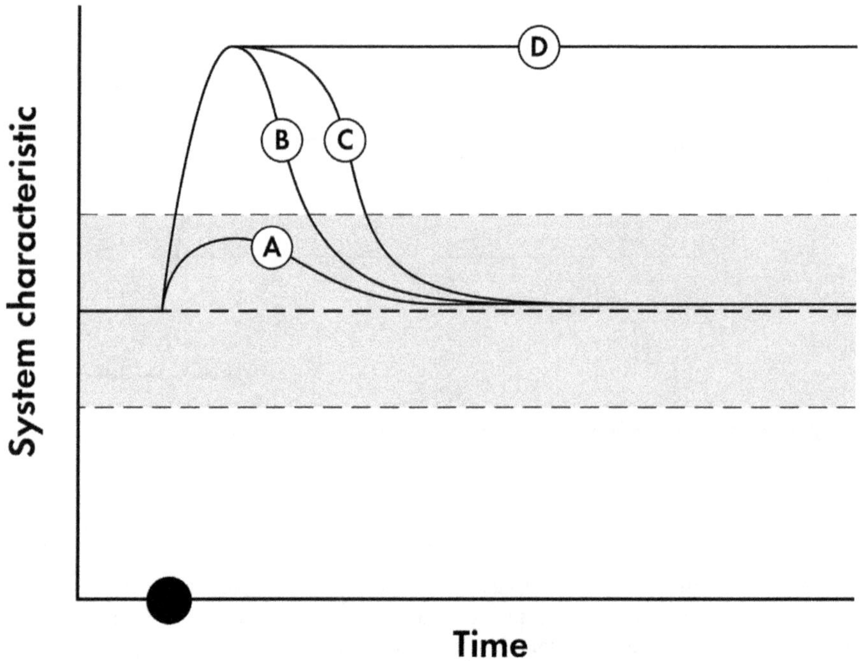

Figure 2.3 Representation of various idealized types of responses to disturbance. Solid lines represent trajectories of response after a disturbance event (solid circle); the long-term baseline conditions in the absence of disturbance are indicated by the gray shading. Response A is the most resistant, as the system characteristics after disturbance never exceed those of baseline. Responses B, C, and D are equally resistant, but they differ in their resilience. Response B is more resilient than response C, as the system characteristics return to baseline more quickly in the former than in the latter. Response D does not exhibit resilience; the disturbance sufficiently alters the system so that it occupies a new state rather than returning to baseline conditions. Adapted from Zimmerman et al. (1996).

patches accumulate native species slowly and might be dominated at some scales by introduced taxa (Aide et al. 1996; see also chapter 8). Although the introduction and establishment of introduced species have increased the total number of species on the island, some introduced species have also endangered endemic animal diversity. The roof rat (*Rattus rattus*), the small Indian mongoose (*Herpestes auropunctatus*), and the earthworm (*Pontoscolex corethrurus*) are examples of introduced species that have affected the distribution and abundance of native fauna (Willig and Gannon 1996; Zou and González 1997). Active management of the biota encourages the persistence of some species and can determine the bounds of their distributions (e.g., the Puerto Rican parrot, *Amazona vittata* [Snyder et al. 1987]).

The biota responds to abiotic gradients in the Luquillo Mountains and interacts with them to produce observable ecological patterns (see chapter 3). The forest canopy moderates temperature in the understory and traps moisture, whereas openings in the canopy lead to increased sunlight at the forest floor, which increases spatial variation in the temperature at the litter layer (Odum et al. 1970; Devoe 1989; Scatena 1990). Fungal mats interconnect dead leaves in the litter and in the canopy, reduce the likelihood of export from terrestrial systems during heavy rains, and enhance the retention of limiting nutrients via their incorporation into biomass (i.e., immobilization) (Lodge and Asbury 1988). Root mats and root grafting among individuals (Basnet et al. 1992) likely stabilize the soil and reduce erosion. Plant species assimilate and concentrate nutrients and trace elements differentially, thereby producing considerable spatial dynamics in biogeochemicals (Scatena et al. 1993). Earthworms mix and aerate the soil and provide routes for the flow of groundwater through the soil (Zou and González 1997). Such mediating influences of the biota are particularly important under the conditions that can result from disturbance (Willig and McGinley 1999).

The distribution of species in the Luquillo Mountains responds to both geological layers and abiotic layers of the current tapestry, and it modifies them in turn. For example, the species composition and physiognomy of plants vary markedly along the elevational gradient in the Luquillo Mountains (Crow and Grigal 1979; Weaver 1991; Gould et al. 2006; also see chapter 3). Cool and wet high elevations support elfin forest, characterized by an intermeshing root mat on the forest floor and small trees festooned with mosses and lichens reaching up to a 3 to 10 m canopy. Lower elevations are warmer and less wet and support tabonuco forest, which is characterized by an extensive litter layer and buttressed trees forming a canopy at 20 to 22 m. Soils at mid- to high elevations that are poorly drained because of the topography support almost pure stands of sierra palm (*Prestoea montana*).

In the Luquillo Mountains, the species richness of most taxa decreases with increasing elevation (Waide et al. 1998), although some groups evince a modal distribution with a maximum in the lowlands (see, e.g., Alvarez 1997). Species appear and disappear along the elevational gradient, and introduced species become less common with distance from human disturbances such as roads (Olander et al. 1998). Variation in the community composition can affect biogeochemical processes, as well as the capacity of the biota to moderate the environment after disturbance. Thus, biotic variability adds a vital layer to the tapestry that is the Luquillo Mountains and increases the complexity of interactions involving the abiotic and biotic environments and the disturbance regime.

Gradients and the Dynamics of Pattern and Process

The proposal that led to the foundation of the Luquillo LTER program (Luquillo LTER 1988) addressed the challenge of linking point and stand data to landscape-scale patterns and processes through simulation modeling: "We propose to develop an explicit scheme for translating geographical information, derived from geographical space, into model parameter space (equivalent to ecological space), using a gradient approach" (p. 14). The term "ecological space," referring to environmental characteristics arrayed along gradients in geographic space, was first used by Minchin (1987) and draws on ideas from Whittaker (1956) and Austin and Cunningham

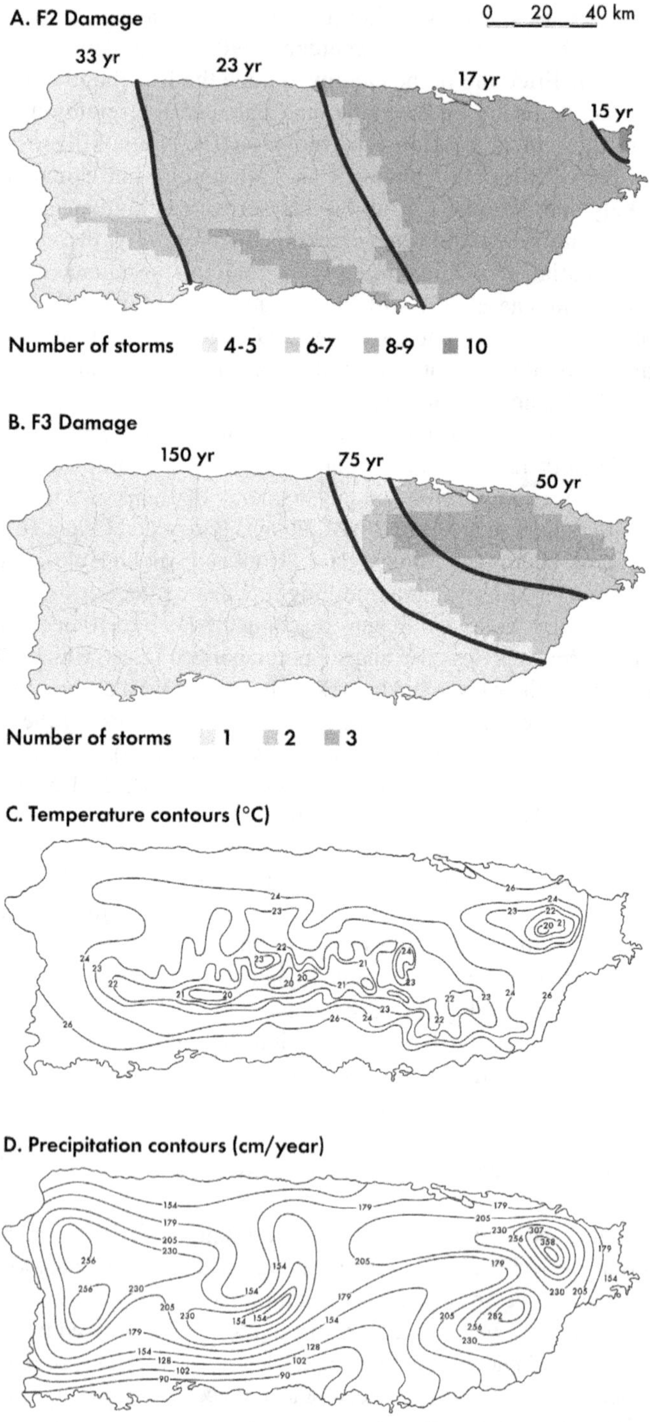
A. F2 Damage
0 20 40 km
33 yr
23 yr
17 yr
15 yr
Number of storms 4-5 6-7 8-9 10
B. F3 Damage
150 yr
75 yr
50 yr
Number of storms 1 2 3
C. Temperature contours (°C)
D. Precipitation contours (cm/year)

(1981). Our initial use of the idea of ecological space did not acknowledge the complications that disturbance might cause with regard to translating geographic information into model parameters. Disturbance can disrupt or create gradients by altering the mapping of ecological space on geographic space. Six years later, after we had experienced the effects of a major hurricane, a more sophisticated explication of the relationship between geographical location, resource gradients, and disturbance formed the rationale of the second LTER proposal (Luquillo LTER 1994). Five premises derived from our own observations (Hall et al. 1992a, 1992b) and from the literature (Keddy 1991; Gosz 1992) formed the basis for our research approach.

1. The distribution of organisms and associated rates of ecosystem processes are related to a limited number of spatial gradients of environmental factors (e.g., temperature, sunlight, soil moisture, and soil nutrient levels) in the landscape.
2. Each physical position on the landscape (i.e., geographical space) has a representation in n-dimensional gradient space (or ecological space). This representation in ecological space is determined by interactions of geography, geology, climate, topography, disturbance history, and the biota, which collectively determine the conditions along each primary gradient.
3. Disturbance affecting a spatial position in the landscape displaces conditions in ecological space. Disturbance modifies characteristics with respect to many or all environmental factors or conditions, resulting in a new spatial configuration of ecological space. Describing displacements in ecological space potentially allows for a mechanistic understanding and predictions of changes in distributions and rates of processes.
4. An explicit consideration of the association between geographic space and ecological space facilitates the comparison of different types of disturbances. Different disturbance types have characteristic directions of displacement in ecological space. The size and, especially, the intensity of the disturbance influence the magnitude of the displacement in that characteristic direction. The frequency of disturbances, in conjunction with the response time, influences the impact of subsequent disturbances. If a subsequent disturbance (or a new type of disturbance) results in a further displacement before the response to an initial disturbance is complete, new and unique positions in ecological space might result.

Figure 2.4 Disturbance frequency decreases from east to west across Puerto Rico, based on the number of storms from 1886 to 1996. The number of storms that were classified as F2 (extensive blowdowns; panel [A]) or F3 (forests leveled; panel [B]) on the Fujita scale defines damage classes and return intervals for each damage class (modified from Boose et al. [2004]; damage classes redrawn into new map projections with new shading). Spatial variation characterizes temperature (panel [C]) and precipitation (panel [D]) across Puerto Rico (modified from Gannon et al. [2005]; values converted from English to metric system and inserted into contours of the map). The Luquillo Mountains, located in the northeast corner of the island, experience considerable elevational variation in temperature and precipitation (panels [C] and [D]) and lie in one of the most frequently affected areas of Puerto Rico (panels [A] and [C]).

5. In this conceptual framework, resistance and resilience to disturbance can be defined and quantified. Resistance is related to the displacement in ecological space for a given disturbance. Resilience is the time required in order to return to the original position (or a position much like the original position) for a given displacement.

Our conceptual approach, and especially its development from an energetic perspective (see below), helped to integrate studies at various levels of biotic organization by providing a framework that was intuitively attractive to population, community, and ecosystem ecologists. The idea of ecological space shared concepts with niche theory (Hutchinson 1958, 1965) and thus provided common intellectual ground across subdisciplines. However, the fundamental niche stays relatively constant over ecological time scales, whereas disturbance can modify ecological space over relatively short periods, leading to community reorganization after disturbance. This conceptual approach integrates studies at various levels of biotic organization and provides a mechanism for synthesis and modeling that is extremely powerful because of its quantitative nature. Understanding gained from this approach is directly applicable to the evaluation of techniques for the ecological management of tropical forests under different disturbance regimes.

Dynamics of Ecological Space and the Biota

The mechanisms by which the abiotic environment determines the distribution of species, the composition of communities, and the nature of ecosystem processes act on individual organisms. The currency of that interaction is energy. More specifically, the existence of an organism under particular environmental conditions depends on the energy balance of the organism (Shelford 1951; Maguire 1976; Dill 1978; McNab 1980; Kitchell 1983; Root 1988; Covich 2000). The dynamic nature of the determinants of energy balance arises from variation in abiotic factors, including those associated with the disturbance regime. In the following sections, we review these processes from the perspective of the species (niche-based), the disturbance regime (disturbance-based), and the community (succession-based).

Energetic Basis of Organismal Responses to the Environment

Organisms respond to gradients of and heterogeneity in environmental characteristics as a consequence of their morphological, physiological, and behavioral attributes. These attributes essentially determine the fitness for organisms at any location. These concepts are rooted in niche theory (Hutchinson 1958, 1965) and have been amplified in the context of a common currency, namely, energy (Hall et al. 1992b). The presence, abundance, and behavior of a species are linked intimately to the energetic costs and benefits associated with living in a particular geographic position. A species can persist in an area only if the long-term energy gains equal or exceed the costs, thereby facilitating growth and reproduction (Hall et al. 1992b and references therein). The effects of abiotic resources (e.g., low levels or high levels of nutrients), as well as biotic interactions (e.g., the presence of consumers or mutualists), can be incorporated into such a conceptual model as energetic costs or benefits. For example, in order to survive and persist in an area characterized by low nitrogen availability, an organism must invest in phenotypic characteristics that allow it to accumulate nitrogen, resulting in reduced energy allocation to reproduction. In the absence of mutualism, this tradeoff might narrow the range of acceptable nitrogen levels to a subset of what exists in ecological space. Similarly, the energy used to escape predation, through chemical defenses (e.g., toxins), morphological structures (e.g., thorns or thick cuticles), or behavioral activities (e.g., lunar phobia enacted to avoid visual predators at night), must be diverted from energetic resources that could otherwise be allocated to growth and reproduction. Nonetheless, areas with abundant resources might not support the persistence of a species if the cost of predation is high. Alternatively, the presence of mutualists such as root mychorrizae can reduce the cost of acquiring essential nutrients such as nitrogen or phosphorus, thereby facilitating the persistence and reproduction of a species in a habitat that otherwise would be impossible to thrive in energetically. Energetic tradeoffs exist because the cost of investment in any set of phenotypic characteristics associated with the soma reduces possible investments in reproductive output.

Energetic tradeoffs are particularly relevant for understanding elevational or latitudinal distributions of species because costs (respiration rates) and benefits (assimilation rates) vary with temperature in nonlinear ways (figure 2-5). Respiration is the metabolic cost of executing vital physiological processes. The respiration rate is positively and often exponentially related to the temperature for extensive portions of the thermal gradient (i.e., a Q_{10}-type response). Assimilation rates also increase with temperature, but they do so in a near-asymptotic manner. That is, the rate of increase decreases with increasing temperature (i.e., saturates), essentially reaching zero. The difference between the rates of assimilation and respiration represents the net profit (or loss) associated with life at any point in the thermal gradient. Because of the general shapes of the response curves for assimilation and respiration, the difference results in a net profit curve that is modal or Gaussian in form (figure 2-5). Such net profits are available for allocation to biomass accumulation (growth) or reproduction (figure 2-6). Thus, individuals of a species might be found in environments in which they can (1) subsist only for short periods of time, (2) survive indefinitely but not reproduce (sink habitats), or (3) persist and reproduce beyond replacement (source habitats) (Pulliam 1996). Other biotic or abiotic factors effectively shift the cost or benefit curves, expanding or contracting the thermal range in which individuals maintain source or sink populations.

Disturbance and Biotic Response

Disturbance is one of the most important factors that elicit changes in the structural or functional aspects of ecosystems. Structural elements include community attributes such as species richness, species diversity, guild diversity, species composition, rarity, and species dominance. Functional elements are related to ecosystem processes such as decomposition and production and include primary productivity, secondary productivity, decomposition rates, mineralization rates, and nutrient fluxes. The sequence of changes in these characteristics that follow disturbance can be visualized as a vector or trajectory of response.

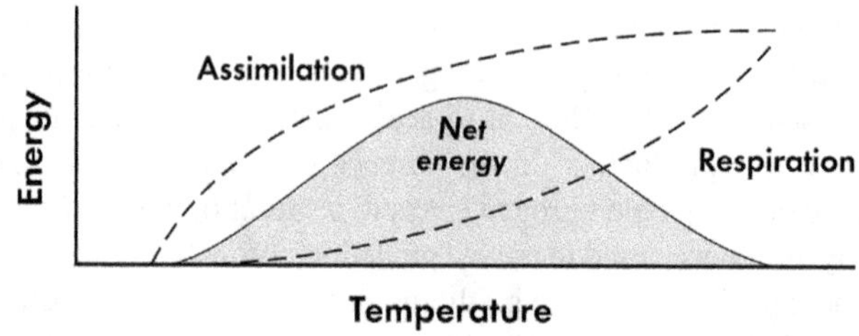

Figure 2.5 The joint effects of assimilation and respiration rates determine whether particular regions of ecological space (here associated with temperature) are occupied by particular species. The exact form and location of rate curves are species-specific and determine the net energy balance (represented by the area under the bell-shaped curve) available for allocation to growth and reproductions. Modified from Hall et al. (1992a).

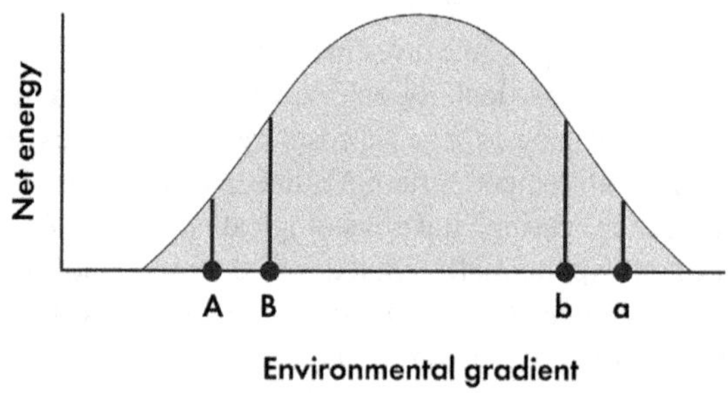

Figure 2.6 The net energetic profit based on the difference between assimilation and respiration is a Gaussian or bell-shaped curve, with the presence of sink and source populations predicated on considerations of energy allocation. Positions along the environmental gradient between points B and b provide sufficient energetic rewards so that populations can produce an excess of individuals (positive growth) and may colonize other areas. Positions along the environmental gradient between points A and B or points a and b do not allow for a population increase, even though individuals can persist there indefinitely. Consequently, populations at these locations in the environmental gradient must be maintained in the long term by immigration from source areas. At positions along the environmental gradient <A or >a, individuals cannot survive indefinitely; thus, a species is represented in those regions by only transient individuals. Modified from Hall et al. (1992a).

Disturbance is caused by an agent or entity (e.g., the winds of a hurricane, the heat of a fire) that initiates changes in the spatiotemporal characteristics of the ecological system of interest, often detected as changes in the amount or distribution of biomass. Most ecological systems are subject to a number of disturbance agents. The combination of agents at a particular place represents the disturbance regime. Any particular disturbance event might alter the frequency, extent, or intensity of other disturbances (figure 2-7). Such interactions can be additive, synergistic, or antagonistic and are important considerations when attempting to understand disturbance and response in ecological systems (Walker and Willig 1999; Willig and Walker 1999).

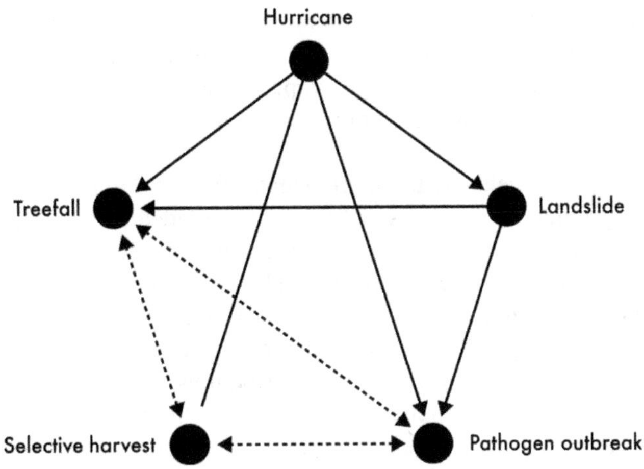

Figure 2.7 Representation of aspects of the disturbance regime for the Luquillo Mountains as embodied by the interactions between disturbance elements (e.g., hurricanes, landslides, treefalls, selective harvest, pathogen outbreak). For simplicity of exposition, only a few of all possible elements are illustrated. Arrows represent the influence of one element of a disturbance regime on another (e.g., the occurrence of a hurricane increases the likelihood of subsequent disturbance from landslides). Solid lines indicate strong influences, whereas dashed lines indicate weak influences. Double-headed arrows represent reciprocal causality or effects. Modified from Willig and Walker (1999).

Because the Luquillo Mountains are situated on an island in the Caribbean with a long history of human settlement and are in the path of the Atlantic Trade Winds, they have a complex disturbance regime. The regime (or a portion of it) may be represented as a number of interacting agents including hurricanes, landslides, treefalls, selective harvest, and pathogen outbreaks (figure 2-7). The occurrence of one agent of disturbance (e.g., a hurricane) might enhance the likelihood of subsequent disturbances (e.g., landslides). Moreover, some agents of disturbance might have reciprocal effects: treefalls enhance the likelihood of pathogen outbreaks, and pathogens enhance the likelihood of treefalls (Goheen and Hansen 1993; Webb 1999). At any point in time, the disturbance regime might enhance or reduce the spatial heterogeneity in local climatic or abiotic characteristics, thereby affecting the abundance and distribution of species across the landscape.

Heterogeneity or variability in the environmental characteristics to which organisms respond can arise in a variety of ways, including as a result of topography, disturbance, and succession. All of these sources of variation interact within a landscape, and in turn they are affected by their spatial context. In the Luquillo Mountains, topographic variation generates gradients in important climatic drivers such as solar insolation, temperature, and precipitation (figure 2-8) and produces environmental heterogeneity of abiotic factors (Cox et al. 2002). At broad spatial extents, we evaluate how elevational variation in key environmental drivers produces gradients that induce variation in populations and communities, as well as in associated biogeochemical processes. At narrower spatial extents, we describe how environmental variation associated with the local topographic position—ridge, slope, upland valley, and riparian valley (i.e., the catena)—translates into population-, community-, and ecosystem-level variability.

The response of organisms to environmental factors defines their niches and facilitates the prediction of species distributions, provided that key environmental factors have a consistent association with geographic space over time. Modeling algorithms (e.g., GARP) (Peterson 2001) can be used to define key factors associated with species occurrences. When combined with spatially explicit environmental data, these algorithms predict the fundamental niches of organisms. Differences between predicted and actual distributions might point to biotic interactions that affect realized niches or to dispersal limitations. However, these models often are based on average conditions that do not reflect temporal extremes, and as a result they might predict overly broad distributions. Consequently, such models might fail to capture the full temporal variability in the spatial distribution of organisms.

Because of their particular niche characteristics, species are predisposed to exist under environmental conditions associated with particular geographic areas. However, points in geographical space do not maintain a constant relationship with ecological space because of disturbances and biotic responses, including succession. Thus, species can persist in a particular area only if they can survive and reproduce under the environmental conditions that occur over long time scales relative to the life of an organism. Because the relationship between geographic space and ecological space is dynamic, the relationship between the physical template and the distribution and abundance of animal, plant, and microbial species cannot be understood without reference to the disturbance regime. As the mapping of ecological space to geographical space changes, species co-occurrences might be affected, with consequent cascading effects on competitive, predatory, or mutualistic relationships (figure 2-9; see also chapter 6). The biota's response to the dynamic relationship between geographic and ecological space is reflected in successional changes that have their origins in disturbance.

Long-term responses to a disturbance are determined by the postdisturbance environment, which includes the character and heterogeneity of the abiotic environment, the composition of the surviving organisms, and the structural legacies of the disturbance event. However, the characteristics of the postdisturbance environment also are influenced by preceding disturbances, so that at any one time the biota generally is responding to multiple historical disturbances. It is this integrated response that determines the trajectory of the community composition, structure, and function (i.e., successional dynamics) over time after any particular disturbance event. Knowledge of the natural variability (Landres et al. 1999) of an ecosystem is critical to an understanding of responses to a specific disturbance, including those involved in human-managed ecosystems (chapter 7). Consequently, the interplay between disturbance and biotic response is best understood within the context of succession.

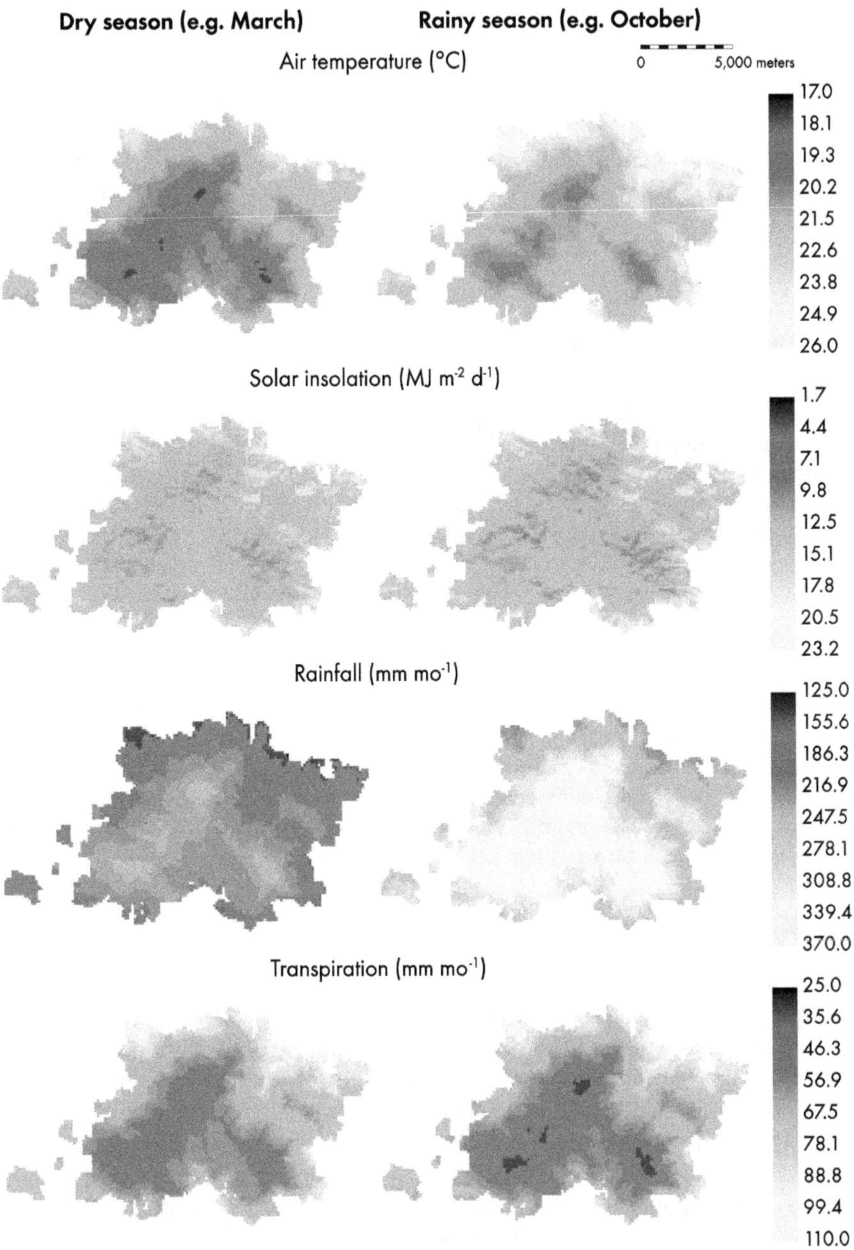

Figure 2.8 Spatial variation in temperature, insolation, precipitation, and transpiration in the Luquillo Mountains based on the spatially explicit model TOPOCLIM (Wooster 1989). Slope, aspect, and elevation are used as input data for the model. Historical climate data are used to parameterize model equations that estimate climatic variables. Simulated air temperature (°C), solar insolation (MJ m^{-2} day^{-1}), rainfall (mm/month), and transpiration (mm/month) are shown for dry and rainy seasons (Wang et al. 2002). Values increase from violet to red in the color spectrum.

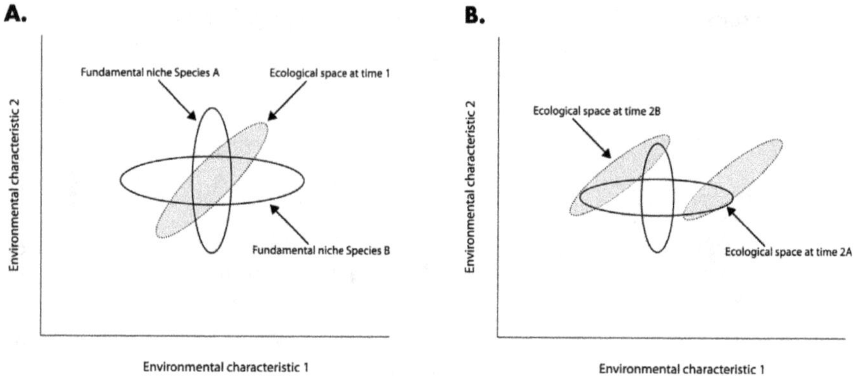

Figure 2.9 In order for a species to persist at a geographic location, it must be able to survive the full range of environmental conditions and resources that occur there over time. Alternatively, the species' behavior (e.g., emigration or migration) can result in the avoidance of unfavorable environmental or resource conditions. Moreover, variations in the ecological attributes of geographic space over time affect species interactions, niche breadth, and co-occurrence at a smaller spatial scale. (A) At a particular point in geographic space, the ranges of values that exist for each of a number of environmental characteristics (e.g., temperature, rainfall) define the ecological space at that point. A species can occur at this geographic point if its fundamental niche overlaps with the corresponding ecological space. Species with fundamental niches that overlap within the existing ecological space (gray shading) can co-occur. (B) As the result of a disturbance, the values of environmental characteristics might change, redefining the ecological space at time 2. If ecological space shifts to position 2A, only one species can persist under the new conditions; if ecological space shifts to position 2B, both species can persist, but they cannot co-occur. In systems in which disturbance creates shifts in ecological space that are frequent compared to species' generation times, broad fundamental niches would be favored (Waide 1996).

Succession

Disturbance initiates succession, influences subsequent trajectories in abiotic and biotic characteristics, and moderates successional rates, endpoints, and durations. We follow a basic conceptual model (Willig and Walker 1999) when attempting to understand how disturbance and succession interact to produce a spatially and temporally dynamic tapestry in the Luquillo Mountains (figure 2-10). Regions of geographic space might be subject to a disturbance, such as a hurricane. A hurricane alters abiotic conditions such as temperature or moisture, as well as the distribution of biomass or necromass and the composition or abundance of species (see Walker et al. 1991, 1996). In essence, the abiotic and biotic conditions of a point in geographic space quickly are altered as a result of the initial disturbance. In the Luquillo Mountains, hurricanes kill and uproot trees (Walker 1991), causing gaps in the forest canopy (Brokaw and Grear 1991). Gaps in the canopy result in higher temperatures and lower humidity throughout a cylindrical area from the top of the canopy to the forest floor (Fernández and Fetcher 1991). Biomass from affected vegetation becomes necromass and is redistributed to the forest floor (Lodge et al. 1991), altering the quality, quantity, and dispersion of resources and substrates. These conditions subsequently influence the abundance and distribution of microbial, plant, and animal species (Walker et al. 1991). The spatial and temporal scales on which organisms integrate or perceive environmental variability in part determine the severity of a disturbance event, as well as subsequent trajectories of change.

Residuals and Legacies

Disturbance directly alters the abiotic and biotic characteristics of geographic space. We distinguish between the immediate manifestations of a disturbance (residuals [Clements 1916]) and the subsequent dynamic nature of the ecosystem as a result of the existence of these residuals (legacies [Vogt et al. 1997]). Residuals can be abiotic (e.g., mineral soil exposed after a landslide, redistribution of rocks and sediment in a stream after a hurricane) or biotic (e.g., coarse woody debris deposited on the forest floor after a hurricane, community composition after selective harvesting; see chapter 4). The relative importance of abiotic and biotic residuals depends on the disturbance's type, frequency, intensity, and extent (see chapter 5). For example, the response to a landslide that exposes mineral soil will be strongly

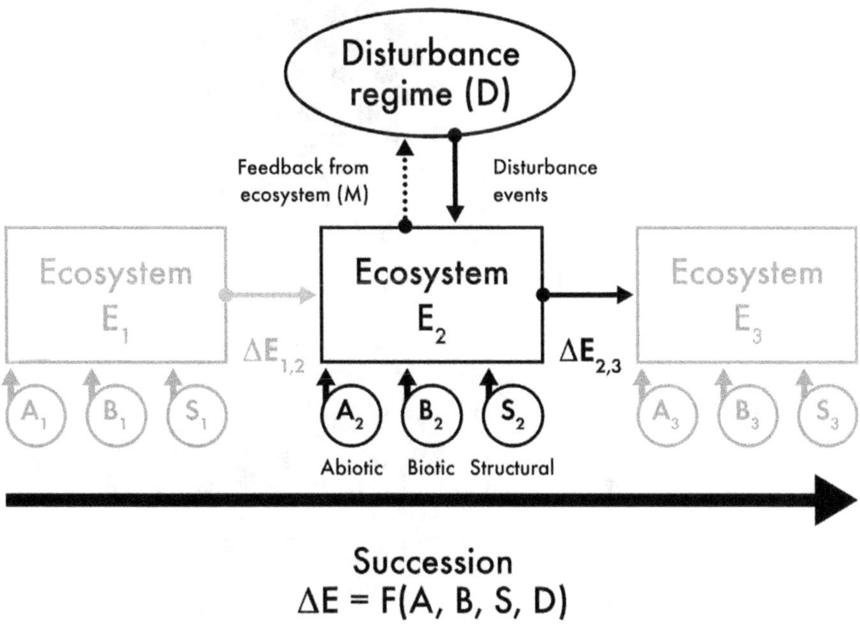

Figure 2.10 Conceptual model linking disturbance and succession as the mechanistic basis for the temporal and spatial complexion of the ecological tapestry of the Luquillo Mountains. Abiotic, biotic, and structural environments (A, B, and S, respectively) interact with one another and the disturbance regime (D) to determine changes in the state of an ecosystem (E). At the same time, the state of an ecosystem can influence the disturbance regime (feedback M). Subscripts indicate the location of the system in time. Modified from Willig and Walker (1999).

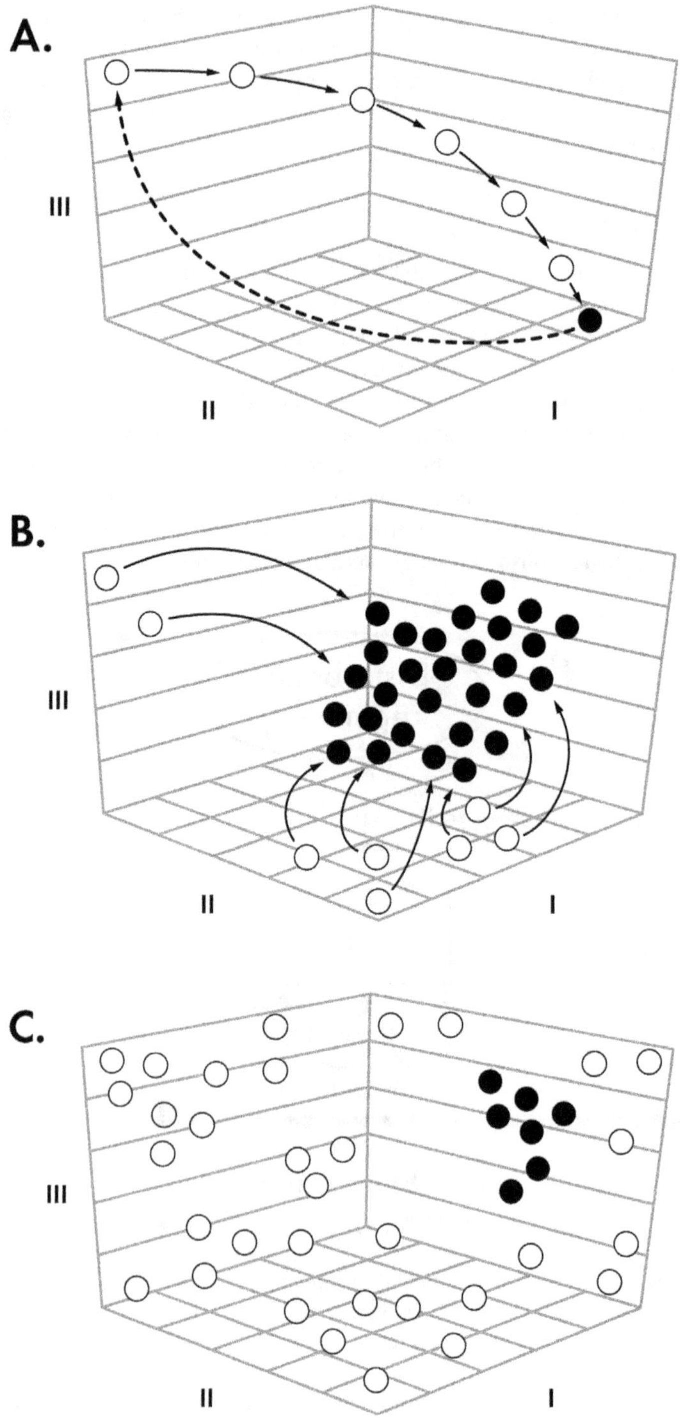
A.
III
II
I
B.
III
II
I
C.
III
II
I

influenced by abiotic residuals, whereas the response to a pathogen outbreak will be strongly influenced by biotic residuals. Moreover, the composition and configuration of the landscape in which a disturbance is situated can affect the importance of biotic and abiotic residuals, because geographic proximity can determine the likelihood of dispersal into an area by native or introduced species.

Residuals influence ecosystem response in many ways (figure 2-11) (Clements 1916; Franklin et al. 2000). Alterations in geomorphology resulting from intense rainfall change hydrologic patterns, as well as soil water and nutrient availability, modifying multiple environmental gradients in geographic space. Organisms that survive a disturbance provide a springboard for succession, and the composition and distribution of the postdisturbance community can have strong effects on the ecosystem that develops under the new abiotic conditions. Organisms that fail to survive a disturbance might alter ecosystem structures and processes (e.g., hydrology) or provide long-term sources of energy and nutrients (Vitousek and Denslow 1986; Zimmerman et al. 1995). For example, a tree blown over and killed by a hurricane (a residual) creates a legacy in the nutrient composition of the soil. Thus, a disturbance event can have strong effects on the abiotic and biotic environments and might even alter the geomorphic template of an ecosystem. The effects of intense disturbances might be apparent in the ecosystem even after subsequent disturbance events. Legacies of previous disturbance events, some of them dating back hundreds of years, contribute to the present-day structure of the Luquillo Mountains (Scatena 1989; García-Montiel and Scatena 1994; Aide et al. 1996; also see chapters 4 and 5).

Figure 2.11 Ecological space may be envisioned as a multidimensional hypervolume that reflects the critical abiotic, biotic, and structural components of a system. Multivariate data reduction methods can be used to reduce these multiple dimensions to a few components (i.e., I, II, and III) that represent the salient features of variation among sites in ecological space. (A) Changes in the ecological characteristics of a site over time facilitate the quantification of the direction and magnitude of successional change. Successional trajectories (solid arrows) are envisioned as the temporally linked ecological conditions of a site (circles) over time in response to some initial disturbance (dashed arrow). In this particular instance, the characteristics of the site return to the predisturbance state in six time increments. (B) Prior to Hurricane Hugo, the tabonuco forest in the vicinity of El Verde Field Station comprised a number of sites, most of which shared similar ecological conditions (solid circles). As a result of numerous minor disturbances (e.g., treefalls), some sites (six open circles in the lower right of the panel) were slightly displaced from the ecological conditions of the "matrix." More intense disturbances, such as landslides, altered the ecological characteristics of sites to a greater degree (open circles in the upper left of the panel). Secondary succession (arrows) occurred as these sites changed over time and converged on the characteristics of the original matrix. (C) As a result of Hurricane Hugo, the ecological configuration of sites in the Luquillo Mountains was altered. Only a few sites (solid circles), generally those protected on the leeward sides of ridges, remained within the range of conditions previously characteristic of the original matrix. Most sites (open circles) were variously displaced in ecological space as a consequence of the hurricane. Modified from Willig and Walker (1999).

Legacies of Natural Disturbances

Residuals from natural disturbances can include characteristics that arise from changes in geomorphology, the modification of environmental conditions and resources, the distribution of surviving organisms and propagules, or alterations in the structural heterogeneity (Franklin et al. 2000). Both abiotic and biotic residuals influence subsequent responses to disturbance, and that influence can manifest as persistent legacies in the forest structure, composition, or function. Spores, seeds, and seedlings that survive a disturbance can initiate succession with minimal delay. However, thick layers of litter might change the rates of germination of surviving seeds (Guzmán-Grajales and Walker 1991), leading to changes in plant species composition that might persist for decades. Modified environmental gradients in geographic space influence rates of productivity and decomposition. Structural residuals can provide critical habitat for other species and moderate change in the microclimate. Living and dead structural elements can persist long after a disturbance and affect the trajectory and rate of succession through legacy effects on soil nutrients and forest structure. Because recurrent natural disturbances, even when infrequent on ecological time scales, occur within the evolutionary experience of organisms in the Luquillo Mountains, the persistence of biotic residuals can increase the resilience of ecosystems after disturbances. However, anthropogenic disturbances, although they sometimes share characteristics with natural disturbances, can have quite different effects on populations, communities, and ecosystems.

Legacies of Anthropogenic Disturbance

The intensities of anthropogenic disturbances differ greatly, from the removal of selected plant parts to the mass harvest of entire populations or communities. In the most intense anthropogenic disturbances, such as deforestation or agriculture, a severe effect is associated with the small quantity of biotic residuals (including structure) that remain in the postdisturbance environment. Anthropogenic disturbances often remove large quantities of the biomass of an ecosystem and leave behind an environment that is greatly altered, nutrient-poor, and often highly homogeneous. Moreover, repeated disturbances that are imposed by design (as in annual tilling and biocide application) forestall natural successional responses (Franklin et al. 2000). Anthropogenic disturbances generally have severe effects on both abiotic and biotic components of the ecosystem, often decoupling the covariation of these components in space and time, and thus moving the ecosystem's characteristics farther from those of the original system in ecological space than would a natural disturbance. The resilience of an ecosystem after anthropogenic disturbance might be low because of severe modifications to the environment that include the absence of biological residuals.

In many instances, the abiotic environment resulting from anthropogenic disturbance constitutes conditions that are beyond the natural variability (Landres et al. 1999) of the system and thus outside the evolutionary experience of native organisms. In these circumstances, the parameters defining the fundamental niches of many native species might not overlap with the ecological space created by a disturbance, and succession might be arrested or co-opted by introduced or immigrating species that are better adapted to the novel combinations of environmental characteristics. The trajectory of response in these cases might be unique and result in new local combinations of species. Nonetheless, intervention by humans (e.g., the establishment and cultivation of selected tree species) can enhance ecosystem resilience. One of the goals of human intervention in this case is to modify the abiotic environment to conditions that occur within the range that is acceptable to native species (Landres et al. 1999). If the abiotic environment returns to predisturbance conditions, biological processes create positive feedbacks that enhance the rate of succession. In some instances, anthropogenic disturbance has created novel ecosystems (sensu Chapin and Starfield 1997) that cannot be restored to conditions within their historical range of environmental variability (Veblen 2003) and which need to be managed using innovative approaches. In the words of Seastedt et al. (2008:548), "In managing novel ecosystems, the point is not to think outside the box, but to recognize that the box itself has moved."

An example of the long-term effect of anthropogenic disturbance comes from the Luquillo Forest Dynamics Plot on the eastern slopes of the Luquillo Mountains. In the past, parts of this 16 ha plot were subjected to different intensities of use, resulting in four distinct categories of canopy cover in 1936 (figure 2-12). The ordination of data from tree surveys conducted in 1989 produced groupings that corresponded closely to the degree of historical anthropogenic disturbance, with secondary relationships to soil type and topography (Thompson et al. 2002). Natural disturbances (hurricanes) and forest development in the intervening period failed to mask the existence of anthropogenic disturbance or the relative severity of disturbance in different parts of the plot. Moreover, the community composition of other organisms (e.g., snails) and functional diversity (e.g., bacteria) showed differences among these same cover classes (Willig et al. 1996, 1998, 2007).

Many human activities that modify ecosystems are usefully viewed from the perspective of regimes of disturbance, rather than as isolated disturbance events (see chapter 4). Indeed, the repeated and systematic application of treatments (i.e., multiple disturbance elements) to prevent recovery might be the salient feature that distinguishes anthropogenic disturbances from natural disturbances. For example, roads in the Luquillo Mountains are initially constructed by clearing corridors of vegetation, bulldozing the land to a convenient configuration, and paving those cleared surfaces with asphalt. The maintenance of roads represents a human-directed disturbance regime designed to sustain ecological conditions at a desirable ecological state and retard succession. Vegetation is cut along the periphery of roads, debris from landslides and landslips is removed, and surfaces are repaired when substrate erosion degrades road surfaces. Similarly, agriculture in the Luquillo Mountains was a human-directed disturbance regime that involved deforestation, plowing, and planting crops. The application of fertilizers and biocides (e.g., fungicides, herbicides, insecticides, and rodenticides), weeding, and replanting are all activities of a human-initiated regime of disturbance that has a profound effect on the ecological state of the system. Conservation, restoration, and reclamation efforts are aspects of human management (see chapter 7) that can be profitably viewed as directed disturbance regimes that attempt to achieve a particular composition and functionality in the targeted ecosystems.

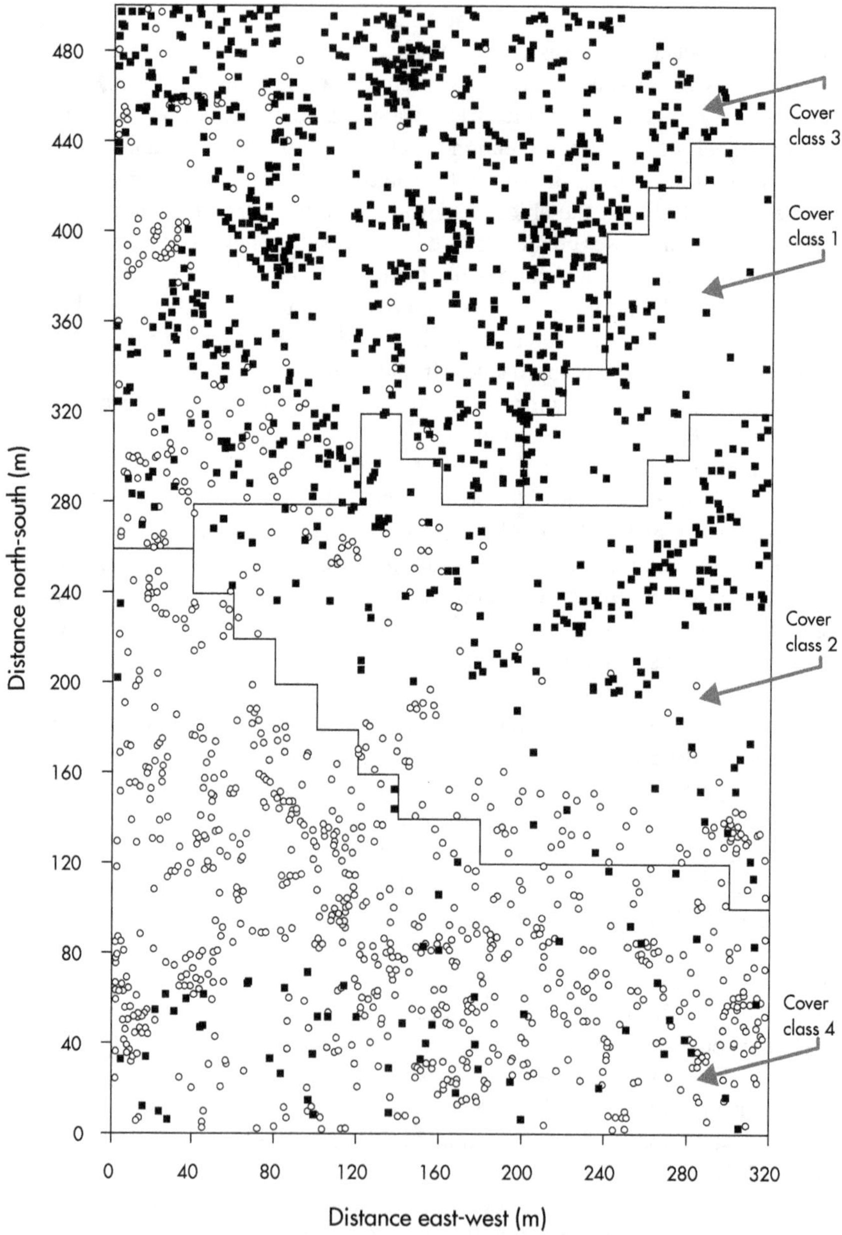

Figure 2.12 Previous land use and distribution of dominant species in the Luquillo Forest Dynamics Plot (after Willig et al. [1996] and Thompson et al. [2002]). Cover classes reflect land uses before 1936, derived from aerial photography. Cover classes 1 through 3 were clear-cut or heavily logged and then used for agriculture or silviculture, whereas cover class 4 was selectively logged. In 1936, canopy cover in classes 1 through 4 was 10 to 20 percent, 20 to 50 percent, 50 to 80 percent, and 80 to 100 percent, respectively. *Dacyodes excelsa* (open circles), a tree of mature forest, dominates the southern half of the plot, whereas *Casearia arborea* (solid squares), a secondary forest species, is more common in the northern section.

Disturbance and the Relationship between Biodiversity and Ecosystem Processes

An environment enriched by the effects of disturbance provides many opportunities to understand the interaction between pattern and process. Disturbance affects both species – area (Grime 1973; Sousa 1984) and species – time (White et al. 2006) relationships and thus influences the diversity and identity of species (the biotic environment) that imbue an ecosystem with structure and functionality. Indeed, in order to predict ecosystem function, it is critical that one understand the interactions among aspects of diversity, species identity, and disturbance (see chapter 6) from a long-term perspective. Disturbance-driven changes in biodiversity influence the abiotic environment through their impact on resource availability and microclimate. Disturbance modifies the community composition, and interactions among new combinations of species alter the effect of species on ecosystem processes (Chapin et al. 2002). Changes in biodiversity feed back to modify the disturbance regime directly through the behavior of species in the community and indirectly through changes in the structural environment. The cumulative effect of all of these factors determines the resistance and resilience of ecosystem processes.

As patch-generating phenomena, disturbances alter spatial heterogeneity at a variety of scales and consequently have the potential to affect beta diversity as well as gamma diversity. Indeed, of the 17 general models posited to represent the relationship between species diversity and productivity (Scheiner and Willig 2005), two directly involve disturbance and five others indirectly involve disturbance to the extent that it creates patches associated with distinctive levels of critical resources (see box 2-1).

Moreover, empirical studies clearly document that not all aspects of diversity (e.g., richness, evenness, diversity, dominance, rarity) are correlated spatially (Wilsey and Potvin 2000; Stevens and Willig 2002; Chalcraft et al. 2004; Wilsey et al. 2005). As a result, the way in which disturbance affects the relation between productivity and biodiversity depends on the particular metric used to characterize biodiversity, as well as the spatial scale at which it is measured in nature (see chapter 8).

Summary and Implications

The ecological tapestry is a vibrant metaphor that captures important aspects of the spatiotemporally dynamic ecosystems of the Luquillo Mountains. Our conceptual approach considers historical factors, as well as contemporary geology, topography, and abiotic conditions, to create spatial variability in ecological space that favors some taxa more than others. This ultimately determines the abundance and distribution of species in the Luquillo Mountains. In addition, interspecific interactions (competition, predation, and mutualism) and heterogeneity arising from a complex disturbance regime (e.g., hurricanes, tropical storms, landslides, treefalls, droughts) that includes anthropogenic elements (e.g., forestry, agriculture, urbanization) combine to add further complexity, variability, and heterogeneity to the warp and weft of the

Box 2.1. Models representing the relationship between species diversity and productivity that involve disturbance directly and indirectly.

Directly involve disturbance

- Disturbance and competition (Huston 1979; Huston and Smith 1987)
- Hump-back model (Grime 1973, 1979)

Indirectly involve disturbance

- Available habitat (Denslow 1980; Rosenzweig and Abramsky 1993)
- Resource competition and resource heterogeneity (Tilman 1982, 1988; Abrams 1988)
- Intertaxon competition (Rosenzweig and Abramsky 1993; Tilman and Pacala 1993)
- Adaptive tradeoffs (Vander Meulen et al. 2001)

fabric composing the Luquillo tapestry. Finally, the various species of the Luquillo Mountains interact with matter and energy to form dynamic ecosystems, with tight coupling between aquatic and terrestrial systems. A number of important implications or insights can be derived from the application of our conceptual framework to the ecosystems of the Luquillo Mountains.

- Understanding present-day functionality requires knowledge of present-day, historical, and ancient processes. These different processes transpire at characteristic rates and interact to produce dynamism in the system.
- Geology, topography, regional climate, and disturbance produce heterogeneity or variation in the abiotic environment.
- The distribution of the biota is influenced by variability created at multiple spatial scales by multiple processes.
- The abundance and distribution of species affect biogeochemical processes and the capacity of the biota to moderate the environment after disturbance events and thus affect successional trajectories.
- The presence, abundance, and behavior of a species are linked intimately to the energetic costs and benefits associated with living in a particular geographic position, provided that key environmental factors have a consistent association with geographic space over time.
- The response to a disturbance is determined by the immediate postdisturbance environment, which includes the character and heterogeneity of the abiotic environment, the composition of the surviving organisms, and the structural legacies of the disturbance event. The integrated response of the biota to these circumstances determines the trajectory of change in the

community composition, structure, and biogeochemical processing (i.e., successional dynamics).

Literature Cited

Abrams, P. A. 1988. Resource productivity-consumer species diversity: Simple models of competition in spatially heterogeneous environments. *Ecology* 69:1418–1433.

Aide, T. M., J. K. Zimmerman, M. Rosario, and H. Marcano. 1996. Forest recovery in abandoned cattle pastures along an elevation gradient in northeastern Puerto Rico. *Biotropica* 28:537–548.

Alvarez, J. 1997. Patterns of abundance, species richness, habitat use and morphology in tropical terrestrial mollusks. Ph.D. dissertation. Texas Tech University, Lubbock, TX.

Andrewartha, H. G., and L. C. Birch. 1954. *The distribution and abundance of animals.* Chicago: University of Chicago Press.

Austin, M. P., and R. B. Cunningham. 1981. Observational analysis of environmental gradients. *Proceedings of the Ecological Society of Australia* 11:109–119.

Basnet, K., F. N. Scatena, G. E. Likens, and A. E. Lugo. 1992. Ecological consequences of root grafting in tabonuco (*Dacryodes excelsa*) trees in the Luquillo Experimental Forest, Puerto Rico. *Biotropica* 25:28–35.

Birdsey, R. A., and P. L. Weaver. 1982. The forest resources of Puerto Rico. Resource Bulletin SO-85. New Orleans, LA: USDA Forest Service.

Boose, E. R., M. I. Serrano, and D. R. Foster. 2004. Landscape and regional impacts of hurricanes in Puerto Rico. *Ecological Monographs* 74:335–352.

Brokaw, N. V. L., and J. S. Grear. 1991. Forest structure before and after Hurricane Hugo at three elevations in the Luquillo Mountains, Puerto Rico. *Biotropica* 23:386–392.

Carpenter, S. R. 2001. Alternate states of ecosystems: Evidence and some implications. Pages 357–383 in M. C. Press, N. J. Huntly, and S. Levin, editors, *Ecology: Achievement and challenge.* Oxford, England: Blackwell.

Chalcraft, D. R., J. W. Williams, M. D. Smith, and M. R. Willig. 2004. Scale dependence in the relationship between species richness and productivity: The role of spatial and temporal turnover. *Ecology* 85:2701–2707.

Chapin, F. S., III, P. A. Matson, and H. A. Mooney. 2002. *Principles of terrestrial ecosystem ecology.* New York: Springer-Verlag.

Chapin, F. S., III, and A. M. Starfield. 1997. Time lags and novel ecosystems in response to transient climatic change in arctic Alaska. *Climatic Change* 35:449–461.

Clements, F. E. 1916. *Plant succession: An analysis of the development of vegetation.* Washington, DC: Carnegie Institute.

Covich, A. P. 2000. Energy flow and ecosystems. Pages 509–523 in S. A. Levin, editor, *Encyclopedia of biodiversity.* Volume 2. San Diego, CA: Academic Press.

Cox, S. B., M. R. Willig, and F. N. Scatena. 2002. Variation in nutrient characteristics of surface soils from the Luquillo Experimental Forest of Puerto Rico—A multivariate perspective. *Plant and Soil* 247:189–198.

Crow, T. R., and D. F. Grigal. 1979. A numerical analysis of arborescent communities in the rain forest of the Luquillo Mountains, Puerto Rico. *Vegetatio* 40:135–146.

Denslow, J. S. 1980. Gap partitioning among tropical rainforest trees. *Biotropica* 12:47–55.

Devoe, N. N. 1989. Differential seeding and regeneration in openings and beneath closed canopy in sub-tropical wet forest. Ph.D. dissertation. Yale School of Forestry and Environmental Science, New Haven, CT.

Dill, L. M. 1978. An energy-based model of optimal feeding-territory size. *Theoretical Population Biology* 14:396–429.

Fernández, D. S., and N. Fetcher. 1991. Changes in light availability following Hurricane Hugo in a subtropical montane forest in Puerto Rico. *Biotropica* 23:393–399.

Franklin, J. F., D. Lindenmayer, J. A. MacMahon, A. McKee, J. Magnuson, D. A. Perry, R. Waide, and D. Foster. 2000. Threads of continuity. *Conservation Biology in Practice* 1:8–16.

Gannon, M. R., M. R. Duran, A. Kurta, and M. R. Willig. 2005. *Bats of Puerto Rico: An island focus and Caribbean perspective*. Lubbock, TX: Texas Tech University Press.

García-Montiel, D., and F. N. Scatena. 1994. The effect of human activity on the structure and composition of a tropical forest in Puerto Rico. *Forest Ecology and Management* 63:57–78.

Goheen, D. J., and E. M. Hansen. 1993. Effects of pathogens and bark beetles on forests. Pages 175–196 in T. D. Schowalter and G. M. Filip, editors, *Beetle-pathogen interactions in conifer forests*. London: Academic Press.

Gosz, J. R. 1992. Gradient analysis of ecological change in time and space: Implications for forest management. *Ecological Applications* 2:248–261.

Gould, W. A., G. González, and G. Carrero Rivera. 2006. Structure and composition of vegetation along an elevational gradient in Puerto Rico. *Journal of Vegetation Science* 17:653–664.

Grime, J. P. 1973. Competition and diversity in herbaceous vegetation. *Nature* 242:344–347.

Grime, J. P. 1979. *Plant strategies and vegetative processes*. Chichester, England: Wiley.

Guzmán-Grajales, S. M., and L. R. Walker. 1991. Differential seedling responses to litter after Hurricane Hugo in the Luquillo Experimental Forest, Puerto Rico. *Biotropica* 23:407–413.

Hall, C. A. S., J. A. Stanford, and F. R. Hauer. 1992a. The distribution and abundance of organisms as a consequence of energy balances along multiple environmental gradients. *Oikos* 65:377–390.

Hall, C. A., M. R. Taylor, and E. Everham. 1992b. A geographically-based ecosystem model and its application to the carbon balance of the Luquillo Forest, Puerto Rico. *Water, Air, and Soil Pollution* 64:385–404.

Huston, M. 1979. A general hypothesis of species diversity. *The American Naturalist* 113:81–101.

Huston, M., and T. Smith. 1987. Plant succession: Life history and competition. *The American Naturalist* 130:168–198.

Hutchinson, G. E. 1958. Concluding remarks. *Cold Spring Harbor Symposium on Quantitative Ecology* 22:415–427.

Hutchinson, G. E. 1965. The niche: An abstractly inhabited hypervolume. Pages 26–78 in *The ecological theater and the evolutionary play*. New Haven, CT: Yale University Press.

Ingegnoli, V. 2004. *Landscape ecology: A widening foundation*. New York: Springer-Verlag.

Keddy, P. A. 1991. Working with heterogeneity: An operator's guide to environmental gradients. Pages 181–201 in J. Kolasa and S. T. A. Pickett, editors, *Ecological heterogeneity*. New York: Springer-Verlag.

Kitchell, J. F. 1983. Energetics. Pages 312–338 in P. Webb and D. Weihs, editors, *Fish biomechanics*. New York: Praeger.

Krebs, C. J. 1985. *Ecology: The experimental analysis of distribution and abundance*, 3rd ed. New York: Harper and Row.

Landres, P. B., P. Morgan, and F. J. Swanson. 1999. Overview of the use of natural variability concepts in managing ecological systems. *Ecological Applications* 9:1179–1188.

Lodge, D. J., and C. E. Asbury. 1988. Basidiomycetes reduce export of organic matter from forest slopes. *Mycologia* 80:888–890.

Lodge, D. J., F. N. Scatena, C. E. Asbury, and M. J. Sánchez. 1991. Fine litter fall and related nutrient inputs resulting from Hurricane Hugo in subtropical wet and lower montane rain forests of Puerto Rico. *Biotropica* 23:336–342.

Luquillo LTER. 1988. Long term ecological research on the Luquillo Experimental Forest. Unpublished proposal to the National Science Foundation. http://luq.lternet.edu/publications/prop1988/Prop1988.pdf. Accessed October 31, 2011.

Luquillo LTER. 1994. Long term ecological research on the Luquillo Experimental Forest II. Unpublished proposal to the National Science Foundation. http://luq.lternet.edu/publications/prop1994/index.html. Accessed October 31, 2011.

Maguire, B., Jr. 1976. Analysis and modeling of evolutionary dynamics with the response structure/evolutionary structure approach. Pages 365–407 in M. K. Hecht, W. C. Steere, and B. Wallace, editors, *Evolutionary biology*. New York: Plenum.

McNab, B. K. 1980. Food habits, energetics, and the population biology of mammals. *The American Naturalist* 116:106–124.

Minchin, P. R. 1987. Simulation of multidimensional community patterns: Towards a comprehensive model. *Vegetatio* 71:145–156.

Odum, H. T., G. Drewry, and J. R. Kline. 1970. Climate at El Verde, 1963–1966. Pages B347–B418 in H. T. Odum and R. F. Pigeon, editors, *A tropical rain forest: A study of irradiation and ecology at El Verde, Puerto Rico*. Oak Ridge, TN: U.S. Atomic Energy Commission.

Olander, L., F. N. Scatena, and W. L. Silver. 1998. Impacts of disturbance initiated by road construction in a subtropical cloud forest in the Luquillo Experimental Forest, Puerto Rico. *Forest Ecology and Management* 109:33–49.

Peterson, A. T. 2001. Predicting species' geographic distributions based on ecological niche modeling. *The Condor* 103:599–605.

Pickett, S. T. A., and P. S. White. 1985. Natural disturbance and patch dynamics: An introduction. Pages 3–13 in S. T. A. Pickett and P. S. White, editors, *The ecology of natural disturbance and patch dynamics*. Orlando, FL: Academic Press.

Pulliam, H. R. 1996. Sources and sinks: Empirical evidence and population consequences. Pages 45–69 in O. E. Rhodes, Jr., R. K. Chesser, and M. H. Smith, editors, *Population dynamics in ecological space and time*. Chicago: University of Chicago Press.

Raffaele, H. A. 1989. *Birds of Puerto Rico and the Virgin Islands*. Princeton, NJ: Princeton University Press.

Root, T. 1988. Energy constraints on avian distributions and abundances. *Ecology* 69: 330–339.

Rosenzweig, M. L., and Z. Abramsky. 1993. How are diversity and productivity related? Pages 52–65 in R. E. Ricklefs and D. Schluter, editors, *Species diversity in ecological communities*. Chicago: University of Chicago Press.

Scatena, F. N. 1989. An introduction to the physiography and history of the Bisley Experimental Watersheds in the Luquillo Mountains of Puerto Rico. General Technical Report SO-72. New Orleans, LA: USDA Forest Service.

Scatena, F. N. 1990. Watershed scale rainfall interception on two forested watersheds in the Luquillo Mountains of Puerto Rico. *Journal of Hydrology* 113:89–102.

Scatena, F. N., W. Silver, T. Siccama, A. Johnson, and M. J. Sanchez. 1993. Biomass and nutrient content of the Bisley Experimental Watersheds, Luquillo Experimental Forest, Puerto Rico before and after Hurricane Hugo, 1989. *Biotropica* 25:15–27.

Scheiner, S. M., and M. R. Willig. 2005. Developing unified theories in ecology as exemplified with diversity gradients. *The American Naturalist* 166:458–469.

Seastedt, T.R., J. Hobbs, and K. N. Suding. 2008. Management of novel ecosystems: Are novel approaches required? *Frontiers in Ecology and the Environment* 6:547–553.

Shelford, V. E. 1951. Fluctuations of forest animal populations in east central Illinois. *Ecological Monographs* 21:183–213.

Smith, T., and M. Huston. 1989. A theory of spatial and temporal dynamics of plant communities. *Vegetatio* 83:49–69.

Snyder, N. F. R., J. W. Wiley, and C. B. Kepler. 1987. *The parrots of Luquillo: Natural history and conservation of the Puerto Rican parrot.* Los Angeles: Western Foundation of Vertebrate Zoology.

Sousa, W. P. 1984. The role of disturbance in natural communities. *Annual Review of Ecology and Systematics* 15:353–391.

Stevens, R. D., and M. R. Willig. 2002. Geographical ecology at the community level: Perspectives on diversity of New World bats. *Ecology* 83:545–560.

Thompson, J., N. Brokaw, J. K. Zimmerman, R. B. Waide, E. M. Everham, III, D. J. Lodge, C. M. Taylor, D. Garcia-Montiel, and M. Fluet. 2002. Land use history, environment, and tree composition in a tropical forest. *Ecological Applications* 12:1344–1363.

Tilman, D. 1982. *Resource competition and community structure.* Princeton, NJ: Princeton University Press.

Tilman, D. 1988. *Plant strategies and the dynamics and structure of plant communities.* Princeton, NJ: Princeton University Press.

Tilman, D., and S. Pacala. 1993. The maintenance of species richness in plant communities. Pages 13–25 in R. E. Ricklefs and D. Schluter, editors, *Species diversity in ecological communities.* Chicago: University of Chicago Press.

Vander Meulen, M. A., A. J. Hudson, and S. M. Scheiner. 2001. Three evolutionary hypotheses for the hump-shaped productivity-diversity curve. *Evolutionary Ecology Research* 3:379–392.

Veblen, T. T. 2003. Historic range of variability of mountain forest ecosystems: Concepts and applications. *The Forestry Chronicle* 79:223–226.

Vitousek, P. M., and J. S. Denslow. 1986. Nitrogen and phosphorus availability in treefall gaps of a lowland tropical rainforest. *The Journal of Ecology* 74:1167–1178.

Vogt, K. A., J. C. Gordon, J. P. Wargo, and D. Vogt. 1997. *Ecosystems: Balancing science with management.* New York: Springer-Verlag.

Waide, R. B. 1996. Birds. Pages 363–398 in D. P. Reagan and R. B. Waide, editors, *The food web of a tropical rain forest.* Chicago: University of Chicago Press.

Waide, R. B., J. K. Zimmerman, and F. N. Scatena. 1998. Controls of primary productivity: Lessons from the Luquillo Mountains in Puerto Rico. *Ecology* 79:31–37.

Walker, L. R. 1991. Tree damage and recovery from Hurricane Hugo in Luquillo Experimental Forest, Puerto Rico. *Biotropica* 23:379–385.

Walker, L. R., D. J. Lodge, N. V. L. Brokaw, and R. B. Waide. 1991. Ecosystem, plant, and animal responses to hurricanes in the Caribbean. *Biotropica* 23:313–521.

Walker, L. R., W. L. Silver, M. R. Willig, and J. K. Zimmerman. 1996. Long-term responses of Caribbean ecosystems to disturbance. *Biotropica* 28:414–614.

Walker, L. R., and M. R. Willig. 1999. An introduction to terrestrial disturbance. Pages 1–16 in L. R. Walker, editor, *Ecosystems of disturbed ground.* Amsterdam: Elsevier.

Wang, H. Q., C. A. S. Hall, F. Scatena, N. Fetcher, and W. Wu. 2002. Modeling the spatial and temporal variability in climate and primary productivity across the Luquillo Mountains, Puerto Rico. *Forest Ecology and Management* 179:69–94.

Weaver, P. L. 1991. Environmental gradients affect forest composition in the Luquillo Mountains of Puerto Rico. *Interciencia* 16:142–151.

Webb, S. L. 1999. Disturbance by wind in temperate-zone forests. Pages 187–222 in L. R. Walker, editor, *Ecosystems of the world 16: Ecosystems of disturbed ground.* Amsterdam: Elsevier.

White, E. P., P. B. Adler, W. K. Lauenroth, R. A. Gill, D. Greenberg, D. M. Kaufman, A. Rassweiler, J. A. Rusak, M. D. Smith, J. R. Steinbeck, R. B. Waide, and J. Yao. 2006. A comparison of the species-time relationship across ecosystems and taxonomic groups. *Oikos* 112:185–195.

Whittaker, R. H. 1956. Vegetation of the Great Smoky Mountains. *Ecological Monographs* 26:1–80.

Willig, M. R., C. P. Bloch, N. Brokaw, C. L. Higgins, J. Thompson, and C. R. Zimmermann. 2007. Cross-scale responses of biodiversity to hurricane and anthropogenic disturbance in a tropical forest. *Ecosystems* 10:824–838.

Willig, M. R., and M. R. Gannon. 1996. Mammals. Pages 399–431 in D. P. Reagan and R. B. Waide, editors, *The food web of a tropical rainforest*. Chicago: University of Chicago Press.

Willig, M. R., and M. A. McGinley. 1999. Animal responses to natural disturbance and roles as patch generating phenomena. Pages 667–689 in L. R. Walker, editor, *Ecosystems of disturbed ground*. Amsterdam: Elsevier.

Willig, M. R., D. L. Moorhead, S. B. Cox, and J. C. Zak. 1996. Functional diversity of soil bacteria communities in the tabonuco forest: The interaction of anthropogenic and natural disturbance. *Biotropica* 28:471–483.

Willig, M. R., M. F. Secrest, S. B. Cox, G. R. Camilo. J. F. Cary, J. Alvarez, and M. R. Gannon. 1998. Long-term monitoring of snails in the Luquillo Experimental Forest of Puerto Rico: Heterogeneity, scale, disturbance, and recovery. Pages 293–322 in F. Dallmeier and J. A. Comiskey, editors, *Forest biodiversity in North, Central and South America, and the Caribbean: Research and monitoring*. Man and Biosphere Series, Volume 21. Paris: UNESCO; New York: Parthenon.

Willig, M. R., and L. R. Walker. 1999. Disturbance in terrestrial ecosystems: Salient themes, synthesis, and future directions. Pages 747–767 in L. R. Walker, editor. *Ecosystems of disturbed ground*. Amsterdam: Elsevier.

Wilsey, B. J., D. R. Chalcraft, C. M. Bowles, and M. R. Willig. 2005. Multidimensional nature of species diversity in grassland communities. *Ecology* 86:1178–1184.

Wilsey, B. J., and C. Potvin. 2000. Biodiversity and ecosystem functioning: Importance of species evenness in an old field. *Ecology* 81:887–892.

Wooster, K. M. 1989. A geographically-based microclimatological computer model for mountainous terrain with application to the Luquillo Experimental Forest in Puerto Rico. Master's thesis. State University of New York College of Environmental Science and Forestry, Syracuse, NY.

Wu, J., and O. L. Loucks. 1995. From balance of nature to hierarchical patch dynamics: A paradigm shift in ecology. *Quarterly Review of Biology* 70:439–466.

Zimmerman, J. K., W. M. Pulliam, D. J. Lodge, V. Quiñones-Orfila, N. Fetcher, S. Guzmán-Grajales, J. A. Parrotta, C. E. Asbury, L. R. Walker, and R. B. Waide. 1995. Nitrogen immobilization by decomposing woody debris and the recovery of tropical wet forest from hurricane damage. *Oikos* 72:314–322.

Zimmerman, J. K., M. R. Willig, L. R. Walker, and W. L. Silver. 1996. Introduction: Disturbance and Caribbean ecosystems. *Biotropica* 28:414–423.

Zou, X., and G. González. 1997. Changes in earthworm density and community structure in abandoned tropical pastures. *Soil Biology Biochemistry* 29:627–629.

3

Geographic and Ecological Setting of the Luquillo Mountains

William H. McDowell, Frederick N. Scatena, Robert B. Waide, Nicholas Brokaw, Gerardo R. Camilo, Alan P. Covich, Todd A. Crowl, Grizelle González, Effie A. Greathouse, Paul Klawinski, D. Jean Lodge, Ariel E. Lugo, Catherine M. Pringle, Barbara A. Richardson, Michael J. Richardson, Douglas A. Schaefer, Whendee L. Silver, Jill Thompson, Daniel J. Vogt, Kristiina A. Vogt, Michael R. Willig, Lawrence L. Woolbright, Xiaoming Zou, and Jess K. Zimmerman

Key Points

- The Luquillo Mountains in northeastern Puerto Rico are geologically dynamic, with recurrent hurricanes, landslides, and earthquakes.
- Puerto Rico has never been physically connected to continents by land bridges, which, together with the island's long distance from North and South America, contributes to its relatively low numbers of native plant and animal species for a tropical location and its high rate of endemism.
- The climate is warm, wet, and relatively aseasonal but shows strong gradients with elevation.
- Soils are deep and highly weathered, with carbon and nutrient concentrations and standing stocks similar to those in many other tropical forests. Soils contain much to most of the available nutrients and total carbon, but plant biomass is a particularly important pool of potassium.
- Nutrient inputs in precipitation are dominated by marine aerosols; these aerosols and rapid weathering contribute to a substantial export of base cations in streams.
- Nitrogen budgets are unbalanced at the watershed scale, suggesting that significant amounts of N fixation are occurring.
- The Luquillo Mountains contain many types of forest, but four are common and particularly well studied: tabonuco, colorado, palm, and elfin.

- Aboveground net primary productivity is high, as it is in many other tropical sites, and aboveground biomass, productivity, and forest stature decrease with elevation.
- Large mammalian herbivores and predators are absent; lizards, frogs, snakes, and a few birds are the top terrestrial predators.
- Stream and river food webs are dominated by freshwater shrimp and fish species that migrate to the estuary; nonmigratory freshwater crabs are also important, but aquatic insects are neither diverse nor abundant.
- Leaf litter decomposition is rapid in both the forest and streams, and detrital pathways provide a major energy source to higher trophic levels.

Introduction

In this chapter, we describe the geologic, geographic, and ecological context in which the Luquillo Mountains (figure 3-1) are situated, with particular emphasis on factors that potentially influence the response of terrestrial and aquatic ecosystems to disturbance. We start with the physical and chemical environment and then discuss the biota. Whenever possible, we address the whole of the Luquillo Mountains, although we know the most about the mid-elevation forests. We address long-term results and ambient conditions as a prelude to our detailed descriptions

Figure 3.1 The Luquillo Mountains of Puerto Rico. (Photograph by William McDowell.)

of disturbances and the biotic response to disturbance in chapters 4, 5, and 8. It is difficult to discuss the Luquillo Mountains without considering the role that previous disturbances have played in shaping the mountains as we know them today (see chapter 1). Consequently, we synthesize the results of several decades of research and reflect on the lessons learned from our research as we place the Luquillo Mountains in the context of other tropical ecosystems, and of forest and stream ecosystems globally.

Geology

Regional Geology

Puerto Rico is located in one of the most dynamic regions on the planet, with severe hurricanes, landslides, tsunamis, and earthquakes all occurring with significant frequency (see chapter 4 for details). Much of this dynamism is related to Puerto Rico's location at the junction of the American and Caribbean crustal plates (Masson and Scanlon 1991; ten Brink et al. 2006). Puerto Rico is the smallest island (figure 3-2) of a large volcanic island-arc (the Greater Antilles) that developed during the Cretaceous, about 100 million years ago (mya), along a broad strike-slip zone between these crustal plates. The Puerto Rico Trench, the deepest spot in the Atlantic, lies about 150 km north of Puerto Rico. Although the islands are primarily volcanic in origin, only dormant volcanoes are currently found in the Greater Antilles. The Lesser Antilles is a younger (created ~35 to 24 mya), predominantly volcanic island arc with active volcanoes lying south and east of Puerto Rico and stretching south to the coast of Venezuela.

Although it is possible to see from one island to the next all the way from Florida to Venezuela, the islands are not thought to have formed continuous land bridges between North and South America (Graham 2003a; Hedges 2006). The geologic history of the Caribbean that is relevant for biogeography is still uncertain and controversial (Graham 2003b). Previously, most of the smaller Caribbean islands, including Puerto Rico, were not thought to have been physically connected to their neighbors (Heatwole and MacKenzie 1967). However, the Virgin Islands, which are part of the Puerto Rico Bank, were contiguous with Puerto Rico during glacial maxima, most recently during the Pleistocene. Furthermore, geologic evidence indicates that the proto-Antilles were once connected below the sea surface, and more recently Heinicke et al. (2007) suggested that Cuba, Hispaniola, and Puerto Rico were connected above sea level during the late Eocene (35 mya) and that subsidence during the Oligocene (~23 to 34 mya) broke these connections. Others have proposed, however, that the emergence of the proto-Antilles above sea level did not occur until the middle Eocene (~47 mya) (Iturralde-Vinent and MacPhee 1999; Graham 2003b), and that Puerto Rico did not emerge until the middle Miocene (~23 to 17 mya), after its separation from Hispaniola in the late Oligocene to early Miocene (~24 mya) had already occurred (Graham 2003b).

Puerto Rico has undergone a full cycle of mountain development subsequent to its emergence in the Miocene and is now relatively stable in terms of volcanic

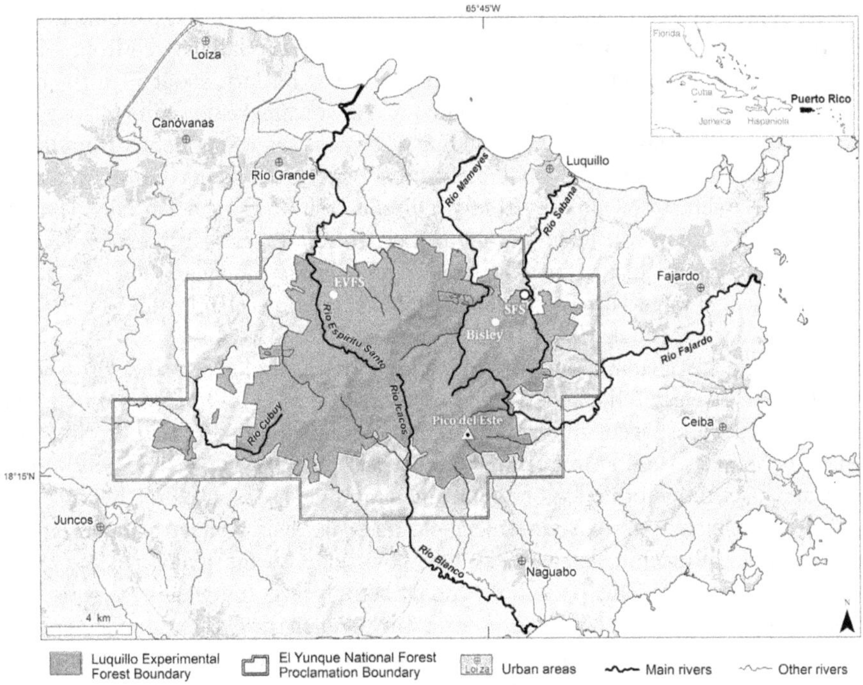

Figure 3.2 Map of northeastern Puerto Rico and the Luquillo Mountains. The dark gray area is the Luquillo Experimental Forest (coterminous with El Yunque National Forest), administered by the U.S. Forest Service. EVFS = El Verde Field Station; SFS = Sabana Field Station (U.S. Forest Service); Bisley = Bisley Experimental Watersheds. (Map by Olga Ramos.)

activity, as the most recent volcanic eruption occurred at least 30 mya (Seiders 1971). There is still, however, considerable seismic activity in the area. The largest earthquake in the past century (7.5 on the Richter scale) occurred in 1918 and originated in the Mona Passage (López-Venegas et al. 2008). This and a subsequent tsunami caused 116 deaths. Earthquakes of magnitude 3.0 or above are common in Puerto Rico.

Local Geology

The island of Puerto Rico is a rugged mountain mass that has been described as a "heap of volcanic debris" (Hodge 1920; Mitchell 1954). The core of Puerto Rico is an east-west trending body that was formed in association with Cretaceous and Tertiary volcanoes. Tilted beds of clastic and carbonate sediments flank this volcanic core and form an apronlike structure that is progressively younger toward both the Caribbean and the Atlantic coasts. The Luquillo Mountains are the eastern terminus of the volcanic core of Puerto Rico and are the dominant geologic feature on the eastern end of the island. Geologically, they are best described as a tilted fault block dominated by northwest-trending faults and northeast-trending folds (Scatena 1989).

The Luquillo Mountains are underlain by volcaniclastic rocks, plutonic quartz diorite intrusions, and contact metamorphic rocks that were all derived from a similar andesitic magma during the Cretaceous and lower Tertiary, about 50 to 100 mya (Seiders 1971) (figure 3-3[B]). The source of the volcaniclastic sediments was an active volcanic complex that was standing at or near sea level. Debris from these volcanoes was deposited in moderately deep water after being transported and reworked by submarine slides, turbidity currents, ash flows, and ash falls. During this period of active volcanism in Puerto Rico, the Caribbean basin experienced a large meteor impact that defines the Cretaceous–Tertiary boundary and which has been implicated in massive regional and global extinctions (Hildebrand and Boynton 1990; Florentin et al. 1991).

Following the accumulation of volcaniclastic debris in the marine environment, late Eocene or early Oligocene tectonic activity about 30 to 40 mya uplifted this material. The subsequent intrusion of plutonic rock (the quartz-rich dioritic Rio Blanco complex) (figure 3-3[B]) marked the last phase of igneous activity in the area and caused the formation of contact metamorphic rocks when the hot igneous intrusion contacted the existing volcaniclastic rock. This contact metamorphism produced the erosion-resistant rocks that now underlie the major peaks of the Luquillo Mountains. This period of tectonic activity was followed by a period of stability until the middle Miocene, about 10 mya, when the Caribbean plate drifted eastward and the Greater and Lesser Antilles began to assume their present configuration. Since the end of the Eocene 34 mya, the regional tectonics of Puerto Rico and the Virgin Islands have been dominated by left-lateral slip between the North American and Caribbean plates (Masson and Scanlon 1991).

Local Topography and Land Forms

The Luquillo Mountains rise to an elevation of 1,074 m and are flanked by a coastal plain to the north, east, and south that is 8 to 16 km wide. Within the Luquillo Mountains, the rugged landscape has three major peaks (El Yunque Peak, East Peak, and El Toro Peak) and four main valleys that correspond to the four major rivers (the Espíritu Santo, the Mameyes, the Fajardo, and the Icacos/Blanco) (figure 3-2). Important research stations are located in several of these major watersheds, including the El Verde Field station in the Espíritu Santo, the Bisley Experimental Watersheds in the Mameyes, and the Sabana Field Station in to the Río Sabana watershed.

Hillslopes in the Luquillo Mountains are steep and form well-defined convex-concave catenas. Ridgetops are typically well-defined *cuchillo* or knife-like divides. Lower hillslope segments are generally concave where they pass into first-order valleys and straight where they join perennial channels. The major physical processes acting on these hillslopes include landslides, slope creep, debris flows, and tree throws (Scatena and Lugo 1995). These processes occur throughout the Luquillo Mountains, but their frequency, magnitude, and ecological significance vary with the bedrock geology, elevation, and forest type (Larsen and Torres Sánchez 1996; Larsen 1997).

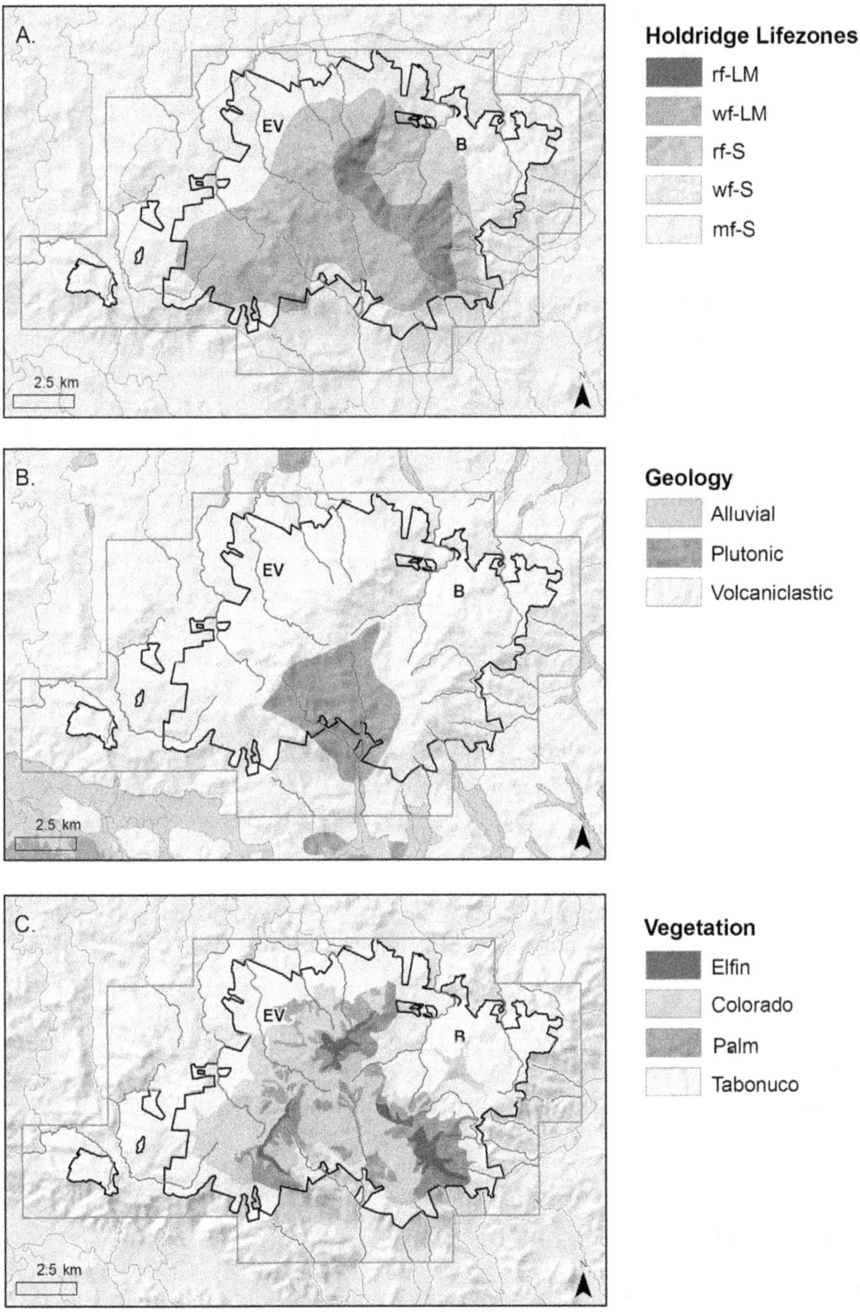

Figure 3.3 Life zones, geology, and vegetation of the Luquillo Mountains. (A) Holdridge Life Zones. rf-LM = lower montane rain forest; wf-LM = lower montane wet forest; rf-S = subtropical rain forest; wf-S = subtropical wet forest; mf-S = subtropical moist forest. (B) Surficial bedrock geology (see text). (C) Vegetation (see text). (Map by Olga Ramos.)

Bedrock under the Icacos Valley is quartz diorite, and volcaniclastic rocks underlie most of the Mameyes and Espíritu Santo valleys, where the Bisley Experimental Watersheds and El Verde Field Station are located (figure 3-3[B]). Bedrock underlying the Bisley Experimental Watersheds is mapped as the Fajardo formation (Briggs and Aquilar-Cortés 1980), whereas the El Verde Field Station research area is underlain by the Hato Puerco formation (Seiders 1971). The two formations have a similar chemistry and origin, but the Hato Puerco formation tends to produce larger boulders when weathered.

Weathering rates on the two bedrock types are rapid, resulting in large hydrologic exports of both dissolved and particulate matter to the sea (McDowell and Asbury 1994). Geochemical techniques suggest that the quartz diorite Río Blanco formation in the Icacos Valley is weathering at the rate of 0.58 cm per millennium, making it one of the fastest weathering granitic terrains that has ever been measured (White and Blum 1995; White et al. 1998). The rate of export of sediment from the Icacos basin is 3,200 kg ha^{-1} y^{-1}, and silica (SiO_2) loss occurs at a rate of 487 kg ha^{-1} y^{-1} (McDowell and Asbury 1994). Work on the chemistry of tributaries to the Icacos suggests that a primary driver of this rapid weathering rate is the rapid physical denudation associated with landslides, which expose fresh mineral surfaces to weathering (Bhatt and McDowell 2007). Landslides occur frequently throughout the Luquillo Mountains, but they are largest and most common in higher-elevation areas where slopes are steep and in areas underlain by quartz diorite bedrock such as the Río Icacos valley (Larsen and Torres- Sanchez 1996). Weathering in the volcaniclastic terrain is also high, with sediment losses of 150 to 330 kg ha^{-1} y^{-1} and SiO_2 losses of 180 to 400 kg ha^{-1} y^{-1} (McDowell and Asbury 1994). The Icacos and Espíritu Santo valleys are U.S. Geological Survey (USGS) Water, Energy, and Biogeochemical Budgets sites that focus on weathering rates (Peters et al. 2006).

The drainage network of the Luquillo Mountains consists of concave valleys and a dense network of intermittent, zero-order swales and gullies draining into first-order channels with steep gradients. Most channels are boulder- and bedrock-lined, with steep sides that tightly confine and structure them. Channels of the upper Icacos valley and some reaches of the Mameyes (Baño de Oro) are notable exceptions, as the quartz diorite bedrock (Río Blanco formation) of the Icacos valley weathers to produce large amounts of quartz sand and broad, sand-filled channels. Waterfalls are common throughout the Luquillo Mountains, and high falls (>3 m) represent an important barrier to the upstream passage of aquatic organisms (figure 3-4) (Covich et al. 2006, 2009; Kikkert et al. 2009; Hein et al. 2011).

Biogeography

The Caribbean Basin is a biogeographically complex region, owing to its complex, and still uncertain, geologic history, described earlier in this chapter. About 100 mya, Puerto Rico and other islands of the proto-Antilles were part of an island arc (Donnelly 1992; Pindell and Kennan 2002) that might have served as a conduit for biotic interchange between North America and South America (Donnelly 1990; Hedges 2006). By the end of the Cretaceous, the island arc had moved 1,000 km to

Figure 3.4 A stream in the The Luquillo Mountains of Puerto Rico. Waterfalls such as this are barriers to the upstream passage of aquatic organisms. (Photograph by William McDowell.)

the northeast and could no longer have served as a conduit between North and South America (Graham 2003b). There were various periods of island emergence and submergence in the period between 50 and 100 mya, but the principal period of sustained emergence began with the compression of the Caribbean Plate against the Bahamas Platform in the middle Eocene, ~49 mya (Iturralde-Vinent and MacPhee 1999; Graham 2003b). The Antilles are only about 1 to 3 crater diameters away from the site of the bolide impact in the Yucatán that defines the Cretaceous–Tertiary boundary (65 mya) (Pindell 1994). The resulting impact waves were hundreds to thousands of meters in height (Maurrassee 1991) and are thought to have extinguished most life in the Caribbean at that time (Hedges et al. 1994; Hedges 2006).

Detailed studies of the flora of Puerto Rico, although extensive, are confined principally to the 20th century. Because the ecosystems of the island had been widely disturbed by at least four hundred years of active human intervention by the time of these studies (Figueroa Colón 1996), the composition of the original flora prior to human presence is not well known. The most recent assessment suggests that a large fraction of species (672 of 3,032) are naturalized or of uncertain origin. Ten percent of the original flora of the island is now extinct, and 38 percent is critically imperiled (Gann and Bradley 2006). Approximately 10 percent of the flora consists of endemics (Liogier and Martorell 1982). Within the remaining areas of native forest, the Luquillo Mountains represent an area of high species richness and endemism, with at least 830 plant species and more than 250 tree species. Nearly half of all of the endemic trees of Puerto Rico (67 of 139) are found in the Luquillo Mountains (Figueroa Colón 1996). A significant number of nonendemic species found in the Luquillo Mountains are endemic to the West Indies, and the remaining trees are widespread in Central and South America. In addition to native species, a large number of species have been introduced. The most significant introductions to the Luquillo Mountains are described later in this chapter ("Effects of Recent Invasions") and in chapter 6.

Puerto Rico's Luquillo Mountains contain fewer tree species than other well-studied tropical sites that bracket the Puerto Rican site in elevation and rainfall (table 3-1). In addition to insularity, many other factors such as life zone, age of the flora, climate, habitat heterogeneity, and disturbance regime can contribute to species richness (Whitmore 1974; Lugo 1987; Lugo et al. 2002). Thus, it is difficult to ascribe these differences in tree species richness among tropical sites to any single causative factor.

In studies of orchid biogeography, Ackerman et al. (2007) found that area (for islands > 750 km^2 in size) and maximum elevation were good predictors of species diversity and endemism in the West Indies. This pattern was primarily driven by the effect of elevation on montane islands, as the species richness of low islands was not associated with land area. Orchid species richness apparently results from an interaction between area and topographic (habitat) diversity (Ackerman et al. 2007). The majority of the 728 species of orchid had a high vagility of seed dispersal, and their occurrence was primarily determined by habitat characteristics (Trejo-Torres and Ackerman 2001). However, the occurrence of 300 single-island endemics, nearly all on high islands, indicates very limited seed dispersal in these species (Ackerman et al. 2007).

Biogeographic patterns in some families of fungi most closely resemble those of the orchid island endemics studied by Ackerman et al. (2007). Two families of agaric (mushroom) fungi that grow primarily on soil, the Hygrophoraceae and Entolomataceae, were analyzed for biogeographic patterns in the Caribbean Basin (Baroni et al. 1997; Lodge et al. 2002). Species in these families were most abundant in moist and wet habitats and were thus most abundant and diverse on high-elevation islands in the Caribbean. One-third to one-half of the species of Hygrophoraceae and Entolomataceae in the Greater Antilles do not occur in the Lesser Antilles (Lodge et al. 2002). Some Greater Antillean species (or their closest relatives) do occur in South or North America, and a few are found in Africa, but pantropical species are very rare (Baroni et al. 1997; Lodge et al. 2002). Long-distance spore dispersal followed by successful colonization appears to be limited in these two families of mushrooms, and speciation appears to be rapid (Lodge et al. 2002).

Compared to tropical mainland areas of similar size and habitat diversity, the biota of Puerto Rico is depauperate for many animal groups (Garrison and Willig 1996; Reagan 1996; Thomas and Kessler 1996), and currently there are no native mammals except bats (Anthony 1918; Reagan and Waide 1996). The islands of the West Indies, including Puerto Rico, lack many families characteristic of mainland avifaunas (Waide 1996). For example, the widespread and diverse avian families of motmots, jacamars, puffbirds, barbets, toucans, woodcreepers, ovenbirds, and manakins are absent from the West Indies. The absence of these groups coupled

Table 3.1 Environmental and stand data for trees ≥ 10 cm dbh in various tropical forest plots with environmental characteristics that span the range of environmental conditions found at the LFDP at El Verde in the Luquillo Mountains

Location	Area (ha)	Latitude (°N)	Elevation (masl)	Temperature (°C)	Rain (mm y^{-1})	Trees (stems ha^{-1})	Basal area (m^2 ha^{-1})	Species (no. ha^{-1})	Spp. $stem^{-1}$	α	Reference
CTFS plots											
LFDP, subtropical wet	16	18	381	22.8	3500	876	34.3	42	0.048	9	Thompson et al. 2004
Thailand, Doi Inthanon	15	18	1700	20.9	1908	519	36.1	67	0.129	21	Kanzaki et al. 2004
Philippines, Palanan	16	17	113	26.1	5000	537	36.1	100	0.186	37	Co et al. 2004
Panama, Barro Colorado Island	50	9	140	26.9	2551	429	27.8	91	0.212	36	Leigh et al. 2004
Colombia, La Planada	25	1	1840	18.3	4087	586	23.8	88	0.150	29	Vallejo et al. 2004
Ecuador, Yasuní	25	0	230	28.4	3081	702	27.3	251	0.358	142	Valencia et al. 2004
One-hectare plots											
Puerto Rico, Bisley, subtropical wet	1	18	301	–	3500	434	28.2	33	0.076	–	Dallmeier et al. 1998
Mexico, Los Tuxtlas, Veracruz	1	18	150	24.6	4639	348	34.9	76	0.218	–	Bongers et al. 1987
Belize, Bladen, subtropical wet	1	16	45	26.7	2490	358	31.5	89	0.249	38	Brewer and Webb 2002
Other plots											
Puerto Rico, subtropical wet	0.81	18	c. 380	c. 22.8	–	936	39.7	45	0.048	–	Wadsworth 1987
Costa Rica, premontane wet	0.80	c. 10	–	–	–	528	33.2	42	0.080	–	Wadsworth 1987

CTFS = Center for Tropical Forest Science, a network of large tropical forest study plots (Condit 1995). The Holdridge Life Zone is given to facilitate comparison of the LFDP with other Neotropical sites and is included in the location description. Holdridge life zone determinations for Puerto Rican sites are from Ewel and Whitmore (1973); those for Costa Rica are from Wadsworth (1987), and those for Belize are from Hartshorn et al. (1984). The low density of stems ≥10 cm dbh at Los Tuxtlas, Mexico, and Bladen, Belize, is partly due to the heavy understory dominance of 5 to 8 cm dbh palms (*Astrocaryum mexicanum*). The Palanan site, in the Philippines, is on a continental (formerly part of the mainland) island and is damaged by strong storms (winds speeds > 100 km h^{-1}) about every 10 years (Co et al. 2004). Stem and species data in the Puerto Rican and Costa Rican sites described by Wadsworth (1987), and calculated per hectare, are means of two c. 0.4 ha plots at each site. α is Fisher's α, which measures species diversity independently of sample size (Condit et al. 1998). Temperature is mean annual.

with the high frequency of natural disturbance in the West Indies promotes the occurrence of a high proportion of habitat generalists in West Indian avifaunas (Waide 1996).

The depauperate nature of the fauna found in Puerto Rico might be in part a consequence of typical island-biogeographic processes related to the size of the island and its distance from pools of colonists. Even compared to other islands in the Greater Antilles, however, Puerto Rico's animal species richness is well below the equilibrium level predicted by its area (MacArthur and Wilson 1967), at least for some taxa such as the bats (Griffiths and Klingener 1988). This is thought to result from Puerto Rico's long isolation from other larger islands and the continents (Hedges 2006), as well as the combined effects of frequent and widespread natural and human disturbances, which likely have extirpated species throughout the Caribbean (Turvey et al. 2007). The effects of disturbance might be especially important in Puerto Rico, which is the only island in the Greater Antilles to have lost all of its native land mammals (six), with some losses occurring following Amerindian or European colonization (Turvey et al. 2007). The concept of "equilibrium" numbers of species as embodied in the MacArthur and Wilson (1967) paradigm is thus difficult to apply to the biota of Puerto Rico (Lazell 2005; Covich 2006).

Climate

Caribbean Paleoclimate and Long-Term Trends

The climate of the Caribbean is warm, has slight but highly predictable seasonal temperature variations, and is subject to a variety of atmospheric systems that influence levels of precipitation. Paleoclimatic data suggest that Puerto Rico's climate, and that of most of the Caribbean, has been relatively stable for many millions of years compared to those of temperate and boreal regions. Studies of flora preserved in sediments from the Oligocene (34 to 24 mya) indicate that in coastal and upland sites, Puerto Rico had a range of tropical to subtropical plant communities similar to those of today (Graham and Jarzen 1969). Of the 44 genera of plants identified by Graham and Jarzen (1969), 31 presently occur in Puerto Rico, 3 occur on other Caribbean islands, and 7 are found in ecologically comparable environments in Latin America. Only three of the genera found were temperate tree species that require a habitat that is not presently available on the island or in the immediate region, suggesting a relatively stable climate for the region over the past 20 million to 30 million years.

There is some evidence that the Caribbean is currently warmer and wetter than it was during the Pleistocene. In the Pleistocene, a permanent snowline might have existed between 2,300 and 2,600 meters above sea level (masl) in Hispaniola, and parts of the Caribbean were more arid than they are today (Schubert and Medina 1982; Schubert 1988). Data from corals in Barbados indicate that temperatures in shallow Caribbean waters were 4°C to 6°C lower during Pleistocene glacial advances to the north (Guilderson et al. 1994). Climatic reconstruction from records of oxygen isotopes in Yucatán (Covich and Stuiver 1974) and Haitian lake sediments

(Hodell et al. 1991) suggests that relatively dry conditions occurred during the early Holocene. This was followed by a wetter mid-Holocene and a return to drier conditions during the late Holocene, several thousand years ago.

Climatic reconstruction in Puerto Rico from flood plain sediments in the coastal plain of the Río Fajardo suggests that the climate of the Luquillo region was humid during the Pleistocene but has become progressively drier during the Holocene (Mellon 2000). Although widespread fires are indicated in the stratigraphy of Holocene charcoal from Laguna Tortuguero on the north-central coast of Puerto Rico, they probably correspond to a peak of human activity around 5,300 years ago, and not to dramatic shifts in climate (Burney and Burney 1994). The Puerto Rican climate has thus been relatively unchanged since the geologic development of the island 30 mya and has not undergone the large shifts from glaciated to unglaciated conditions seen in many other regions.

Climatic data collected over the past 100 years suggest that the region is currently undergoing a period of minor drying and warming. A detailed study of precipitation trends at 24 stations throughout Puerto Rico from 1931 to 1966 shows a statistically significant decrease for most stations of –0.6 to –2.3 mm y^{-1} for the period of May–October (Bisselink 2003). In contrast, precipitation increased by 0.3 to 1.7 mm y^{-1} from November through April (see also chapter 4). The eight stations with the longest monitoring records (approximately one hundred years) all have negative trends in total annual precipitation, with decreases ranging between 1.59 and 4.90 mm y^{-1}. Furthermore, 1997, 1994, and 1991 were the second, third, and sixth driest years in the 20th century (Larsen 2000). These data suggest a trend of decreasing precipitation over most of Puerto Rico during the past century.

Regional temperatures have also changed over time in the Caribbean. In Cuban soils, changes in vertical temperature profiles indicate that climatic warming has increased surface temperatures at a rate of 1.0°C to1.2°C per century over the past 200 to 300 years. In the past 100 to 200 years, deforestation has also contributed to the recorded increases in soil temperature (Cermak et al. 1992).

Large-Scale Climate Drivers

Four major types of atmospheric systems affect the Luquillo Mountains: (1) intratropical systems; (2) extratropical frontal systems; (3) cyclonic systems; and (4) large scale, coupled ocean-atmospheric events (North Atlantic Oscillation [NAO], El Niño-Southern Oscillation [ENSO]). All of these systems can result in large-scale disturbances that generate significant rain and wind (see chapter 4). Neither monsoonal rains nor the Inter-Tropical Convergence Zone, however, affect the climate of the Luquillo Mountains.

Intratropical atmospheric systems are those that originate and generally remain within the tropics and include micro- and meso-scale convective systems and orographic rains. Owing to pronounced orographic effects, rainfall is unevenly distributed over Puerto Rico and ranges from less than 1,000 mm y^{-1} on the southwest (leeward) side of the island to 1,400 mm y^{-1} in the coastal plains on the northeastern (windward) side of the island, and up to 5,000 mm y^{-1} in the mountains (see figure 2-4). These orographic effects on rainfall are largely responsible for the steep environmental gradients that occur in the Luquillo Mountains (García-Martinó et al. 1996). Environmental gradients in the Luquillo Mountains are discussed in detail later in this chapter.

Extratropical frontal systems, locally known as cold fronts, occur during the temperate zone winter and spring with the arrival of polar lows from the northeastern United States. Cyclonic systems (large masses of air that rotate about a low-pressure center) occur from May to November. When the closure of a cyclonic system is incomplete, it is known as a tropical wave; when it is complete, it is termed a tropical storm or a hurricane. During the peak months of cyclonic activity (June–September), an average of two tropical waves pass by the Luquillo Mountains weekly (van der Molen 2002). Rainfall intensities and wind regimes associated with different atmospheric systems are described in more detail in chapter 4.

Large-scale ocean-atmospheric systems like the NAO and the ENSO are a principal cause of global interannual climate variability and have been linked to ecological processes in other tropical forests (Scatena et al. 2005). Although ENSO events have been linked to an increase in hurricane activity in Puerto Rico (Donnelly and Woodruff 2007), the NAO, rather than the ENSO, has the strongest relationship to the Puerto Rican climate (Malmgren et al. 1998). The NAO index is the normalized sea level difference in barometric pressure between the Azores and Iceland. It is significantly related to variations in annual rainfall in Puerto Rico; during years with a high northern winter NAO index, precipitation is generally lower than average. Annual rainfall in the Luquillo Mountains is only weakly correlated with either NAO or ENSO indices (Schaefer 2003), but Greenland (1999) did find a correlation between ENSO events and temperature in the Luquillo Mountains.

Climate of the Luquillo Mountains

The Luquillo Mountains have a humid tropical maritime climate. Water enters the ecosystem as rain and cloud drip, and on rare occasions as hail; snow and frost have never been recorded in the Luquillo Mountains. There is no pronounced dry season as found in monsoonal climates. The regular and predictable seasonal droughts (<50 mm monthly precipitation) or dry periods (<100 mm monthly precipitation) that are found in other tropical sites such as those in Barro Colorado Island, Panama (Zimmerman et al. 2007), or in a variety of Asian forests (Richards 1996) do not occur in the Luquillo Mountains. Episodic drought and rainy periods do occur throughout the year in the Luquillo Mountains, however.

The amount of rainfall increases with elevation, but during a 2-year study it showed similar patterns throughout the year at low, middle, and high elevations (figure 3-5). Long-term records from a mid-elevation site (El Verde Field Station, 365 masl; figure 3-2) show little seasonal variation over a 30-year record (figure 3-6). The highest rainfall tends to occur in May or September through December, but any month can be very wet (over 400 mm) or relatively dry (below 125 mm; figure 3-6). No month averages below 200 mm of precipitation. In general, more rain falls during the day than during the night (in terms of the total precipitation depth), but the frequency of

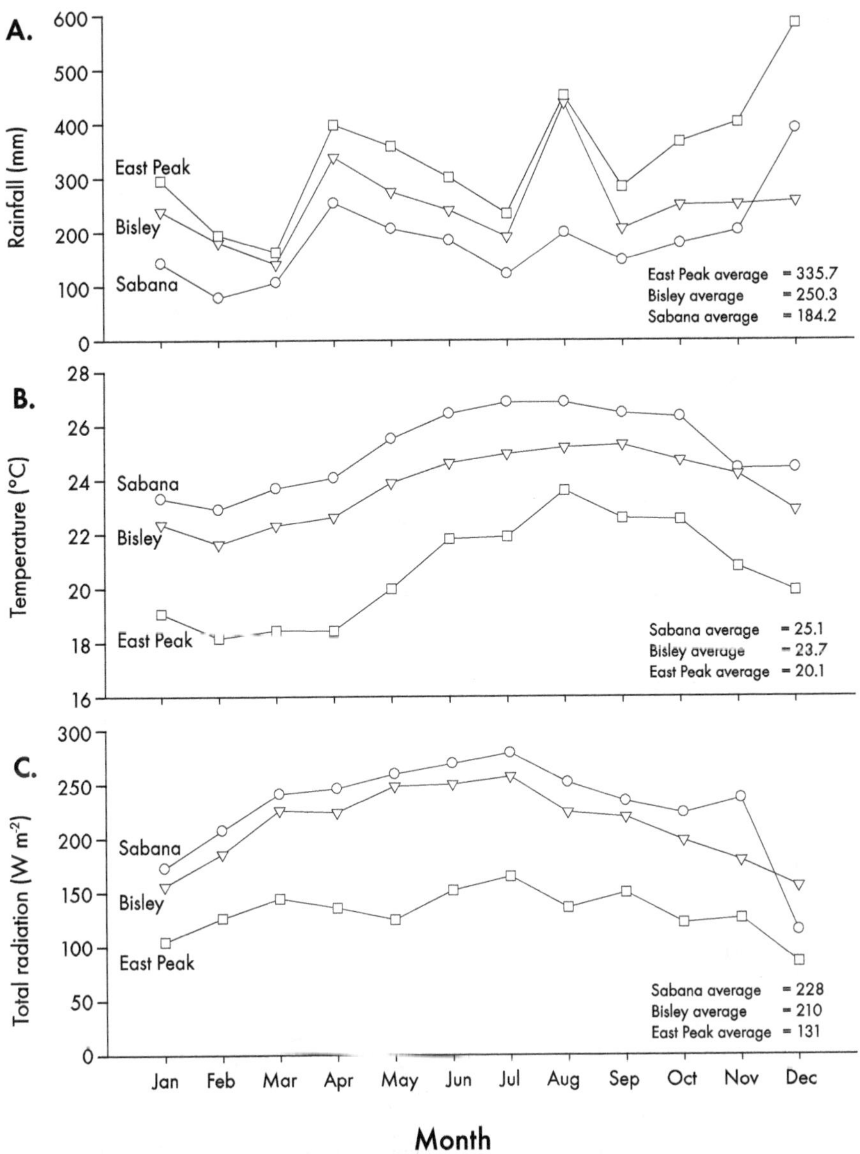

Figure 3.5 Variation in climatic conditions from 2000 to 2002 at three sites spanning the elevation gradient in the Luquillo Mountains: Sabana Field Station (100 masl), Bisley Experimental Watersheds (359 masl), and East Peak (Pico del Este) (1051 masl). (A) Average monthly precipitation (mm). (B) Average hourly air temperature by month (°C). (C) Average monthly total radiation (W m^{-2}).

rainfall events is highest at night and in the early morning (Schellekens et al. 1999). This pattern reflects the occurrence of smaller, low-intensity events at night and in the early morning, and larger events during the day.

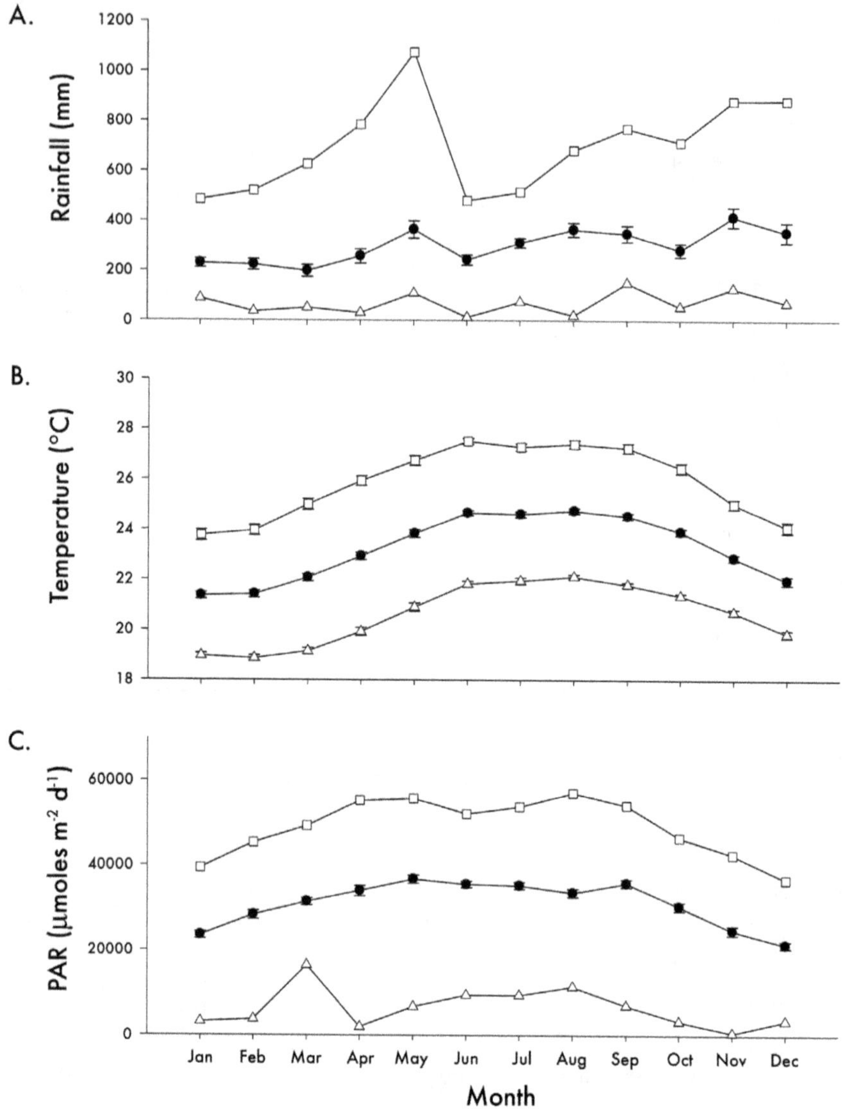

Figure 3.6 Long-term seasonal variation in climatic conditions at the El Verde Field Station in the Luquillo Mountains (350 masl). (A) Monthly precipitation in 1975–2004 (● = mean ± SEM; □ = maximum for 1975–2004; △ = minimum for 1975–2004). (B) Mean daily air temperature (± SEM) by month in 1975–2004 (● = monthly mean; □ = hottest day of the month; △ = coolest day of the month). (C) Monthly photosynthetically available radiation (PAR = μmol m^{-2} day^{-1}) at canopy level at the El Verde Field Station (350 masl). ● = mean daily values ± SEM; □ = maximum light level recorded during a day, by month (mean ± SEM); △ = minimum light level recorded during a day, by month (mean ± SEM), in 2000–2004.

Daily average air temperatures at all elevations in the Luquillo Mountains show small (3°C to 4°C) but highly predictable seasonal patterns, with the highest temperatures in June–July and the lowest temperatures in January (figures 3-5 and 3-6). The average annual air temperature varies with location and elevation in the Luquillo Mountains, with a difference of about 5°C between the temperature measured at the base of the mountains at Sabana Field Station (100 masl) and that at the weather station on East Peak (1,051 masl). Similar results were reported earlier by Briscoe (1966). Over the course of an average day, the temperature changes by 5°C at the base of the mountain but by only about 1°C at East Peak (figure 3-7), which is almost constantly bathed in cloud and fog. Humidity is high throughout the year in the Luquillo Mountains, with monthly minima (65 to 70 percent) and maxima (95 percent) showing no strong seasonal patterns (figure 3-8). Changes in the relative humidity over the course of a day strongly mirror changes in the air temperature at all elevations in the Luquillo Mountains, with humidity decreasing as air temperatures increase (figure 3-7). Windspeed is constant throughout the year, at about 1.3 m s^{-1} at the El Verde Field Station and 1.2 m s^{-1} at the Bisley Experimental Watersheds (figure 3-9). The average daily maximum wind speed is 6.4 m s^{-1} at the Bisley Experimental Watersheds and shows no seasonal patterns (figure 3-9). The average monthly total radiation shows strong seasonal patterns at lower elevations in the Luquillo Mountains, peaking in July, but seasonal patterns are much less distinct and the average total radiation is about 40 percent lower at East Peak than at lower elevations (figure 3-5). Photosynthetically active radiation (PAR) shows broad maxima during the period from June to August and has its lowest values in January (figure 3-6). Meteorological data for the Bisley Experimental Watersheds are available online from USGS National Weather Information Service Meteorological Station 50065549. Additional summary data on the climate of the Luquillo Mountains can be found at http://luq.lternet.edu/data/lterdb90/metadata/BisleyTowergraphs-Rad.htm for the Bisley Experimental Watersheds, and raw data files can be found for data throughout the Luquillo Mountains at http://luq.lternet.edu/data/databasesbycategory.html#Meteorology.

Climatic Gradients

Rainfall, humidity, wind, and cloudiness all tend to increase with elevation, whereas irradiance, air temperature, and soil temperature decrease (figures 3-5 and 3-7) (Briscoe 1966; Brown et al. 1983; García-Martinó et al. 1996). Rainfall shows a particularly strong pattern with elevation (figure 3-10). Taken together, these climatic variables all contribute to the decreased evapotranspiration and higher runoff observed at higher elevations (García-Martinó et al. 1996). The highest elevations of the Luquillo Mountains are covered in clouds for weeks at a time, resulting in significant interception of cloud moisture. Cloud cover is estimated to occur at least 75 percent of the time at East Peak, with an average cloud water deposition of 1 to 4 mm day^{-1}, which is higher than the deposition at most other sites globally where significant cloud inputs occur (Asbury et al. 1994; Eugster et al. 2006; Holwerda et al. 2006). Because rainfall is also very high at East Peak (15 to 30 mm day^{-1}), however, cloud water inputs represent a small part of the total hydrologic input.

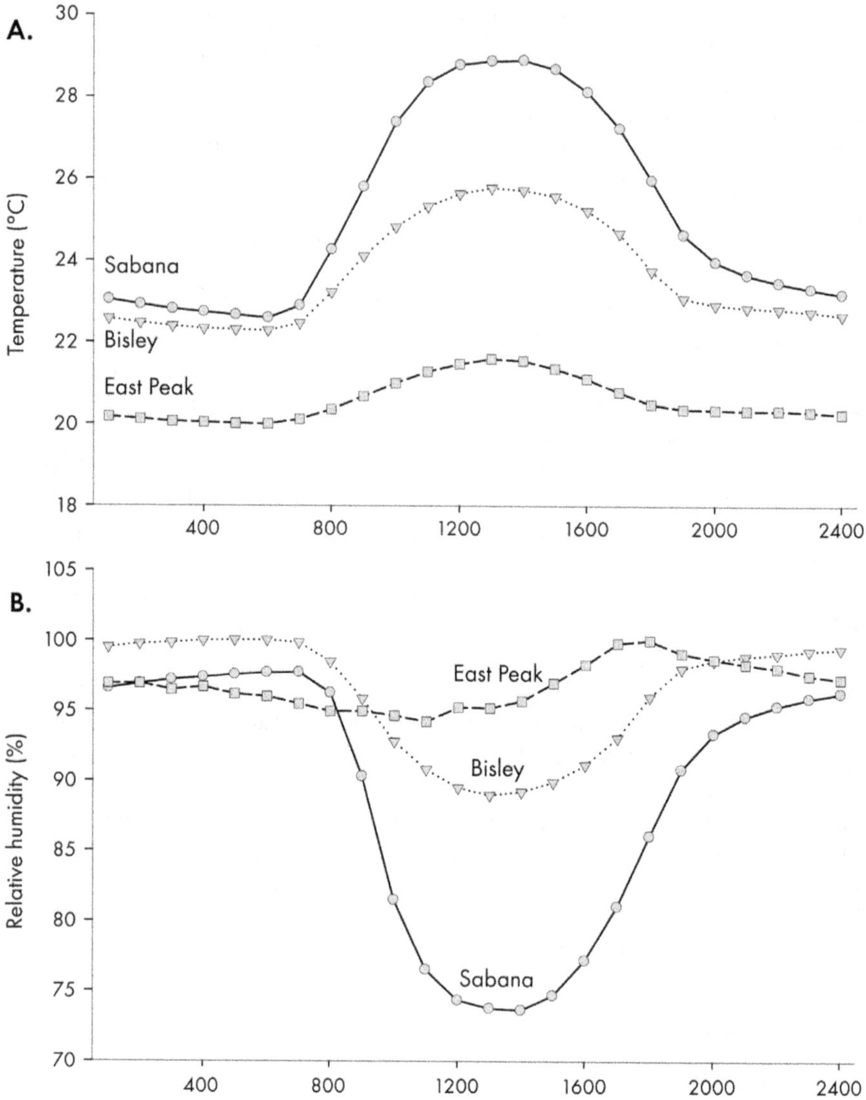

Figure 3.7 (A) Diel variation in mean hourly temperature (°C) and (B) relative humidity (percent) during 2000–2002 at three sites spanning the elevation gradient in the Luquillo Mountains: Sabana Field Station (100 masl), Bisley Experimental Watersheds (359 masl), and East Peak (1051 masl).

Local geographic position is particularly important in determining environmental conditions in the Luquillo Mountains, owing to the strong and steady trade winds. Seasonal patterns in rainfall, for example, vary spatially. The Luquillo summit casts a moderate rain shadow on its downwind flank, and the location of this shadow varies seasonally with the prevailing wind direction. Rainfall at the

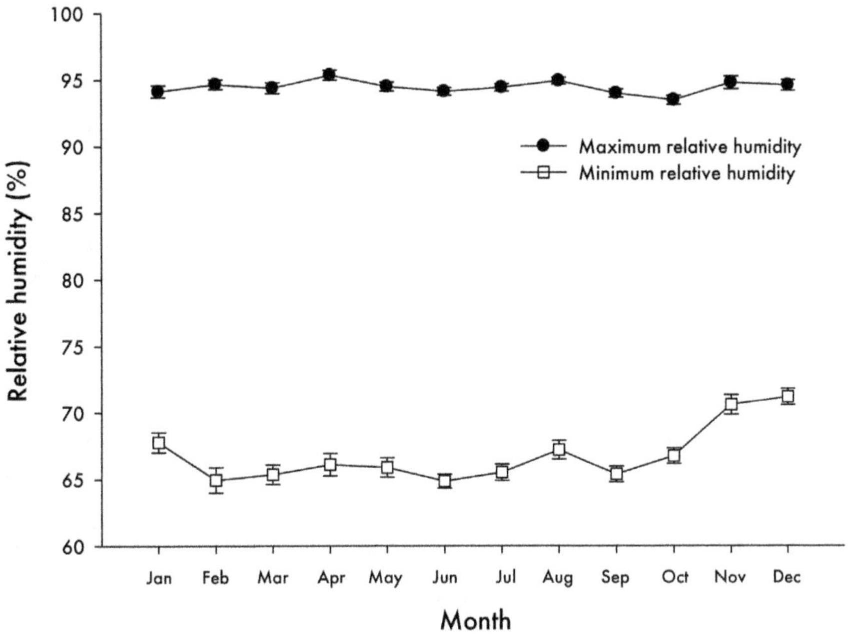

Figure 3.8 Variation in relative humidity by month during 2000–2004 at the El Verde Field Station (350 masl). □ = mean daily maximum humidity ± SEM; △ = mean daily minimum relative humidity ± SEM.

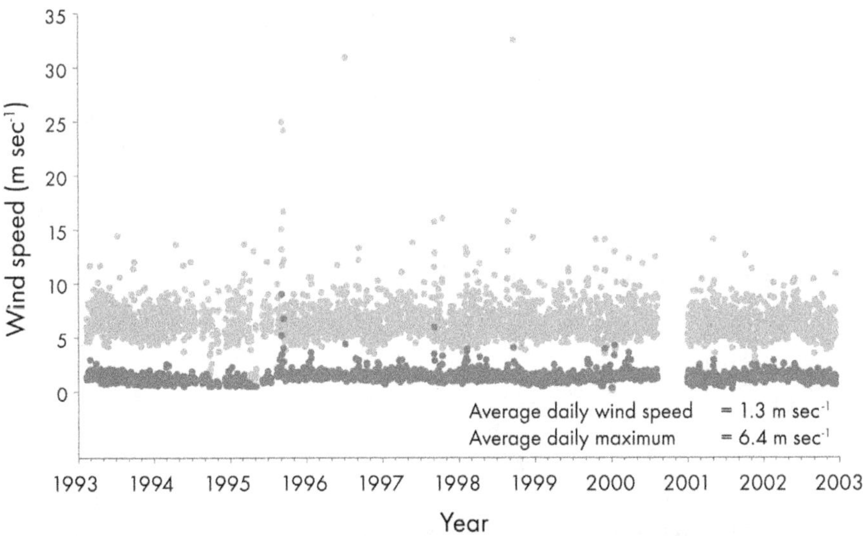

Figure 3.9 Long-term variation in wind velocity (m s^{-1}) at the Bisley Experimental Watersheds (359 masl) for the period of 1993–2002. • = daily average wind velocity; = daily maximum wind velocity. For the period of record, the mean daily wind velocity averaged 1.3 m s^{-1}, and the daily maximum wind velocity averaged 6.4 m s^{-1}.

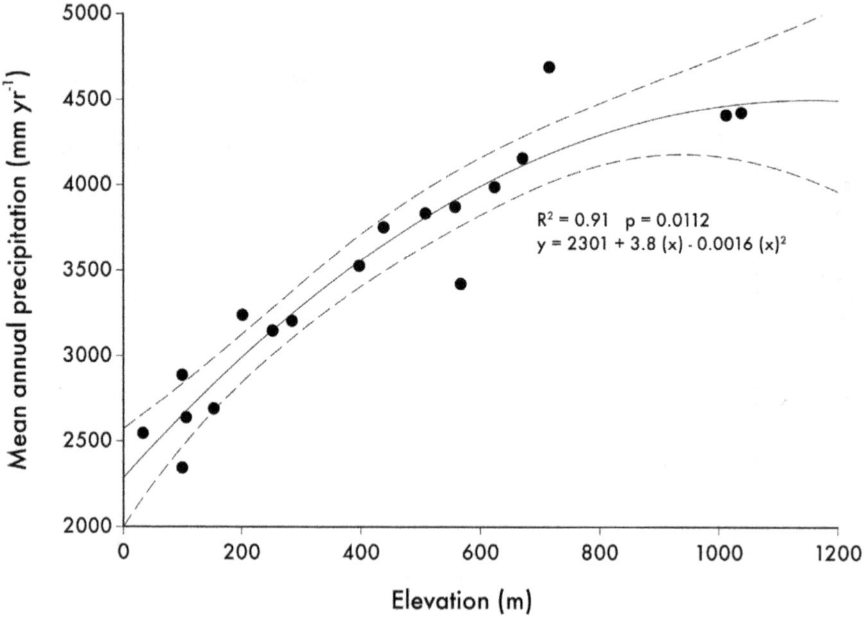

Figure 3.10 Variation in annual precipitation with elevation in the Luquillo Mountains. Redrawn from data presented in García-Martinó et al. (1996).

Bisley Experimental Watersheds and the El Verde Field Station, which are at similar elevations, is comparable throughout much of the year but is up to 30 percent lower at El Verde in May and June (Heartsill-Scalley et al. 2007). Wang et al. (2003) modeled the spatial and temporal variability of air temperature, solar insolation, rainfall, and transpiration (see figure 2.8) within the Luquillo Mountains. Their results show a complex pattern of spatial variability in climatic variables. The combined effects of elevation and geographic position on rainfall and temperature result in five different subtropical Holdridge life zones in the Luquillo Mountains: moist forest, wet forest, lower montane wet forest, lower montane rain forest, and rain forest (figure 3-3) (Ewel and Whitmore 1973).

Nutrient Cycling

Atmospheric Inputs

In the Luquillo Mountains, the biogeochemical cycles of most elements are driven by high rainfall, rapid river runoff, and the proximity of the mountains to the sea. Rains are intense and frequent, averaging three showers daily, and they bring large amounts of sea salt aerosols with them. Marine aerosols are the source of nearly all the sodium, chloride, magnesium, and potassium in rain (McDowell et al. 1990), and because of this, inputs of sodium are much higher than those typically seen in other humid tropical forests (figure 3-11). Calcium (Ca), nitrogen (N), and phosphorus (P) in rain come from predominantly nonmarine sources. For calcium, dust from the Sahara is a likely source of non–sea salt inputs (McDowell et al. 1990; Heartsill-Scalley et al. 2007). Because rain chemistry changes relatively little with elevation, the wet deposition of various elements is proportional to the increases in rainfall with elevation (Asbury et al. 1994). Cloud deposition also adds to nutrient deposition at high elevations (Asbury et al. 1994).

Saharan dust is a common occurrence throughout the Caribbean (Prospero and Nees 1986; Shinn et al. 2000; Muhs et al. 2007). It is found most often from June to August, when it can cause atmospheric haze, is readily visible as orange particles in rain collectors, and is the subject of local newspaper articles because of its nuisance value for residents of Puerto Rico. Dust inputs to the Caribbean coincide with North African droughts and are correlated with the NAO (Moulin et al. 1997). Beyond the effects of Saharan dust on soluble Ca concentrations in rainfall, little is known of its ecological significance in Puerto Rico, although recent work suggests that it might be a significant source of P to watersheds of the Luquillo Mountains (Pett-Ridge 2009). Evidence from the Hawaiian Islands, where dust from Asian deserts is important, suggests that atmospherically transported dust can be a major

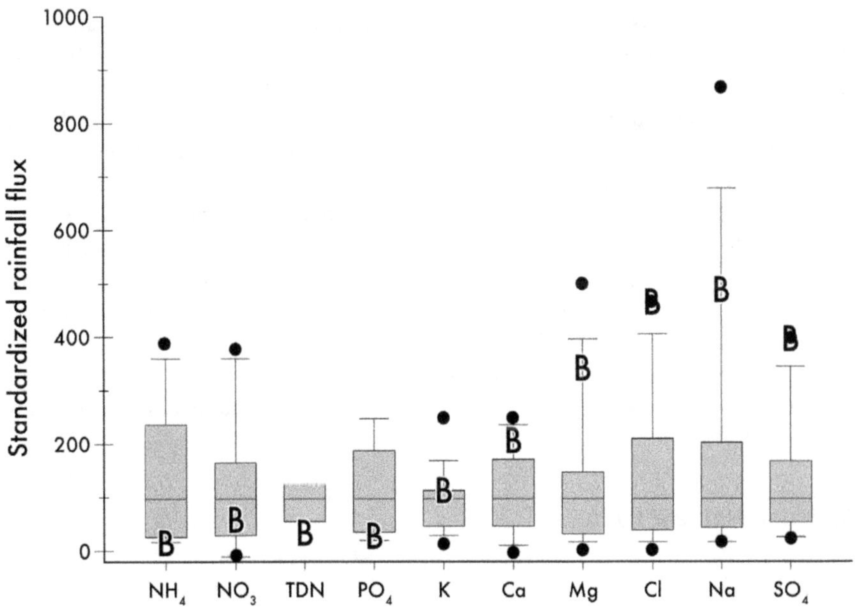

Figure 3.11 Box plots of standardized values for input of ammonium, nitrate, phosphate, potassium, calcium, magnesium, chloride, sodium, and sulfate (NH_4, NO_3, PO_4, K, Ca, Mg, Cl, Na, and SO_4) in rainfall at the Bisley Experimental Watersheds (B) compared to other humid tropical forests. Standardization is done by expressing the values from each site as a percentage of the median for all sites. Sample size ranges from 7 to 23, with most being at least 14. Box shows 25th through 75th percentiles; error bars show 10th and 90th percentiles; solid circles are outliers. Adapted and redrawn from Scatena (1998).

source of elements such as Ca and P in highly weathered landscapes (Chadwick et al. 1999). On some Caribbean islands (Barbados and the Florida Keys), Saharan dust plays an important role in soil formation (Muhs et al. 2007), but its significance for elemental cycles in the Luquillo Mountains is not as well understood. Atmospheric deposition can be a critical source of growth-limiting nutrients for the reestablishment of vegetation following disturbance. Zarin and Johnson (1995) found that rainfall inputs were sufficient to provide nearly all of the Ca required for the growth of vegetation on landslide scars in the Luquillo Mountains, and that they provided significant quantities of other nutrients needed for regrowth.

With regard to elements not found in significant quantities in sea salt (e.g., N, P, and potassium [K]), the chemistry of rain in the Luquillo Mountains is relatively dilute, making the overall deposition rates of these elements modest and at or below levels recorded at other tropical sites (figure 3-11). The rates of inorganic N deposition are typical of those found throughout rural Central and South America (Sanhueza and Santana 1994). In comparison with rain from remote areas of the world, Puerto Rican rainfall was only slightly enriched in non-sea-salt sulfate and nitrate in the period from 1984 to 1987, indicating that there was little anthropogenic influence on the precipitation chemistry (McDowell et al. 1990). Since then, however, nitrate deposition in wet-only precipitation samples analyzed as part of the National Atmospheric Deposition Program has steadily increased (Ortiz-Zayas et al. 2006).

The sources of this increase in nitrate deposition are unknown. Urbanization has increased in all directions around the Luquillo Mountains, and this might be contributing to the observed increases in nitrate deposition. Increased volcanic activity since 1995 in the Soufrière Hills, Montserrat, has increased the deposition of total dissolved N, but not nitrate, in bulk precipitation collected at the Bisley Experimental Watersheds during periods of volcanic activity (Heartsill-Scalley et al. 2007). There are few anthropogenic pollution sources to the northeast of Puerto Rico, the direction from which the dominant trade winds originate. Because air masses from North America do reach the island, they might also be contributing to nitrate and sulfate inputs above global background levels (McDowell et al. 1990). Air masses from North America can reach Puerto Rico throughout the year, but most of the North American air reaching Puerto Rico arrives in early spring, when the rainfall pH is somewhat reduced, dropping from its typical pH of 5.5 to values around 5.0.

Soils and Nutrient Pools

Soil Characteristics

Soils in the Luquillo Mountains are deep, highly weathered iron (Fe) and aluminum (Al) clay soils with nutrient concentrations that are typical for the tropics (Sánchez 1976; Silver et al. 1994). The surface organic layer, or forest floor, above the clayey soil is generally poorly developed or intermittent, so it represents a relatively minor pool of nutrients at most locations in the Luquillo Mountains. Nonetheless, most of the carbon flowing through the food web passes through the detrital system on the forest floor (Lodge 1996), where it is processed rapidly (Ostertag et al. 2003).

Soils of the Luquillo Mountains are primarily classified as Ultisols. Less commonly found, but present in small areas, are other soil orders (Oxisols, Entisols, and Inceptisols). Oxisols are mostly found in the well-drained upland areas in the tabonuco forest, whereas Inceptisols and Entisols are located in the drainage areas along streams and in the upper elevations near quartz diorite intrusions. Detailed descriptions of soils have been published by Silver et al. (1994, 1996), Scatena and Lugo (1995), the Soil Survey Staff (1995), and Cox et al. (2002). Ultisols of the Luquillo Mountains have a clay content that ranges from 35 percent to 88 percent; the average clay content is (52 ± 5) percent (McGroddy and Silver 2000). Unlike the riparian Inceptisols of the Icacos basin, where clay concentrations (ranging from 49 percent to 15 percent) tend to decrease with depth, the Ultisols of the Bisley Experimental Watersheds have higher concentrations of clay (51 percent to 54 percent), and clay concentrations do not vary with depth (McDowell et al. 1992).

Soil carbon (C) concentrations are typically 2 percent to 4 percent in surface soils (0–10 cm) of the Luquillo Mountains, and they decline to less than 1 percent at depths of 35 to 60 cm. Soil N is typically found in concentrations of 0.1 percent to 0.4 percent in surface soils and declines to 0.1 percent or less at depth (Fox 1982; McDowell et al. 1992; Silver et al. 1994; Scatena and Lugo 1995). In the tabonuco forest, soil C and N concentrations can be as high as 7.6 percent and 0.67 percent, respectively, in surface soils (0–10 cm) and up to 4.1 percent and 0.29 percent in subsurface (10–25 cm) soils (Li 1998). Light-fraction C and N, which are thought to represent more biologically available material than the denser fractions, was measured as 2.9 and 0.17 mg g^{-1}, respectively, in the surface soils studied by Li (1998). Light-fraction C shows large increases with elevation in the Luquillo Mountains and accounts for almost all C storage in soils above 900 m elevation (McGroddy and Silver 2000). The soil N concentration (averaging 0.31 percent in surface soils) is somewhat lower in the Bisley Experimental Watersheds than in soils found in other tropical montane sites (figure 3-12). Soil C is higher under decaying logs than elsewhere (Zalamea et al. 2007).

The amount of extractable phosphorus typically decreases sharply with depth in the soil profile (Silver et al. 1994). Soils from mid-elevation tabonuco forests have levels of potassium chloride-extractable P averaging (26 ± 2) $\mu g\ g^{-1}$ in surface soils (0–10 cm), and these levels declined to 3 $\mu g\ g^{-1}$ at depths of 35 to 60 cm. In samples from colorado forests, at higher elevations, Frizano et al. (2002) found that concentrated hydrochloric acid extraction recovered 27 $\mu g\ g^{-1}$ P at both the surface and depths of 35 to 60 cm. They also examined a wide range of extractants and found very high variability in the patterns of extractable P with depth. Extractable nutrient cations from mid-elevation soils in the Bisley Experimental Watersheds average 0.42, 1.83, and 1.37 cmol kg^{-1} for K^+, Ca^{2+}, and magnesium (Mg^{2+}), respectively (Scatena and Lugo 1995) (table 3-2).

The soil chemistry shows strong variation with topographic position in tabonuco and colorado forests, but the patterns vary. In the volcaniclastic soils associated with tabonuco forests, organic matter and nitrogen levels are highest on the ridges and upper slopes, but in the colorado forest soils the pattern is reversed, with the highest C and N levels in riparian soils (tables 3-2 and 3-3). In both forest types, greater amounts of extractable Ca, Mg, and P are found in riparian than in upland

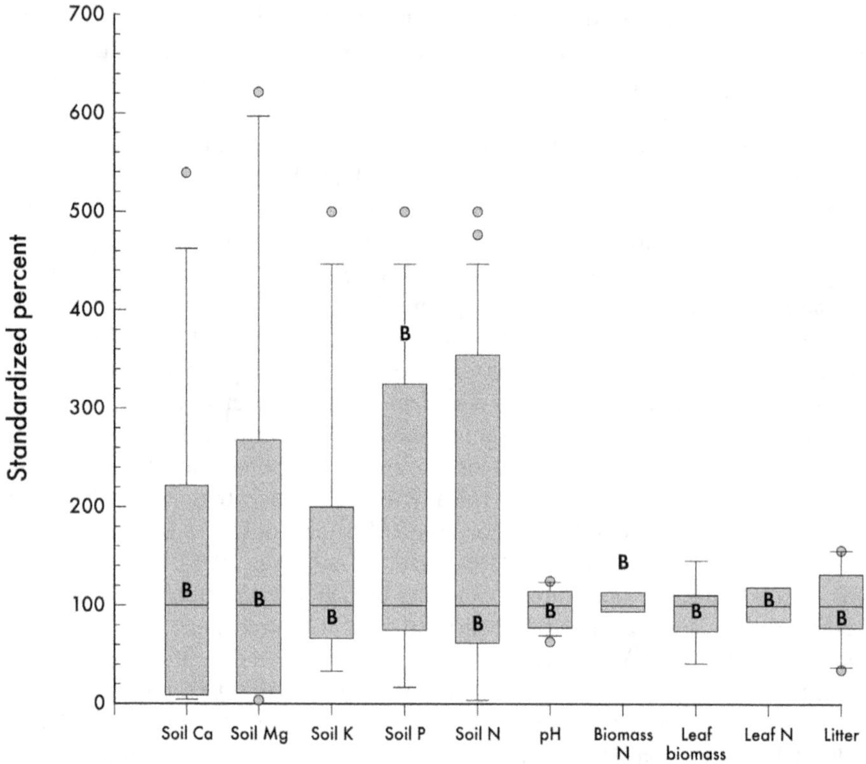

Figure 3.12 Box plots of soil chemistry, biomass, and chemistry of leaf and total biomass in the Bisley Experimental Watersheds (B) of the Luquillo Mountains compared to other humid tropical forests. Box plots show standardized percentage values for soil chemistry (Soil Ca, Mg, K, P, N, and pH), aboveground biomass nitrogen stock (Biomass N), leaf biomass, leaf nitrogen concentration (Leaf N), and litterfall (Litter). Standardization is done by expressing the values from each site as a percentage of the median for all sites. Sample size ranges from 7 to 23 tropical forest sites, with at least 14 sites for most parameters. Box shows 25th through 75th percentiles; error bars show 10th and 90th percentiles; solid circles are outliers. Adapted and redrawn from Scatena (1998).

soils (tables 3-2 and 3-3). Organic matter tends to accumulate on mid-elevation ridges and slopes dominated by *Dacryodes excelsa* (figure 3-13), a tree that produces thick surface root mats, but the exact mechanisms of soil carbon accumulation on ridges are not known. The higher organic matter content leads to greater exchangeable acidity and lower soil pH on ridges (4.8) as compared to riparian valleys (5.4) (table 3-2). High densities of basidiomycete litter mats on slopes contribute to the retention of leaf litter and the protection of surface soil from the erosive effects of rain and overland flow, thereby conserving soil carbon on slopes (Lodge and Asbury 1988; Lodge et al. 2008).

The patterns in soil chemical properties along the catena can be partially explained by redox processes and their effects on the amount and form of Fe in the

Table 3.2 Variation of soil chemistry (organic matter, nutrients, exchangeable cations, and pH) in the Bisley Experimental Watersheds by topographic position and for the entire watershed

Group	*n*	SOM	N	P	K	Ca	Mg	Na	Fe	Mn	pH
		Mg ha^{-1}		kg ha^{-1}			cmol kg^{-1}				
Ridge	22	210a	9.0a	42.0a	0.39ab	0.88a	0.76a	0.31a	4.74a	0.29a	4.8a
		(23)	(0.7)	(4.8)	(0.04)	(0.28)	(0.26)	(0.03)	(0.76)	(0.09)	(0.1)
Slope	40	163a	7.7a	39.8a	0.44a	1.56a	1.51ab	0.29a	2.83ab	0.52ab	5.0ab
		(11)	(0.5)	(3.8)	(0.05)	(0.40)	(0.44)	(0.03)	(0.31)	(0.09)	(0.1)
Upland valley	12	143ab	7.0ab	47.1a	0.36b	3.18a	1.93b	0.31a	2.68b	0.97b	5.2bc
		(18)	(0.7)	(5.6)	(0.04)	(1.93)	(0.75)	(0.05)	(0.72)	(0.16)	(0.1)
Riparian valley	9	131b	7.2b	71.7b	0.48a	3.33b	1.59b	0.24a	2.65b	0.91ab	5.4c
		(9)	(1)	(10.3)	(0.04)	(0.64)	(0.35)	(0.47)	(0.4)	(0.2)	(0.1)
Entire watershed	83	170	8.5	45.4	0.42	1.83	1.37	0.29	3.33	0.572	5
		(9)	(0.3)	(2.7)	(0.03)	(0.37)	(0.24)	(0.02)	(0.3)	(0.06)	(0)

Modified from Scatena and Lugo 1995. Mean chemistry of surface soils (0 to 60 cm) for individual geomorphic settings (ridge, slope, upland valley, and riparian valley) is shown with the standard deviation on the line below. The estimation of the average nutrient standing stocks for the combined Bisley watersheds 1 and 2 is based on the frequency of each geomorphic setting within the combined watersheds. Sample *n* refers to the number of sites sampled within the watershed or the geomorphic setting. SOM = soil organic matter plus forest floor. For each column, means with the same letters are not different at the 0.05 level according to Duncan's multiple range test.

soil. In well-aerated soils, oxidized Fe (Fe^{3+}) can coat exchange sites, essentially blocking the retention of other base cations that are normally associated with mineral exchange complexes (Abruna and Smith 1953; Fox 1982). Ridges and slopes are generally well aerated, whereas upland valleys and riparian zones experience frequent low-oxygen events associated with rainfall and high stream flow (Silver et al. 1999). Soil exchange sites become available for other base cations such as Ca and Mg when Fe becomes reduced, resulting in higher levels of extractable cations in riparian soils (table 3-2). Conversely, scouring of riparian valleys in tabonuco forest during overland flow events leads to lower carbon stocks (Weaver et al. 1987; Lodge et al. 2008).

Soil Carbon and Nutrient Pools

Nutrient pools in soils of the Luquillo Mountains are similar to those reported for other lower montane wet tropical forests (Silver et al. 1994; Scatena 1998). Soil organic matter averages 170 Mg ha^{-1} to a depth of 60 cm in the Bisley Experimental Watersheds (table 3-4). Soil P is found at somewhat higher standing stocks in the Luquillo Mountains (45 kg ha^{-1}) than at other tropical forests (Scatena and Lugo 1995; Scatena 1998).

Pools of available nutrients in the soils of the Luquillo Mountains are typically as great as or greater than those in the aboveground biomass (table 3-4). The exception is K, with up to an order of magnitude more K being stored in the plant biomass than in the labile soil pool (Scatena et al. 1993; Silver et al. 1994).

Table 3.3 Variation of soil chemistry (organic matter, nutrients, exchangeable cations, and pH) and bulk density (BD) in the Icacos watershed by topographic position and depth

Position	SOM	C	N	P	K	Ca	Mg	Na	Fe	Mn	BD
		%		mg kg^{-1}	mg g^{-1}					mg kg^{-1}	g cm^{-3}
Ridge											
0–10 cm	13.62 (1.12)	5.19 (0.51)	0.23 (0.02)	6.70 (0.53)	0.06 (0.003)	0.12 (0.03)	0.06 (0.07)	0.04 (0.003)	2.06 (0.15)	5.32 (0.88)	0.69 (0.02)
10–25 cm	9.13 (0.51)	2.88 (0.25)	0.14 (0.01)	3.78 (0.54)	0.04 (0.003)	0.02 (0.00)	0.02 (0.002)	0.03 (0.004)	1.54 (0.11)	1.93 (0.16)	0.86 (0.03)
25–50 cm	7.15 (0.41)	1.18 (0.09)	0.07 (0.01)	0.68 (0.21)	0.01 (0.002)	0.01 (0.00)	0.01 (0.001)	0.01 (0.001)	0.55 (0.07)	0.76 (0.1)	1.3 (0.05)
Slope											
0–10 cm	12.15 (0.56)	4.75 (0.37)	0.23 (0.02)	8.18 (0.71)	0.05 (0.002)	0.06 (0.01)	0.03 (0.003)	0.04 (0.003)	1.92 (0.17)	3.61 (0.41)	0.67 (0.03)
10–25 cm	9.64 (0.48)	3.53 (0.30)	0.17 (0.01)	6.24 (0.68)	0.04 (0.003)	0.02 (0.00)	0.02 (0.002)	0.03 (0.002)	1.58 (0.11)	2.17 (0.18)	0.9 (0.03)
25–50	6.37 (0.33)	1.33 (0.15)	0.06 (0.01)	1.30 (0.33)	0.01 (0.002)	0.01 (0.00)	0.01 (0.001)	0.01 (0.001)	0.70 (0.09)	1.52 (0.4)	1.25 (0.04)
Riparian											
0–10 cm	21.46 (0.94)	7.33 (0.65)	0.33 (0.02)	14.57 (0.95)	0.08 (0.005)	0.26 (0.05)	0.08 (0.009)	0.05 (0.004)	1.00 (0.11)	41.80 (11.2)	0.58 (0.02)
10–25 cm	20.07 (0.77)	6.39 (0.57)	0.29 (0.02)	11.90 (0.87)	0.06 (0.004)	0.23 (0.09)	0.05 (0.006)	0.03 (0.003)	0.83 (0.08)	35.78 (10.7)	0.62 (0.02)
25–50 cm	18.79 (1.85)	5.46 (0.65)	0.20 (0.02)	8.04 (0.78)	0.04 (0.006)	0.18 (0.07)	0.05 (0.009)	0.04 (0.004)	0.67 (0.09)	29.24 (8.99)	0.65 (0.04)

Modified from McSwiney (1999). Chemistry of surface soils by depth for individual geomorphic settings (ridge, slope, and riparian zone) is shown with standard deviation on the line below. Three different catenas were sampled at lower-elevation sites (approx 725 m) in the Rio Icacos, and three samples were taken at each position and depth in each catena. SOM = soil organic matter plus forest floor.

Figure 3.13 Group of tabonuco trees (*Dacryodes excelsa*), which dominate especially on ridge tops, with indivduals usually interconnected by roots. (Photograph by Jerry Bauer.)

Table 3.4 Variation in nutrient capital and organic matter (OM) in plant biomass and soils of the tabonuco forest type.

	OM Mg ha^{-1}	N	P	K	Ca	Mg
				kg ha^{-1}		
Overstory	221	614	33.4	514	464	126
Roots (C)	72.4	203	10.9	79.6	156	50.7
Roots (F)	2.20	34.6	1.2	2.4	17.0	2.2
Understory	4.31	55.8	3.4	48.6	16.0	13.8
Total biomass	300	907	48.9	644	653	192
Soil	170	8,500	45.4	70	600	1,100
Percent total standing stock as biomass	64	10	52	90	52	15

Values represent watershed mean values of the Bisley Experimental Watersheds, tabonuco forest type. In soils, total N is reported, but other nutrients represent the extractable fraction only. Plant biomass is divided into overstory trees with dbh > 2.5 cm (including boles, leaves, and bark), coarse roots (C), fine roots (F), and understory plants. Data from Scatena et al. (1993), Silver et al. (1994), and Scatena and Lugo (1995).

Nutrient pools in aboveground biomass are similar to those in other montane tropical forests in New Guinea (Grubb and Edwards 1982) and Venezuela (Grimm and Fassbender 1981), but they are higher than those in upper montane forest sites in Jamaica (Tanner 1985) and Hawaii (Mueller-Dombois et al. 1984). A comparison with a broad range of montane tropical sites shows that soils in Bisley are typical for base cations, elevated in extractable P, and lower than the median in N (figure 3-12).

Although mineral nutrients are present in soils, they are not necessarily available to plants. The soils of the Luquillo Mountains, as in many tropical areas, have phosphorus-fixing clays dominated by iron and aluminum oxides that bind tightly to phosphorus and make it less available to plants and microbes. Oxygen levels in soils can also affect the nutrient availability through a number of mechanisms. Low oxygen (O_2) availability in the flooded soils of high-elevation elfin forests in the Luquillo Mountains (Silver et al. 1999) might result in decreased decomposition rates and the decreased efficiency of nutrient uptake by roots or their mycorrhizal fungi. On the other hand, low O_2 concentrations can also increase the availability of P, as periods of soil anoxia result in the reduction of Fe^{3+} to Fe^{2+}, which releases the P held in $FePO_4$ bonds (Silver et al. 1999). Oxygen concentrations in soils decrease significantly as the annual rainfall at a location increases, and they can reach very low levels (<3 percent) at individual sampling points for periods of up to 25 consecutive weeks (Silver et al. 1999). Soil O_2 concentrations of <3 percent are frequently cited as being below the critical threshold for the survival of some herbs and wetland plants (Drew 1990).

Variation in Soils with Elevation

Soils of the Luquillo Mountains show considerable variation in C, N, and P with elevation and vegetation type (table 3-5). Carbon and nitrogen are highest in high-elevation soils, and the C:N ratio (mass:mass) increases from 12 to 24 along the elevational

Table 3.5 Variation in nutrient pools of surface soils (0–10 cm depth) from pasture and three forest types in the Luquillo Mountains.

Vegetation (elevation, masl)	C	N	P	K	Ca	Mg	Na	Fe	Mn	pH
	Mg ha^{-1}		kg ha^{-1}							
Pasture	34.9	2.78	6.34	60.0	217	136	30.3	482	65.1	3.72
(100)	(6.54)	(0.62)	(3.11)	(14.0)	(47.4)	(57)	(6.1)	(146)	(23.7)	(0.07)
Tabonuco	47.1	3.51	7.49	75.3	472	242	49.7	676	41.2	3.81
(300)	(11.3)	(0.53)	(1.28)	(31.3)	(499)	(151)	(19.0)	(304)	(28.5)	(0.28)
Palo colorado	67.3	3.49	4.92	46.5	293	95	38.5	813	24.9	3.87
(650)	(23.7)	(1.40)	(2.08)	(6.12)	(212)	(33)	(8.96)	(114)	(21.7)	(0.23)
Elfin	220	9.30	6.01	37.3	202	67	43.6	501	14.2	3.79
(950)	(54.7)	(2.68)	(1.59)	(12.4)	(134)	(14)	(6.04)	(171)	(15.6)	(0.22)

Data from Cox et al. (2002). Each sample represents the mean (SD) of two ridge sites and two valley sites per forest type, with multiple auger samples composited from each site.

gradient (table 3-5) (Cox et al. 2002). Available cations tend to be highest in the mid-elevation tabonuco forest, but wide variability in elemental contents and in the interactions between elevation and topographic position within each forest type results in few statistically significant differences for individual elements (Cox et al. 2002).

Internal Nutrient Fluxes

Litterfall, throughfall, and the movement of soil solution and groundwater represent large internal transfers of nutrients within the forest ecosystems of the Luquillo Mountains. The importance of each pathway differs by element. Throughfall, for example, provides more than twice the flux of K to the forest floor than does litterfall (McDowell 1998). For nitrogen, however, the opposite situation occurs, with 20 times as much N transferred in litterfall as in throughfall (figure 3-14).

Internal nitrogen dynamics have been particularly well studied in tabonuco forests (see the summary by Chestnut et al. [1999]). In the forests of the Luquillo Mountains, the sum of N export and net biomass N accumulation exceeds N inputs

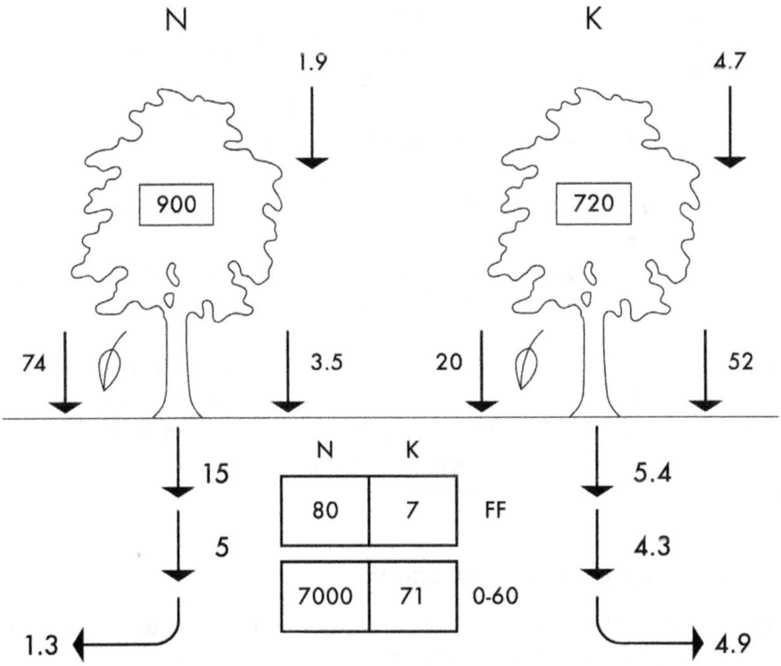

Figure 3.14 Standing stocks and internal fluxes of nitrogen (N) and potassium (K) in the tabonuco forest type, Luquillo Experimental Forest, Puerto Rico. Standing stocks (boxes) are kg ha^{-1}; fluxes (arrows) are kg ha^{-1} y^{-1}. Fluxes shown include precipitation, throughfall, litterfall, soil solution at 40 and 80 cm, and streamflow. FF = forest floor; 0–60 = soil pools from 0 to 60 cm in depth. Standing stock of potassium in soil is the exchangeable pool only. Modified from McDowell (1998). Reprinted with permission from Cambridge University Press.

in rainfall (McDowell and Asbury 1994). This imbalance in the nitrogen budget might be due to unmeasured inputs from biological nitrogen fixation or unmeasured changes in the total standing stock of soil N. Because standing stocks are very large relative to inputs and outputs, it is hard to rule out changes in soil N storage as a potential explanation for the imbalance in the N budget (Chestnut et al. 1999). Nitrogen fixation is also very likely to occur, as it is typically high in tropical environments (Cleveland et al. 1999), and various N fixers are known to occur in the Luquillo Mountains (Edmisten 1970). Some variability in the rates of N fixation appears to be associated with patterns of past land use. Trees planted with coffee (e.g., *Inga*) are common at many sites in the Luquillo Mountains, and in the Bisley Experimental Watersheds soils under *Inga* have a higher N content than those under other species (Beard et al. 2005). Free-living soil microbes in the Luquillo Mountians contribute the most to N-fixation rates on an areal basis, but nitrogenase activity is highest on a per-gram basis in mosses (Cusack et al. 2009). An estimate of the rate of N fixation in watershed-scale studies of N inputs and outputs (McDowell and Asbury 1994; Chestnut et al. 1999) suggests that it might be up to 16 kg ha^{-1} y^{-1}. No systematic survey of N fixation has been conducted in the Luquillo Mountains.

The aquatic habitat within bromeliad tanks is particularly nutrient rich, with the concentrations of many elements being orders of magnitude higher than those found in rain or streamwater (Richardson et al. 2000). These tanks serve as spatially distributed, high-nutrient aquatic microcosms in the terrestrial ecosystem. Depending on the elevation, the average total dissolved N (TDN) and phosphate (PO_4^{3-}) concentrations can exceed 3 and 0.4 mg l^{-1}, respectively, and dissolved organic carbon (DOC) concentrations can exceed 50 mg l^{-1}. In contrast, the average concentrations of TDN, PO_4^{3-}, and DOC in stream water rarely exceed 0.25, 0.02, and 3 mg l^{-1}, respectively (McDowell et al. 1990; McDowell and Asbury 1994). Annual nutrient budgets indicate that these bromeliad microcosms are nutrient rich because of their high inputs of both throughfall and litter from canopy trees. In general, tank bromeliads in all forest types accumulate <5 percent of the nutrients that pass through them; the exception is in high-elevation elfin forest, where bromeliads accumulate about 25 percent of P and K inputs (Richardson et al. 2000). The relative importance of bromeliad phytotelmata (tanks) as storage compartments increases with elevation, as the bromeliad density increases, along with their efficiency of nutrient retention (Richardson et al. 2000).

Interfaces as Biogeochemical Hot Spots

Interfaces where two distinct parts of an ecosystem meet are often important in defining ecological processes at multiple spatial and temporal scales. Interfaces play an important role in biogeochemical transformations because the transformations often occur at higher rates per surface area or volume at interfaces than at adjacent, homogeneous units of the landscape. McClain et al. (2003) have summarized the situations in which interfaces are more active, and they have proposed a formal set of definitions, including "hot spots" and "hot moments," which often occur at interfaces. Hot spots are points in the landscape at which the rates of biogeochemical processes are disproportionately high relative to the surrounding area.

They are often the result of converging hydrologic flowpaths, where the reactants needed for a reaction (e.g., the low-oxygen water, nitrate, and organic matter needed for denitrification) are delivered to a point in the landscape. Hot moments occur when the same confluence of reactants and conditions are present for a specific moment in time, generating intense biological activity.

Drought-induced crashes in microbial populations result in pulses of nutrient availability when the forest floor is rewetted and the microbial biomass killed off by the drought is mineralized (Lodge et al. 1994), representing an excellent example of the "hot moment" concept. Local weather systems such as large storms and prolonged droughts (see chapter 4) can have a significant effect on the short-term variation in internal nutrient cycling rates in tropical forests (Lodge et al. 1994). These pulses are often quantitatively significant, because microbes represent a significant fraction of the total labile nutrient pool. Mean fungal biomass accounted for 22 percent of the total phosphorus in the litter layer at El Verde, and for between 3 percent and 85 percent of the litter P at different sites (Lodge 1993, 1996). In soil at El Verde, fungal biomass accounted for 0.8 percent to 20 percent of the labile (Olson extractable) P and 24 percent of the Ca (Lodge 1993, 1996). The rapid growth and nutrient immobilization by microbes under favorable moisture conditions helps to retain nutrients against leaching loss; crashes in fungal populations in response to drying release nutrients from the microbial biomass, making them available to plants (Lodge 1993; Lodge et al. 1994).

In the Luquillo Mountains, the oxic–anoxic interface and the stream–groundwater interface represent two hot spots for biogeochemical transformations. These two can be related; changes in oxygen status are often associated with the groundwater–streamwater interface in the Luquillo Mountains (see, e.g., McDowell et al. 1992). But oxic–anoxic interfaces can also be found in upland soils, far from the stream's edge (Silver et al. 1999). In these soils, the maintenance of low-oxygen conditions is typically related to inputs of rainfall that drive the metabolic processes that deplete molecular oxygen in the soil matrix and prevent the resupply of atmospheric oxygen by reducing open pore space.

The low soil O_2 concentrations that can occur in Luquillo Mountain soils affect a variety of biogeochemical processes in upland soils, in addition to having effects on riparian biogeochemistry. Elfin forest soils, for example, have extremely high soil methane (CH_4) concentrations (3 percent to 24 percent), indicating the strong influence of anaerobic processes. These high soil CH_4 concentrations result in net CH_4 emission into the atmosphere in the elfin forest ([98 ± 50] mg m^{-2} d^{-1}), and net emission is also seen in lower elevation valleys ([5 ± 1] mg m^{-2} d^{-1}), but soils in other parts of the forest are net CH_4 consumers (McSwiney 1999; Silver et al. 1999). Nitrous oxide (N_2O) flux responds little to rainfall in chronically wet soils but appears to be related to differences in oxygen concentrations across topographic gradients, with maximal N_2O production occurring at intermediate oxygen levels (McSwiney et al. 2001).

At the stream–water interface, or riparian zone, the interplay between oxygen, carbon, and nitrogen drives most biogeochemical processes (see, e.g., Hedin et al. 1998). The riparian zone, as a topographic low point, tends to have high C inputs that are often stored in layers of buried organic matter along the banks of larger streams. These high C inputs often result in the O_2 depletion of riparian groundwater and increased rates of dissimilatory pathways in the nitrogen cycle. These potential pathways include the dissimilatory reduction of nitrate to ammonium (Silver et al. 2001) and denitrification, which results in the conversion of nitrate to gaseous end products (N_2O or N_2) that eventually leave the soil and return to the atmosphere.

The importance of the riparian zone in regulating losses of N from the Luquillo Mountains was first examined by McDowell et al. (1992) and Bowden et al. (1992), and their two companion papers were the first to examine the significance of the riparian zone for N biogeochemistry in a tropical environment. They compared riparian zone function in the two bedrock types and found that the differences in bedrock resulted in geomorphological differences in riparian zones that had major impacts on N retention and loss. In soils derived from volcaniclastic parent material (as in the Bisley Experimental Watersheds), groundwater flow paths were shallow, oxygen status varied along the flow path (e.g., patches of oxidized and reduced soils), and little nitrogen in any form was delivered to the riparian zone (McDowell et al. 1992; Schellekens et al. 2004). In contrast, in soils derived from igneous parent material (e.g., the quartz diorite of the Icacos watershed), flow paths were much deeper, and the nitrogen dynamics were characterized by sharp transitions along the flow path. Upslope groundwater was entirely oxic and showed a significant accumulation of nitrate (up to 1 mg l^{-1} as N). This nitrate was lost, in part to N_2O production at the slope–floodplain interface (Bowden et al. 1992; McSwiney et al. 2001), and ammonium (NH_4) accumulated in the extremely reduced groundwater of the floodplain. The amounts of ammonium and total dissolved N subsequently decreased as groundwater passed through the stream bank and into the stream, suggesting the importance of coupled nitrification-denitrification in the variably oxygenated soils of the stream's edge (McSwiney et al. 2001).

Subsequent work has examined the importance of riparian processes in regulating N flux at the reach and basin scale in the Río Icacos valley. Chestnut and McDowell (2000) intensively monitored groundwater inputs along a 100 m reach of a tributary to the Río Icacos. By directly measuring the groundwater inputs and groundwater chemistry, they determined that the N export would be 6 to 10 times greater in the absence of riparian and hyporheic N retention or denitrification. Madden (2004) expanded this approach to the main stem of the Río Icacos, using a variety of direct and indirect measurements of groundwater inputs to the main stem. She estimated that the hydrologic losses of N from the entire Icacos valley would be double or triple the observed values if not for denitrification in the riparian zone.

Stream and Atmospheric Outputs

The export of nutrients from Luquillo Mountain watersheds in stream flow is comparable to that reported from other humid tropical watersheds (table 3-6). One of the most detailed comparisons among tropical watersheds was published by Lewis et al. (1999), who synthesized data on the export of nitrogen in organic, inorganic, and particulate forms from large and small basins throughout South America and the Caribbean. They found that losses of N as dissolved organic and particulate N were significant, and that Luquillo Mountain sites were typical of those found throughout the Neotropics. They also found that nitrogen export in tropical streams and rivers is greater than nitrogen export from temperate watersheds with a similar degree of human effects. Phosphorus and K losses are moderate and typical of those from humid tropical forests (table 3-6). The hydrologic export of DOC from watersheds of the Luquillo Mountains is within the range reported for other tropical forests (25 to 100 kg ha^{-1} y^{-1}) (McDowell and Asbury 1994). The adsorption of DOC on mineral soil appears to limit the loss of DOC in runoff, as it does in temperate forests (McDowell 1998). Losses of base cations in stream water show clear variation with bedrock geology. In the quartz diorite lithology (Río Icacos), K concentrations and fluxes are at least double those from the volcaniclastic lithology, but the opposite is true for Mg (McDowell and Asbury 1994). Losses of N show no clear patterns in relation to the watershed size or forest type (McDowell and Asbury 1994; Schaefer et al. 2000).

Instream processes, as well as watershed processes, can be important in regulating nutrient losses. Studies with stable isotopic tracers show that the rates of nitrification are very high in Bisley (stream 3) relative to temperate streams. The fraction of total NH_4 uptake converted to nitrate (NO_3) is 60 percent, higher than at any other site studied by Webster et al. (2003), and the uptake length (the distance an average molecule travels before uptake or assimilation) is only 26 m (Merriam et al. 2002). Nitrate uptake is relatively slow in comparison, occurring over many hundreds of meters (Merriam et al. 2002).

The flashy nature of stream flow in the Luquillo Mountains produces high temporal variability in the rates of watershed nutrient output. Streams in the Luquillo Mountains respond quickly to rainfall (figure 3-15), and discharge can change a hundred-fold in a few hours. In a comparison of Long Term Ecological Research (LTER) sites, Post and Jones (2001) found that streams of the Luquillo Mountains are among the quickest to respond to rainfall, owing to their shallow flow paths through macropores in the dense clay soils. To use the terminology of Olden and Poff (2003), stream flows in the Luquillo Mountains are typically "perennial flashy or runoff." Flow paths and stream base flow differ considerably between the two bedrock types in the Luquillo Mountains, with the quartz diorite bedrock of the Icacos basin producing deeper flow paths and more stable base flows than the volcaniclastic bedrock (McDowell et al. 1992; McDowell and Asbury 1994; Schellekens et al. 2004). The mass of dissolved and particulate matter exported from watersheds in stream water typically increases with increased stream discharge, and streams in the Luquillo Mountains are no exception. The export of sediments is particularly sensitive to discharge, because sediment concentrations increase with increased flow (figure 3-16). The export of dissolved nutrients is less sensitive to increases in stream flow, as the concentrations of most elements decrease with increased flow (e.g., Shanley et al. 2011; figure 3-16). There is little evidence of seasonality in stream chemistry, and long-term trends appear to be driven by hurricanes (figure 3-17).

For carbon, nitrogen, and sulfur, trace gas fluxes can be an important watershed-scale export term. In their synthesis of the nitrogen budget for tabonuco forests, for example, Chestnut et al. (1999) estimated that losses of N resulting

Table 3.6 Nutrient flux from tropical and temperate forested watersheds with relatively little anthropogenic disturbance.

Name	Location	NO_3-N	NH_4-N	DON	TDP	Na	K	Ca	Mg	DOC	SiO_2	Runoff
		kg ha^{-1} y^{-1}										cm
W3	Malaysia[a]	0.38	0.29	1.93	0.19	30.7	12.4	19.7	17.3	86	104	196
W6	Malaysia[a]	0.10	0.22	4.15	0.07	7.1	3.2	3.1	3.2	194	13	195
Kali Mondo	Indonesia[b]				0.70	27.4	22.0	29	30.5		538	359
Pedra Preta	Amapa, BR[c]	0.71	0.24			15.1	4.7	22.1	7.5			135
Calado	Manaus, BR[d]	2.67	0.18	0.76	0.05	2.6	0.5	0.5	0.3			165
Tempisquito	Costa Rica[e]	6.10		3.00	0.57	114	26	192	70	37		290
Temp. Sur	Costa Rica[e]	4.90		2.10	0.33	339	64	442	137	27		430
Kathia	Costa Rica[e]	5.60		1.50	0.34	110	21	151	57	19		310
Marilin	Costa Rica[e]	4.00		1.90	0.46	123	41	174	67	24		160
El Jobo	Costa Rica[e]	4.30		2.00	0.34	104	35	133	53	26		140
Zompopa	Costa Rica[e]	6.00		3.40	0.43	164	31.0	199	73	43		280
Icacos	Puerto Rico[f]	2.54	0.68	4.79	0.08	161	17.3	96	35	94	486	368
Sonadora	Puerto Rico[f]	1.39	0.30	3.74	0.05	96	5.5	44.1	28.1	74	180	438
Toronja	Puerto Rico[f]	0.90	0.26	2.80	0.03	113	4.9	87.6	62.7	33	325	175
Orinoco	Venezuela[g]	0.95	0.41	1.92	0.24	17.5	7.8	30.9	7.9	52	70	119
Hubbard Brk.	New Hampshire, USA[h]	0.60	0.10	1.30	0.02	7.2	1.9	13.7	3.1	29	38	83
Coweeta	N. Carolina, USA[i]	0.05	0.04		0.02	8.7	4.1	5.3	3.3	15	59	119
HJ Andrews	Oregon, USA[j]	0.81		0.69	0.70	52.8	9.5	123	8.6		114	237

From McDowell (2002). DON = dissolved organic nitrogen; TDP = total dissolved phosphorus; DOC = dissolved organic carbon.

[a] Grip et al. 1994.
[b] Bruijnzeel 1983.
[c] Forti et al. 2000.
[d] Williams and Melack 1997.
[e] Newbold et al. 1995.
[f] McDowell and Asbury 1994.
[g] Lewis and Saunders 1989.
[h] Campbell et al. 2000; Likens et al. 1977.
[i] Swank and Waide 1988.
[j] Sollins et al. 1980.

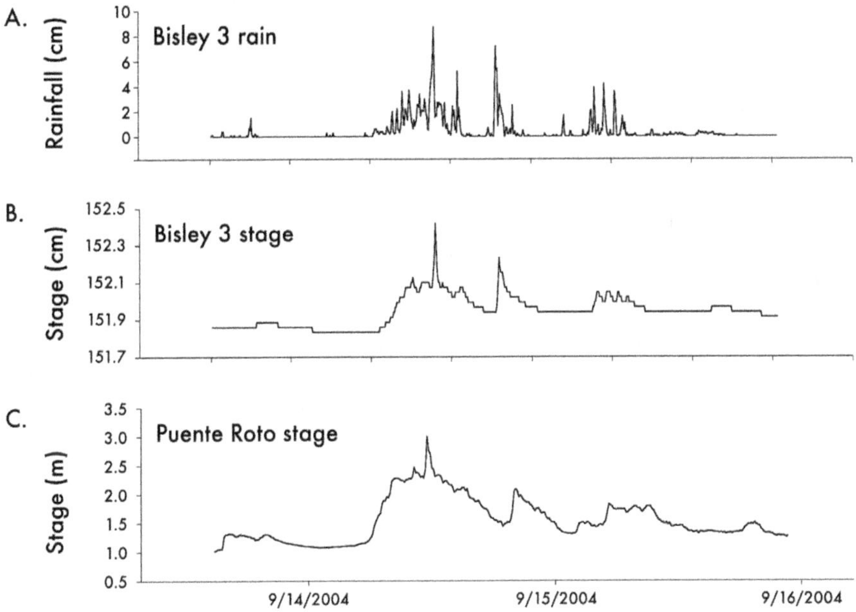

Figure 3.15 Response of stream stage height to rainfall in Bisley Experimental Watershed (BEW) 3 during several days in September 2004. Stream discharge typically increases as the cube of the stage height. (A) Rainfall in BEW 3 (mm day^{-1}) collected during a 2-week sampling period. (B) Stage height at the gauge on BEW 3 (cm above an arbitrary datum). (C) Stage height (m above an arbitrary datum) on the Río Mameyes at Puente Roto, near the edge of the Luquillo Experimental Forest.

from denitrification accounted for 1 to 4 kg ha^{-1} y^{-1}; the upper end of this range is equal to inputs of nitrogen in rainfall. Trace gas fluxes and trace gas concentrations tend to be highly sensitive to the topographic position. Nitrous oxide fluxes tend to be highest at topographic breaks in the colorado forest (Bowden et al. 1992; McSwiney et al. 2001). In a year-long study of soils from ridgetops to the stream bank, McSwiney et al. (2001) found that highest fluxes of N_2O were typically found in the topographic break where the ridge meets the riparian floodplain, and that available manganese (Mn) was a good predictor of high N_2O flux. Only soils containing available Mn produced significant N_2O fluxes. In contrast, CH_4 flux was less clearly related to the topography, with significant rates of CH_4 consumption found at all topographic positions (McSwiney et al. 2001). Concentrations of CH_4 and N_2O at depths of 10 to 80 cm in the colorado forest are also sensitive to topographic position, with the highest concentrations in riparian and streambank soils and the lowest in ridge soils (McSwiney 1999; Silver et al. 1999). The response of biogeochemical conditions to an environmental variable such as rainfall also can vary with topographic position. For example, soil O_2 concentrations in valley soils are correlated with rainfall from the previous day, but at ridge sites they are correlated with cumulative rainfall inputs over the previous 4 weeks.

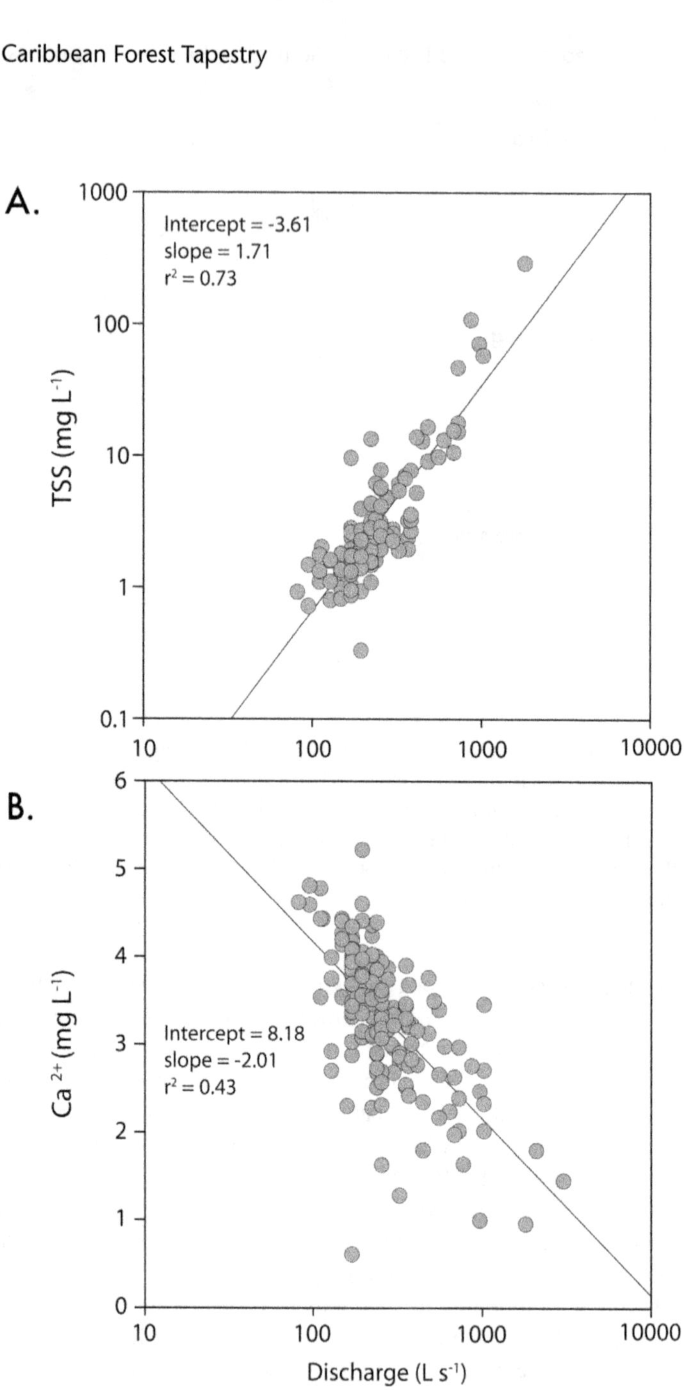

Figure 3.16 Variation in the concentration of (A) total suspended sediments (TSS) and (B) calcium (Ca^{2+}) with stream flow in the Río Icacos in the Luquillo Mountains in 1983–1986. Data from McDowell and Asbury (1994).

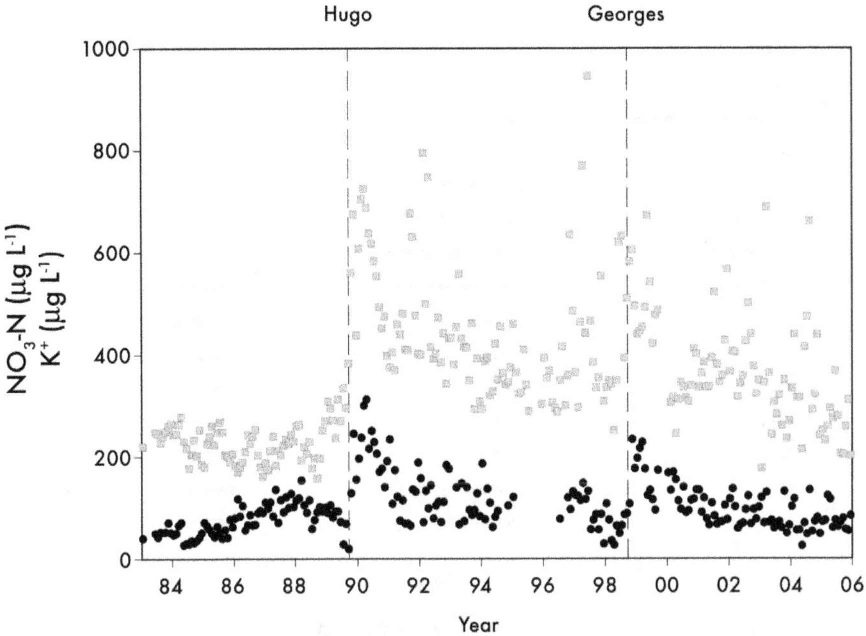

Figure 3.17 Long-term variation in monthly average concentrations of nitrate (NO_3-N; solid circles) and potassium (K^+; boxes) in water in Quebrada Sonadora in the Luquillo Mountains in relation to major hurricanes Hugo and Georges. Data from McDowell and Asbury (1994), Schaefer et al. (2000), and unpublished work of the authors.

Terrestrial Biota and Ecosystem Processes

Primary Producers

Species Composition

Forests in the Luquillo Mountains have been classified into four major forest types: tabonuco, colorado, elfin, and palm brake forests (Gleason and Cook 1927; Wadsworth 1987). Other minor forest types include *Pterocarpus* forest, palm floodplain forests, palm brakes, and bogs (Brown et al. 1983). The tabonuco forest type (occurring in the subtropical moist forest and subtropical wet forest life zones) (Ewel and Whitmore 1973) is named for the tabonuco (*Dacryodes excelsa*), which is the dominant tree species growing from the lower slopes near sea level to elevations of about 600 m. In well-developed stands of this forest type, the taller trees exceed 30 m in height, there is a fairly continuous canopy at 20 m, and the shaded understory is moderately dense (figure 3-18). The shape of the canopy profile varies following hurricane disturbance, with reduced cover at the highest points in the profile. The most common tree species in this forest type are *Casearia arborea*, *Dacryodes excelsa*, *Manilkara bidentata*, *Inga laurina*, and *Sloanea berteriana* (Thompson et al. 2002). These tree species and the sierra palm *Prestoea montana* (previously

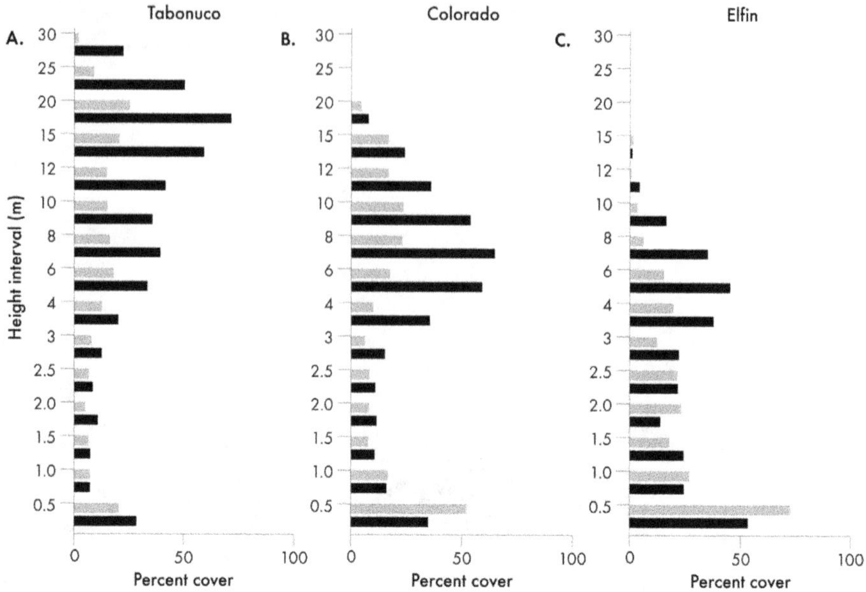

Figure 3.18 Vegetation height profiles in 1 ha plots in (A) tabonuco (350 m elevation, 475 sample points), (B) colorado (750 m, 451 points), and (C) elfin (900 m, 451 points) forest plots in the Luquillo Mountains in 1989, before (dark bars) and after (light bars) Hurricane Hugo. Horizontal scale shows total points with cover as a percentage of the total number of grid points in each plot. Vertical scale is graduated and shows the upper limit for each height interval. From Brokaw and Grear (1991). Reprinted with permission of the Association for Tropical Biology and Conservation.

Euterpe globosa and named *P. acuminata* in Henderson et al. [1995]; figure 3-19) account for 65 percent of all stems ≥ 10 cm diameter at a breast height of 1.3 m from the ground (dbh) in the intensively studied Luquillo Forest Dynamics Plot (LFDP) near El Verde Field Station (figure 3-2). *Prestoea montana* contributes the greatest number of stems to the total stem count in the LFDP. The most common shrubs in the tabonuco forest are *Palicourea riparia*, *Psychotria berteroana*, and *Piper glabrescens*. Grasses, ferns, and forbs are frequent on the ground, especially in canopy gaps. Epiphytes are common, but vines are uncommon (Rice et al. 2004). Both the El Verde Field Station and the Bisley Experimental Watersheds, principal research sites for the Luquillo LTER Program, are in tabonuco forest (see Lugo and Scatena [1995] for a synthesis).

Biomass in the tabonuco forest type ranges from 122 to 300 Mg ha^{-1} on average (Ovington and Olson 1970; Scatena et al. 1993; Beard et al. 2005). Live fine root biomass ranges from 1.5 to 8.0 Mg ha^{-1}, and total live fine and structural root biomass totals 20 to 74 Mg ha^{-1} (Parrotta and Lodge 1991; Kangas 1992; Lugo 1992; Scatena et al. 1993; Silver and Vogt 1993; Vogt et al. 1995, 1997). Fine root biomass changed significantly with weather events (Parrotta and Lodge 1991), with the higher biomasses recorded before Hurricane Hugo and the lowest recorded during the 1994 drought (see chapter 5; Beard et al. 2005). These fine root biomass

Figure 3.19 Wind in a palm (*Prestoea montana*) forest, Luquillo Mountains of Puerto Rico. (Photograph by Jerry Bauer.)

levels are about one-third of those recorded in tropical lowland forests in South America (Cuevas et al. 1991; Vogt et al. 1997) but are similar to those in other montane tropical forests (see, e.g., Grubb 1977). Much of the tabonuco forest type in the Luquillo Mountains was logged for valuable tree species such as tabonuco or partially cleared for coffee or other crops prior to the purchase of most of the Luquillo Mountains by the United States Forest Service in the 1930s (García-Montiel and Scatena 1994). The impacts of past land use on the distribution and abundance of trees were still evident in 1989 (Thompson et al. 2002).

The structure of tabonuco forest shows clear variation with topography. Tabonuco trees are most common on ridges and least common in riparian valleys, where sierra palms are common (Basnet 1992; Johnston 1992; Thompson et al. 2002). Root biomass also varies by topographic position in the tabonuco forest type, with higher fine root biomass on ridges than on slopes or valleys (Vogt et al. 1995, 1997). This might be due to differences in species composition, with shallow-rooted palms found growing most commonly in riparian areas and deeper-rooted dicots being more common on ridges.

Higher in elevation, extending up to about 900 m in the subtropical rain forest and lower montane wet forest life zones (Ewel and Whitmore 1973), is the colorado forest type, named for the dominant tree, palo colorado (*Cyrilla racemiflora*). Species also found in this forest type include *Magnolia splendens*, *Matayba domingensis*, *Micropholis garciniaefolia*, *M. chrysophylloides*, *Calycogonium squamulosum*,

Ocotea moschata, and *Croton poecilanthus* (Brown et al. 1983). Most of the colorado forest type is found growing on quartz diorite bedrock; it is not known to what extent this bedrock favors a shift to colorado forest type at higher elevations, although changing environmental conditions alone drive the shift to colorado forest in the Sonadora watershed of the Río Espíritu Santo drainage (Barone et al. 2008), which does not contain quartz diorite. In the colorado forest type, soils are often saturated, and root mats on the soil surface are common. The canopy of the colorado forest type reaches about 15 m in height (figure 3-18), and its biomass is 130 Mg ha^{-1} (table 3-7) (Weaver and Murphy 1990).

Sierra palms (*Prestoea montana*) are frequently found at the same elevation and in the same life zones as tabonuco and palo colorado forests, but they achieve maximum dominance as palm brakes in especially steep and wet areas (Lugo et al. 1995). Depending on the degree of soil satuation and aspect, the number of associated tree species can vary between 24 and 35 species per 0.4 ha. Like palo colorado forest, palm forest is about 15 m in height. The aboveground biomass in palm-dominated floodplain forest can be as high as 223 Mg ha^{-1}, with 54 Mg ha^{-1} of palm biomass alone (Frangi and Lugo 1985). Palms are also found in riparian forest (palm floodplain forest) and on very steep slopes at low elevation (palm brakes) (Frangi 1983; Lugo et al. 1995). Because individual palm trees are found throughout the forest, and because patches of palm forest are found in a variety of wet or steep environments at most elevations in the Luquillo Mountains, it is difficult to make generalizations about the palm forest type, although palms are usually associated with saturated soils and disturbance.

Elfin forest type, a dense forest growing on saturated soils derived from bedrock formed by contact metamorphism, is found above 900 masl (Weaver 1995) in the subtropical rain forest life zone as defined by Ewel and Whitmore (1973). The canopy height is typically 3 to 5 m (figure 3-18), although a variant of elfin forest growing in more protected sites such as small valleys near mountain peaks can

Table 3.7 Aboveground biomass, litterfall, and net primary productivity from four forest types of the Luquillo Mountains found in three subtropical life zones.

	Subtropical wet forest (Tabonuco)	Lower montane wet forest		Lower montane rain forest (Elfin)
		(Palm)	(Colorado)	
Biomass (Mg ha^{-1})	190	174	130	80
Litterfall (Mg ha^{-1} y^{-1})				
Leaf	4.94	6.26	5.05	2.45
Wood	1.38	0.86	1.22	0.28
Flower	0.17	0.18	[0.23]	—
Fruit	0.34	1.14		—
Miscellaneous	1.78	0.36	0.30	0.37
Total litterfall	8.6	8.8	6.8	3.1
Total aboveground NPP (Mg ha^{-1} y^{-1})	10.5	19.5	7.60	3.70

From Weaver and Murphy (1990). Aboveground net primary productivity is estimated as the sum of annual litterfall plus stem increment.

reach a height of 10 m. In elfin forest, the trees and ground are covered with mosses and epiphytes, and the sierra palm can be common in some areas. The dominant species of tree in the elfin forest is *Tabebuia rigida*, and the community also includes *Calycogonium squamulosum, Ocotea spathulata, Calyptranthes krugii*, and *Miconia pachyphylla* (Brown et al. 1983; Weaver and Murphy 1990). The biomass in the elfin forest is 80 Mg ha^{-1} (table 3-7) (Weaver et al. 1986; Olander et al. 1998). The elfin forest has also been referred to as cloud forest, because it typically occurs at elevations with persistent cloud cover, or as dwarf forest. We prefer the term "elfin forest" to describe the particularly small-stature forest growing on peaks and high ridges (Boynton 1968; Howard 1968) over "cloud forest," because the cloud level frequently is as low as 600 m elevation in the Luquillo Mountains and thus enshrouds both colorado and palm forest types in addition to the elfin forest. Other studies in the Caribbean (e.g., Tanner 1977) have used the term "cloud forest" to describe forests that we refer to as colorado forest type.

For fuller descriptions of the species composition in the forest types, see Wadsworth (1951), Odum and Pigeon (1970), Brown et al. (1983), Lugo and Scatena (1995), Lugo et al. (1995), Weaver (1995), and Thompson et al. (2002, 2004). All of these forest communities continue to be dominated by native species while existing in a variable matrix of human and natural disturbance, species invasion, and forest regeneration following agricultural abandonment at lower elevations (Gould et al. 2006).

Tank bromeliads (mainly *Vriesia* and *Guzmania* spp.) are important components of lower canopy and understory plant communities throughout the Luquillo Mountains. They are most abundant at the highest elevations, where the forest canopy becomes more open and rainfall increases, and the same species can be both epiphytic and saxicolous (ground living). In the elfin forest on East Peak, the dominant bromeliad is *Vriesea sintenisii*, which can have a density of up to 3.2 plants m^{-2} (Richardson et al. 2000).

The phenology of vascular plants in the Luquillo Mountains follows patterns in annual solar insolation, as has been suggested for other tropical forests (van Schaik et al. 1993). Seasonal drought has most often been assumed to be the primary abiotic factor controlling the timing of leaf flush and reproduction in tropical forests, but van Schaik et al. (1993) and Wright and van Schaik (1994) questioned that conclusion for all but the driest forests. Results from ever-wet tabonuco forest support the conclusion that leaf flush and flowering are driven by light levels in the absence of seasonal drought (figure 3-20). Leaf flush is highly synchronous in eight dominant species of understory trees and shrubs and is highest in June, when light levels are highest (figure 3-20) (Angulo-Sandoval and Aide 2000). Zimmerman et al. (2007) showed that seasonal patterns in flowering are also tied to periods of maximal light levels, with highest flowering in June, July, and August (figure 3-20). In general, flowering peaks are broad in most tabonuco forest species, with 75 percent of flowering observations in a given species spread over a 3- to 6-month period (summary of 10 years of data on flower parts falling into litter traps) (Zimmerman et al. 2007). Summed over all species, peak flowering occurs during the period of June through August, and relatively few species have peak flowering in October through March (figure 3-20). In palo colorado forest, peak flower and fruit

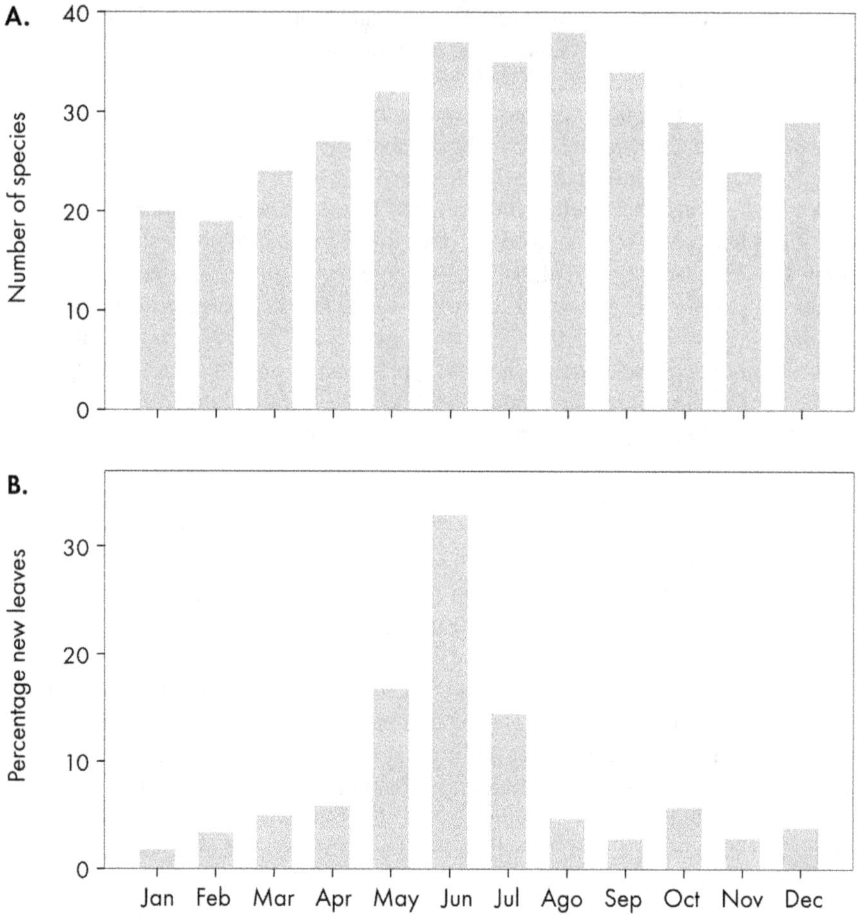

Figure 3.20 Seasonal variation in (A) flowering and (B) leaf production in the tabonuco forest type. The extent of flowering is shown as the monthly sum of the number of species in peak flower, with peak flowering for a species defined by the months containing 75 percent of all observations (Zimmerman et al. 2007). Leaf production is the mean percentage of leaf area sampled that is newly produced when sampled each month, averaged over eight understory species (Angulo-Sandoval and Aide 2000).

production also occurs in June (Weaver and Murphy 1990). Biotic factors such as herbivory and seed predation might also play a role in promoting leaf flushing and peaks of reproductive effort (Angulo-Sandoval and Aide 2000; Angulo-Sandoval et al. 2004).

Primary Productivity

In forests of the Luquillo Mountains, the measurement of primary productivity is complicated by frequent disturbance, leading to a mosaic of forest patches at different successional stages. A synthesis of early data across forest types collected before Hurricane Hugo struck the Luquillo Mountains in 1989 indicates that the aboveground net primary productivity (ANPP) ranged from 2.7 to 19.5 Mg ha^{-1} y^{-1} (table 3-7) (Weaver and Murphy 1990; Lugo 1992). Net primary productivity peaks at mid-elevation in the Luquillo Mountains (Waide et al. 1998; Harris 2006). Forest-wide, ANPP has been estimated at 9.4 Mg ha^{-1} y^{-1} (Wang et al. 2003). The primary productivity of bromeliads increases with elevation and makes up 12.8 percent of ANPP in the elfin forest (Richardson et al. 2000).

The production of leaves, fruit, flowers, and small wood in litterfall often represents the majority of aboveground primary productivity (table 3-7), and thus can be used as an indicator of patterns in productivity. Litterfall in the Luquillo Mountains follows a pattern similar to that for the ANPP, declining with elevation except for the mid-elevation palm stand. In the tabonuco forest type, there is an extensive data set on leaf fall (e.g., Wiegert and Murphy 1970; Lodge et al. 1991; Zou et al. 1995; Vogt et al. 1996; Zalamea and González 2008), which averages about 57 to 80 percent of total litterfall (table 3-7). Values for annual leaf fall were consistent (ranging from 1.29 to 1.38 g m^{-2} d^{-1}) in measurements taken over several decades before Hurricane Hugo and among stands with different disturbance histories (Odum 1970a; Lodge et al. 1991; Zou et al. 1995). The total litterfall has varied dramatically following major named storms and hurricanes, however, with litterfall equivalent to a year or more of daily background rates occurring as a result of single hurricanes (figure 3-21). In the absence of major storms, leaf fall patterns in the tabonuco forest are correlated primarily with solar radiation, day length, and air temperature (Zalamea and González 2008). Litterfall also increases from the riparian zones to ridgetops in the tabonuco forest (Vogt et al. 1996; Beard et al. 2005).

The control of primary productivity in forests of the Luquillo Mountains is a complicated and still-unresolved issue (Waide et al. 1998). In general, wet tropical forests are thought to be limited by phosphorus or trace elements rather than nitrogen (Vitousek 1982, 1984; Martinelli et al. 1999). In the Luquillo Mountains, limitation by phosphorus or nitrogen seems unlikely, as soil phosphorus and the nitrogen concentration of the foliage are as high as or higher than in other tropical sites (figure 3-12). However, following Hurricane Hugo, the experimental addition of nutrients resulted in increased productivity in tabonuco and elfin forest plots (Zimmerman et al. 1995; Waide et al. 1998). The control of primary productivity might be more complicated than a single-factor limitation, especially at higher elevations. Episodic water shortage, frequent inundation in clouds, root inhibition because of low oxygen levels in periodically waterlogged soil, exposure to strong winds, and reduced leaf temperatures and photosynthesis at higher elevations (as discussed in other sections of this chapter) can all contribute to a limiting of the primary productivity.

Herbivores and Herbivory

Herbivorous insects and a wide range of invertebrates living near the forest floor are the dominant primary consumers, as there are no large mammalian herbivores in the Luquillo Mountains. Small animals, including many birds and the omnivorous

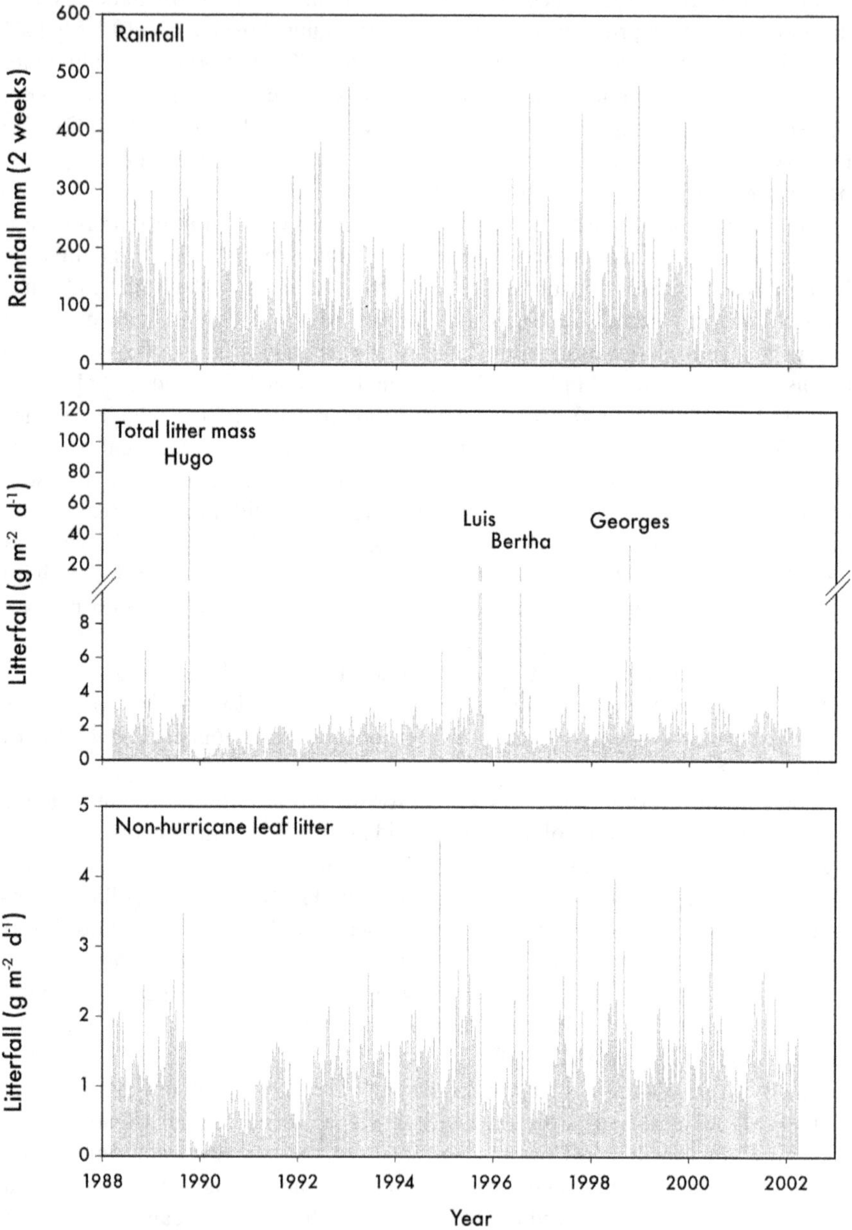

Figure 3.21 Long-term variation in (A) rainfall, (B) total litterfall, and (C) nonhurricane leaf fall collected every 2 weeks in the Bisley Experimental Watersheds of the Luquillo Mountains. Hurricanes are shown by name above the associated litterfall. Litterfall is total (leaf litter plus fruits and woody material collected in litter traps). Nonhurricane leaf fall excludes collection periods immediately following major hurricanes and does not include wood or fruits. From Scatena et al. (1996) and unpublished data.

black or roof rat (*Rattus rattus*) and Indian mongoose (*Herpestes auropunctatus*), also consume fruits and seeds (Willig and Gannon 1996). In the colorado forest type, herbivory ranges from 0.6 to 11 percent of leaf area, depending on tree species (Weaver and Murphy 1990). Canopy herbivory in the tabonuco forest type, measured as the frequency of herbivore-caused damage (6 percent) (Odum and Ruiz-Reyes 1970), is comparable to rates in other forest ecosystems (Pfeiffer 1996) and is highly correlated with the density of roaches and orthopterans (Dial and Roughgarden 1995). Schowalter and Ganio (1999) showed that canopy herbivory increased with greater canopy closure.

Many of the herbivores in the tabonuco forest type are polyphagous, eating a variety of plant species. Bark beetles (Scolytidae) are relatively common, and some are known to eat seed pods of the tree *Inga vera*, as well as dead and decaying wood; a few are known to eat live trees (Garrison and Willig 1996). Several species of grasshoppers are common in the tabonuco forest and can reach lengths of 45 mm (Garrison and Willig 1996). Synchronous leaf production among plant species in the Luquillo Mountains appears to significantly lower rates of herbivory during maximal leaf production in June (Angulo-Sandoval and Aide 2000; Angulo-Sandoval et al. 2004).

Snails and walking sticks are two well-studied groups of herbivores that are found primarily in the understory and on the forest floor in the Luquillo Mountains. Seventeen species of gastropods representing 14 genera, 12 families, and 3 subclasses have been identified (Willig et al. 1998; Bloch 2004). Eight species are arboreal grazers, six are forest floor grazers (or detritivores), and three are carnivores. Three grazer/detritivores (*Caracolus caracolla*, *Nenia tridens*, and *Gaeotis nigrolineatus*) are the most abundant snails, with mean densities in 1994–2003 (mean ± SD) of 0.20 ± 0.06, 0.19 ± 0.12, and 0.07 ± 0.04 individuals m^{-2}, respectively. They can reach local densities of up to one to three individuals m^{-2}, depending on the extent of the disturbance and microhabitat characteristics (Willig et al. 1998; Bloch 2004). In general, *N. tridens* is most often associated with treefall gaps, whereas *C. caracolla* (figure 3-22) is more often found in undisturbed forest (Alvarez and Willig 1993). The slug *G. nigrolineatus* is one of the few species that are strongly associated with a particular plant species; it is commonly found on the leaflets of the sierra palm *Prestoea montana*. Because snails and slugs are not particularly mobile, they can be affected strongly by disturbances that affect microclimate (especially temperature and humidity) and the availability of detritus (see chapter 5). Crabs of the genus *Epilobocera* (figure 3-23) eat fruits and flowers on the forest floor. Although these crabs have an obligate freshwater life-history phase, they forage widely on the forest floor (Covich and McDowell 1996). More details on crabs are given further on in the section "Aquatic Biota and Ecosystem Processes."

The Luquillo Mountains harbor five species of walking stick, but only one (*Lamponius portoricensis*) is common (Garrison and Willig 1996; Tilgner et al. 2000; Van Den Bussche et al. 1988). *Lamponius* is found primarily in the forest understory and is most common in areas containing one of its important food plants, *Piper glabrescens* (Willig et al. 1993). Prior to Hurricane Hugo, the density of *L. portoricensis* in a 100 m^2 area dominated by a treefall gap was between 0.4 and 1 individual m^{-2} (Willig et al. 1986). Because of its size, habitat associations, and

Figure 3.22 Terrestrial snails, such as *Caracolus caracolla*, are abundant generalist consumers in the Luquillo Mountains, eating fresh leaves, leaf litter, fruits, fungi, and bacteria. (Photograph by Paul Klawinski.)

food preferences, *L. portoricensis* might act as a keystone species by affecting species survivorship and nutrient cycling within light gaps (Willig et al. 1986).

Birds and bats consume fruit, seeds, and nectar, as do some insects. Birds ranging in size by over two orders of magnitude (3 to 300 g) consume a variety of plant parts. The most notable frugivorous specialists are the Scaly-naped Pigeon (*Columba tagioenas squamosa*), the Puerto Rican Spindalis (*Spindalis portoricencis*), the Puerto Rican Parrot (*Amazona vittata*), and the Antillean Euphonia (*Euphonia musica*), but fruit is an important element of the diet of many other species (e.g., the Red-legged Thrush [*Turdus plumbeus*], the Black-whiskered Vireo [*Vireo altiloquus*], the Pearly-eyed Thrasher [*Margarops fuscatus*], and the Puerto Rican Bullfinch [*Loxigilla portoricensis*]). The range of food types consumed differs among species; the Puerto Rican Parrot feeds on at least 58 plant species (Snyder et al. 1987), and the Antillean Euphonia is a mistletoe specialist (Waide 1996). Species whose diet consists primarily of seeds include the Ruddy Quail-Dove (*Geotrygon montana*), the Zenaida Dove (*Zenaida aurita*), and the Black-faced Grassquit (*Tiaris bicolor*). Nectarivores include two hummingbirds (Puerto Rican Emerald [*Chlorostilbon maugaeus*] and Green Mango [*Anthracothorax viridis*]) and the Bananaquit (*Coereba flaveola*), which forages among flowers for nectar and insects.

Bats appear to be an important part of the nocturnal food web in the tabonuco forest, although their ecological functions are not well known (Willig and Gannon

Figure 3.23 The freshwater crab *Epilobocera sinuatifrons* (Pseudothelphusidae) on the forest floor with a palm seed (*Prestoea montana*) in its left chela. (Photograph by Paul Klawinski.)

1996). Thirteen species of bats occur on the island of Puerto Rico. Of the species found in the Luquillo Mountains, one (*Monophyllus redmani*) is a nectarivore, and four (*Brachyphylla cavernarum*, *Artibeus jamaicensis*, *Stenoderma rufum*, and *Erophylla sezekorni*) are frugivores. Densities are not known for any bat species in the Luquillo Mountains, but relative abundances suggest that *A. jamaicensis* and *S. rufum* are numerically the most important species (Willig and Gannon 1996).

Detritivores

Detritivores found in the litter layer of Luquillo Mountain soils include mites, millipedes, centipedes, collembolans, ants, flies, beetles, isopods, termites, and earthworms. Faunal inventories at the El Verde Field Station demonstrated that about half of the total faunal biomass was concentrated in a relatively thin layer of soil and litter (Odum 1970a; Pfeiffer 1996). Mites are the numerically dominant taxon in the litter layer of the tabonuco forest, accounting for 33 to 69 percent of all arthropods extracted from litter samples, with densities of 1,000 to 2,700 individuals m^{-2} (Pfeiffer 1996; Richardson et al. 2005). The dominance of litter invertebrates by mites is typical for tropical sites (Pfeiffer 1996). Because of their small size, however, mites account for only 1 percent of the total invertebrate biomass

(6.4 mg m^{-2}) (Richardson et al. 2005). Larger, less common taxa such as Isoptera (termites) and Diplopoda (millipedes) are found at densities of a few hundred m^{-2}, but they account for nearly 40 percent of the total invertebrate biomass in the litter layer of the Luquillo Mountains (table 3-8) (Richardson et al. 2005). Ants are also an important component of the litter invertebrate community, with densities of 500 to 1,200 m^{-2} (Pfeiffer 1996; Richardson et al. 2005), but the army and leaf cutter ants found in many other tropical regions are not present in Puerto Rico. The densities of macroarthropods (such as millipedes, isopods, cockroaches, and crickets) are higher than those found in other tropical sites (Pfeiffer 1996). The population levels of various taxa found in the litter layer vary over the course of the year but do not show synchronous or strong seasonal variations. In a detailed study of monthly changes in forest floor leaf litter invertebrates at El Verde, Pfeiffer (1996) found that the numbers of Diptera and Lepidoptera increased 5- or 10-fold in June, and isopod numbers declined 4-fold. No long-term studies have been conducted to determine whether these variations in abundance represent robust seasonal trends or the response to a particular event that occurred during the study year.

Litter invertebrates are thought to be particularly important as agents of litter decomposition in tropical relative to temperate forests (Heneghan et al. 1999). Experimental manipulations restricting invertebrate access to litter suggest that up to 66 percent of litter decomposition in the Luquillo Mountains is due to forest floor invertebrates (González and Seastedt 2001). Most oribatid mites and collembolans have well-developed mouth parts capable of fragmenting organic matter while feeding on the microflora adhering to this detritus (Seastedt 1984), and they are thought to be one of the key invertebrate groups responsible for litter decomposition in the forest floor (González and Seastedt 2001).

Termites are specialist consumers of cellulose in litter and wood. There are four species of termites in the tabonuco forest (McMahan 1996), and all are xylophagous on dead standing wood, downed boles, or smaller twigs and branches. In Puerto Rico, there are no termites that cultivate fungi and their reproductive structures (i.e., mushrooms) or feed on soil, and the termite species richness is much lower in the Luquillo Mountains than the 40 to 85 species reported in Malaysia, Cameroon, or Guiana (McMahan 1996). Thus, the overall contribution of termites to plant decomposition is thought to be less in the Luquillo Mountains than in other tropical forests (McMahan 1996).

Nasutitermes costalis is the most evident and most widely studied termite in the Luquillo Mountains. Nests constructed by the worker caste can be up to 40 cm in diameter, have a half-life of about 4 y, and occur at a density of approximately 4.5 ha^{-1}. The density of *N. costalis* individuals in tabonuco forest litter ranges from 86 to 95 m^{-2}, with a dry mass biomass of 74 to 207 mg m^{-2} (Wiegert 1970; Richardson et al. 2005). The polymorphism of workers (which are found as distinctive large and small lines; McMahan 1996) might account for the similar densities but wide differences in total biomass reported in the two studies.

Earthworms are a key component of the soil fauna and play an important role in litter decomposition, as well as in the maintenance of the physical structure and porosity of soils in the Luquillo Mountain (Lyford 1969; Camilo and Zou 1999;

Table 3.8 Average abundance and biomass of major taxonomic groups of invertebrates in the litter layer of the forest floor collected from both palm and nonpalm forests at elevations ranging from 350 to 1000 masl in the Luquillo Mountains.

Abundance			Biomass		
Individuals m^{-2}		Percent of total	mg m^{-2}		Percent of total
Acari	979	33.9	Isoptera	130.9	23.3
Formicidae	525	18.2	Diplopoda	82.0	14.6
Collembola	285	9.9	Coleoptera (adults)	60.3	10.7
Isoptera	247	8.6	Mollusca	55.7	9.9
Coleoptera (adults)	180	6.2	Araneae	39.5	7.0
Hemiptera and Homoptera	112	3.9	Onychophora	33.3	5.9
Diptera (adults)	105	3.6	Formicidae	30.9	5.5
Diptera (immature)	99	3.4	Isopoda	23.6	4.2
Isopoda	79	2.7	Chilopoda	20.9	3.7
Coleoptera (immature)	73	2.5	Hemiptera and Homoptera	19.8	3.5
Diplopoda	67	2.3	Diptera (adults)	13.1	2.3
Pseudoscorpio nes	51	1.8	Lepidoptera (adults)	9.8	1.7
Araneae	29	1.0	Blattodea	8.0	1.4
All other taxa[a]	27	0.9	All other taxa[a]	7.6	1.3
Chilopoda	8	0.3	Collembola	6.9	1.2
Lepidoptera (adults)	7	0.3	Acari	6.4	1.1
Lepidoptera (larvae)	6	0.2	Coleoptera (immature)	5.2	0.9
Opiliones	4	0.2	Pseudoscorpio nes	4.9	0.9
Blattodea	4	0.1	Diptera (immature)	4.6	0.8
Mollusca	1	<0.1	Opiliones	2.9	0.5
Onchophora	0	<0.1	Lepidoptera (larvae)	2.6	0.5
Total	2,889		Total	562.9	

Modified from Richardson et al. (2005).

[a] "All other taxa" combines those individual taxa that either occurred more infrequently than those in the table or made up only a small biomass.

Liu and Zou 2002). Approximately 30 species of terrestrial oligochaetes have been described in Puerto Rico, and about half of them are present in the Luquillo Mountains (González et al. 2007). At least two species of Puerto Rican earthworms are endemic: *Estherella montana* (figure 3-24) and *Neotrigaster complutensis* (Borges 1996). González et al. (2007) described earthworm communities along an elevation gradient of eight forest types in northeastern Puerto Rico and found that the density, biomass, and diversity of worms varied significantly among forest types, with the highest earthworm density in the *Pterocarpus* forest. The total earthworm biomass is highest in the elfin and *Pterocarpus* forest types. The number of earthworm species increases with elevation and is predicted by soil pH and fine root density.

In tabonuco forest, the introduced *Pontoscolex corethrurus* dominates the total earthworm density and biomass. The average density and biomass of *Pontoscolex corethrurus* in the tabonuco forest type is 95 individuals and 21.6 g of fresh biomass m^{-2} (González et al. 1996; González and Zou 1999b). This earthworm increases N availability and rates of N mineralization in soils (González and Zou 1999a). The introduced *Ocnerodrilus parki* dominates the total earthworm density in the colorado, palm, and elfin forest types (Borges and Alfaro 1997), with unknown effects on the soil structure or biogeochemical processes. Earthworm abundance in the Luquillo Mountains varies with plant species composition and soil properties. Densities and biomass are nearly twice as high in soil beneath *Dacryodes excelsa*, for example (109 worms m^{-2} and 31 g of fresh biomass m^{-2}), than beneath *Heliconia caribea* (64 individuals and 17 g of fresh biomass m^{-2}; González et al. 1999). Carbon and nitrogen concentrations in the top 25 cm of the soil profile also vary with the two plant communities, suggesting interactions among earthworms, vegetation, and soil carbon and nutrient status.

Figure 3.24 *Estherella* sp. is a native earthworm commonly found in the Luquillo Mountains of Puerto Rico. (Photograph by Grizelle González.)

Microbes and Litter Decomposition

Much of the carbon in the food web of the tabonuco forest passes directly from plants through the detrital system rather than through herbivores (Lodge 1996). Fungi, bacteria, and a variety of invertebrates are important decomposers of detrital inputs to the forest floor (figure 3-25). Terrestrial fungi in the Luquillo Mountains typically prefer particular types of substrates, such as roots, leaves, petioles, twigs, branches, and logs (Holler and Cowley 1970; Lodge and Cantrell 1995; Lodge 1996, 1997). Furthermore, most terrestrial wood decomposers and other wood-inhabiting fungi in the Luquillo Mountains have preferences for a particular diameter of branch or bole, and sometimes for whether the bole lies on the ground or is suspended above it (Lodge and Cantrell 1995; Lodge 1996, 1997; Huhndorf and Lodge 1997). Basidiomycetes and slime molds also have preferences for whether the decomposing leaves are on the ground or suspended in the understory (Lodge and Cantrell 1995; Stephenson et al. 1999). Likewise, the majority of microfungi in decomposing leaves have strong preferences for a particular leaf species or leaf type, resulting in low similarities in microfungal communities on different leaf species that are located on the same patch of forest floor (Cowley 1970; Polishook et al. 1996; Lodge 1997; Santana et al. 2005). Microfungi that were dominant in a particular leaf species decomposed their preferred substrate faster than did microfungi feeding on "nonpreferred" leaves (Santana et al. 2005). In contrast, very few of the fungi that inhabit decaying wood show strong host preferences or specificity (Huhndorf and Lodge 1997; Lodge 1997). Microbial biomass in soil tracks litterfall, with peak microbial biomass occurring one month prior to peak litterfall (Ruan et al. 2004).

Rates of leaf, wood, and fine root decomposition are rapid in the Luquillo Mountains, as they are in many other tropical forests (La Caro and Rudd 1985; Bloomfield et al. 1993; Zou et al. 1995; Vogt et al. 1996; Sullivan et al. 1999). Leaves decompose rapidly, with 75 to 80 percent mass loss of mixed litter assemblages (tabonuco forest type) in 1 year (Zou et al. 1995). Turnover rates vary approximately twofold among species, with the sierra palm having the slowest decomposition rate among species that have been tested (La Caro and Rudd 1985; Vogt et al. 1996). In the Luquillo Mountains, as is typical elsewhere, a high lignin content is associated with slow leaf decomposition (La Caro and Rudd 1985; Bloomfield et al. 1993; Sullivan et al. 1999; Santana et al. 2005). Rates of mass loss in leaves with a high lignin content were increased by about 20 percent when the leaves were decomposed on litter mats formed by ligninolytic basidiomycete fungi (Lodge et al. 2008). Although the litter mat density is highest on steep slopes, especially in tabonuco forest (Lodge et al. 2008), leaf litter decomposition does not vary with topographic position in either the tabonuco forest type (Wiegert and Murphy 1970; Bloomfield et al. 1993; Sullivan et al. 1999; Ruan et al. 2005) or the colorado forest type (Sullivan et al. 1999).

The decomposition of plant materials is controlled by climatic conditions on a global scale (Meentemeyer 1978; Coûteaux et al. 1995; Parton et al. 2007). At the local scale, under similar climatic conditions, the litter chemistry can also regulate decomposition rates (Melillo 1982). Studies from the tabonuco forest show that although the litter chemistry clearly affects rates of decomposition, soil arthropods

Figure 3.25 *Collybia johnstonii* is a litter decomposer fungus that binds fresh litterfall into mats, thereby reducing erosion on steep slopes; translocates phosphorus from decomposing leaves into freshly fallen leaves to build biomass; uses lignin-degrading enzymes that accelerate decomposition; alters subsequent microbial communities and processes; and is abundant in tabonuco forest under closed canopy but is sensitive to canopy opening from disturbance. (Photograph by Jean Lodge.)

and earthworms have particularly strong influences on the rates of litter decomposition in the Luquillo Mountains. González and Seastedt (2001) reported that soil arthropods were responsible for up to 66 percent of the total decomposition of *Cecropia schreberiana*. Earthworms also accelerated the decomposition of mixed-species litterbags that represented the natural species composition of the tabonuco forest (Liu and Zou 2002), and the addition of debris that facilitated fungal and invertebrate colonization resulted in increased rates of leaf decomposition (Ruan et al. 2005). Especially after hurricanes, when the number of habitats available for the amphibian *Eleutherodactylus coqui* increases, decomposition rates of litter increase within the 1 m^2 area used by coqui to call for mates at night, because of nutrient inputs from their feces (Beard et al. 2003).

During the decomposition process, leaf litter of all but the highest N concentration tends to increase in N concentration during the early stages of decomposition, often twofold (Lodge 1993; Parton et al. 2007). This global tendency has also been observed in the tabonuco forest in mixed-species litterbags by Zou et al. (1995), who found that an initial 1.1 percent N concentration in leaf litter increased to a final value of 2.2 percent after 300 days of decomposition. The total N mass in absolute terms can also increase during the early stages of leaf decomposition (Zou et al. 1995) due to N immobilization by microorganisms active in the decomposition process. Absolute increases in the N content can also occur during the decomposition of wood, and this N accumulation in decaying wood and leaves is thought to be important in whole-ecosystem C and N dynamics in tabonuco forest (Lodge et al. 1994; Zimmerman et al. 1995; Walker et al. 1996; Miller and Lodge 1997; Beard et al. 2005). Similar to N, other nutrients such as P, Mg, and Ca can also increase in concentration, especially when the nutrient is in short supply in the substrate that is being decomposed. As is typical of tropical forests, phosphorus appears to be highly conserved and is translocated by basidiomycete fungi that decompose leaf litter in tabonuco forest, resulting in P contents that exceed initials during the early stages of leaf decomposition, whereas nitrogen contents only rarely exceed 100 percent of the initial amount (Lodge 1993, 1996; Lodge et al. 1994, 2008).

Predators

Frogs and anoline lizards are the dominant predators in the canopy and understory of the tabonuco forest (Garrison and Willig 1996; Reagan et al. 1996), with lizards dominating the daytime food web and frogs dominating the nocturnal food web. The densities of frogs are among the highest recorded anywhere (Stewart and Woolbright 1996). Frogs are generalist predators that take mainly invertebrate prey, and they are prey to numerous vertebrate and invertebrate predators (Stewart and Woolbright 1996). Most of the frogs in the Luquillo Mountains are terrestrial breeding members of the genus *Eleutherodactylus* (family Leptodactylidae) that range throughout the forest and are generally not restricted to the vicinity of standing water. The most common is *E. coqui* (figure 3-26), which occurs across the island from lowlands to mountain tops, and which attains very high densities of two frogs m^{-2} in the mid-elevations of the Luquillo Mountains (Stewart and Woolbright 1996). Reproduction in *E. coqui* is highest in the summer, and the total population numbers peak with high juvenile densities in the winter. *Eleutherodactylus coqui* is limited by retreat and nest sites (Stewart and Pough 1983). Locally high densities are associated with concentrations of suitable retreat sites including fallen leaves of *Prestoea montana* and *Cecropia schreberiana*, and density within the forest can be patchy because of changes in the plant community following disturbances such as treefalls (Woolbright 1996).

Of the other 15 species of Puerto Rican *Eleutherodactylus*, 12 were historically found in the vicinity of the Luquillo Mountains, although all were less numerous and more restricted in range than *E. coqui* (Rivero 1978). At least four of these have undergone widespread local extinctions in the past 40 years, consistent with the global pattern of amphibian declines, and at least two of these species are probably extirpated from the Luquillo Mountains (Woolbright 1997). The remaining frog community generally varies with elevation and cover type (Drewry 1970; Rivero

Figure 3.26 A coquí (*Eleutherodactylus coqui*), the abundant tree frog in the Luquillo Mountains of Puerto Rico. (Photograph by Jerry Bauer.)

1978). One aquatic breeding native, *Leptodactylus albilabris*, is also common in the Luquillo Mountains, building bubble nests in puddles. The only nonnative amphibian commonly seen in the Luquillo Mountains is the cane toad, *Bufo marinus*, which moves into the forest along roads.

Anoline lizards are conspicuous, abundant, and well-studied predators of insects on Caribbean islands, and they are particularly important in daytime food webs in the Luquillo Mountains. Seven species of anole are found in the Luquillo Mountains, three of which (*Anolis occultus*, *A. krugi*, and *A. cristatellus*) inhabit edges and openings in the forest (Reagan 1996). The Puerto Rican giant anole (*A. cuvieri*) is a rare, canopy-dwelling species that feeds on snails, butterfly and moth larvae, beetles, walking sticks, plant material, and other anoles (Reagan 1996). Three other species are common within tabonuco forest and are specialized foragers found on small branches and twigs (*A. stratulus*), tree trunks into the crown (*A. evermanni*), and tree trunks and the ground (*A. gundlachi*). On Caribbean islands where insectivorous mammals and birds are more rare than in continental sites, anoles are among the most important higher-order consumers and have significant effects on the structure of terrestrial food webs (Schoener and Toft 1983; Schoener and Spiller 1987; Reagan 1996).

Surveys from canopy towers at El Verde found that *A. stratulus* was extremely abundant in the forest canopy and *A. evermanni* used the canopy frequently (Reagan 1996). Individual *A. stratulus* occupy small ellipsoidal home ranges/territories (males only) layered within the forest canopy. This three-dimensional habitat partitioning is unique among terrestrial vertebrates and allows *A. stratulus* to achieve the highest population densities of any lizard species (Reagan 1992). *Anolis gundlachi* was most abundant near the ground (Reagan 1992). The combined abundance of these three species is approximately 2.5 individuals m^{-2} (Reagan 1996), with *A. stratulus* contributing more than 80 percent of the individuals. Repeated population estimates of *A. stratulus* in different seasons and years found relatively stable population numbers, consistent with data for other West Indian anoles (Schoener 1985).

The diets of the four common forest anoles included 34 animal orders, 20 of which were insects (Reagan 1996). The most common prey in *A. gundlachi* stomachs were ants, but Lepidoptera larvae, crickets, and earthworms constituted the largest volume of prey consumed. *Anolis gundlachi* consumed several taxa of arthropods inhabiting soil litter that were not eaten by the other two species. *Anolis evermanni* is a generalist, foraging on tree trunks, in the canopy, and on rocks in streams beds. All *A. evermanni* consumed ants, Homoptera, and spiders regardless of where they foraged, but individuals foraging in streams also ate significant numbers of seeds, as well as insects that dwell on the surface of water. Ants were also the most common prey for *A. stratulus*, followed by Homoptera and Diptera, but, by volume, planthoppers (Homoptera) constituted nearly 50 percent of their stomach contents. Stomachs of *A. stratulus* held fewer insects during the drier part of the year (February–May), suggesting the possibility of food limitation for this species (Reagan 1996; see also Licht 1974; Andrews 1976; Sexton et al. 1976; Lister 1981). Reagan (1996) estimated the total daily intake of insects for these three species at around 450,000 individuals ha^{-1}.

Nine nonanoline reptile species are found throughout Puerto Rico in moist forest at elevations of up to about 600 m (Thomas and Kessler 1996). This assemblage includes two typhlopid blind snakes (*Typhlops platycephalus* and *T. rostellatus*), one amphisbaenian (*Amphisbaenia caeca*), an anguid lizard (*Diploglossus pleei*), two geckos (*Sphaerodactylus macrolepis* and *S. klauberi*), the Puerto Rican boa (*Epicrates inornatus*), and two colubrid snakes (*Alsophis portoricensis* and *Arrhyton exiguum*; see Thomas and Kessler [1996] for photographs). Except for the boa and one of the colubrids (*Alsophis portoricensis*), which forage in trees, these species prey on arthropods in the soil and leaf litter. The effect of nonanoline reptiles on their prey species is unknown, as neither abundances nor foraging rates are known for these species (Thomas and Kessler 1996). However, information on their diet indicates some degree of specialization, especially for *Typhlops* (termites and ants), *Diploglossus* (millipedes), and *Alsophis* and *Arrhyton* (lizards).

Spiders are the dominant predators on the forest floor, with a mean annual density (356 m^{-2}) that is much higher than in most other temperate or tropical forests (Pfeiffer 1996). Predaceous beetles, bugs, and centipedes are also found on the forest floor (table 3-8), but relatively little is known of their densities or feeding habits.

By their diversity and abundance, birds are among the most important consumers in the Luquillo Mountains. The avifauna of Puerto Rico, including on the islands of Vieques, Culebra, Mona, Monito, and Desecheo and on smaller cays and islands, includes a total of 275 extant species, 36 of which are introduced (Raffaele et al.

1998). Approximately 136 bird species, including 14 endemics and 31 introduced species, breed in Puerto Rico. Sixty-six species of land birds occur in the Luquillo Mountains, and two extinct (*Aratinga chloroptera*, Psittacidae; *Corvus leucognaphalus*, Corvidae) and three fossil (*Tyto cavatica*, Tytonidae; *Geotrygon larva*, Columbidae *Corvus nasicus*, Corvidae) species might also have occurred there (Waide 1996).

In the tabonuco forest type of the Luquillo Mountains, long-term studies at the El Verde Field Station provide information about the structure and dynamics of the avian community and about the importance of birds as consumers (Waide 1996). The most common species in mature tabonuco forest include the Bananaquit, the Black-whiskered Vireo, the Ruddy Quail-Dove, the Scaly-naped Pigeon, the Puerto Rican Tanager (*Nesospingus speculiferus*), the Puerto Rican Tody (*Todus mexicanus*; figure 3-27), and the Puerto Rican Emerald (Waide 1996). Seven of the fifteen most common species are endemic to Puerto Rico. Comparisons of counts conducted in 1964–1966 and 1981–1982 found both increases (Black-whiskered Vireo,

Figure 3.27 The Puerto Rican tody (*Todus mexicanus*), an understory insectivore and representative of the family Todidae, endemic to the West Indies. (Photograph by Jerry Bauer.)

Ruddy Quail-Dove, Scaly-naped Pigeon, Puerto Rican Spindalis) and decreases (Pearly-eyed Thrasher) in abundance during a period of relatively stable forest composition and structure. The distribution of species among feeding guilds differs from continental tropical avifaunas by having a smaller proportion of insectivores (16.7 percent) and a larger proportion of frugivores (33.3 percent) (Waide 1996). The reduced number of insectivorous species might reflect competition from abundant frogs and lizards in tabonuco forest. High numbers of lizards also lead to the occurrence in the Luquillo Mountains of an endemic predatory bird specializing on lizards as prey (the Puerto Rican Lizard Cuckoo [*Coccyzus vieilloti*]).

In tabonuco forest, the introduced black rat is common and can reach densities of up to 40 individuals ha^{-1} (Snyder et al. 1987). This contrasts with data from the palo colorado forest type, where rats can attain densities of 281 individuals ha^{-1} (Willig and Gannon 1996). Black rats feed on the forest floor as well as in the trees, consuming fruits from a variety of early (e.g., *Cecropia*) and late (e.g., *Dacryodes*) successional trees, and they also eat snails, fungi, insects, lizards, and frogs. Although more rare than the black rat, the small Indian mongoose is similarly omnivorous (Willig and Gannon 1996). Because of their abundance, size, metabolic rate, and omnivorous food habits, both of these introduced mammals likely have altered the structure and dynamics of food webs and are now integral components of the animal community in the Luquillo Mountains and the entire island of Puerto Rico.

Bromeliads as Specialized Terrestrial Habitats

In the terrestrial environment, bromeliads act as widely dispersed aquatic microcosms with both terrestrial and aquatic animal communities. Within a single bromeliad, habitats range from accumulations of dry leaf litter to the truly aquatic phytotelmata at the base of the bromeliad leaves. Bromeliads are colonized by a variety of detritivorous animals, including isopods, millipedes, cockroaches, and beetles. Dipteran larvae such as crane flies and mosquitoes are found in the pools of water trapped by the bromeliads. Some of the animals found in bromeliads, such as the pseudoscorpion *Macrochernes attenuatus* and the hydrophilid beetle *Omicrus ingens*, are endemic to bromeliads and to Puerto Rico (Hansen and Richardson 1998; Richardson 1999). Bromeliads and their associated fauna are tightly linked to atmospheric processes and thus can be particularly sensitive to climate change (Lugo and Scatena 1992).

Terrestrial Food Web

Trophic interactions among terrestrial species are best understood for the tabonuco forest, about which Reagan and Waide (1996) summarized 4 decades of research at El Verde Field Station. More than 2,600 animal species are known from El Verde, and more than 2,500 of these are invertebrates (Garrison and Willig 1996; Pfeiffer 1996). This number of invertebrates is likely an underestimate, as not all species have been described. Five interrelated features characterize the terrestrial food web at El Verde (figure 3-28) and distinguish it from food webs in similar, continental tropical forests. These features are the absence of large herbivores and predators, low faunal richness, a superabundance of frogs and lizards, discontinuities within the food web, and frequent disturbances. The first three of these distinctive features arise either directly or indirectly from Puerto Rico's geographic position as an island in the Caribbean basin and its past history of isolation.

Because most animal species occurring in Puerto Rico arrived through the process of overwater dispersal from South America, the present fauna of the island lacks those groups that have poor dispersal capabilities, including large mammalian herbivores (e.g., deer, tapirs) and predators (e.g., jaguar) and large frugivorous birds (e.g., toucans, guans, curassows, chachalacas, and turkeys). The absence of these taxa has significant effects on the structure of the food web, in which the largest predators are relatively small and include birds, introduced mammals, and a reptile (figure 3-28). The proportion of top predators is smaller than in continental food webs, which might reduce the top-down control of consumer populations.

The Luquillo Mountains have fewer animal species overall than mainland tropical forests do (Waide 1987; Reagan et al. 1996). Precise comparisons are difficult because of the lack of data from mainland sites representing the same life zones as in the Luquillo Mountains. The relatively small number of species affects the structure of the terrestrial food web in at least two ways. Reduced interspecific competition leads to habitat generalization in the existing species (Waide 1996). Moreover, the relatively small number of species limits the number of possible feeding interactions within the food web, with potential effects on food chain length and connectivity that would not be found in more species-rich communities.

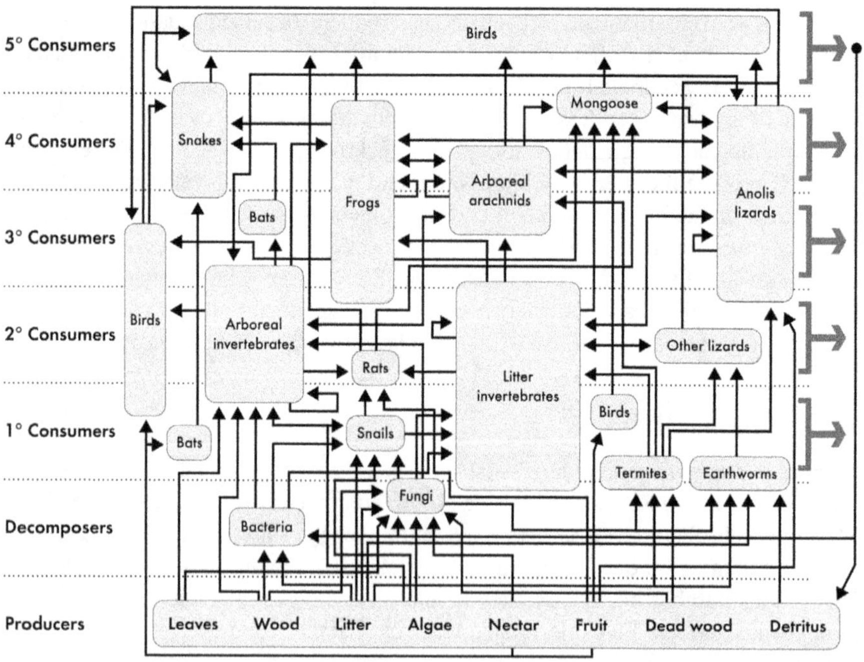

Figure 3.28 Terrestrial food web of the subtropical wet forest in the Luquillo Mountains. Redrawn from Reagan and Waide (1996).

The absence of large predators in Puerto Rico is thought to be responsible for the extremely high densities of small frog (*Eleutherodactylus*) and lizard (*Anolis*) predators that are found in tabonuco forest (Reagan et al. 1996). The abundance of these small predators has a number of potential consequences for the structure of the animal community and for functional attributes of these forest ecosystems such as rates of herbivory and nutrient cycling. Small, ectothermic predators have higher conversion efficiencies, can potentially support longer food chains and more levels of predators, and can potentially promote reciprocal predation (feeding loops) (Formanowicz et al. 1981; Reagan et al. 1996). Angulo-Sandoval et al. (2004) have also proposed that the high densities of frogs and lizards in the tabonuco forest type have led to a suppression of invertebrate herbivores and lower rates of herbivory than in continental communities where larger predators occur (e.g., Panama). Dial and Roughgarden (1995) provided support for this hypothesis by showing that the exclusion of lizards increased the density of invertebrate herbivores and the frequency of herbivory. Similarly, Beard et al. (2002, 2003) showed that the experimental manipulation of coqui populations directly affects herbivory rates by invertebrates.

Both spatial and temporal discontinuities in the animal community lead to compartmentalization of the food web. Distinct groups of animals inhabit aquatic (see below) and terrestrial habitats, which minimizes consumption between groups. The vertical stratification of foraging by species in the terrestrial community structures connections within the food web. The dependence on live or dead sources of energy separates the food web into predatory and detrital compartments, with the vast majority of carbon flowing through the detrital compartment and mycorrhizal fungi (Lodge 1996; Pfeiffer 1996). The food web at El Verde is distinguished by differences in the activity times of the most abundant predatory taxa that separate the roles that they each play within the food web. *Eleutherodactylus* are primarily nocturnal, and *Anolis* are diurnal (Reagan et al. 1996; Stewart and Woolbright 1996). This is reflected in their respective diets and, to an even greater degree, in the diets of the bird and snake predators that feed on them. This separation in activity times divides the food web into compartments (subwebs) and increases the complexity of the trophic structure in tabonuco forest.

Frequent disturbance in the Luquillo Mountains might also structure the food web. The relative scarcity of large predators (which are more likely to go extinct than small predators in a dynamic disturbance environment [Pimm 1982]), the prevalence of omnivory, and the tendency toward donor-controlled predator–prey systems are all characteristics of the El Verde food web that might be affected by frequent disturbance (Reagan et al. 1996). Frequent disturbance-driven changes in the habitat structure and microenvironment work in favor of habitat generalists. All of these factors suggest a strong relationship between the disturbance regime of the Luquillo Mountains and the structure of the food web.

Terrestrial Elevational Gradient

One of the most striking features of the elevational gradient in the Luquillo Mountains is the sharp decline in tree stature from the base of the mountain to the summit. In the elfin forest of the Luquillo Mountains, trees seldom exceed 5 m in height, but they are commonly 25 m or taller at lower elevations (figure 3-18) (Waide et al. 1998). Aboveground biomass also tends to decrease with elevation in dicotyledonous communities in the Luquillo Mountains, as does net primary productivity (see the section "Terrestrial Primary Producers" above; table 3-7).

The underlying mechanisms that reduce forest stature with elevation have been the subject of considerable debate but little consensus (Bruijnzeel and Veneklaas 1998). Mineral nutrient deficiencies, low rates of transpiration, waterlogged soils, wind stress, and reduced nutrient uptake and root damage from polyphenolic inhibition at the soil–root interface have all been suggested as possible causes (Odum 1970b; Grubb 1977; Tanner 1977; Lawton 1982; Weaver and Murphy 1990; Bruijnzeel et al. 1993). Grubb (1977) suggested that although anoxia can affect tropical montane forest plant communities, the primary stress would be induced by the low pH and low soil nutrient levels caused by high rates of nutrient leaching.

Our data do not provide a convincing explanation for the stunted vegetation on the peaks of the Luquillo Mountains. There is no compelling evidence of direct nutrient or pH limitation in the elfin forest. Levels of N and P in the soil are somewhat higher than at lower elevations, the pH is unchanged, and base cation levels are only marginally lower (table 3-5). Even though standing stocks of N are higher in the elfin forest soils, concentrations (mg g^{-1}) of N and Ca in foliage are significantly lower in the elfin forest than in the colorado forest type. When expressed on a unit area basis, however, the nutrient levels are comparable. Understanding the extent to which nutrient availability limits the productivity and stature of elfin vegetation is further complicated by the fact that there can be significant foliar uptake of nutrients from precipitation or uptake by fine roots found in the canopy (Nadkarni 1981).

The number of tree species decreases with increasing elevation, with about 170 species in the tabonuco forest type, 90 in the colorado forest type, and 40 in elfin forest (Weaver and Murphy 1990). The mean height, dbh, and basal area per hectare also tend to decrease with increasing elevation, whereas the stem density increases (White 1963; Weaver and Murphy 1990).

The composition of tree communities along the elevational gradient in the Luquillo Mountains suggests that they have a complicated origin and do not match either continuous or community unit distributional models. A recent study examined changes in the vegetation community structure with elevation by sampling along three transects (0.1 ha plots every 50 m in elevation) in different watersheds of the Luquillo Mountains (Barone et al. 2008). Based on an analysis of the clustering of the elevational ranges and modes of tree species, the data showed that the upper boundaries of species ranges were significantly clustered on the two longest transects, whereas lower boundaries were not. These changes in the community structure corresponded roughly to the broad forest types discussed above (tabonuco, palo colorado, palm, and elfin), but there was also significant nestedness among the plots because some species had broad elevational ranges. These patterns thus do not match either continuous or community unit distributional models along the elevational gradient, as has been seen in other tropical mountains (Ashton 2003).

Elevation has a marked indirect effect on termite abundance in the Luquillo Mountains through its effects on plant communities and litter type. Termites are absent from the thinly distributed litter of the high-elevation elfin forest, but they are abundant (442 m^{-2}) in litter under palm stands at all elevations (Richardson et al. 2005). Termites are vulnerable to predation and dehydration, and the physical conditions in moist layers of palm litter might provide the necessary protection. Ants and most other taxonomic groups show similar patterns of decline in abundance with increasing elevation, but they consistently have their highest densities in palm litter in all forests. These comparisons of palm and nonpalm litter invertebrate communities up the elevation gradient suggest that community composition is determined more by the forest type than by the direct climatic effects of decreasing temperature and increasing rainfall (Richardson et al. 2005).

The distribution of birds with elevation in the Luquillo Mountains has not been studied systematically. However, some species, such as the Puerto Rican Parrot and the Elfin Woods Warbler (*Dendroica angelae*), do favor higher-elevation forests.

The species richness and abundance of decomposer basidiomycetes and pyrenomycetes decline at higher elevations. The few basidiomycete species that are found at high elevations show an interesting biogeographic affinity with North American taxa, and in some cases the same species is found in both the Luquillo Mountains and North America (Baroni et al. 1997). As noted above, the overall tree species richness also declines with elevation, and the decline in fungal diversity might be a reflection of the declining numbers of potential hosts. Many ascomycetous fungi and their asexual stages are restricted to colonization of the dead leaves of particular trees (Laessøe and Lodge 1994; Lodge et al. 1995; Polishook et al. 1996; Lodge 1997; Santana et al. 2005). Although some decomposer basidiomycetes are widespread among the ecological zones of the Luquillo Mountains, many species are largely restricted to a particular life zone, as confirmed by terminal restriction fragment length polymorphism analysis (Lodge et al. 2008; Cantrell et al., in press). In contrast, a greater proportion of bacteria are shared among forest types (Cantrell et al., in press). Although total soil microbial C does not differ between the elfin and tabonuco forests (Zou et al. 2005), the total soil C does increase with elevation (Wang et al. 2002), and soil microbial communities also differ significantly among forest types along the elevation gradient (Cantrell et al., in press).

The invertebrate community living in the phytotelmata of bromeliads shows striking shifts in diversity with elevation, with the highest diversity at mid-elevation in the palo colorado forest type (Richardson 1999; Richardson et al. 2000). Species richness is high in tabonuco forest (167 species), peaks in mid-elevational colorado forest (198), and is significantly lower in elfin forest (97) (Richardson et al. 2000). Typical litter detritivores, such as isopods, millipedes, and cockroaches, were reduced in abundance in the elfin forest, as were larvae of the tipulid fly *Trentepohlia dominicana*. Scirtid beetle larvae (*Scirtes* sp.), the most abundant species in the two lower-elevation forest types, were absent from the elfin forest, as were hydrophilid beetles, *Omicrus ingens*, the naidid worm *Aulophorus superterrenus*, and larvae of the large predatory elaterid beetle *Platycrepidius* sp. In general, invertebrate predators were absent or few in number in the elfin forest (Richardson 1999). Changes in microclimate and nutrient conditions in the bromeliad phytotelmata are probably responsible for the changes in animal diversity with elevation. The colorado forest might provide the most favorable conditions for the survival of both larval and adult invertebrates, as it has lower wind velocities than the elfin forest, higher rainfall than the tabonuco forest, and an intermediate level of anoxia in the phytotelmata. Bromeliads in the colorado forest type are thus less likely to dry out, allowing species with lower abundance and drought intolerance a greater chance of persistence during dry periods.

Aquatic Biota and Ecosystem Processes

Primary Producers and Stream Energy Budgets

Primary production by benthic algae and inputs of leaves, fruits, and other material from the terrestrial landscape form the basis of stream food webs in the Luquillo Mountains. Where light limits primary productivity, as is often the case, organic matter of terrestrial origin (e.g., dissolved organic carbon in groundwater, leaf litter from the adjoining forest) fuels much of the stream metabolism (Ortiz-Zayas et al. 2005). Algae present in streams of the Luquillo Mountains most commonly include diatoms, green algae, and blue-green algae (Pringle 1996). Macrophytes are typically absent, except *Elodea*, which is found at low elevations. Long strands of filamentous green and blue-green algae are observed periodically, but typically few algae are visible in the streambed (Pringle 1996; Pringle et al. 1999). Frequent high-discharge events scour the streambed and remove algae from rock surfaces. In between high-discharge events, herbivory by atyid shrimps (middle- to high-elevation streams) or snails (lower-elevation streams) plays a key role in maintaining the algal standing crop at low levels (Pringle and Blake 1994; Pringle et al. 1999; March et al. 2002).

The net primary productivity is low in small streams of the Luquillo Mountains, and it is often undetectable with whole-stream measures of respiration and productivity (Buzby 1998; Merriam et al. 2002; Ortiz-Zayas et al. 2005). Ortiz-Zayas et al. (2005) conducted an extensive study of the primary productivity and respiration in the Río Mameyes, with 8 to 10 measurements at each of multiple sites over 2 years. They found that the rates of oxygen production were low in headwaters of the Río Mameyes (<70 g O_2 m^{-2} y^{-1}) throughout the year, but they were higher (453 to 634 g O_2 m^{-2} y^{-1}) in the middle and lower reaches. Ratios of productivity:respiration (P/R) were typically about 0.2, with only one station exceeding a P/R of 1 for only a few of the dates sampled (Ortiz-Zayas et al. 2005). The Río Icacos and other streams in areas with quartz diorite bedrock support particularly few attached algae, owing to the very sandy and unstable streambed, but throughout the Luquillo Mountains there is little evidence of significant primary production in streams.

Atyid shrimps are key grazers in stream ecosystems of the Luquillo Mountains, significantly affecting the algal standing crop, the community structure, and the spatial heterogeneity of algal communities (Pringle 1996; Pringle et al. 1999).

Because of their size relative to insects, shrimp can affect the algal standing crop and community structure on spatial and temporal scales that are quite different from those of smaller invertebrates. Algal biovolumes can be as high as 26 $cm^3\ m^{-2}$ in sunny spots where consumption rates by atyid shrimp are low or in streams with few *Atya*, but they are typically 0.03 to 0.18 $cm^3\ m^{-2}$ in other environments (Pringle 1996). Grazing by shrimp and light levels interact to determine this heterogeneity in the algal biovolume. In Quebrada Toronja, a stream with high shrimp densities near the El Verde Field Station, the algal standing crop in the margins of pools with direct sunlight was 140-fold greater than that in deeper areas where atyids foraged; in shaded pools, the standing crop in pool margins was only five times that in deeper areas (Pringle 1996). Shrimps also influence the algal community composition, maintaining low-diversity diatom-dominated communities where they graze; ungrazed pool margins have significantly greater taxonomic richness and structural complexity (figure 3-29).

Different phenological patterns of leaf fall among native and nonnative riparian species provide a spatially and temporally heterogeneous series of alternative energy sources for stream microbes and detritivores. Relatively little is known about how qualitative differences in the nutrient content and leaf chemistry might drive variability in the food quality of different species of riparian leaves. In the Luquillo Mountains, more than 40 species of riparian trees can supply leaf litter at various times (Reed 1998). Native riparian species such as tabonuco, *Cecropia schreberiana*, and sierra palm are commonly distributed along stream banks in the Luquillo Mountains.

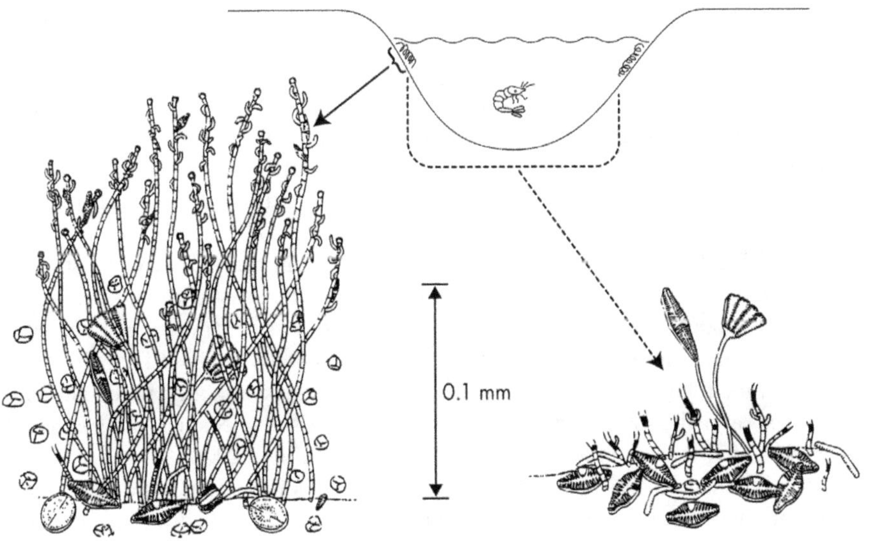

Figure 3.29 Effects of shrimp on benthic algal community composition. Where shrimp have access to the streambed, most of the algal flora is small prostrate diatoms. On pool edges where shrimp do not forage, filamentous algae dominate. Modified and redrawn from Pringle (1996).

Aquatic Consumers and Decomposers

Decapod crustaceans (shrimps and crabs) are the most important group of consumers in streams of the Luquillo Mountains. They include four atyid shrimps (including the common *Atya lanipes*), the common xiphocaridid (*Xiphocaris elongata*), various predatory palaemonid shrimp of the genus *Macrobrachium*, and the crab *Epilobocera sinuatifrons* (Covich and McDowell 1996; Zimmerman and Covich 2003). Each of these crustaceans is omnivorous and displays multiple methods of feeding as an adult. For example, *Atya lanipes* switches between filtering fine particles from flowing water and scraping/gathering benthic algae and fine particulate organic matter (FPOM), and *Xiphocaris elongata* shreds coarse organic matter and also scrapes and gathers benthic algae and FPOM (Pringle et al. 1993; Pringle 1996; Crowl et al. 2001; March et al. 2001). The feeding habits of atyid and xiphocaridid juveniles are poorly known but are thought to be similar to those of adults. Stable isotopic analysis indicates that algal-based resources, as well as detrital food sources, are important to stream consumers, even in small forested headwater streams (March and Pringle 2003). The results of a two-source mixing model suggest that shrimps relied more on algal-based carbon resources than terrestrially derived resources at three sites along the Río Espíritu Santo (March and Pringle 2003).

Fishes and snails are also important consumers in many of the streams of the Luquillo Mountains, particularly the algivorous goby (*Sicydium plumeri*), the predatory mountain mullet (*Agonostomus monticola*), and herbivorous neritid snails (*Neritina* spp.) (Erdman 1986; Nieves 1998; Blanco 2005). Aquatic invertebrates other than decapods and snails include a low diversity of aquatic insects (e.g., baetid and leptophlebiid mayflies, hydroptilid caddisflies, and libellulid dragonflies), as well as miscellaneous invertebrates such as aquatic worms, copepods, and mites (Buzby 1998; Greathouse and Pringle 2006). There are no species of stoneflies in the streams of the Luquillo Mountains, and the total known richness of aquatic insects is approximately 60 to 70 species (Covich and McDowell 1996).

Typical mean densities and biomass of aquatic invertebrates within the Luquillo Mountains range from ~200 to 6,000 individuals m^{-2} and ~0.3 to 10 g ash-free dry mass m^{-2} (Greathouse and Pringle 2006). Densities and biomass reach higher values (~25,000 individuals m^{-2} and ~25 g ash-free dry mass m^{-2}) when streams draining the Luquillo Mountain enter the lowlands of the coastal plain (Greathouse and Pringle 2006). Typically, aquatic invertebrate biomass is dominated by shrimps, crabs, and snails. Insects and other invertebrates generally account for only a few percent of the total standing stock of aquatic invertebrates, although particular habitats, such as riffles, sometimes have high densities of nondecapod, nongastropod invertebrates (Merriam et al. 2002; Greathouse and Pringle 2006).

All of the native shrimps, fishes, and neritid snails of the Luquillo Mountains have a marine stage in their life cycle, and thus migrations up and down the drainage basin are an important feature of the stream community (Pringle 1997; Holmquist et al. 1998; March et al. 1998; Nieves 1998; Benstead et al. 2000; Pyron and Covich 2003; Blanco and Scatena 2005). The life cycles of shrimps, neritid snails, and *Sicydium* (gobies) are categorized as freshwater amphidromous (adults breed in freshwater, larvae passively drift to estuary before returning to freshwater as juveniles). The American eel (*Anguilla rostrata*) is catadromous (migratory to the sea for breeding). Detailed life cycles of other fishes, such as mountain mullet, are poorly known, but they are thought to be freshwater amphidromous (Nieves 1998). Other invertebrates (e.g., *Epilobocera*, aquatic insects) lack a marine stage but "migrate" between freshwater and land. The semiaquatic crab (*Epilobocera*) has direct development in fresh water. Juveniles then feed and develop in fresh water, and adults move between fresh water and the forest floor (Covich and McDowell 1996; Zimmerman and Covich 2003).

Effects of Shrimp Foraging

The foraging activities of atyid and *Xiphocaris* shrimps have large effects on benthic sediment, algae, and insects. Field observations and numerous experimental studies using exclosure/enclosure techniques have documented that shrimp reduce benthic algal biomass, reduce the standing stock of benthic organic matter and nitrogen, and alter algal and insect communities (Pringle et al. 1993; Pringle and Blake 1994; Pringle 1996; Pringle et al. 1999; March et al. 2002; Greathouse et al. 2006b; see also chapter 6). When shrimp were excluded from Quebrada Sonadora, a shrimp-rich river, for example, benthic organic material increased 10-fold (from 1.1 to 10.6 g ash-free dry mass m^{-2}), and benthic nitrogen increased 5-fold (from 0.04 g m^{-2} to 0.2 g m^{-2}) (Pringle et al. 1999). Pringle et al. (1993) suggested that the differences in the abundance of atyid shrimp seen among streams of the Luquillo Mountains result in changes in the distribution and abundance of relatively sessile benthic invertebrates. Their hypothesis is supported by several lines of evidence. Enclosure/exclosure experiments show that foraging by atyid shrimp and *Xiphocaris* reduces the numbers of retreat-dwelling chironomid (midge) larvae (e.g., Pringle et al. 1993; March et al. 2002), and field observations indicate that particle-feeding benthic insects such as black flies are restricted to fast-flowing riffles and pool margins outside of shrimp foraging areas (Pringle et al. 1993; Buzby 1998). Other benthic invertebrates that are negatively affected by shrimp include odonate dragonflies, caenid mayflies, ceratopogonid midges, limpets, and aquatic worms (Greathouse et al. 2006b). Shrimp can have positive effects on motile mayflies such as Baetidae (Buzby 1998; Greathouse et al. 2006b).

Leaf Decomposition

Leaf decomposition in the streams of the Luquillo Mountains is rapid, with most species of leaves fully decomposed in less than 9 months (Padgett 1976; Vogt et al. 1996). Shrimp and fungi dominate the decomposition process. The aquatic hyphomycetes *Campylospora chaetocladia*, *Triscelophorus monosporus*, and *Pyramidospora casuarinae* were the most abundant of 16 fungal species found to colonize leaves during an experimental study of leaf decomposition (Padgett 1976). Whole-pool manipulations of shrimp abundance suggest that the presence of both *Xiphocaris elongata* and *Atya* species is necessary for the efficient processing of leaf material. *Xiphocaris* shred the leaves, and *Atya* filter the resulting particles from the water column (Crowl et al. 2001; also see chapter 6). Changes in shrimp communities along the elevational gradient are reflected in changes in the rates of litter decomposition (March et al. 2001). Leaf-shredding insects are uncommon; only *Phanocerus* elmid beetles and *Phyllocius pulchrus* (a calamoceratid caddisfly) are present, and they exist in low numbers (Buzby 1998). Leaf-mining Chironomidae also occur in Puerto Rican streams (Greathouse and Pringle 2006; Greathouse et al. 2006c).

Food Webs along the Aquatic Elevational Gradient

Because of the nature of streams and flowing waters, stream communities at a point in geographic space are inextricably linked to upstream and downstream communities. This is particularly so in the Luquillo Mountains, where many stream biota have direct connections to the sea at some point in their life cycle. The species composition of aquatic communities and the influence of aquatic consumers on ecosystem-level processes (e.g., decomposition) in the Luquillo Mountains vary with elevation and the position of natural and anthropogenic barriers such as waterfalls and dams (Greathouse et al. 2006a, 2006c; Covich et al. 2009). Longitudinal distributions of shrimps, fishes, and snails are particularly influenced by their migratory life cycles between fresh and salt water and by variation among taxa in their abilities to migrate past barriers (Covich and McDowell 1996; Covich et al. 1996).

Geomorphic breaks are central to understanding the community structure and food webs in streams of the Luquillo Mountains. Predatory fishes such as mountain mullet are typically limited to elevations below 400 m, because waterfalls limit their passage upstream. Although neritid snails can climb steep slopes, they also are limited to lower elevations below waterfalls. Their distribution is thought to represent the tradeoffs among predation risk, the energetic demands of migrating upstream, and life span (Pyron and Covich 2003). Gobies are found at elevations of up to ~700 m because they have the ability to move upstream against high-velocity currents using sucking discs evolved from pectoral fins (Erdman 1961, 1986). Shrimp reach the highest-elevation headwater streams, beyond the upstream limits of *Sicydium*. Xiphocaridid and atyid shrimps also occur at much higher abundances upstream from waterfalls. These high abundances above waterfalls are thought to be due to the release from predation by fish and/or competition from neritid snails (Covich 1988; March et al. 2002; Greathouse and Pringle 2006; Covich et al. 2009; Hein et al. 2011).

Distributions of functional feeding groups along the elevational gradient have been well studied in the Río Mameyes and Río Espíritu Santo (figure 3-2). In the Río Mameyes drainage, from the headwaters of the Río de La Mina (720 masl) to within 2.5 km of the Río Mameyes mouth (5 masl), several patterns are observed with increasing catchment area/decreasing elevation (Greathouse and Pringle 2006). Xiphocaridid and atyid shrimps reach their highest densities and standing stocks upstream from the upper limit of predatory fishes (figure 3-30). In contrast, high densities and standing stocks of gastropods (primarily neritid snails) occur at sites where predatory fishes are present (figure 3-30). *Macrobrachium* shrimps have high densities of small juveniles at lower-elevation sites, but no clear patterns in

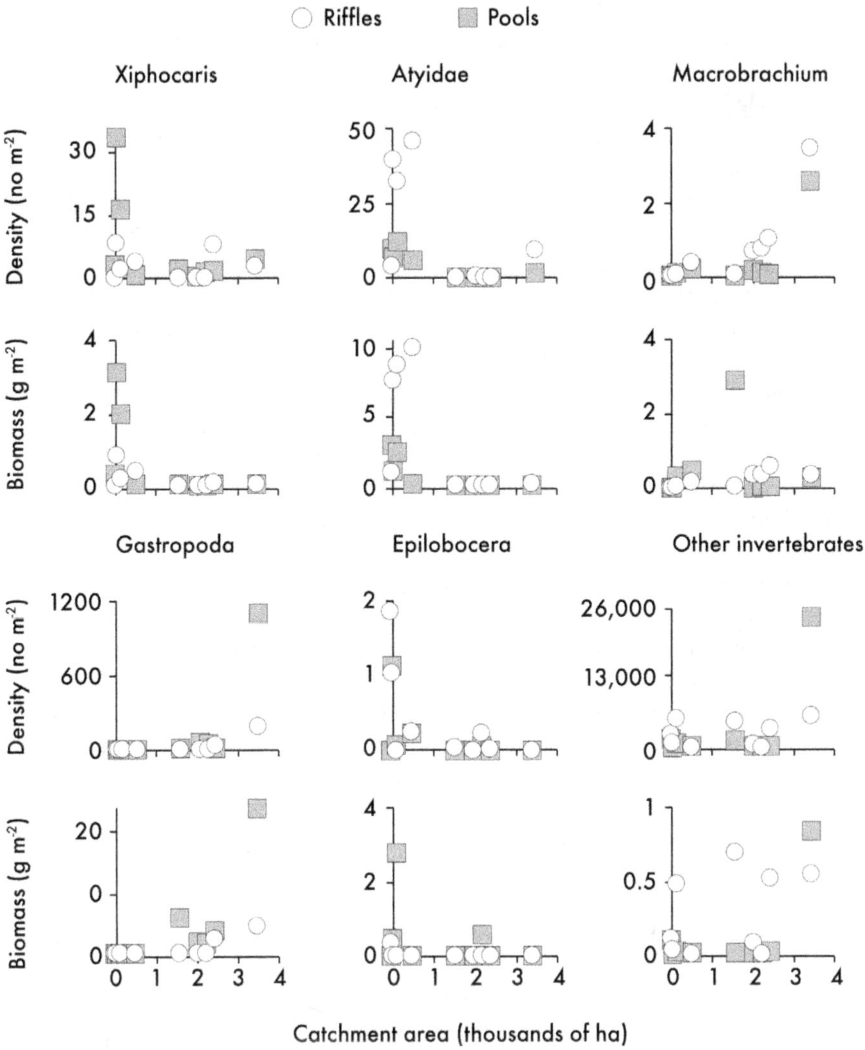

Figure 3.30 Densities and standing stocks (biomass as ash-free dry mass) of aquatic invertebrates in riffles and pools along a stream continuum from the headwaters of the Río de La Mina to within 2.5 km of the mouth of the Río Mameyes. Data from Greathouse and Pringle (2006). Invertebrate groups are shrimps (*Xiphocaris elongata*, Atyidae, *Macrobrachium*), Gastropoda (primarily neritid snails, but also the snail *Thiara granifera*, and limpets), crabs (*Epilobocera sinuatifrons*), and other invertebrates (e.g., aquatic insects, Oligochaeta, Copepoda). Shrimps and crabs were sampled via depletion electroshocking over a known area. Gastropods and other invertebrates were sampled using standard quantitative methods appropriate to each habitat (e.g., Surber net in riffles, cores in pools). Horizontal black bars below the bottom *x*-axes indicate sites at which predatory fishes (e.g., *Agonostomus monticola*, *Anguilla rostrata*, *Eleotris pisonis*) are present. Samples taken from riffles are represented by open circles, and those taken from pools by solid squares. Note axes of different scales.

biomass. Crabs reach their highest densities and biomass at high-elevation sites (figure 3-30) above high (>10 m), steep waterfalls (Covich et al. 2006; Covich et al. 2009; Hein et al. 2011). The remaining invertebrates (including insects) are grouped into a single "other" category that shows the highest density and biomass in low-elevation pools where abundant *Elodea* provides a complex habitat (figure 3-30).

Similar elevational patterns in shrimp and snail densities in the neighboring Espíritu Santo drainage drive variation in leaf decomposition rates (March et al. 2001) and algal biomass (March et al. 2002). When shrimp were excluded from a mid-elevation site lacking predatory fishes (at ~300 masl), leaf decomposition rates decreased by almost 50 percent ($k = 0.067$ day^{-1} vs. 0.036 day^{-1}, $p = 0.019$; March et al. 2001). In contrast, at both mid- and low-elevation sites where predatory fishes were present (~90 and 10 masl, respectively), the exclusion of macrobiota had no significant effect on rates of leaf breakdown. Subsequent laboratory experiments confirmed that the shrimp *Xiphocaris elongata* was the dominant consumer of leaf material but that it consumed significantly less when in the presence of predatory shrimps (*Macrobrachium* spp.). The combined results of laboratory and field experiments indicate that interference competition/predation between these two taxa accounts for the differences in leaf breakdown rates observed between sites. The role of *X. elongata* in detrital processing is context dependent, with strong effects occurring only in stream headwaters, where predatory fishes and *Macrobrachium* spp. are less abundant, and where *Macrobrachium* spp. make up a smaller proportion of the shrimp biomass.

The effects of shrimp exclusion on epilithic communities in the Espíritu Santo drainage also varied with elevation (March et al. 2002). At two mid-elevation sites (300 and 90 masl) where snails were absent or low in abundance, shrimp exclusion had strong effects on the accrual of inorganic and organic material, chlorophyll *a*, algal biovolume, and biomass of Chironomidae. At the low-elevation site (10 masl), snails were abundant, and shrimp exclusion had no effect on benthic organic matter, algae, or Chironomidae.

Algae appear to be important food resources for shrimp along the elevational gradient despite the relatively low primary productivity of these streams. Shrimp appear to rely primarily on algal carbon for growth in larger streams with sufficient sunlight, with no strong patterns in the importance of terrestrial versus algal food sources along the elevational gradient (March and Pringle 2003).

Understanding the context-dependent effects of stream biota along river continua is critical, owing to the migratory life cycle of shrimps and fishes of streams draining the Luquillo Mountains (Covich and McDowell 1996; March et al. 1998; Greathouse and Pringle 2006). With the increasing number of large man-made dams limiting the access of shrimps and fishes to upper-elevation sites, changes in a variety of ecological processes are likely (Pringle 1997; Holmquist et al. 1998; Benstead et al. 1999), and these changes are expected to vary with elevation (March et al. 2001, 2002; Greathouse et al. 2006c) (see chapter 7).

Migration patterns also differ with elevation and stream size. During base flows, the densities of larval shrimp drifting downstream to the estuary increase exponentially with increasing stream size (as measured by cumulative stream length) (March et al. 1998; Kikkert et al. 2009). Whether this relationship holds true during storm flows is unknown. The diel periodicity of larval shrimp drift also appears to vary with elevation in response to the risk of predation. Larval shrimp drift was strongly nocturnal at five low- and mid-elevation sites where predatory fishes were present but showed no diel periodicity at a mid-elevation site lacking predatory fishes due to its position above a waterfall. Upstream migration by juvenile shrimp and neritid snails also shows elevational patterns (see, e.g., Pyron and Covich 2003).

The River Continuum Concept (RCC) formalizes predicted changes in stream food webs with stream order and position in the drainage network (Vannote et al. 1980; Minshall et al. 1985). At high elevations, streams are small and shaded. Coarse particulate organic matter (CPOM) from terrestrial sources is predicted to dominate basal resources, and functional feeding groups that rely on this CPOM (shredders, collector-gatherers, and collector-filterers) are predicted to dominate macroinvertebrate biomass. As the elevation decreases and streams widen, these medium-sized streams with higher light levels are predicted to support more algae and the functional feeding groups (scrapers, collector-gatherers, and collector-filterers) that feed on algae and benthic biofilms. At the lowest-elevation sites in a large river system, high turbidity is predicted to result in low light levels and low algal productivity, and thus a macroinvertebrate community dominated by collector-filterers, which utilize transported FPOM. Macroinvertebrate predators are predicted to show no consistent change with stream order.

These RCC predictions for temperate streams largely hold true for the Río Mameyes, a system that spans small headwater streams to medium-sized channels within the Luquillo Mountains before entering the ocean as a fourth-order stream in the urbanized lowland floodplain downstream of the Luquillo Mountains (Ortiz-Zayas et al. 2005; Greathouse and Pringle 2006). The P/R in the Río Mameyes increases from headwaters to lowlands, as predicted (Ortíz-Zayas et al. 2005). The relative dominance of macroinvertebrate biomass also follows predictions for most functional groups: shredders decreased, scrapers increased, collector-gatherers decreased, and predators showed no change from headwaters to lowlands (Greathouse and Pringle 2006). Filterers, represented by shrimp of the genus *Atya*, decreased with distance downstream, rather than increasing as predicted by the RCC.

Stream chemistry reflects both terrestrial and aquatic biogeochemical processes, and thus the changes in terrestrial and aquatic ecosystems documented in the preceding paragraphs might be expected to cause changes in the stream chemistry and nutrient export with elevation. Contrary to this expectation, however, data published to date show no striking differences in the stream chemistry among watersheds with different mean elevations. McDowell and Asbury (1994), for example, found that nitrate-N concentrations in the high-elevation Río Icacos (700 to 1,000 masl; 66 μg l^{-1}) were similar to those in the low-elevation Toronja watershed (62 μg l^{-1}). Fluxes of NO_3-N from the Icacos (2.5 kg ha^{-1} y^{-1}) were much greater, however, than those from the Toronja (0.9 kg ha^{-1} y^{-1}), owing to the much higher runoff at higher elevations (McDowell and Asbury 1994). Temporal variation in N flux is greater than the spatial variation alone, as NO_3-N export varied from 0.7 to 8 kg ha^{-1} y^{-1} among six watersheds in the years before and after Hurricane Hugo (Schaefer et al. 2000).

Effects of Recent Invasions

Although the biotic assemblage in a given locale or region is frequently thought of as resulting from ecological and evolutionary processes occurring over thousands to millions of years, the rapid pace of biotic introductions and successful invasions in the past few centuries have resulted in significant changes in the biota of many regions. The flora and fauna of Puerto Rico have been affected in many ways by introduced species. Bamboos (*Bambusa vulgaris*, *B. longispiculata*, *B. tulda*, *B. tuldoides*, *Dendrocalamus strictus*) are common along the roads of the Luquillo Mountains, where they were originally planted by the U.S. Forest Service to assist in erosion control. Bamboos have spread along stream channels (O'Connor et al. 2000) but are not widespread in the Luquillo Mountains. Pomarrosa (*Syzygium jambos*) is another nonnative species common in riparian zones, but it too is not common elsewhere in the tabonuco forest type. Nonnative plants are common in lower-elevation forests that have undergone extensive human modification or suffered significant hurricane effects, but as the native overstory returns, these introduced trees lessen in importance (Lugo 2004; Thompson et al. 2007). Black rats are not native to Puerto Rico and likely reached the island with Ponce de Léon in 1508 (Snyder et al. 1987). Rats and the Indian mongoose (intentionally introduced to control the rats) threaten a variety of native fauna, including four bird species (Puerto Rican Parrot, Short-eared Owl [*Asio flammeus*], Puerto Rican Whip-poor-will [*Caprimulgus noctitherus*], and Key West Quail Dove [*Geotrygon chrysia*]) and two snake species (Puerto Rican boa and Puerto Rican racer [*Alsophis portoricencis*]) (Raffaele et al. 1973). The endangered birds construct nests in which eggs and nestlings are vulnerable to predation by rats and mongooses. The cane toad (*Bufo marinus*) was introduced in order to control pests in sugar cane and is now found in much of the Luquillo Mountains. Disturbance due to anthropogenic practices seems to be the major factor causing the spread of introduced earthworms in the tropics (González et al. 2006). Introduced earthworms can establish their populations in sites modified after deforestation (e.g., forest-pasture conversion), tree plantations, and cultivation activities, and also follow human migrations (González et al. 1996; Zou and González 2002). Conversely, native species can return upon the regrowth of forest in abandoned pastures (Sánchez et al. 2003).

Most major functional groups of plants and animals have one or more important introduced species that play a significant role in community dynamics and ecosystem processes. Introduced earthworms, rats, mongooses, and the cane toad each play an important role in terrestrial food webs, and introduced bamboo and pomarrosa are plant species that have a significant role in stream food webs.

Aquatic invasions have occurred in lowland Puerto Rican streams, including those that originate in the Luquillo Mountains. The freshwater snail *Thiara granifera* is a conspicuous introduced species. In the main stem of the Río Mameyes, *T. granifera* reaches standing stocks of ~2 g m^{-2}, but its biomass is generally one to two orders of magnitude lower than the biomass of native *Neritina* snails (Greathouse and Pringle 2006). *Thiara granifera* occurs only at lower elevations (its island-wide upstream limit is ~480 masl) (Chaniotis et al. 1980) and is low in biomass in most streams of the Luquillo Mountains. Competition from *T. granifera* is thought to have caused an island-wide decline in lotic populations of *Biomphalaria glabrata*, the native snail that serves as host to the liver fluke that causes schistosomiasis (Butler et al. 1980; DeJong et al. 2001). The invasion of *T. granifera* in the 1950s appears to have been accidental (Butler et al. 1980; Chaniotis et al. 1980), and its impact on native snails thus represents a positive unintended impact on human health. In contrast, the snail *Marisa cornuarietis* was introduced intentionally for the biological control of *B. glabrata* in standing waters such as farm ponds (Butler et al. 1980). The prevalence of schistosomiasis is now very low in the streams of Puerto Rico, as are densities of *B. glabrata* (Giboda et al. 1997).

Although fisheries introductions across the island have primarily focused on reservoirs, introduced reservoir fishes do invade running waters. The abundance of these introduced fishes is high in streams above large reservoirs but low in streams below reservoirs and in streams with no reservoirs. These patterns indicate that the near extirpation of native fishes and shrimps from streams above dams that are large enough to block migrations results in stream communities with reduced biotic resistance to invasion (Holmquist et al. 1998). This biotic resistance of the native fauna might explain why the aquatic fauna of the Luquillo Mountains is remarkably lacking in introduced species.

Aquatic habitats across the island, including those in the Luquillo Mountains, are at risk for future invasions by a variety of aquarium and aquaculture species that are poorly regulated (Williams et al. 2001). Australian redclaw (*Cherax quadricarinatus*) is a particular threat to the Luquillo Mountains. A population of this crayfish has become established in the Carraizo Reservoir on the Río Grande de Loíza, a river that drains the Luquillo Mountains, and this species appears to be capable of outcompeting native shrimps (Williams et al. 2001).

Introduced plants appear to be altering or supplementing stream food webs in the Luquillo Mountains in ways that are not necessarily negative. Asian species such as bamboo and pomarrosa provide some of the ecosystem functions provided by native species (e.g., leaf litter food sources, woody debris, and shade) (Covich et al. 1999). More freshwater shrimp (both *Atya* and *Macrobrachium*) were found in pools with riparian bamboo than in adjacent pools of similar size that lacked bamboo, and laboratory studies showed that shrimp prefer nonnative bamboo when offered either bamboo or native leaves as cover (O'Connor 1998). The microhabitat created by bamboo litter in streams thus appears to be very well suited for use by these shrimp.

Luquillo Mountains from a Tropical Perspective

Understanding the drivers of spatial and temporal variability in ecosystem structure and function is a long-standing goal in ecology. Within the Luquillo Mountains, one of our primary research foci has been to examine the importance of gradients in driving spatial variability in community structure and ecosystem processes. The broad gradients in rainfall and temperature associated with elevation provide the primary abiotic drivers of variation in community structure and ecosystem processes in the Luquillo Mountains. Patches of different bedrock and the disturbance history (landslide, hurricane damage, intensive past land use) provide complexity along the elevational gradient. With the high rainfall and runoff found in the Luquillo Mountains, aquatic–terrestrial interfaces occur frequently and are "hot spots" of biogeochemical activity (McDowell et al. 1992; McDowell 2001). With such spatial and temporal variation in abiotic drivers and ecosystem properties across the Luquillo Mountains, it is difficult to describe the variation across the landscape in purely spatial terms; a more dynamic temporal component is also needed in order to capture the ways in which a site varies depending on the legacy of past disturbance events and the biotic responses to them.

The concept of ecological space (see chapter 2) provides a useful way to organize our understanding of how environmental characteristics change over time in response to underlying landscape features and the disturbance regime. The heterogeneity of ecological characteristics in geographical space is dictated by a combination of geographic circumstances (e.g., leeward/windward vs. elevation to drive rainfall; elevation/cloudiness vs. aspect to drive PAR at the canopy), the underlying geologic substrate (quartz diorite vs. volcaniclastic bedrock), and the legacies of past disturbances. The biota both respond to ecological space and help create it. Seedling germination and growth, for example, require specific conditions for various species (Guzmán-Grajales and Walker 1991), and successful recruitment of the seedling causes changes in the light and moisture characteristics that are important elements in the ecological space at the site.

Soils provide important nutrient pools in terrestrial ecosystems, and their chemical and physical properties are highly variable in tropical forests (figure 3-12). Wet tropical forests were once thought to have soils containing low concentrations of mineral nutrients. Unfortunately this concept has become embedded in the popular and scientific literature, even though it is not generally applicable (Sánchez 1976; Richter and Babbar 1991; Lal and Sánchez 1992). Although some tropical forests, such as those on Amazonian white sands, conform to this model, many areas of wet tropical forest have soils with considerable mineral pools, including the Luquillo Mountains (Silver et al. 1994). Many tropical forests also have large nitrogen pools, which are presumably the result of high rates of nitrogen fixation (Cleveland et al. 1999; Cusack et al. 2009) and the legacies of past land uses (Beard et al. 2005).

A corollary to this general misconception regarding the nutrient content of tropical soils is that plant biomass is the primary nutrient store in tropical rainforests. For many years, tropical forests were characterized as nutrient-poor ecosystems with low nutrient-holding capacity, with the nutrient content of the aboveground biomass greatly exceeding that of labile nutrient storage in soils (see Whitmore [1989] for an overview of the genesis of this concept). Such characterizations led many to believe that tropical forests were extremely fragile ecosystems, and that plant biomass was much more important than soils for nutrient cycling and conservation. These generalizations were derived from many sources, but the strongest empirical evidence for this view came from studies on forest C and nutrient distribution and cycling in San Carlos de Río Negro in the Venezuelan Amazon (see the review by Jordan [1985]). Although these studies represented some of the most careful and complete early ecosystem research conducted in the tropics, the results were not necessarily generalizable to a wide range of tropical environments because of the unusual soil mineralogy. The soils of the San Carlos site, white sands or psamments, are a relatively rare tropical soil type with some of the lowest cation and P availability of any tropical environment (Jordan 1985; Cuevas and Medina 1986, 1988; Medina and Cuevas 1989). Results from the Luquillo Mountains and other tropical forests suggest that a low nutrient content is not particularly characteristic of tropical forests, as many sites are rich in nutrients owing to their geologic history and soil depth. This has important implications for ecosystem behavior following a disturbance, which is addressed in chapter 5.

Summary

The Luquillo Mountains contain insular ecosystems that have never been connected to a continental land mass and which are subject to severe disturbances, including hurricanes, landslides, and earthquakes. Soils are deep, weathered, and not particularly nutrient poor, and they support a moderately diverse flora and fauna with high endemism. Strong gradients in rainfall, temperature, and insolation, driven by elevation and aspect, help structure the forests found at different elevations and topographic positions in the Luquillo Mountains. Forest productivity does not appear to be limited by nitrogen and declines with increasing elevation. Soil carbon and nitrogen concentrations in surface horizons increase with increasing elevation, and the topographic position also causes substantial variation in the soil chemistry. Concentrations of inorganic nitrogen in streams are high relative to those in montane temperate sites and are stable over time, except for brief periods following hurricanes. One of the greatest ecological distinctions between Puerto Rico and mainland tropical forests is the complete lack of large mammals and the absence of many families of birds, reptiles, and amphibians, which are the result of Puerto Rico's biogeographic insularity and disturbance history. The low species richness for several vertebrate taxa relative to that in otherwise similar continental montane forests is typical of Caribbean islands. No large mammalian herbivores are found, and herbivory is dominated by insects and birds. Top predators in the forest include lizards, frogs, and a few species of birds; large mammalian predators are absent. Stream food webs are dominated by shrimps and crabs, with food webs being fueled by both detrital and algal resources. Shrimps play a key role in stream ecosystems by maintaining a low algal standing crop and benthic insect abundance, altering the algal species composition, regulating benthic inorganic sediments and the quality and quantity of benthic organic matter, and driving rates of litter decomposition. Invasive plants and animals are prominent in both aquatic and terrestrial ecosystems and appear to be most successful following disturbance.

Literature Cited

Abruna, F., and R. M. Smith. 1953. Clay mineral types and related soil properties in Puerto Rico. *Soil Science* 75:411–420.

Ackerman, J. D., J. C. Trejo-Torres, and Y. Crespo-Chuy. 2007. Orchids of the West Indies: Predictability of diversity and endemism. *Journal of Biogeography* 34:779–876.

Alvarez, J., and M. R. Willig. 1993. Effects of treefall gaps on the density of land snails in the Luquillo Experimental Forest of Puerto Rico. *Biotropica* 25:100–110.

Andrews, R. M. 1976. Growth rate in island and mainland anoline lizards. *Copeia* 1976:477–482.

Angulo-Sandoval, P., and T. M. Aide. 2000. Leaf phenology and leaf damage of saplings in the Luquillo Experimental Forest, Puerto Rico. *Biotropica* 32:415–422.

Angulo-Sandoval, P., H. Fernández-Marín, J. K. Zimmerman, and T. M. Aide. 2004. Changes in patterns of understory leaf phenology and herbivory following hurricane damage. *Biotropica* 36:60–67.

Anthony, H. E. 1918. Indigenous land mammals of Puerto Rico, living and extinct. *Memoirs of the American Museum of Natural History, New Series* 2:331–435.

Asbury, C. E., W. H. McDowell, R. Trinidad-Pizarro, and S. Berrios. 1994. Solute deposition from cloud water to the canopy of a Puerto Rican montane forest. *Atmospheric Environment* 28:1773–1780.

Ashton, P. S. 2003. Floristic zonation of tree communities on wet tropical mountains revisited. *Perspectives in Plant Ecology, Evolution and Systematics* 6:87–104.

Barone, J. A., J. Thomlinson, P. A. Cordero, and J. K. Zimmerman. 2008. Metacommunity structure of tropical forest along an elevation gradient in Puerto Rico. *Journal of Tropical Ecology* 24:525–534.

Baroni, T. J., D. J. Lodge, and S. A. Cantrell. 1997. Tropical connections: Sister species and species in common between the Caribbean and the Eastern United States. *McIlvainea* 13:4–19.

Basnet, K. 1992. Effect of topography on the pattern of trees in a tabonuco (*Dacryodes excelsa*) forest of Puerto Rico. *Biotropica* 24:31–42.

Beard, K. H., A. K. Eschtruth, K. A. Vogt, D. J. Vogt, and F. N. Scatena. 2003. The effects of the frog *Eleutherodactylus coqui* on invertebrates and ecosystem processes at two scales in the Luquillo Experimental Forest, Puerto Rico. *Journal of Tropical Ecology* 19:607–617.

Beard K. H., K. A. Vogt, and A. Kulmatiski. 2002. Top-down effects of a terrestrial frog on nutrient concentrations in a subtropical wet forest. *Oecologica* 133:583–593.

Beard, K. H., K. A. Vogt, D. J. Vogt, F. N. Scatena, A. P. Covich, R. Sigurdardottir, T. G. Siccama, and T. A. Crowl. 2005. Structural and functional responses of a subtropical forest to 10 years of hurricanes and droughts. *Ecological Monographs* 75:345–361.

Benstead, J. P., J. G. March, and C. M. Pringle. 2000. Estuarine stages and upstream migration of freshwater shrimps in two tropical rivers, Puerto Rico. *Biotropica* 32:545–547.

Benstead, J. P., J. G. March, C. M. Pringle, and F. N. Scatena. 1999. Effects of a low-head dam and water abstraction on migratory tropical stream biota. *Ecological Applications* 9:656–668.

Bhatt, M. P., and W. H. McDowell. 2007. Controls on major solutes within the drainage network of a rapidly weathering tropical watershed. *Water Resources Research* 43:W11402.

Bisselink, B. 2003. Precipitation trends in Puerto Rico: Quantification and explanation of complex patterns. M.S. thesis. Vrije Universiteit, Amsterdam, The Netherlands.

Blanco, J. F. 2005. Physical habitat, disturbances, and the population ecology of the migratory snail *Neritina virginea* (Gastropoda: Neritidae) in streams of Puerto Rico. Ph.D. dissertation. University of Puerto Rico, Río Piedras, PR.

Blanco, J. F., and F. N. Scatena. 2005. Floods, habitat hydraulics and upstream migration of *Neritina virginea* (Gastropoda: Neritidae) in Northeastern Puerto Rico. *Caribbean Journal of Science* 41:55–74.

Bloch, C. P. 2004. Response of snail populations and communities to large-scale disturbance: Long-term and landscape perspectives. Ph.D. dissertation. Texas Tech University, Lubbock, TX.

Bloomfield, J., K. A. Vogt, and D. J. Vogt. 1993. Decay rate and substrate quality of fine roots and foliage of two tropical tree species in the Luquillo Experimental Forest, Puerto Rico. *Plant and Soil* 150:233–245.

Bongers, F., J. Popma, J. Meave del Castillo, and J. Carabias. 1987. Structure and floristic composition of the lowland rain forest of Los Tuxtlas, Mexico. *Vegetatio* 74:55–80.

Borges, S. 1996. Taxonomic key and annotated bibliography to the earthworms of Puerto Rico. Scientific survey of Puerto Rico and the Virgin Islands: An eighty-year reassessment of the islands' natural history. *Annals of the New York Academy of Sciences* 776:249–256.

Borges, S., and M. Alfaro. 1997. The earthworms of Baño de Oro, Luquillo Experimental Forest, Puerto Rico. *Soil Biology and Biochemistry* 29:231–234.

Bowden, W. B., W. H. McDowell, C. E. Asbury, and A. M. Finley. 1992. Riparian nitrogen dynamics in two geomorphologically distinct tropical rain forest watersheds-nitrous oxide fluxes. *Biogeochemistry* 18:77–99.

Boynton, H. W. 1968. The ecology of an elfin forest in Puerto Rico. 2. The microclimate of Pico del Oeste. *Journal of the Arnold Arboretum* 49:419–430.

Brewer, S. W., and M. A. H. Webb. 2002. A seasonal evergreen forest in Belize: Unusually high tree species richness for northern Central America. *Biological Journal of the Linnean Society* 138:275–296.

Briggs, R. P., and E. Aquilar-Cortes. 1980. Geological map of the Fajardo and Cayo Icacos Quadrangles, Puerto Rico. Miscellaneous Geological Investigations of the U.S. Geological Survey, Map I-1153, Washington, D.C.

Briscoe, C. B. 1966. Weather in the Luquillo Mountains of Puerto Rico. Forest Service Research Paper ITF-3. Río Piedras, PR: U.S. Department of Agriculture, Forest Service, Institute of Tropical Forestry.

Brokaw, N. V. L., and J. S. Grear. 1991. Forest structure before and after Hurricane Hugo at three elevations in the Luquillo Mountains, Puerto Rico. *Biotropica* 23:386–392.

Brown, S., A. E. Lugo, S. Silander, and L. Liegel. 1983. Research history and opportunities in the Luquillo Experimental Forest. General Technical Report SO-44. New Orleans, LA: USDA Forest Service.

Bruijnzeel, L. A. 1983. The chemical mass balance of a small basin in a wet monsoonal environment and the effect of fast-growing plantation forest. Pages 229–239 in B. W. Webb, editor, *Dissolved loads of rivers and surface water quantity/quality relationships*. IAHS Pub. No. 141. Paris: International Association of Hydrological Sciences.

Bruijnzeel, L. A., and J. Veneklaas. 1998. Climatic conditions and tropical montane forest productivity: The fog has not lifted yet. *Ecology* 79:3–9.

Bruijnzeel, L. A., M. J. Waterloo, J. Proctor, A. T. Kuiters, and B. Kotterink. 1993. Hydrological observations in montane rain forests on Gunung Silam, Sabah, Malaysia, with special reference to the Massenerhebung effect. *Journal of Ecology* 81:145–167.

Burney, D. A., and L. P. Burney. 1994. Holocene charcoal stratigraphy from Laguna Tortugero, Puerto Rico and the timing of human arrival on the island. *Archaeological Science* 21:273–281.

Butler, J. M., F. F. Ferguson, J. R. Palmer, and W. R. Jobin. 1980. Displacement of a colony of *Biomphalaria glabrata* by an invading population of *Tarebia granifera* in a small stream in Puerto Rico. *Caribbean Journal of Science* 16:73–80.

Buzby, K. M. 1998. The effect of disturbance on the ecological efficiency of a small tropical stream. Ph.D. dissertation. College of Environmental Science and Forestry, State University of New York, Syracuse, NY.

Camilo, G. R., and X. Zou. 1999. Soil fauna in managed forests: Lessons from the Luquillo Experimental Forest, Puerto Rico. Chapter 13 in A. Grajal, R. Fimble, and J. Robinson, editors, *Effects of logging in the tropics*. Oxford, England: Oxford University Press.

Campbell, J. L., J. W. Hornbeck, W. H. McDowell, D. C. Buso, J. B. Shanley, and G. E. Likens. 2000. Dissolved organic nitrogen budgets for upland, forested ecosystems in New England. *Biogeochemistry* 49:123–142.

Cantrell, S. A., D. J. Lodge, C. A. Cruz, L. M. García, J. R. Pérez-Jiménez, and M. Molina. In press. Differential abundance of microbial functional groups along the elevation gradient from the Luquillo Mountains to the coast. Special issue on the elevation gradient in the Luquillo Mountains of Puerto Rico. *Ecological Bulletins*.

Cermak, V., L. Bodri, and J. Safanda. 1992. Underground temperature fields and changing climate: Evidence from Cuba. *Global and Planetary Change* 97:325–337.

Chadwick, O. A., L. A. Derry, P. M. Vitousek, B. J. Huebert, and L. O. Hedin. 1999. Changing sources of nutrients during four million years of ecosystem development. *Nature* 397:491–497.

Chaniotis, B. N., J. M. Butler, Jr., F. F. Ferguson, and W. R. Jobin. 1980. Bionomics of Tarebia granifera (Gastropoda: Thiaridae) in Puerto Rico, an Asiatic vector of Paragonimiasis westermani. *Caribbean Journal of Science* 16:81–90.

Chestnut, T. J., and W. H. McDowell. 2000. C and N dynamics in the riparian and hyporheic zones of a tropical stream, Luquillo Mountains, Puerto Rico. *Journal of the North American Benthological Society* 19:199–214.

Chestnut, T. J., D. J. Zarin, W. H. McDowell, and M. Keller. 1999. A nitrogen budget for late-successional hillslope tabonuco forest, Puerto Rico. *Biogeochemistry* 46:85–108.

Cleveland, C. C., A. R. Townsend, D. S. Schimel, H. Fisher, R. W. Howarth, L. O. Hedin, S. S. Perakis, E. F. Latty, J. C. Von Fischer, A. Elseroad, and M. F. Wasson. 1999. Global patterns of terrestrial biological nitrogen (N_2) fixation in natural ecosystems. *Global Biogeochemical Cycles* 13:623–645.

Co, L. I., D. A. Lagunzad, J. V. LaFrankie, N. A. Bartolome, J. E. Molina, S. I. Yap, H. G. Garcia, J. P. Bautista, E. C. Grumpal, R. R. Araño, and S. J. Davies. 2004. Palanan Forest Dynamics Plot, Philippines. Pages 574–584 in E. C. Losos and E. G. Leigh, Jr., editors, *Tropical forest diversity and dynamism: Results from a long-term tropical forest network*. Chicago: University of Chicago Press.

Condit, R. 1995. Research in large, long-term tropical forest plots. *Trends in Ecology and Evolution* 10:18–22.

Condit, R., R. B. Foster, S. P. Hubbell, R. Sukumar, E. G. Leigh, N. Manokoran, S. Loo de Lao, J. V. LaFrankie, and P. S. Ashton. 1998. Assessing forest diversity on small plots: Calibration using species-individual curves from 50-ha plots. Pages 247–268 in F. Dallmeier and J. A. Comiskey, editors, *Forest biodiversity research, modeling and monitoring*. UNESCO Man and the Biosphere Series, Vol. 20. Pearl River, NY: Parthenon.

Coûteaux, M. M., P. Bottner, and B. Berg. 1995. Liter decomposition, climate and litter quality. *Trends in Ecology and Evolution* 10:63–66.

Covich, A. P. 1988. Atyid shrimp in the headwaters of the Luquillo Mountains, Puerto Rico: Filter feeding in natural and artificial streams. *Verhandlungen der Internationale Vereinigung fur Theoretische und Angewandte Limnologie* 23:2108–2113.

Covich, A. P. 2006. Dispersal-limited biodiversity of tropical insular streams. *Polish Journal of Ecology* 54:523–547.

Covich, A. P., T. A. Crowl, and T. Heartsill-Scalley. 2006. Effects of drought and hurricane disturbances on headwater distributions of palaemonid river shrimp (*Macrobrachium* spp.) in the Luquillo Mountains, Puerto Rico. *Journal of North American Benthological Society* 25:99–107.

Covich, A. P, T. A. Crowl, C. L. Hein, M. J. Townsend, and W. H. McDowell. 2009. Predator-prey interactions in river networks: Comparing shrimp spatial refugia in two drainage basins. *Freshwater Biology* 54:450–465.

Covich, A. P., T. A. Crowl, S. L. Johnson, and M. Pyron. 1996. Distribution and abundance of tropical freshwater shrimp along a stream corridor: Response to disturbance. *Biotropica* 28:484–492.

Covich, A. P., and W. H. McDowell. 1996. The stream community. Pages 433–459 in D. P. Reagan and R. B. Waide, editors, *The food web of a tropical rain forest*. Chicago: University of Chicago Press.

Covich, A. P., M. A. Palmer, and T. A. Crowl. 1999. The role of benthic invertebrate species in freshwater ecosystems. *BioScience* 49:119–127.

Covich, A. P., and M. Stuiver. 1974. Changes in oxygen 18 as a measure of long-term fluctuations in tropical lake levels and molluscan populations. *Limnology and Oceanography* 19:682–691.

Cowley, G. T. 1970. Fleshy fungi in relation to irradiation and cutting in the Luquillo Experimental Forest. Pages F9–F13 in H. T. Odum and R. F. Pigeon, editors, *A tropical rain forest: A study of irradiation and ecology at El Verde, Puerto Rico*. Oak Ridge, TN: U.S. Atomic Energy Commission.

Cox, S. B., M. R. Willig, and F. N. Scatena. 2002. Variation in nutrient characteristics of surface soils from the Luquillo Experimental Forest of Puerto Rico: A multivariate perspective. *Plant and Soil* 247:189–198.

Crowl, T. A., W. H. McDowell, A. P. Covich, and S. L. Johnson. 2001. Freshwater shrimp effects on detrital processing and nutrients in a tropical headwater stream. *Ecology* 82:775–783.

Cuevas, E., S. Brown, and A. E. Lugo. 1991. Above- and belowground organic matter storage and production in a tropical pine plantation and a paired broadleaf secondary forest. *Plant and Soil* 135:257–268.

Cuevas, E., and E. Medina. 1986. Nutrient dynamics within Amazonian forest ecosystems. I. Nutrient flux in fine litter fall and efficiency of nutrient utilization. *Oecologia* 68:466–472.

Cuevas, E., and E. Medina. 1988. Nutrient dynamics within Amazonian forests. II. Fine root growth, nutrient availability and leaf litter decomposition. *Oecologia* 76:222–235.

Cusack, D. F., W. Silver, and W. H. McDowell. 2009. Biological nitrogen fixation in two tropical forests: Ecosystem-level patterns and effects of nitrogen fertilization. *Ecosystems* 12:1299–1315.

Dallmeier, F., J. A. Comiskey, and F. N. Scatena. 1998. Five years of forest dynamics following Hurricane Hugo in Puerto Rico's Luquillo Experimental Forest. Pages 231–248 in F. Dallmeier and J. A. Comiskey, editors, *Forest Biodiversity in North, Central and South America, and the Caribbean*. UNESCO Man and the Biosphere Series, Vol. 21. Pearl River, NY: Parthenon.

DeJong, R. J., J. A. T. Morgan, W. L. Paraense, J.-P. Pointier, M. Amarista, P. F. K. Ayeh-Kumi, A. Babiker, C. S. Barbosa, P. Brémond, A. P. Canese, C. P. de Souza, C. Dominguez, S. File, A. Gutierrez, R. N. Incani, T. Kawano, F. Kazibwe, J. Kpikpi, N. J. S. Lwambo, R. Mimpfoundi, F. Njiokou, J. N. Poda, M. Sene, L. E. Velásquez, M. Yong, C. M. Adema, B. V. Hofkin, G. M. Mkoji, and E. S. Loker. 2001. Evolutionary relationships and biogeography of *Biomphalaria* (Gastropoda: Planorbidae) with implications regarding its role as host of the human bloodfluke, *Schistosoma mansoni*. *Molecular Biology and Evolution* 18:2225–2239.

Dial, R., and J. Roughgarden. 1995. Experimental removal of insectivores from rain forest canopy: Direct and indirect effects. *Ecology* 76:1821–1834.

Donnelly, J. P., and J. D. Woodruff. 2007. Intense hurricane activity over the past 5,000 years controlled by El Niño and the West African monsoon. *Nature* 447:465–468.

Donnelly, T. W. 1990. Caribbean biogeography: Geological considerations on the problem of vicariance vs dispersal. Pages 595–609 in A. Azarolli, editor, *Biogeographical aspects of insularity*. Rome: Accademia Nazionale dei Linecei.

Donnelly, T. W. 1992. Geological setting and tectonic history of Mesoamerica. Pages 1–13 in D. Quintero and A. Aiello, editors, *Insects of Panama and Mesoamerica.* Oxford, England: Oxford University Press.

Drew, M. C. 1990. Sensing soil oxygen. *Plant, Cell, and Environment* 13:681–693.

Drewry, G. E. 1970. The role of amphibians in the ecology of a Puerto Rican rain forest. Pages 16–54 in R. G. Clements, G. E. Drewry, and R. J. Lavigne, editors, *The Rain Forest Project annual report, No. 147.* San Juan, PR: Puerto Rico Nuclear Center.

Edmisten, J. 1970. Preliminary studies of the nitrogen budget of a tropical rain forest. Pages H211–H215 in H. T. Odum and R. F. Pigeon, editors, *A tropical rain forest: A study of irradiation and ecology at El Verde, Puerto Rico.* Oak Ridge, TN: U.S. Atomic Energy Commission.

Erdman, D. S. 1961. Notes on the biology of the gobiid fish *Sicydium plumieri* in Puerto Rico. *Bulletin of Marine Science of the Gulf and Caribbean* 11:448–456.

Erdman, D. S. 1986. The green stream goby, *Sicydium plumieri. Tropical Fish Hobbyist* 34:70–74.

Eugster, W., R. Burkard, F. Holwerda, F. N. Scatena, and L. A. Bruijnzeel. 2006. Characteristics of fog and fogwater fluxes in a Puerto Rican elfin cloud forest. *Agricultural and Forest Meteorology* 139:288–306.

Ewel, J. J., and J. L. Whitmore. 1973. The ecological life zones of Puerto Rico and the U.S. Virgin Islands. Research Paper ITF-18. Río Piedras, PR: USDA Forest Service, Institute of Tropical Forestry.

Figueroa Colón, J. C. 1996. Phytogeographical trends, centers of high species richness and endemism, and the question of extinctions in the native flora of Puerto Rico. Pages 89–102 in J. C. Figueroa Colón, editor, *The scientific survey of Puerto Rico and the Virgin Islands: An eight-year reassessment of the islands' natural history.* Annals of the New York Academy of Sciences, Vol. 776. New York: New York Academy of Sciences.

Florentin, J.-M., R. Maurrasse, and S. Gautam. 1991. Impacts, tsunamis, and the Haitian Cretaceous-Tertiary Boundary layer. *Science* 252:1690–1692.

Formanowicz, D. R., Jr., M. M. Stewart, K. Townsend, F. H. Pough, and P. F. Brussard. 1981. Predation by giant crab spiders on the Puerto Rican frog Elutherodactylus coqui. *Journal of Herpetology* 37:125–129.

Forti, M. C., R. Boulet, A. J. Melfi, and C. Neal. 2000. Hydrogeochemsitry of a small catchment in northeastern Amazonia: A comparison between natural with deforested parts of the catchment (Serra Do Navio, Amapa State, Brazil). *Water, Air, and Soil Pollution* 118:263–279.

Fox, R. L. 1982. Some highly weathered soils of Puerto Rico, 3. Chemical properties. *Geoderma* 27:139–176.

Frangi, J. L. 1983. Las tierras pantanosas de la montaña Puertorriqueña. Pages 233–247 in A. E. Lugo, editor, *Los bosques de Puerto Rico* [*The forests of Puerto Rico*]. San Juan, PR: USDA Forest Service Institute of Tropical Forestry and Puerto Rico Department of Natural Resources.

Frangi, J. L., and A. E. Lugo. 1985. Ecosystem dynamics of a subtropical floodplain forest. *Ecological Monographs* 55:351–369.

Frizano, J., A. H. Johnson, D. R. Vann, and F. N. Scatena. 2002. Soil phosphorus fractionation during forest development on landslides scars in the Luquillo Mountains, Puerto Rico. *Biotropica* 34:17–26.

Gann, G. D., and K. A. Bradley. 2006. Floristic assessment of the northern karst belt of Puerto Rico, phase three. Final report submitted to the International Institute of Tropical Forestry by The Institute for Regional Conservation. Río Piedras, PR: USDA Forest Service, International Institute of Tropical Forestry.

García-Martinó, A. R., G. S. Warner, F. N. Scatena, and D. L. Civco. 1996. Rainfall, runoff and elevation relationships in the Luquillo Mountains of Puerto Rico. *Caribbean Journal of Science* 32:413–424.

García-Montiel, D. C., and F. N. Scatena. 1994. The effect of human activity on the structure and composition of a tropical forest in Puerto Rico. *Forest Ecology and Management* 63:57–78.

Garrison, R. W., and M. R. Willig. 1996. Arboreal invertebrates. Pages 183–246 in D. P. Reagan and R. B. Waide, editors, *The food web of a tropical rain forest.* Chicago: University of Chicago Press.

Giboda, M., E. A. Malek, and R. Correa. 1997. Human schistosomiasis in Puerto Rico: Reduced prevalence rate and absence of Biomphalaria glabrata. *American Journal of Tropical Medicine and Hygiene* 57:564–568.

Gleason, H. A., and M. T. Cook. 1927. Plant ecology of Porto Rico. *Scientific Survey of Porto Rico and the Virgin Islands* 7:1–173. New York: New York Academy of Sciences.

González, G., E. García, V. Cruz, S. Borges, M. Zalamea, and M. M. Rivera. 2007. Earthworm communities along an elevation gradient in Northeastern Puerto Rico. *European Journal of Soil Biology* 43:S24–S32.

González, G., and T. R. Seastedt. 2001. Soil fauna and plant litter decomposition in tropical and subalpine forests. *Ecology* 82:955–964.

González, G., C. Yu Huang, X. Zou, and C. Rodríguez. 2006. Earthworm invasions in the tropics. *Biological Invasions* 8:1247–1256.

González, G., and X. Zou. 1999a. Plant and litter influences on earthworm abundance and community structure in a tropical wet forest. *Biotropica* 31:486–493.

González, G., and X. M. Zou. 1999b. Earthworm influence on N availability and the growth of *Cecropia schreberiana* in tropical pasture and forest soils. *Pedobiologia* 43:1–6.

González, G., X. Zou, and S. Borges. 1996. Earthworm abundance and species composition in abandoned tropical croplands: Comparisons of tree plantations and secondary forests. *Pedobiologia* 40:385–391.

González, G., X. M. Zou, A. Sabat, and N. Fetcher. 1999. Earthworm abundance and distribution pattern in contrasting plant communities within a tropical wet forest in Puerto Rico. *Caribbean Journal of Science* 35:93–100.

Gould, W., G. González, and G. Carrero. 2006. Native, endemic, non-native and critical plants in eight forest types along an elevational gradient in Puerto Rico: Implications for conservation. *Journal of Vegetation Science* 17:653–664.

Graham, A. 2003a. Geohistory models and Cenozoic paleoenvironments of the Caribbean region. *Systematic Botany* 28:378–386.

Graham, A. 2003b. Historical phytogeography of the Greater Antilles. *Brittonia* 55:357–383.

Graham, A., and D. M. Jarzen. 1969. Studies in neotropical paleobotany I. The Oligocene communities of Puerto Rico. *Annals of the Missouri Botanical Gardens* 56:308–357.

Greathouse, E., and C. M. Pringle. 2006. Does the river continuum concept apply on a tropical island? Longitudinal variation in a Puerto Rican stream. *Canadian Journal of Fisheries and Aquatic Sciences* 63:134–152.

Greathouse, E. A., C. M. Pringle, and J. G. Holmquist. 2006a. Conservation and management of migratory fauna: Dams in tropical streams of Puerto Rico. *Aquatic Conservation-Marine and Freshwater Ecosystems* 16:695–712.

Greathouse, E. A., C. M. Pringle, and W. H. McDowell. 2006b. Do small-scale exclosure/enclosure experiments predict effects of large-scale extirpation of freshwater migratory fauna? *Oecologia* 149:709–717.

Greathouse, E. A., C. M. Pringle, W. H. McDowell, and J. G. Holmquist. 2006c. Indirect upstream effects of dams: Consequences of migratory consumer extirpation in Puerto Rico. *Ecological Applications* 16:339–352.

Greenland, D. 1999. ENSO-related phenomena at Long Term Ecological Research sites. *Physical Geography* 20:491–507.

Griffiths, T. A., and D. Klingener. 1988. On the distribution of Greater Antillean bats. *Biotropica* 20:240–251.

Grimm, U., and H. W. Fassbender. 1981. Ciclos bioquímicas en un ecosistema forestal de los Andes Occidentales de Venezuela I. Inventario de las reservas orgánicas e minerales (N, P, K, Ca, Mg, Fe, Al, Na) [Biogeochemical cycles in a forested ecosystem of the Western Andes of Venezuela. I. Inventory of organic matter and nutrient standing stocks (N, P, K, Ca, Mg, Fe, Al, Na)]. *Turrialba* 31:27–37.

Grip, H., A. Malmer, and F. K. Wong. 1994. Converting tropical rain forest to forest plantation in Sabah, Malaysia 1. Dynamics and net losses of nutrients in control catchment streams. *Hydrological Processes* 8:179–194.

Grubb, P. J. 1977. Control of forest growth and distribution on wet tropical mountains: With special reference to mineral nutrition. *Annual Review of Ecology and Systematics* 8:83–107.

Grubb, P. J., and P. J. Edwards. 1982. Studies of mineral cycling in montane rain forest in New Guinea. III. The distribution of mineral elements in the above-ground material. *Journal of Ecology* 70:623–648.

Guilderson, T. P., R. G. Fairbanks, and J. L. Rubenstone. 1994. Tropical temperature variations since 20,000 years ago: Modulating interhemispheric climate change. *Science* 263:663–665.

Guzmán-Grajales, S. M., and L. R. Walker. 1991. Differential seedling responses to litter after Hurricane Hugo in the Luquillo Experimental Forest, Puerto Rico. *Biotropica* 23:407–413.

Hansen, M., and B. A. Richardson. 1998. A new species of *Omicrus* Sharp (Coleoptera: Hydrophilidae) from Puerto Rico and its larva, the first known larva of Omicrini. *Systematic Entomology* 23:1–8.

Harris, N. L. 2006. The carbon balance of a tropical forested ecosystem along an elevational gradient in the Luquillo Mountains, Puerto Rico. Ph.D. dissertation. State University of New York, Syracuse, NY.

Hartshorn, G., L. Nicolait, L. Hartshorn, G. Bevier, R. Brightman, J. Cal, A. Cawich, W. Davidson, R. Dubois, C. Dyer, J. Gibson, W. Hawley, J. Leonard, R. Nicolait, D. Weyer, H. White, and C. Wright. 1984. *Belize country environmental profile: A field study*. Belize City, Belize: Robert Nicolait and Associates, Ltd.

Heartsill-Scalley, T., F. N. Scatena, C. Estrada, W. H. McDowell, and A. E. Lugo. 2007. Disturbance and long-term patterns of rainfall and throughfall nutrient fluxes in a subtropical forest in Puerto Rico. *Journal of Hydrology* 333:472–485.

Heatwole, H., and F. MacKenzie. 1967. Herpetogeography of Puerto Rico. IV. Paleogeography, faunal similarity and endemism. *Evolution* 21:429–438.

Hedges, S. B. 2006. Paleogeography of the Antilles and origin of West Indian terrestrial vertebrates. *Annals of the Missouri Botanical Garden* 93:231–244.

Hedges, S. B., C. A. Hass, and L. R. Maxoson. 1994. Towards a biogeography of the Caribbean. *Cladistics* 10:43–55.

Hedin, L. O., J. C. von Fischer, N. E. Ostrom, B. P. Kennedy, M. G. Brown, and G. P. Robertson. 1998. Thermodynamic constraints on nitrogen transformations and other biogeochemical processes at soil-stream interfaces. *Ecology* 79:684–703.

Hein, C. L., A. S. Pike, J. F. Blanco, A. P. Covich, F. N. Scatena, C. P. Hawkins, and T. A. Crowl. 2011. Effects of coupled natural and anthropogenic factors on the community structure of diadromous fish and shrimp species in tropical island streams. *Freshwater Biology* 56:1002–1015.

Heinicke, P. M., W. E. Duellman, and S. B. Hedges. 2007. Major Caribbean and Central American frog faunas originated by ancient oceanic dispersal. *Proceedings of the National Academy of Sciences* 104:10092–10097.

Henderson, A., G. Galeano, and B. Bernal. 1995. *Field guide to the palms of the Americas*. Princeton, NJ: Princeton University Press.

Heneghan, L., D. C. Coleman, X. Zou, D. A. Crossley, Jr., and B. L. Haines. 1999. Soil microarthropod contributions to decomposition dynamics: Tropical-temperate comparisons of a single substrate. *Ecology* 80:1873–1882.

Hildebrand, A. R., and W. V. Boynton. 1990. Proximal Cretaceous-Tertiary boundary impact deposits in the Caribbean. *Science* 248:843–846.

Hodell, D. A., J. H. Curtis, G. A. Jones, A. Higuera-Gundy, M. Brenner, M. W. Binford, and K. T. Dorsey. 1991. Reconstruction of Caribbean climate change over the past 10,500 years. *Nature* 352:790–793.

Hodge, E. T. 1920. Geology of the Coamo-Guyama district: Introduction to the geology of Porto Rico. Pages 111–228 in *Scientific survey of Puerto Rico and the Virgin Islands*. Volume 1. New York: New York Academy of Sciences.

Holler, J. R., and G. T. Cowley. 1970. Response of soil, root, and litter microfungal populations to radiation. Pages F35–F39 in H. T. Odum and R. F. Pigeon, editors, *A tropical rain forest: A study of irradiation and ecology at El Verde, Puerto Rico*. Oak Ridge, TN: U.S. Atomic Energy Commission.

Holmquist, J. G., J. M. Schmidt-Gengenbach, and B. B. Yoshioka. 1998. High dams and marine-freshwater linkages: Effects on native and introduced fauna in the Caribbean. *Conservation Biology* 12:621–630.

Holwerda, F., R. Burkard, W. Eugster, F. N. Scatena, A. G. C. A. Meesters, and L. A. Bruijnzeel. 2006. Estimating fog deposition at a Puerto Rican elfin cloud forest site: Comparison of the water budget and eddy covariance methods. *Hydrological Processes* 20:2669–2692.

Howard, R. A. 1968. The ecology of an elfin forest in Puerto Rico. 1. Introduction and composition studies. *Journal of the Arnold Arboretum* 49:381–418.

Huhndorf, S. M., and D. J. Lodge. 1997. Host specificity among wood inhabiting pyrenomycetes (Fungi, Ascomycetes) in a wet tropical forest. *Tropical Ecology* 38:307–315.

Iturralde-Vinent, M. A., and R. D. E. MacPhee. 1999. Paleogeography in the Caribbean region: Implications for Cenozoic biogeography. Bulletin No. 238. New York: American Museum of Natural History.

Johnston, M. H. 1992. Soil-vegetation relationships in a tabonuco forest community in the Luquillo Mountains of Puerto Rico. *Journal of Tropical Ecology* 8:253–263.

Jordan, C. F. 1985. *Nutrient cycling in tropical forest ecosystems*. Chichester, England: John Wiley and Sons.

Kangas, P. 1992. Root regrowth in a subtropical wet forest in Puerto Rico. *Biotropica* 24:463–465.

Kanzaki, M., M. Hara, T. Yamakura, T. Ohkubo, M. N. Tamura, K. Sringernyuang, P. Sahunalu, S. Teejuntuk, and S. Bunyavejchewin. 2004. Doi Inthanon Forest Dynamics Plot, Thailand. Pages 474–481 in E. C. Losos and E. G. Leigh, Jr., editors, *Tropical forest diversity and dynamism: Results from a long-term tropical forest network*. Chicago: University of Chicago Press.

Kikkert, D. A., T. A. Crowl, and A. P. Covich. 2009. Upstream migration of amphidromous shrimps in the Luquillo Experimental Forest, Puerto Rico: Temporal patterns and environmental cues. *Journal of the North American Benthological Society* 28:233–246.

La Caro, F., and R. L. Rudd. 1985. Leaf litter disappearance rates in Puerto Rican montane rain forest. *Biotropica* 17:269–276.

Laessøe, T., and D. J. Lodge. 1994. Three host-specific *Xylaria* species. *Mycologia* 86:436–446.

Lal, R., and P. S. Sanchez, editors. 1992. *Myths and science of soils of the tropics.* SSSA Special Publication No. 29. Madison, WI: Soil Science Society of America.

Larsen, M. C. 1997. Tropical geomorphology and geomorphic work: A study of geomorphic processes and sediment and water budgets in montane humid-tropical forested and developed watersheds, Puerto Rico. Ph.D. dissertation. University of Colorado, Boulder, CO.

Larsen, M. C. 2000. Analysis of 20th century rainfall and streamflow to characterize drought and water resources in Puerto Rico. *Physical Geography* 21:494–421.

Larsen, M. C., and A. J. Torres-Sanchez. 1996. Geographic relations of landslide distribution and assessment of landslide hazards in the Blanco, Cibuco, and Coamo basins, Puerto Rico. Water Resources Investigations Report 95–4029. San Juan, PR: U.S. Geological Survey.

Lawton, R. O. 1982. Wind stress and elfin stature in a montane rain forest tree: An adaptive explanation. *American Journal of Botany* 69:1224–1230.

Lazell, J. 2005. *Island: Fact and theory in nature.* Berkeley, CA: University of California Press.

Leigh, E. G., Jr., S. Loo de Lao, R. Condit, S. P. Hubbell, R. B. Foster, and R. Pérez. 2004. Barro Colorado Forest Dynamics Plot, Panama. Pages 451–463 in E. C. Losos and E. G. Leigh, Jr., editors, *Tropical forest diversity and dynamism: Results from a long-term tropical forest network.* Chicago: University of Chicago Press.

Lewis, W. M., Jr., J. M. Melack, W. H. McDowell, M. McClain, and J. E. Richey. 1999. Nitrogen yields from undisturbed watersheds in the Americas. *Biogeochemistry* 46:149–162.

Lewis, W. M., Jr., and J. F. Saunders III. 1989. Concentration and transport of dissolved and suspended substances in the Orinoco River. *Biogeochemistry* 7:203–240.

Li, Y. 1998. Soil carbon dynamics in tropical wet forests in Puerto Rico. Ph.D. dissertation. University of Puerto Rico, Río Piedras, PR.

Licht, P. 1974. Response of *Anolis* lizards to food supplementation in nature. *Copeia* 1974:215–221.

Likens, G. E., F. H. Bormann, R. S. Pierce, J. S. Eaton, and N. M. Johnson. 1977. *Biogeochemistry of a forested ecosystem.* New York: Springer-Verlag.

Liogier, A., and G. Martorell. 1982. Flora of Puerto Rico and adjacent islands: A systematic synopsis. Río Piedras, PR: University of Puerto Rico Press.

Lister, B. C. 1981. Seasonal relationships of rain forest anoles. *Ecology* 62:1548–1560.

Liu, Z. G., and X. M. Zou. 2002. Exotic earthworms accelerate plant litter decomposition in a Puerto Rican pasture and a wet forest. *Ecological Applications* 12:1406–1417.

Lodge, D. J. 1993. Nutrient cycling by fungi in wet tropical forests. Pages 37–57 in S. Isaac, J. C. Frankland, R. Watling, and A. J. S. Whalley, editors, *Aspects of Tropical Mycology.* British Mycological Society Symposium Series 19. Cambridge, England: Cambridge University Press.

Lodge, D. J. 1996. Microorganisms. Pages 53–108 in D. P. Regan and R. B. Waide, editors, *The food web of a tropical forest.* Chicago: University of Chicago Press.

Lodge, D. J. 1997. Factors related to diversity of decomposer fungi in tropical forests. *Biodiversity and Conservation* 6:681–688.

Lodge, D. J., and C. E. Asbury. 1988. Basidiomycetes reduce export of organic matter from forest slopes. *Mycologia* 80:888–890.

Lodge, D. J., T. J. Baroni, and S. A. Cantrell. 2002. Basidiomycetes of the Greater Antilles project. Pages 45–60 in R. Watling, J. C. Frankland, A. M. Ainsworth, S. Isaac, and

C. Robinson, editors, *Tropical Mycology, Vol. 1, Macromycetes*. Egham, England: CABI Publishing.

Lodge, D. J., and S. Cantrell. 1995. Fungal communities in wet tropical forests: Variation in time and space. *Canadian Journal of Botany* 73:1391–1398.

Lodge, D. J., I. Chapela, G. Samuels, F. A. Uecker, D. Desjardin, E. Horak, O. K. Miller, Jr., G. L. Hennbert, C. A. Decock, J. Ammirati, H. H. Burdsall, Jr., P. M. Kirk, D. W. Minter, R. Halling, T. Laessøe, G. Mueller, F. Oberwinkler, D. N. Pegler, B. Spooner, R. H. Petersen, J. D. Rogers, L. Ryvarden, R. Watling, E. Turnbull, and A. J. S. Whalley. 1995. A survey of patterns in fungal diversity. Pages 157–173 in C. Scheldegger and P. Wolseley, editors, Lichen conservation. Proceedings of the symposium "Lichens—A strategy for conservation." *Mitteilungen der Eidgenossischen Forschungsanstalt fur Wald, Schnee und Landschaft*, Vol. 70.

Lodge, D. J., W. H. McDowell, J. Macy, S. K. Ward, R. Leisso, C. Campos, and K. Kühnert. 2008. Distribution and role of mat-forming saprobic basidiomycetes in a tropical forest. Pages 195–208 in L. Boddy, J. C. Frankland, and P. Van West, editors, *Ecology of saprotrophic basidiomycetes*. London: Elsevier.

Lodge, D. J., W. H. McDowell, and C. P. McSwiney. 1994. The importance of nutrient pulses in tropical forests. *Trends in Ecology and Evolution* 9:384–387.

Lodge, D. J., F. N. Scatena, C. E. Asbury, and M. J. Sanchez. 1991. Fine litterfall and related nutrient inputs resulting from Hurricane Hugo in subtropical wet and lower montane rain forests of Puerto Rico. *Biotropica* 23:364–372.

López-Venegas, A. M., U. S. ten Brink, and E. L. Geist. 2008. Submarine landslide as the source for the October 11, 1918 Mona Passage tsunami: Observations and modeling. *Marine Geology* 254:35–46.

Lugo, A. E. 1987. Are island ecosystems different from continental ecosystems? *Acta Científica* 1:48–54.

Lugo, A. E. 1992. Comparison of tropical tree plantations with secondary forests of similar age. *Ecological Monographs* 62:1–41.

Lugo, A. E. 2004. The outcome of alien tree invasions in Puerto Rico. *Frontiers in Ecology and the Environment* 2:265–273.

Lugo, A. E., A. Bokkestijn, and F. N. Scatena. 1995. Structure, succession, and soil chemistry of palm forests in the Luquillo Experimental Forest. Pages 142–177 in A. E. Lugo and C. Lowe, editors, *Tropical forests: Management and ecology*. New York: Springer-Verlag.

Lugo, A. E., and F. N. Scatena. 1992. Epiphytes and climate change research in the Caribbean: A proposal. *Selbyana* 13:123–130.

Lugo, A. E., and F. N. Scatena. 1995. Ecosystem-level properties of the Luquillo Experimental Forest with emphasis on the tabonuco forest. Pages 59–108 in A. E. Lugo and C. Lowe, editors, *Tropical forests: Management and ecology*. New York: Springer-Verlag.

Lugo, A. E., F. N. Scatena, W. Silver, S. Molina Colón, and P. G. Murphy. 2002. Resilience of tropical wet and dry forests in Puerto Rico. Pages 195–225 in L. H. Gunderson and L. Pritchard, Jr., editors, *Resilience and the behavior of large-scale systems*. Washington, DC: Island Press.

Lyford, W. H. 1969. The ecology of an elfin forest in Puerto Rico, soil root, and earthworm relationships. *Journal of the Arnold Arboretum* 50:210–225.

MacArthur, R. H., and E. O. Wilson. 1967. *The theory of island biogeography*. Princeton, NJ: Princeton University Press.

Madden, N. T. 2004. Scaling up: Nitrogen flux at the stream-soil interface in the Rio Icacos, PR. M.S. thesis. University of New Hampshire, Durham, NH.

Malmgren, B. A., A. Winter, and D. Chen. 1998. El Niño—Southern Oscillation and North Atlantic Oscillation control of climate in Puerto Rico. *Journal of Climate* 11:2713–2717.

March, J. G., J. P. Benstead, C. M. Pringle, and M. W. Ruebel. 2001. Linking shrimp assemblages with rates of detrital processing along an elevational gradient in a tropical stream. *Canadian Journal of Fisheries and Aquatic Sciences* 58:470–478.

March, J. G., J. P. Benstead, C. M. Pringle, and F. N. Scatena. 1998. Migratory drift of larval amphidromous shrimps in two tropical streams, Puerto Rico. *Freshwater Biology* 40:1–14.

March, J. G., and C. M. Pringle. 2003. Food web structure and basal resource utilization along a tropical island stream continuum, Puerto Rico. *Biotropica* 35:84–93.

March, J. G., C. M. Pringle, M. J. Townsend, and A. I. Wilson. 2002. Effects of freshwater shrimp assemblages on benthic communities along an altitudinal gradient of a tropical island stream. *Freshwater Biology* 47:377–390.

Martinelli, L. A., M. C. Piccolo, A. R. Townsend, P. M. Vitousek, E. Cuevas, W. McDowell, G. P. Robertson, O. C. Santos, and K. Treseder. 1999. Nitrogen stable isotopic composition of leaves and soil: Tropical versus temperate forests. *Biogeochemistry* 46:45–65.

Masson, D. G., and K. M. Scanlon. 1991. The neotectonic setting of Puerto Rico. *Geological Society of America Bulletin* 103:144–154.

Maurrassee, F. J.-M. R. 1991. A plate-tectonic model for the Mesozoic and Early Cenozoic history of the Caribbean plate. *Tectonophysics* 296:269–291.

McClain, M. E., E. W. Boyer, C. L. Dent, S. E. Gergel, N. B. Grimm, P. M. Groffman, S. C. Hart, J. W. Harvey, C. A. Johnston, E. Mayorga, W. H. McDowell, and G. Pinay. 2003. Biogeochemical hot spots and hot moments at the interface of terrestrial and aquatic ecosystems. *Ecosystems* 6:301–312.

McDowell, W. H. 1998. Internal nutrient fluxes in a tropical rain forest. *Journal of Tropical Ecology* 14:521–536.

McDowell, W. H. 2001. Hurricanes, people, and riparian zones: Controls on nutrient losses from forested Caribbean watersheds. *Forest Ecology and Management* 154:443–451.

McDowell, W. H. 2002. Nutrient flux from tropical rain forests. Pages 149–163 in R. S. Ambasht and N. K. Ambasht, editors, *Modern trends in applied terrestrial ecology*. New York: Kluwer Academic.

McDowell, W. H., and C. E. Asbury. 1994. Export of carbon, nitrogen and major ions from three tropical montane watersheds. *Limnology and Oceanography* 39:111–125.

McDowell, W. H., W. B. Bowden, and C. E. Asbury. 1992. Riparian nitrogen dynamics in two geomorphologically distinct tropical rain forest watersheds-subsurface solute patterns. *Biogeochemistry* 18:53–75.

McDowell, W. H., C. Gines-Sanchez, C. E. Asbury, and C. R. Ramos-Perez. 1990. Influence of seasalt aerosols and long range transport on precipitation chemistry at El Verde, Puerto Rico. *Atmospheric Environment* 24A:2813–2821.

McGroddy, M., and W. L. Silver. 2000. Variations in belowground carbon storage and soil CO_2 flux rates along a wet tropical climate gradient. *Biotropica* 32:614–624.

McMahan, E. A. 1996. Termites. Pages 109–135 in D. P. Reagan and R. B. Waide, editors, *The food web of a tropical rain forest*. Chicago: University of Chicago Press.

McSwiney, C. P. 1999. Controls on spatial and temporal variability in nitrous oxide fluxes across a tropical rainforest ecosystem in the Luquillo Experimental Forest, Puerto Rico. Ph.D. dissertation. University of New Hampshire, Durham, NH.

McSwiney, C. P., W. H. McDowell, and M. Keller. 2001. Distribution of nitrous oxide and regulators of its production across a tropical rainforest catena in the Luquillo Experimental Forest, Puerto Rico. *Biogeochemistry* 56:265–286.

Medina, E., and E. Cuevas. 1989. Patterns of nutrient accumulation and release in Amazonian forests of the upper Rio Negro basin. Pages 217–240 in J. Proctor, editor, *Mineral nutrients in tropical forest and savanna ecosystems*. Special Publications Series of the British Ecological Society, No. 9. London: Blackwell Scientific Publications.

Meentemeyer, V. 1978. Macroclimate and lignin control of litter decomposition rates. *Ecology* 59:465–472.

Melillo, J. M. 1982. Nitrogen and lignin control of hardwood leaf litter decomposition dynamics. *Ecology* 63:621–626.

Mellon, M. T. 2000. Records of Pleistocene terrestrial climate change in Puerto Rico archived in stacked paleosols. M.S. thesis. University of Pennsylvania, Philadelphia, PA.

Merriam, J. L., W. H. McDowell, J. L. Tank, W. M. Wollheim, C. L. Crenshaw, and S. L. Johnson. 2002. Characterizing nitrogen dynamics, retention and transport in a tropical rainforest stream using an in situ ^{15}N addition. *Freshwater Biology* 47:143–160.

Miller, R. M., and D. J. Lodge. 1997. Fungal responses to disturbance: Agriculture and forestry. Pages 65–84 in K. Esser, P. A. Lemke, and D. T. Wicklow, editors, *The mycota*. Berlin: Springer-Verlag.

Minshall, G. W., K. W. Cummins, R. C. Petersen, C. E. Cushing, D. A. Bruns, J. R. Sedell, and R. L. Vannote. 1985. Developments in stream ecosystem theory. *Canadian Journal of Fisheries and Aquatic Sciences* 42:1045–1055.

Mitchell, R. C. 1954. A survey of the geology of Puerto Rico. Technical Paper 13. Río Piedras, PR: University of Puerto Rico Agricultural Experiment Station.

Moulin, C., C. E. Lambert, F. Dulac, and U. Dayan. 1997. Control of atmospheric export of dust from North Africa by the North Atlantic Oscillation. *Nature* 387:691–694.

Mueller-Dombois, D., P. M. Vitousek, and K. W. Bridges. 1984. *Canopy dieback and ecosystem processes in Pacific forests*. Hawaii Botanical Science Paper 44. Honolulu, HI: University of Hawaii.

Muhs, D. R., J. R. Budahn, J. M. Prospero, and S. N. Carey. 2007. Geochemical evidence for African dust inputs to soils of western Atlantic islands: Barbados, the Bahamas, and Florida. *Journal of Geophysical Research-Earth Surface* 112:F02009.

Nadkarni, N. M. 1981. Canopy roots: Convergent evolution in nutrient cycles. *Science* 214:1023–1024.

Newbold, J. D., B. W. Sweeney, J. K. Jackson, and L. A. Kaplan. 1995. Concentrations and export of solutes from six montane streams in northwestern Costa Rica. *Journal of the North American Benthological Society* 14:21–37.

Nieves, L. O. 1998. Ecological study of the freshwater stream fishes of the upper Mameyes River (Bisley) in the Luquillo Mountains of Puerto Rico. Ph.D. dissertation. George Mason University, Fairfax, VA.

O'Connor, P. J. 1998. Habitat selection in insular tropical streams: Macroinvertebrate responses to a riparian invasion by non-indigenous bamboos. M.S. thesis. Colorado State University, Ft. Collins, CO.

O'Connor, P. J., A. P. Covich, F. N. Scatena, and L. L. Loope. 2000. Non-indigenous bamboo along head water streams of the Luquillo Mountain, Puerto Rico: Leaf fall, aquatic leaf decay and patterns of invasion. *Journal of Tropical Ecology* 16:499–516.

Odum, H. T. 1970a. Rain forest structure and mineral cycling homeostasis. Pages IH3–H52 in H. T. Odum and R. F. Pigeon, editors, *A tropical rain forest: A study of irradiation and ecology at El Verde, Puerto Rico*. Oak Ridge, TN: U.S. Atomic Energy Commission.

Odum, H. T. 1970b. Summary: An emerging view of the ecological system at El Verde. Pages I191–I281 in H. T. Odum and R. F. Pigeon, editors, *A tropical rain forest: A study of irradiation and ecology at El Verde, Puerto Rico*. Oak Ridge, TN: U.S. Atomic Energy Commission.

Odum, H. T., and R. F. Pigeon. 1970. A tropical rain forest: A study of irradiation and ecology at El Verde, Puerto Rico. Oak Ridge, TN: U.S. Atomic Energy Commission.

Odum, H. T., and J. Ruiz-Reyes. 1970. Holes in leaves and the grazing control mechanism. Pages I69–I70 in H. T. Odum and R. F. Pigeon, editors, *A tropical rain forest: A study of irradiation and ecology at El Verde, Puerto Rico*. Oak Ridge, TN: U.S. Atomic Energy Commission.

Olander, L. P., F. N. Scatena, and W. L. Silver. 1998. Impacts of disturbance initiated by road construction in a subtropical cloud forest in the Luquillo Experimental Forest, Puerto Rico. *Forest Ecology and Management* 109:33–49.

Olden, J. D., and N. L. Poff. 2003. Redundancy and the choice of hydrologic indices for characterizing streamflow regimes. *River Research and Applications* 19:101–121.

Ortiz-Zayas, J. R., E. Cuevas, O. L. Mayol-Bracero, L. Donoso, I. Trebs, D. Figueroa-Nieves, and W. H. McDowell. 2006. Urban influences on the nitrogen cycle in Puerto Rico. *Biogeochemistry* 79:109–133.

Ortiz-Zayas, J. R., W. M. Lewis, J. F. Saunders, J. H. McCutchan, and F. N. Scatena. 2005. Metabolism of a tropical rainforest stream. *Journal of the North American Benthological Society* 24:769–783.

Ostertag, R., F. N. Scatena, and W. L. Silver. 2003. Forest floor decomposition following hurricane litter inputs in several Puerto Rican forests. *Ecosystems* 6:261–273.

Ovington, J. D., and J. S. Olson. 1970. Biomass and chemical content of El Verde lower montane rain forest plants. Pages H53–H78 in H. T. Odum and R. F. Pigeon, editors, *A tropical rain forest: A study of irradiation and ecology at El Verde, Puerto Rico*. Oak Ridge, TN: U.S. Atomic Energy Commission.

Padgett, D. E. 1976. Leaf decomposition by fungi in a tropical rainforest stream. *Biotropica* 8:166–178.

Parrotta, J. A., and D. J. Lodge. 1991. Fine root dynamics in a subtropical wet forest following hurricane disturbance in Puerto Rico. *Biotropica* 23:343–347.

Parton, W., W. L. Silver, I. C. Burke, L. Grassens, M. E. Harmon, W. S. Currie, J. Y. King, E. C. Adair, L. A. Brandt, S. C. Hart, and B. Fasth. 2007. Global-scale similarities in nitrogen release patterns during long-term decomposition. *Science* 315:361–364.

Peters, N. E., J. B. Shanley, B. T. Aulenbach, R. M. Webb, D. H. Campbell, R. Hunt, M. C. Larsen, R. F. Stallard, J. Troester, and J. F. Walker. 2006. Water and solute mass balance of five small, relatively undisturbed watersheds in the US. *Science of the Total Environment* 358:221–242.

Pett-Ridge, J. C. 2009. Contributions of dust to phosphorus cycling in tropical forests of the Luquillo Mountains, Puerto Rico. *Biogeochemistry* 94:63–80.

Pfeiffer, W. J. 1996. Litter invertebrates. Pages 137–182 in D. P. Reagan and R. B. Waide, editors, *The food web of a tropical rain forest*. Chicago: University of Chicago Press.

Pimm, S. L. 1982. *Food webs*. New York: Chapman and Hall.

Pindell, J. L. 1994. Evolution of the Gulf of Mexico and the Caribbean. Pages 13–39 in S. K. Donovan and T. A. Jackson, editors, *Caribbean geology: An introduction*. Kingston, Jamaica: The University of the West Indies Publishers' Association.

Pindell, J. L., and L. Kennan. 2002. Plate model for the Caribbean. http://www.ig.utexas.edu/CaribPlate/forum/pindell_kennan.htm. Accessed February 22, 2008.

Polishook, J. D., G. F. Bills, and D. J. Lodge. 1996. Microfungi from decaying leaves of two rain forest trees in Puerto Rico. *Journal of Industrial Microbiology* 17:284–294.

Post, D. A., and J. A. Jones. 2001. Hydrologic regimes of forested, mountainous, headwater basins in New Hampshire, North Carolina, Oregon, and Puerto Rico. *Advances in Water Resources* 24:1195–1210.

Pringle, C. M. 1996. Atyid shrimps (Decapoda: Atyidae) influence the spatial heterogeneity of algal communities over different scales in tropical montane streams, Puerto Rico. *Freshwater Biology* 35:125–140.

Pringle, C. M. 1997. Exploring how disturbance is transmitted upstream: Going against the flow. *Journal of the North American Benthological Society* 16:425–438.

Pringle, C. M., and G. A. Blake. 1994. Quantitative effects of atyid shrimp (Decapoda: Atyidae) on the depositional environment in a tropical stream: Use of electricity for experimental exclusion. *Canadian Journal of Fisheries and Aquatic Sciences* 51:1443–1450.

Pringle, C. M., G. A. Blake, A. P. Covich, K. M. Buzby, and A. M. Finley. 1993. Effects of omnivorous shrimp in a montane tropical stream: Sediment removal, disturbance of sessile invertebrates, and enhancement of understory algal biomass. *Oecologia* 93:1–11.

Pringle, C. M., N. Hemphill, W. H. McDowell, A. Bednarek, and J. G. March. 1999. Linking species and ecosystems: Different biotic assemblages cause interstream differences in organic matter. *Ecology* 80:1860–1872.

Prospero, J. M., and R. T. Nees. 1986. Impact of the North African drought and El Niño on mineral dust in the Barbados trade winds. *Nature* 320:735–738.

Pyron, M., and A. P. Covich. 2003. Migration patterns, densities, and growth of *Neritina punctulata* snails in Río Espíritu Santo and Río Mameyes, Northeastern Puerto Rico. *Caribbean Journal of Science* 39:338–347.

Raffaele, H., J. Wiley, O. Garrido, A. Keith, and J. Raffaele. 1998. *A guide to the birds of the West Indies*. Princeton, NJ: Princeton University Press.

Raffaele, H. A., M. J. Velez, R. Cotte, J. J. Whelan, E. R. Keil, and W. Cupiano. 1973. Rare and endangered animals of Puerto Rico: A committee report. Hyattsville, MD: U.S. Soil Conservation Service.

Reagan, D. P. 1992. Congeneric species distribution and abundance in a three-dimensional habitat: The rain forest anoles of Puerto Rico. *Copeia* 1992:392–403.

Reagan, D. P. 1996. Anoline lizards. Pages 321–345 in D. P. Reagan and R. B. Waide, editors, *The food web of a tropical rain forest*. Chicago: University of Chicago Press.

Reagan, D. P., G. R. Camilo, and R. B. Waide. 1996. The community food web: Major properties and patterns of organization. Pages 461–510 in D. P. Reagan and R. B. Waide, editors, *The food web of a tropical rain forest*. Chicago: University of Chicago Press.

Reagan, D. P., and R. B. Waide. 1996. *The food web of a tropical rain forest*. Chicago: University of Chicago Press.

Reed, A. M. 1998. Scale-dependent influences of riparian processes on the dominant detritivore of a headwater stream in Puerto Rico. M.S. thesis. Colorado State University, Ft. Collins, CO.

Rice, K., N. Brokaw, and J. Thompson. 2004. Liana abundance in a Puerto Rican forest. *Forest Ecology and Management* 190:33–41.

Richards, P. W. 1996. *The tropical rain forest*, 2nd edition. Cambridge, England: Cambridge University Press.

Richardson, B. A. 1999. The bromeliad microcosm and the assessment of faunal diversity in a Neotropical forest. *Biotropica* 31:321–336.

Richardson, B. A., M. J. Richardson, F. N. Scatena, and W. H. McDowell. 2000. The effects of nutrient availability and other elevational changes on bromeliad populations and the phytotelm invertebrate communities in a humid tropical forest in Puerto Rico. *Journal of Tropical Ecology* 16:167–188.

Richardson, B. A, M. J. Richardson, and F. N. Soto-Adames. 2005. Separating the effects of forest type and elevation on the diversity of litter invertebrate communities in a humid tropical forest in Puerto Rico. *Journal of Animal Ecology* 74:926–936.

Richter, D. D., and L. I. Babbar. 1991. Soil diversity in the tropics. *Advances in Ecological Research* 21:315–389.

Rivero, J. A. 1978. Los anfibios y reptiles de Puerto Rico [The amphibians and reptiles of Puerto Rico]. San Juan, PR: Editorial Universitaria, Universidad de Puerto Rico.

Ruan, H. H., Y. Q. Li, and X. M. Zou. 2005. Soil communities and plant litter decomposition as influenced by forest debris: Variation across tropical riparian and upland sites. *Pedobiologia* 49:529–538.

Ruan, H. H., X. M. Zou, F. N. Scatena, and J. K. Zimmerman. 2004. Asynchronous fluctuation of soil microbial biomass and plant litterfall in a tropical wet forest. *Plant and Soil* 260:147–154.

Sánchez, P. A. 1976. *Properties and management of soils in the tropics*. New York: John Wiley and Sons.

Sánchez, Y., X. M. Zou, S. Borges, and H. H. Ruan. 2003. Recovery of native earthworms in abandoned tropical pastures. *Conservation Biology* 17:999–1006.

Sanhueza, E., and M. Santana. 1994. Atmospheric wet depositions in tropical America. *Israel Journal of Chemistry* 34:327–334.

Santana, M., D. J. Lodge, and P. Lebow. 2005. Relationships of host recurrence in fungi to rates of tropical leaf decomposition. *Pedobiologia* 49:549–564.

Scatena, F. N. 1989. An introduction to the physiography and history of the Bisley Experimental Watersheds in the Luquillo Mountains of Puerto Rico. General Technical Report SO-72. New Orleans, LA: USDA Forest Service Southern Forest Experiment Station.

Scatena, F. N. 1998. A comparative ecology of the Bisley Biodiversity Plot and Experimental Watersheds, Luquillo Experimental Forest, Puerto Rico. Pages 213–230 in F. Dallmeier and J. A. Comisky, editors, *Forest Biodiversity in North, Central and South America and the Caribbean: Research and monitoring*. New York: Parthenon.

Scatena, F. N., and A. E. Lugo. 1995. Geomorphology, disturbance, and the soil and vegetation of two subtropical wet steepland watersheds of Puerto Rico. *Geomorphology* 13:199–213.

Scatena, F. N., S. Moya, C. Estrada, and J. D. Chinea. 1996. The first five years in the reorganization of aboveground biomass and nutrient use following Hurricane Hugo in the Bisley experimental watersheds, Luquillo Experimental Forest, Puerto Rico. *Biotropica* 28:424–440.

Scatena, F. N., E. Planos, and J. Schellekens. 2005. Natural disturbances and the hydrology of humid tropical forests. Pages 489–512 in M. Bonell and L. A. Bruijnzeel, editors, *Forests, water and people in the humid tropics: Past, present and future hydrological research for integrated land and water management*. Cambridge, England: Cambridge University Press.

Scatena, F. N., W. Silver, T. Siccama, A. Johnson, and M. J. Sanchez. 1993. Biomass and nutrient content of the Bisley Experimental Watersheds, Luquillo Experimental Forest, Puerto Rico, before and after Hurricane Hugo, 1989. *Biotropica* 25:15–27.

Schaefer, D. A. 2003. Watershed hydrological and chemical responses to precipitation variability in the Luquillo Mountains of Puerto Rico. Pages 141–157 in D. Greenland, D. G. Goodin, and R. C. Smith, editors, *Climate variability and ecosystem response at long-term ecological research sites*. Oxford, England: Oxford University Press.

Schaefer, D. A., W. H. McDowell, F. N. Scatena, and C. E. Asbury. 2000. Effects of hurricane disturbance on stream water concentrations and fluxes in eight tropical forest watersheds of the Luquillo Experimental Forest. *Journal of Tropical Ecology* 16:189–207.

Schellekens, J., F. N. Scatena, L. A. Bruijnzeel, A. I. J. M. Van Dijk, M. M. A. Groen, and R. J. P. Van Hogezand. 2004. Stormflow generation in a small rainforest catchment in the Luquillo Experimental Forest, Puerto Rico. *Hydrological Processes* 18:505–530.

Schellekens, J., F. N. Scatena, L. A. Bruijnzeel, and A. J. Wickel. 1999. Modelling rainfall interception by a lowland tropical rain forest in northeastern Puerto Rico. *Journal of Hydrology* 225:168–184.

Schoener, T. W. 1985. Some comments on Connell's and my reviews of field experiments on interspecific competition. *American Naturalist* 125:730–740.

Schoener, T. W., and D. Spiller. 1987. Effect of lizards on spider populations: Manipulative reconstruction of a natural experiment. *Science* 236:949–952.

Schoener, T. W., and C. Toft. 1983. Spider populations: Extraordinarily high densities on islands without top predators. *Science* 219:1353–1355.

Schowalter, T. D., and L. M. Ganio. 1999. Invertebrate communities in a tropical rain forest canopy in Puerto Rico following Hurricane Hugo. *Ecological Entomology* 24:191–201.

Schubert, C. 1988. Climatic changes during the last glacial maximum in Northern South America and the Caribbean: A review. *Interciencia* 13:128–137.

Schubert, C., and E. Medina. 1982. Evidence of Quaternary glaciation in the Dominican Republic: Implications for Caribbean paleoclimatology. *Palaeogeography, Palaeoclimatology and Palaeoecology* 39:282–294.

Seastedt, T. R. 1984. The role of microarthropods in decomposition and mineralization. *Annual Review of Ecology and Systematics* 29:25–46.

Seiders, V. M. 1971. Cretaceous and lower Tertiary stratigraphy of the Gurabo and El Yunque Quadrangles, Puerto Rico. Bulletin 1294–4. Washington, DC: U.S. Geological Survey.

Sexton, O. J., J. Bauman, and E. Ortleb. 1976. Seasonal food habits of *Anolis limifrons*. *Ecology* 53:182–186.

Shanley, J. B., W. H. McDowell, and R. F. Stallard. 2011. Long-term patterns and short-term dynamics of stream solutes and suspended sediment in a rapidly weathering tropical watershed. *Water Resources Research* 47:W07515.

Shinn, E. A., G. W. Smith, J. M. Prospero, P. Betzer, M. L. Hayes, V. Garrison, and R. T. Barber. 2000. African dust and the demise of Caribbean coral reefs. *Geophysical Research Letters* 27:3029–3032.

Silver, W. L., D. J. Herman, and M. K. Firestone. 2001. Dissimilatory nitrate reduction to ammonium in tropical forest soils. *Ecology* 82:2410–2416.

Silver, W. L., A. E. Lugo, and M. Keller. 1999. Soil oxygen availability and biogeochemistry along rainfall and topographic gradients in upland wet tropical forest soils. *Biogeochemistry* 44:301–328.

Silver, W. L., F. N. Scatena, A. H. Johnson, T. G. Siccama, and M. J. Sánchez. 1994. Nutrient availability in a montane wet tropical forest: Spatial patterns and methodological considerations. *Plant & Soil* 164:129–145.

Silver, W. L., F. N. Scatena, A. H. Johnson, T. G. Siccama, and F. Watt. 1996. At what temporal scales does disturbance affect belowground nutrient pools? *Biotropica* 28:441–457.

Silver, W. L., and K. A. Vogt. 1993. Fine root dynamics following single and multiple disturbances in a subtropical wet forest ecosystem. *Journal of Ecology* 81:729–738.

Snyder, N. F. R., J. W. Wiley, and C. B. Kepler. 1987. The parrots of Luquillo: Natural history and conservation of the Puerto Rican Parrot. Los Angeles: Western Foundation of Vertebrate Zoology.

Soil Survey Staff. 1995. Order 1 soil survey of the Luquillo long-term ecological research grid, Puerto Rico. Lincoln, NE: U.S. Department of Agriculture, Natural Resources Conservation Service.

Sollins, P., C. C. Grier, F. M. McCorison, K. Cromack, Jr., R. Fogel, and R. L. Fredriksen. 1980. The internal element cycles of an old-growth Douglas-fir ecosystem in western Oregon. *Ecological Monographs* 50:261–285.

Stephenson, S. L., J. C. Landolt, and D. L. Moore. 1999. Protostelids, dictyostelids, and myxomycetes in the litter microhabitat of the Luquillo Experimental Forest, Puerto Rico. *Mycological Research* 103:209–214.

Stewart, M. M., and F. H. Pough. 1983. Population density of tropical forest frogs: Relation to retreat sites. *Science* 221:570–572.

Stewart, M. M., and L. L. Woolbright. 1996. Amphibians. Pages 273–320 in D. P. Reagan and R. B. Waide, editors, *The food web of a tropical rain forest*. Chicago: University of Chicago Press.

Sullivan, N. H., W. B. Bowden, and W. H. McDowell. 1999. Short-term disappearance of foliar litter in three species before and after a hurricane. *Biotropica* 31:382–393.

Swank, W. T., and Waide, J. B. 1988. Characterization of baseline precipitation and stream chemistry and nutrient budgets for control watersheds. Pages 57–79 in W. T. Swank and D. A. Crossley, Jr., editors, *Forest hydrology and ecology at Coweeta*. New York: Springer-Verlag.

Tanner, E. V. J. 1977. Four montane rain forests of Jamaica: A quantitative characterization of the floristics, the soils and the foliar mineral levels, and a discussion of the interrelations. *Journal of Ecology* 65:883–918.

Tanner, E. V. J. 1985. Jamaican montane forests: Nutrient capital and cost of growth. *Journal of Ecology* 73:553–568.

ten Brink, U. S., E. L. Geist, and B. D. Andrews. 2006. Size distribution of submarine landslides and its implication to tsunami hazard in Puerto Rico. *Geophysical Research Letters* 33:L11307.

Thomas, R., and A. G. Kessler. 1996. Nonanoline reptiles. Pages 347–362 in D. P. Reagan and R. B. Waide, editors, *The food web of a tropical rain forest*. Chicago: University of Chicago Press.

Thompson, J., N. Brokaw, J. K. Zimmerman, R. B. Waide, E. M. Everham, D. J. Lodge, C. M. Taylor, D. García-Montiel, and M. Fluet. 2002. Land use history, environment, and tree composition in a tropical forest. *Ecological Applications* 12:1344–1363.

Thompson, J., N. Brokaw, J. K. Zimmerman, R. B. Waide, E. M. Everham III, and D. A. Schaefer. 2004. Luquillo Forest Dynamics Plot. Pages 540–550 in E. Losos and E. G. Leigh, Jr., editors, *Tropical forest diversity and dynamism: Results from a long-term tropical forest network*. Chicago: University of Chicago Press.

Thompson, J., A. E. Lugo, and J. Thomlinson. 2007. Land use history, hurricane disturbance, and introduced species survival in a subtropical wet forest in Puerto Rico. *Plant Ecology* 192:289–301.

Tilgner, E. H., G. R. Camilo, and C. F. Moxey. 2000. A new species of *Lamponius* (Phasmida: Phasmatidae) from Puerto Rico. *Journal of Orthoptera Research* 9:21–26.

Trejo-Torres, J. C., and J. D. Ackerman. 2001. Biogeography of the Antilles based on a parsimony analysis of orchid distributions. *Journal of Biogeography* 28:775–794.

Turvey, S. T., J. R. Oliver, Y. M. Narganes Storde, and P. Rye. 2007. Late Holocene extinction of Puerto Rican native land mammals. *Biology Letters* 3:193–196.

Valencia, R., R. Condit, R. B. Foster, K. Romoleroux, G. V. Muñoz, J.-C. Svenning, E. Magård, M. Bass, E. C. Losos, and H. Balslev. 2004. Yasuní Forest Dynamics Plot, Ecuador. Pages 609–620 in E. Losos and E. G. Leigh, Jr., editors, *Tropical forest diversity and dynamism: Results from a long-term tropical forest network*. Chicago: University of Chicago Press.

Vallejo, M. I., C. Samper, H. Mendoza, and J. T. Otero. 2004. La Planada Forest Dynamics Plot, Colombia. Pages 517–526 in E. Losos and E. G. Leigh, Jr., editors, *Tropical forest diversity and dynamism: Results from a long-term tropical forest network*. Chicago: University of Chicago Press.

Van Den Bussche, R. A., M. R. Willig, R. K. Chesser, and R. B. Waide. 1988. Genetic variation and systematics of four taxa of Neotropical walking sticks (Phasmatodea: Phasmatidae). *Proceedings of the Entomological Society of Washington* 90:422–427.

van der Molen, M. K. 2002. Meteorological impacts of land use change in the maritime tropics. Ph.D. dissertation. Vrije Universiteit, Amsterdam, The Netherlands.

Vannote, R. L., G. W. Minshall, K. W. Cummins, J. R. Sedell, and C. E. Cushing. 1980. The river continuum concept. *Canadian Journal of Fisheries and Aquatic Sciences* 37:130–137.

van Schaik, C. P., J. W. Terborgh, and S. J. Wright. 1993. The phenology of tropical forests: Adaptive significance and consequences for primary consumers. *Annual Review of Ecology and Systematics* 24:353–377.

Vitousek, P. M. 1982. Nutrient cycling and nutrient-use efficiency. *American Naturalist* 199:553–572.

Vitousek, P. M. 1984. Litterfall nutrient cycling and nutrient limitation in tropical forests. *Ecology* 65:285–298.

Vogt, K., H. Asbjornsen, A. Ercelawn, F. Montagnini, and M. Valdes. 1997. Roots and mycorrhizas in plantation ecosystems. Pages 247–296 in E. K. Sadanandan Nambiar and A. G. Brown, editors, *Management of soil, nutrients and water in tropical plantation forests*. ACIAR Monograph No. 43. Bruce, ACT, Australia: Australian Centre for International Agricultural Research.

Vogt, K. A., D. J. Vogt, H. Asbjornsen, and R. A. Dahlgren. 1995. Roots, nutrients and their relationship to spatial patterns. *Plant and Soil* 168–169:113–123.

Vogt, K. A., D. J. Vogt, P. Boom, A. Covich, F. N. Scatena, H. Asbornsen, J. L. O'Hara, J. Pérez, T. G. Siccama, J. Bloomfield, and J. F. Ranciato. 1996. Litter dynamics along stream, riparian, and upslope areas following Hurricane Hugo, Luquillo Experimental Forest, Puerto Rico. *Biotropica* 28:458–470.

Wadsworth, F. H. 1951. Forest management in the Luquillo Mountains. I. The setting. *Caribbean Forester* 12:93–132.

Wadsworth, F. H. 1987. Composition of trees in insular forest ecosystems. *Acta Cientifíca* 1:77–80.

Waide, R. B. 1987. The fauna of Caribbean island ecosystems: Community structure and conservation. *Acta Cientifica* 1:64–71.

Waide, R. B. 1996. Birds. Pages 362–398 in D. P. Reagan and R. B. Waide, editors, *The food web of a tropical rain forest*. Chicago: University of Chicago Press.

Waide, R. B., J. K. Zimmerman, and F. N. Scatena. 1998. Controls of primary productivity: Lessons from the Luquillo Mountains in Puerto Rico. *Ecology* 79:31–37.

Walker, L. R., D. J. Zarin, N. Fetcher, R. W. Myster, and A. H. Johnson. 1996. Ecosystem development and plant succession on landslides in the Caribbean. *Biotropica* 28:566–576.

Wang, H., J. D. Cornell, C. A. S. Hall, and D. Marley. 2002. Spatial and seasonal dynamics of surface soil carbon in the Luquillo Experimental Forest, Puerto Rico. *Ecological Modeling* 147:105–122.

Wang, H. Q., C. A. S. Hall, F. N. Scatena, N. Fetcher, and W. Wu. 2003. Modeling the spatial and temporal variability in climate and primary productivity across the Luquillo Mountains, Puerto Rico. *Forest Ecology and Management* 179:69–94.

Weaver, P. L. 1995. The colorado and dwarf forests of Puerto Rico's Luquillo Mountains. Pages 109–141 in A. E. Lugo and C. Lowe, editors, *Tropical forests: Management and ecology*. New York: Springer-Verlag.

Weaver, P. L., R. A. Birdsey, and A. E. Lugo. 1987. Soil organic matter in secondary forests in Puerto Rico. *Biotropica* 19:17–23.

Weaver, P. L., E. Medina, D. Pool, K. Dugger, J. Gonzales-Liboy, and E. Cuevas. 1986. Ecological observations in the dwarf elfin forest of the Luquillo Mountains in Puerto Rico. *Biotropica* 18:79–85.

Weaver, P. L., and P. G. Murphy. 1990. Forest structure and productivity in Puerto Rico's Luquillo Mountains. *Biotropica* 22:69–82.

Webster, J. R., P. J. Mulholland, J. L. Tank, H. M. Vallet, W. K. Dodds, B. J. Peterson, W. B. Bowden, C. N. Dahm, S. Findlay, S. V. Gregory, N. B. Grimm, S. K. Hamilton, S. L. Johnson, E. Marti, W. H. McDowell, J. L. Meyer, D. D. Morrall, S. A. Thomas, and W. W. Wollheim. 2003. Factors affecting ammonium uptake in streams—An inter-biome perspective. *Freshwater Biology* 48:1329–1352.

White, H. H. 1963. Variation of stand structure correlated with altitude in the Luquillo Mountains. *Caribbean Journal of Forestry* 1:46–52.

White, A. F., and A. E. Blum. 1995. Effects of climate on chemical weathering in watersheds. *Geochimica et Cosmochimica Acta* 59:1729–1747.

White, A. F., A. E. Blum, M. S. Schulz, D. V. Vivit, D. A. Stonestrom, M. Larsen, S. F. Murphy, and D. Eberl. 1998. Chemical weathering in a tropical watershed, Luquillo Mountains, Puerto Rico: I. Long-term versus short-term weathering fluxes. *Geochimica Cosmochimica Acta* 62:209–226.

Whitmore, T. C. 1974. Change with time and the role of cyclones in tropical rain forest on Kolombangara, Solomon Islands. Institute Paper No. 46. Oxford, England: Commonwealth Forestry Institute, University of Oxford.

Whitmore, T. C. 1989. Tropical forest nutrients: Where do we stand? A tour de horizon. Pages 1–13 in J. Proctor, editor, *Mineral nutrients in tropical forest and savanna ecosystems*. London: Blackwell Scientific.

Wiegert, R. G. 1970. Energetics of the nest-building termite, *Nasutitermes costalis* (Holmgren), in a Puerto Rican forest. Pages I-57–I-64 in H. T. Odum and R. F. Pigeon, editors, *A tropical rain forest: A study of irradiation and ecology at El Verde, Puerto Rico*. Oak Ridge, TN: U.S. Atomic Energy Commission.

Wiegert, R. G., and P. Murphy. 1970. Effects of season, species and location on the disappearance rate of leaf litter in a Puerto Rican rain forest. Pages H101–H104 in H. T. Odum and R. F. Pigeon, editors, *A tropical rain forest: A study of irradiation and ecology at El Verde, Puerto Rico*. Oak Ridge, TN: U.S. Atomic Energy Commission.

Williams, E. H., L. Bunkley-Williams, C. G. Lilyestrom, and E. A. R. Ortiz-Corps. 2001. A review of recent introductions of aquatic invertebrates in Puerto Rico and implications for the management of nonindigenous species. *Caribbean Journal of Science* 37:246–251.

Williams, M. R., and J. M. Melack. 1997. Solute export from forested and partially deforested catchments in the central Amazon. *Biogeochemistry* 38:67–102.

Willig, M. R., and M. R. Gannon. 1996. Mammals. Pages 399–431 in D. P. Reagan and R. B. Waide, editors, *The food web of a tropical rain forest*. Chicago: University of Chicago Press.

Willig, M. R., R. W. Garrison, and A. Bauman. 1986. Population dynamics and natural history of a Neotropical walking stick, *Lamponius portoricensis* Rehn (Phasmatodea, Phasmatidae). *Texas Journal of Science* 38:121–138.

Willig, M. R., E. A. Sandlin, and M. R. Gannon. 1993. Structural and taxonomic components of habitat selection in the Neotropical folivore, *Lamponius portoricensis* (Phasmatodea: Phasmatidae). *Environmental Entomology* 22:634–641.

Willig, M. R., M. F. Secrest, S. B. Cox, G. R. Camilo, J. F. Cary, J. Alvarez, and M. R. Gannon. 1998. Long-term monitoring of snails in the Luquillo Experimental Forest of Puerto Rico: Heterogeneity, scale, disturbance, and recovery. Pages 293–322 in F. Dallmeier and J. Comisky, editors, *Forest biodiversity in North, Central, and South America and the Caribbean: Research and monitoring*. Man and the Biosphere Series, Vol. 21. Carnforth, Lancashire, England: UNESCO and Parthenon Press.

Woolbright, L. L. 1996. Disturbance influences long-term population patterns in the Puerto Rican frog, *Eleutherodactylus coqui* (Anura: Leptodactylidae). *Biotropica* 28:493–501.

Woolbright, L. L. 1997. Local extinctions of anuran amphibians at El Verde, Puerto Rico. *Journal of Herpetology* 31:572–576.

Wright, S. J., and C. P. van Schaik. 1994. Light and the phenology of tropical trees. *American Naturalist* 143:192–199.

Zalamea, M., and G. González. 2008. Leaffall phenology in a subtropical wet forest in Puerto Rico: From species to community patterns. *Biotropica* 40:295–304.

Zalamea, M., G. González, C. L. Ping, and G. Michaelson. 2007. Soil organic matter dynamics under decaying wood in a subtropical wet forest: Effect of tree species and decay stage. *Plant Soil* 296:173–185.

Zarin, D. J., and A. H. Johnson. 1995. Nutrient accumulation during primary succession in a montane tropical forest, Puerto Rico. *Soil Science Society of America Journal* 59:1444–1452.

Zimmerman, J. K., W. M. Pulliam, D. J. Lodge, V. Quiñones-Orfila, N. Fetcher, S. Guzmán-Grajales, J. A. Parrotta, C. E. Asbury, L. R. Walker, and R. B. Waide. 1995. Nitrogen immobilization by decomposing woody debris and the recovery of tropical wet forest from Hurricane damage. *Oikos* 72:314–322.

Zimmerman, J. K, S. J. Wright, O. Calderón, M. Aponte-Pagan, and S. Paton. 2007. Flowering and fruiting phenologies of seasonal and aseasonal neotropical forests: The role of annual changes in irradiance. *Journal of Tropical Ecology* 23:231–251.

Zimmerman, J. K. H., and A. P. Covich. 2003. Distribution of juvenile crabs (*Epilobocera sinuatifrons*) in two Puerto Rican headwater streams: Effects of pool morphology and past land-use legacies. *Archiv für Hydrobiologie* 158:343–357.

Zou, X. M., and G. González. 2002. Earthworms in tropical tree plantations: Effects of management and relations with soil carbon and nutrient use efficiency. Pages 289–301 in M. V. Reddy, editor, *Management of tropical plantation forests and their soil litter system*. Enfield, NH: Science Publishers Inc.

Zou, X. M., H. H. Ruan, Y. Fu, X. D. Yang, and L. Q. Sha. 2005. Estimating soil labile organic carbon and turnover rates using a sequential fumigation-incubation procedure. *Soil Biology and Biochemistry* 37:1923–1928.

Zou, X., C. P. Zucca, R. B. Waide, and W. H. McDowell. 1995. Long-term influence of deforestation on tree species composition and litter dynamics of a tropical rain forest in Puerto Rico. *Forest Ecology and Management* 78:147–157.

4

Disturbance Regime

Frederick N. Scatena, Juan Felipe Blanco, Karen H. Beard, Robert B. Waide, Ariel E. Lugo, Nicholas Brokaw, Whendee L. Silver, Bruce L. Haines, and Jess K. Zimmerman

Key Points

- The Luquillo Mountains are affected by a wide array of environmental processes and disturbances. Events that concurrently alter the environmental space of several different areas of the Luquillo Mountains occur every 2 to 5 years. Events such as hurricanes that cause widespread environmental modification occur once every 20 to 60 years.
- The most common disturbance-generating weather systems that affect the Luquillo Mountains are (1) cyclonic systems, (2) noncyclonic intertropical systems, (3) extratropical frontal systems, and (4) large-scale coupled ocean-atmospheric events (e.g., North Atlantic Oscillation, El Niño-Southern Oscillation). Unlike some tropical forests, disturbances associated with the passage of the Inter-Tropical Convergence Zone or monsoonal rains do not occur.
- Hurricanes are considered the most important natural disturbance affecting the structure of forests in the Luquillo Mountains. Compared to other humid tropical forests, Luquillo has a high rate of canopy turnover caused by hurricanes but a relatively low rate caused by tree-fall gaps. Historically, pathogenic disturbances have not been common.
- Human-induced disturbances have historically included tree harvesting for timber and charcoal, agriculture, and agroforestry. In the past few decades, water diversions, fishing and hunting, and road building have been important disturbances. Present and future human-induced disturbances are related to regional land use change, the disruption of migratory corridors, and forest drying related to coastal plain deforestation and regional climate change.

- Hurricane-related storm discharges can cause significant geomorphic modifications to Luquillo stream channels, and stream water concentrations of sediments and nutrients can be elevated for months to years following a major hurricane. However, the largest floods are not necessarily associated with hurricanes, and the annual peak discharge can occur in any month of the year but is most common in the late summer and fall.
- Over the entire island of Puerto Rico, 1.2 landslide-producing storms occur each year. In the Luquillo Mountains, landslides are typically covered with herbaceous vegetation within 2 years, have closed canopies of woody vegetation in less than 20 years, and have aboveground biomass of the adjacent forest after several decades.

Introduction

The Luquillo Mountains, like many humid tropical environments, is a dynamic ecosystem that is affected by a wide array of environmental processes and disturbances (figures 4-1 and 4-2). Quantifying the magnitude, frequency, and impact of these natural disturbances on both geographical and ecological space is essential to understanding and managing these forests. This chapter reviews the causes, frequencies, and discrete and cumulative impacts of disturbances on

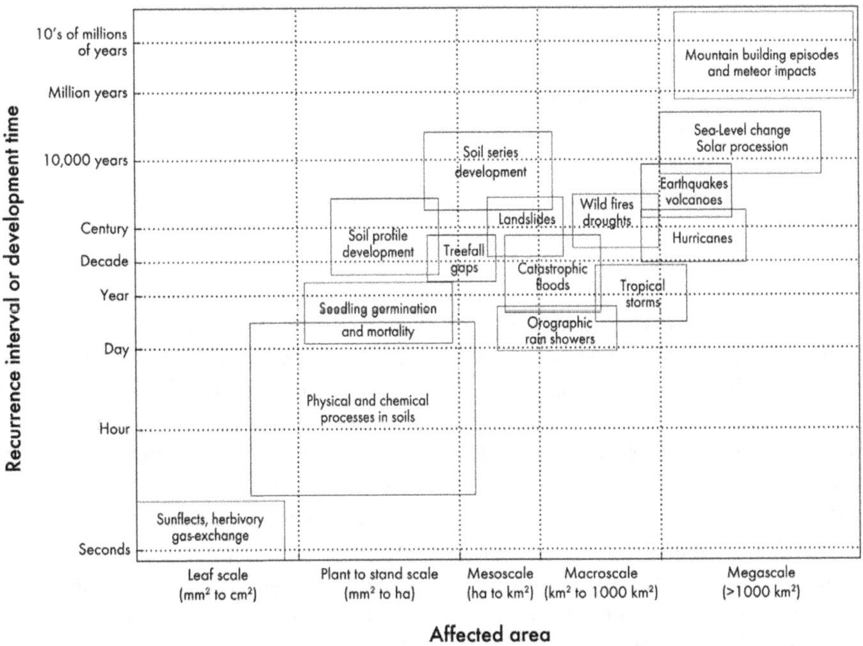

Figure 4.1 Spatial and temporal relationships of natural disturbances and processes affecting the Luquillo Mountains. (Modified from Scatena 1995.)

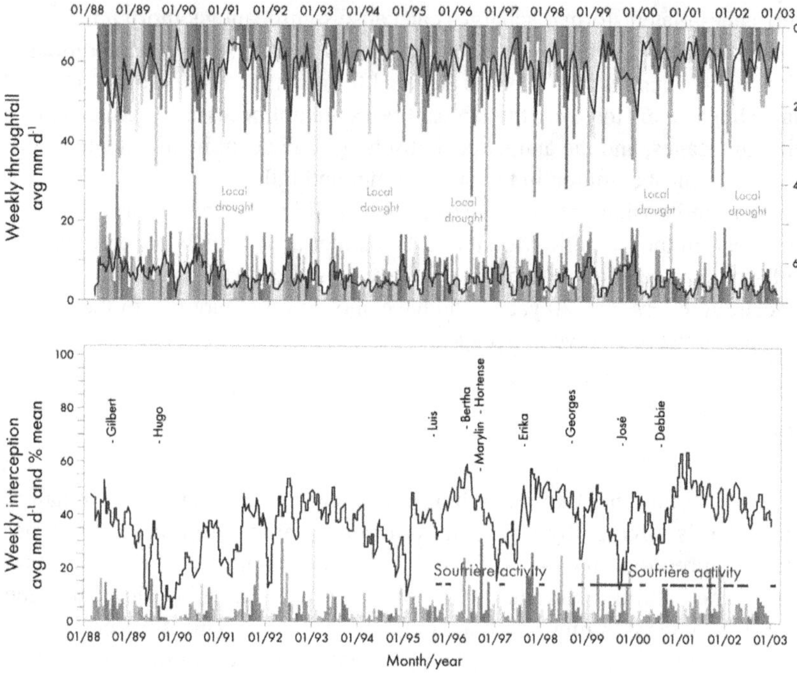

Figure 4.2 Weekly rainfall and throughfall and significant climatic events in the Bisley Research Watersheds, 1988 to 2003. (From Heartsill-Scalley et al. 2007.)

the Luquillo ecosystem. Subsequent chapters discuss the ecosystem's recovery after disturbance.

Disturbances can be defined as relatively discrete events that alter the structure of populations, communities, and ecosystems (see chapter 2) (White and Pickett 1985; Lugo and Scatena 1996; Walker and Willig 1999). "Disturbance regime" refers to the sum of disturbances acting on a particular location. The natural disturbances specified by the United Nations in the International Decade of Natural Disaster Reduction were earthquakes, windstorms, tsunamis, floods, landslides, volcanic eruptions, wildfires, insect infestations, drought, and desertification. Treefalls, pathogens, exotic invasions, and meteor impacts are also known to affect humid tropical forests. Of these 14 types of disturbances, 10 are known to have caused community-level impacts in northeastern Puerto Rico during the past century. These disturbances have also acted on a landscape that has undergone dramatic land use changes associated with forest harvesting and clearing, agriculture, urbanization, water diversions, and other modifications to hydrologic and nutrient cycles.

Quantifying the effects of disturbances on landform morphology and ecosystem development has been a traditional theme in geomorphology and ecology (Wolman and Miller 1960; Connell 1978). It is now generally recognized (Lugo and Scatena 1996) that the effect of a disturbance on the morphology of a landscape or the structure of an ecosystem depends on the following:

- The **type** of disturbance (i.e., flood, fire, landslide, biologic, anthropogenic, etc.)
- The **intensity** of the force exerted (i.e., wind velocity and duration, rainfall magnitude and intensity, earthquake magnitude, etc.)
- The ecosystem **component** that is directly impacted by the forces exerted (i.e., soil, biomass, leaf area, etc.)
- The spatial extent and the spatial distribution of impacts
- The return period or **frequency** of the event
- The **initial condition** and **resistance** (see chapter 2) of the system
- The **resilience** (see chapter 2) of the system and the magnitude of the constructive or restorative processes that occur between disturbances

Mortality is also a complex process that occurs over many spatial and temporal scales. Mortality events can range from "background events" to large-scale "catastrophic events" (Lugo and Scatena 1996). Background mortality is typically associated with senescence, competition, and succession. Catastrophic mortality occurs when a forest is mechanically or chemically impacted by an external force such as a hurricane, a landslide, or toxic waste. When expressed as percentage of stems or biomass per year, the background mortality is typically less than 3 percent per year. The median value of the background mortality in 68 pantropical moist, wet, and rain forest stands was 1.6 percent per year; this is similar to values reported from the Luquillo Mountains, as well as from temperate and boreal forests (Lugo and Scatena 1996). In contrast, catastrophic events can cause 100 percent mortality in small areas.

Tectonic Drivers of Disturbance

The Luquillo Mountains were formed from shallow marine deposits and the material produced by ancestral volcanoes that existed to the south of the present mountains (see chapter 3) (Scatena 1989a). This volcaniclastic bedrock formed from the debris of these volcanoes is of late to upper Cretaceous age (70 to 112 million years ago) and is intruded by a quartzdioritic batholith that underlies the Rio Blanco watershed (Seiders 1971a, 1971b). Existing geochronology suggests that the Rio Blanco batholith is 47 million years old and of Eocene age (Cox et al. 1977). It is also the only major Eocene addition of felsic magma to the Greater Antilles (Smith et al. 1998).

The island is currently located on the edge of a continental-type tectonic block that is rotating in a counter-clockwise fashion (Masson and Scanlon 1991). The tectonic forces associated with this block are ultimately responsible for the volcanic activity and mountain building that has formed the Luquillo Mountains. During its evolution, the island has also undergone considerable erosion and at one time might have had mountains tall enough to support a cold temperatelike flora (Graham and Jarzen 1969). During the Quaternary, uplift of the island has outpaced sea level rise and is estimated from subaerial coral reefs to have a rate of 0.055 mm y^{-1} (Taggart 1992). Over the Holocene, the net rate of uplift was probably higher and is estimated to have been between 0.125 and 0.25 mm y^{-1} (Clark and Wilcock 2000).

Although the origin of the Luquillo Mountains is closely linked to volcanic activity, presently volcanic activity is not an important disturbance in the area. Nevertheless, the Luquillo Mountains do receive occasional ash falls from volcanoes in the lower Caribbean. Between 1987 and 2003, at least two volcanic ash falls from the lower Caribbean island of Montserrat blanketed Puerto Rico with enough fine ash to be noticeable to the public and temporarily close the San Juan International Airport. Thick, catastrophic ash falls similar to those that have occurred in other tropical forests (Whittaker and Walden 1992) are unknown in the historical or recent geologic record of the Luquillo Mountains. Given Puerto Rico's distance from active volcanoes, such catastrophic ash falls are extremely rare, if not impossible, events. Nevertheless, volcanic events and Saharan dust deposition are detectable in Luquillo rainfall and throughfall and might account for up to 9 percent of the inputs of some constituents (McDowell et al. 1990; Heartsill-Scalley et al. 2007). Although there is no evidence that this dust is causing a major health problem in Puerto Rico, during intense events the concentrations of respirable dust do affect some island residents, and U.S. Environmental Protection Agency standards have probably been exceeded occasionally in the Caribbean (Prospero and Lamb 2004).

Although the Luquillo Mountains are not volcanically active, they are within a tectonically active zone, and multiple earthquakes are measured on the island each year. In 1918, an earthquake caused landslides throughout the mountainous regions of the island (Reid and Taber 1919). Devastating earthquake-generated tsunamis occurred in the northern Caribbean in 1867, 1918, and 1946 (Dillon and Brink 1999). However, owing to their elevation, the Luquillo Mountains have never been directly affected by tsunamis. The exact frequency of forest-modifying earthquake-induced disturbances in the Luquillo Mountains is unknown. However, based on these three historic events, a rate for the island can be conservatively estimated at one or two major events per century. Because the majority of the earthquake activity is located between Puerto Rico and the island of Hispaniola, the rate for northeast Puerto Rico and the Luquillo Mountains is probably less than that for the western part of the island. The residuals (see chapter 2) of these tectonic activities include landslides, fault scarps, and other topographic features. Their legacies (see chapter 2) include the location and morphology of stream channels (Ahmad et al. 1993), the location of palm brakes (Lugo et al. 1995), and the Luquillo Mountains themselves.

Meteor Impacts

Catastrophic meteor impacts are events that can have local, continental, and global consequences (Toon et al. 1997). The catastrophic disturbances associated with meteor impacts include earthquakes, blast waves, tsunamis, and fires. Effects from dust, smoke, and acid rain might have longer-term effects on the global climate and biota. Meteor impacts in the Luquillo Mountains have not been recorded in historical times. However, a meteor impact in the Caribbean basin has been implicated in the global Cretaceous-Tertiary age mass extinction of dinosaurs and apparently generated a free-standing ocean wave in the Caribbean that was over 500 m high (Hildebrand and Boynton 1990; Florentin et al. 1991) that would have impacted the ancestral Luquillo region. Because the flora of the Luquillo Mountains developed shortly after this impact (Graham and Jarzen 1969), the widespread extinction it caused might have had a fundamental, but yet unquantified, impact on the ecology of the Luquillo ecosystem.

Atmospheric Drivers of Disturbance in the Luquillo Mountains

The Luquillo Mountains have a humid tropical maritime climate that has rainfall and runoff every month of the year (see chapter 3 and figure 4-2). At mid-elevations, the median daily rainfall is low (3 mm/day), but rain events are numerous (267 rain days per year) and of relatively low intensity (<5 mm/h) (Schellekens et al. 2004). Nevertheless, individual storms with rainfall greater than 125 mm/day occur annually, and daily rainfalls greater than 600 mm have been recorded. The most common disturbance-generating weather systems that affect the Luquillo Mountains are (1) cyclonic systems, (2) noncyclonic intertropical systems, (3) extratropical frontal systems, and (4) large-scale coupled ocean-atmospheric events (e.g., North Atlantic Oscillation [NAO], El Niño-Southern Oscillation [ENSO]).

Unlike some tropical forests, the Luquillo Mountains do not commonly have disturbances associated with the annual passage of the Inter-Tropical Convergence Zone (ITCZ) or monsoonal rains (Walsh 1997). In general, Puerto Rico is too far north to directly experience the seasonal rainfalls that are associated with the ITCZ. Likewise, the relatively small size of Puerto Rico and its orientation relative to the prevailing winds prevent monsoonal systems from developing and sculpting the Luquillo landscape.

Cyclonic Systems

Cyclonic systems are large masses of air that rotate about a low-pressure center, and they include tropical waves and hurricanes. Tropical waves have incompletely closed circulations, whereas hurricanes have completely closed circulations. The occurrence of these systems is mainly confined to the period from May through November, when an average of two waves pass by the island every week (van der Molen 2002). Globally, approximately 82 hurricanes occur in a typical year, 12 percent of which pass through the Caribbean (table 4-1). However, the return period for a hurricane passing directly over the Luquillo Mountains is between 50 and 60 years (Scatena and Larsen 1991). In general, the frequency of hurricanes varies with the season (table 4-2, figure 4-3), decade (figure 4-4), and regional physiography (Boose et al. 1994). Multidecade variation in cyclone activity has been linked to variations in thermohaline oceanic circulation, global sea surface temperatures, West African monsoons, African droughts, and ENSO events (Gray et al. 1997).

Hurricanes are considered the most important natural disturbance affecting the structure of forests in the Luquillo Mountains (Crow 1980; Scatena and Lugo 1995). However, because cyclonic systems develop north and south of the ITCZ and travel poleward, many humid tropical forests are unaffected by hurricanes. In addition to the northern Caribbean, hurricanes are frequent disturbances in Madagascar,

Table 4.1 Mean annual named tropical storm and hurricane frequency for the Caribbean and other tropical regions. Adapted from Walsh (1997) and Planos Gutiérrez (1999)

Region	Tropical storms y^{-1}	Hurricanes y^{-1}	All named storms	Percentage of total	Principal humid tropical forests
North Indian Ocean	3.5	2.2	5.7	7.0	Andama Islands
North Atlantic/ Caribbean	4.2	5.2	9.4	11.5	Caribbean
Southwest Indian Ocean	7.4	3.8	11.2	13.7	Mauritius, Reunion, Madagascar
Southwest Pacific	10.9	3.8	14.8	18.1	Queensland, Fiji, Solomons, Vanuatu
Eastern North Pacific	9.3	5.8	15.2	18.6	None
Western North Pacific	7.5	17.8	25.3	31.0	Philippines, Taiwan, S. China, Borneo
Total	42.8	38.6	81.6	100.0	

Table 4.2 Monthly distribution of Atlantic cyclones between 1890 and 1990. (After Planos Gutiérrez 1999.)

	June	July	August	September	October	November	Other	Total
Percentage of total	6	8	25	34	20	5	2	100
Number for century	50	64	206	288	171	38	15	832
Maximum recorded in each month	3	4	7	7	6	2	2	NA

Mauritius, Reunion, the southwest Indian Ocean, the northern Philippines, Sabah, Taiwan, parts of Indo-China, the Pacific islands, and tropical Queensland (Walsh 1997). Hurricanes are rare to nonexistent in the humid tropical forests of South America, Africa, and northern Malaysia.

There is no simple, direct relationship between the magnitude and the destructive powers of Caribbean hurricanes (Planos Gutiérrez 1999). In general, windspeeds depend on the path of the hurricane and the local aspect and exposure. When hurricanes pass directly over the Luquillo Mountains, ground-level windspeeds can surpass 140 km h^{-1}. When hurricanes pass over or near other parts of the island, the Luquillo Mountains typically have sustained winds near canopy level of 60 km $h^{-1}$and gusts of over 150 km h^{-1} (table 4-3). The total amount of rain that falls in a given area for a given hurricane depends on (a) the intake of humid air into the

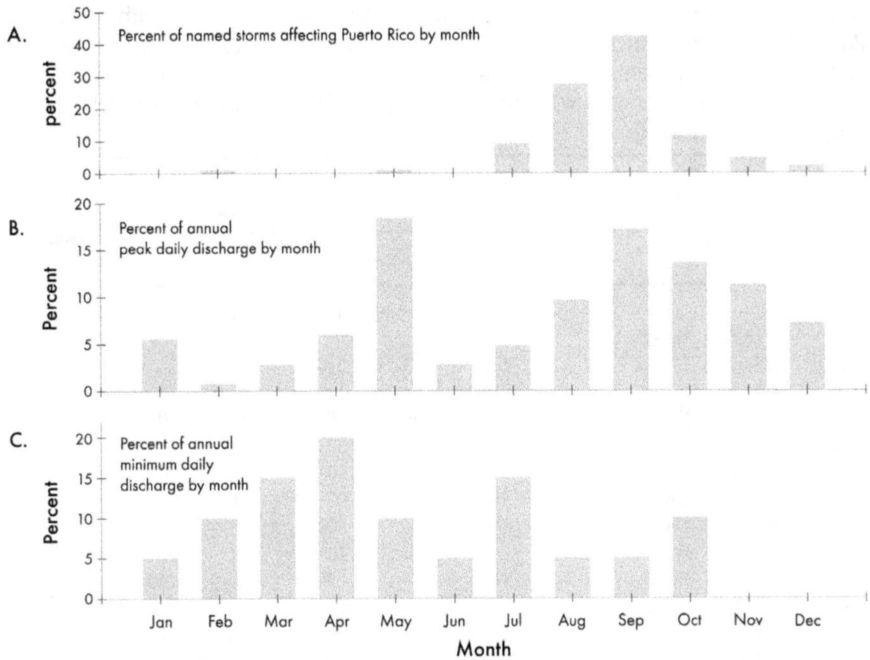

Figure 4.3 Named storms affecting Puerto Rico by month between 1899 and 1999 and percent occurrence of annual peaks and low flows by month for streams draining the Luquillo Mountains of Puerto Rico. (Modified from Scatena 2001.)

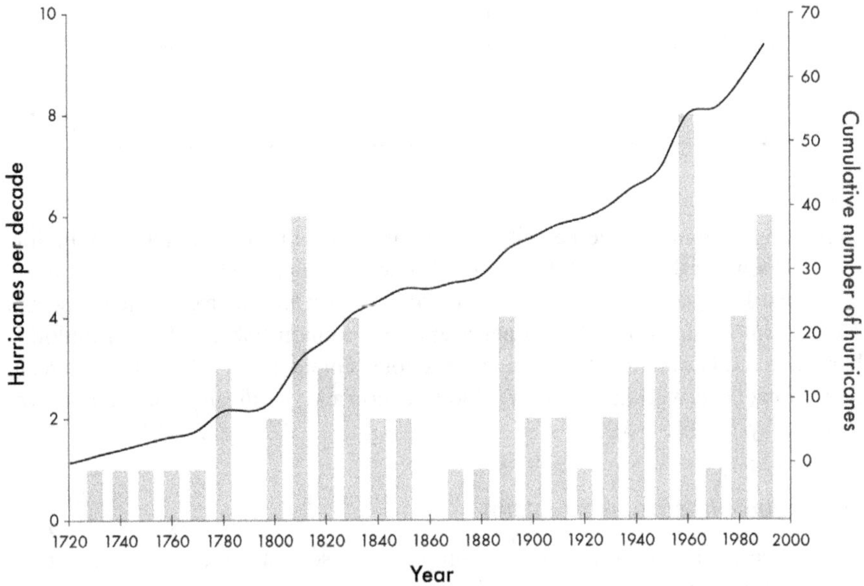

Figure 4.4 Number of hurricanes passing within 2 degrees of Puerto Rico between 1720 and 2000.

Table 4.3 Windspeeds at Roosevelt Roads and San Juan associated with named Puerto Rican storms since 1950. Based on data from the National Weather Service National Hurricane Center (http://www.nhc.noaa.gov), the National Climatic Data Center (http://ncdc.noaa.gov), and the USGS Caribbean Water Resources District (http://pr.water.usgs.gov)

Year	Date	Name	Landfall Y/N	Location of major impact	Roosevelt 2-minute gusts		San Juan 2-minute gusts		Storm duration over island
					Speed	*Direction*	*Speed*	*Direction*	
					km h^{-1}	deg	km h^{-1}	deg	h
1956	8/12/56	Betsy	Y	SE					3
1960	9/5/60	Donna	N	NE	93	210			
1963	9/26/63	Edith	N	SW	54	120	50	140	
1964	8/23/64	Cleo	N	S	56	50	63	50	
1966	8/26/66	Faith	N	NE	59	30	76	360	
1966	9/28/66	Inez	N	S	52	90	67	50	
1967	9/9/67	Beulah	N	SW	43	90	44	90	
1979	8/30/79	David	N	SW	74	100	76	50	
1979	2/17/79	Edith	N		24	70	30	90	
1987	9/22/87	Emily	N	SE	63	150	33	140	
1988	9/11/88	Gilbert	N	S	69	100	56	90	
1989	9/18/89	Hugo	Y	NE	nd	nd	148	320	4
1995	9/5/95	Luis	N	NE	64	350	65	360	
1996	9/15/96	Marilyn	N		37	90	nd	nd	
1996	7/9/96	Bertha	N	NE	37		nd	nd	
1996	9/10/96	Hortense	Y	SW	74	80	nd	nd	2
1997	9/5/97	Erika	N	NE	nd	nd	nd	nd	
1998	9/22/98	Georges	Y	SE	102	140	nd	nd	7
1999	10/20/99	Jose	N	NE	64	80	nd	nd	
2000	8/23/00	Debby	N	NE	60	180	nd	nd	
Average					60.3	120	64.4	158	4
Median					60.0	95	63.0	90	3.5
Maximum					102	350	148	360	7

circulating system, (b) the velocity of the winds within the hurricane, (c) the forward velocity of the eye, (d) the length of time for which the hurricane directly affects a particular area, and (e) the position of the storm and site relative to the ocean. Total storm rainfalls of 100 mm per event are common (table 4-4), and multiday hurricane totals of over 1500 mm are possible (Gupta 1975, 1988). Daily stream flows associated with hurricanes in Puerto Rico vary significantly but can be over 50 mm day^{-1} (table 4-5).

Noncyclonic Intertropical Systems

This group of atmospheric systems comprises those that originate and generally remain within the tropics and include micro- and meso-scale convective systems and orographic rains. Land-sea breezes that result from the differential heating of land and water surfaces are a dominant process that drives these systems, and they

Table 4.4 Daily maximum rainfall associated with named Puerto Rican storms since 1950. Based on data from the National Weather Service National Hurricane Center (http://www.nhc.noaa.gov), the National Climatic Data Center (http://ncdc.noaa.gov), and the USGS Caribbean Water Resources District (http://pr.water.usgs.gov)

Year	Date	Name	San Juan	Fajardo	Roosevelt Roads	Canovanas	East Peak
			mm d^{-1}	mm d^{-1}	mm d^{-1}	mm d^{-1}	mm d^{-1}
1956	8/12/56	Betsy	81	104		135	
1960	9/5/60	Donna	40	209	188	nd	
1963	9/26/63	Edith	22	24	9	18	
1964	8/23/64	Cleo	18	31	nd	18	
1966	8/26/66	Faith	13	19	17	10	
1966	9/28/66	Inez	30	27	28	23	
1967	9/9/67	Beulah	8	11	5	25	
1979	8/30/79	David	67	117	117	233	48
1979	2/17/79	Edith	3	9	28	152	68
1987	9/22/87	Emily	3	14		14	24
1988	9/11/88	Gilbert	47	33	40	36	91
1989	9/18/89	Hugo	225			79	
1995	9/5/95	Luis	54	51	28	46	97
1996	9/15/96	Marilyn	86		10	47	51
1996	7/9/96	Bertha	40		39	61	145
1996	9/10/96	Hortense	208		73	182	
1997	9/5/97	Erika	7		13	8	23
1998	9/22/98	Georges	103		90	215	
1999	10/20/99	Jose	18		15	46	68
2000	8/23/00	Debby	118		44	132	nd
Average			59.6	54.1	46.5	77.9	68.3
Median			40.0	29.0	28.0	46.0	68.0
Maximum			225	209	188	233	145

are responsible for many of the short rainfalls that are common throughout the day. Disturbances generated by these systems are most common in the summer months, when rainfalls of 100 mm in 24 h or less can occur (Planos Gutiérrez 1999). The most intense rains occur over relatively small areas, and rainfalls of greater than 200 mm per event are known to occur. Landslides, uprooted trees, and localized and coastal plain floods are often associated with these events.

Extratropical Frontal Systems

Disturbance-generating rainfalls that occur from November to April are typically associated with cold fronts that originate in extratropical areas to the north. Rains associated with these systems are usually of low intensity but can last for several days. Storm totals are usually less than 150 mm. Nevertheless, intense rainfalls can be associated with these frontal systems, and landslides and flooding are common when intense downpours follow several days of persistent, soil-saturating rain. The

Table 4.5 Peak stream flows associated with named Puerto Rican storms since 1950. Based on data from the National Weather Service National Hurricane Center (http://www.nhc.noaa.gov), the National Climatic Data Center (http://ncdc.noaa.gov), and the USGS Caribbean Water Resources District (http://pr.water.usgs.gov)

Year	Date of max. rain	Name	Mameyes daily peak	Espiritu Santo daily peak	Fajardo daily peak	Icacos daily peak
			mm d^{-1}	mm d^{-1}	mm d^{-1}	mm d^{-1}
1963	9/26/63	Edith			0.46	
1964	8/23/64	Cleo			0.51	
1966	8/26/66	Faith		1.35	0.46	
1966	9/28/66	Inez		3.66	0.95	
1967	9/9/67	Beulah	0.54	0.83	0.33	
1979	8/30/79	David		15.29	10.01	
1979	2/17/79	Edith		16.62	1.15	
1987	9/22/87	Emily	1.55	0.95	0.24	1.53
1988	9/11/88	Gilbert	3.16	4.49	1.58	4.46
1989	9/18/89	Hugo	35.90	22.23	51.84	32.74
1995	9/5/95	Luis	2.49	2.56	2.37	6.06
1996	9/15/96	Marilyn	1.87	1.12	1.21	4.32
1996	7/9/96	Bertha	5.84	7.58	3.71	14.84
1996	9/10/96	Hortense	27.64	20.39	15.37	25.0
1997	9/5/97	Erika	0.61	0.49	0.12	0.49
1998	9/22/98	Georges	19.11	24.3	4.49	17.70
1999	10/20/99	Jose	5.29	8.04	4.29	7.04
2000	8/23/00	Debby	9.09	12.64	8.36	10.38
Average			9.4	8.9	6.0	13.7
Median			4.2	6.0	1.4	7.0
Hurricane maximum			35.9	24.3	51.8	32.7
Record maximum			35.9	32.3	55.7	37.6
Date of record			9/18/89	8/13/90	1/5/92	4/21/93

record discharge and floods of the Río Fajardo were caused by a 1992 cold front. Similar extratropical fronts are important disturbance-generating systems in tropical forests in Central America, the South Pacific, the South Atlantic, South Africa, and Australia.

These systems might have had a larger influence in the past, as cooler tropical climates during the late Quaternary glacial period and the last glacial-interglacial transition have been linked to an increased frequency of polar air masses reaching the tropics (Servant et al. 1993).

Coupled Ocean-Atmospheric Systems

Large-scale ocean-atmospheric systems like the NAO and the ENSO are principal causes of global interannual climate variability and have been linked to disturbances in other tropical forests (Scatena et al. 2005). During El Niño events, the entire Caribbean region is relatively dry from September to October (Chen and Taylor 2002). During declining ENSO phases, the Caribbean is relatively wet during April and July. These ENSO events have also been linked to Caribbean Sea surface temperature anomalies (Spence et al. 2004) and to an increase in global hurricane activity and disturbances in other tropical forests. Nevertheless, the NAO has a stronger relationship with Puerto Rico's annual climate than the ENSO does (Malmgren and Winter 1998). This index is the normalized sea-level pressure difference between the Azores and Iceland and is significantly related to annual rainfall. In general, during years with a high winter NAO, the precipitation in Puerto Rico is lower than average. However, correlations between annual rainfall and NAO or ENSO indices are generally weak. Likewise, the relationships between these indices and the specific occurrence of hurricanes or other large-scale disturbances in Puerto Rico are poor.

Biotic Drivers of Disturbances in the Luquillo Mountains

Population and Land Use Change

Petroglyphs and scattered archeological remains suggest that Luquillo ecosystems were affected by indigenous populations (see chapter 1). The analysis of preserved plant parts from archeological sites also indicates that Pre-Columbian inhabitants had measurable impacts on the island's vegetation and were responsible for local extinctions and species introductions (Newsom 1993). Nevertheless, the two periods of the greatest human-induced transformations of the Luquillo landscape occurred immediately after European settlement in 1498 and after the Spanish Crown opened the island to immigration in the early 1800s (Scatena 1989a). Most of the Luquillo Mountains below 400 m have undergone the following sequence of land use: selective logging and agroforestry, clearing and agriculture, farm abandonment and reforestation, and construction and urban buildup of the surrounding areas (Thomlinson et al. 1996). Coincident with this change in land use was an increase in per capita energy use and a switch from internal (e.g., solar, biomass) to external (e.g., fossil fuel) sources of energy.

The municipal centers that surround the Luquillo Mountains, including Río Grande, Luquillo, Fajardo, and Ceiba, were incorporated between 1772 and 1840. By 1895, large parts of the coastal plain and foothills were planted with sugar cane (Thomlinson et al. 1996). During this period, most of the agricultural activity within the present Luquillo Experimental Forest (LEF) consisted of small subsistence farms and coffee plantations. Both of the areas that now encompass the El Verde and Bisley research areas supported shade coffee plantations at that time. However, most of the commercial coffee plantations in Luquillo were abandoned after a major hurricane in 1898 (Scatena 1989a). Overall, the Luquillo coffee industry could not compete with plantations in the interior of the island because of low yields related to hurricane damage, high rainfall, and relatively acidic and less productive soils. Nevertheless, both the El Verde and Bisley areas, like most of the lower LEF, supported small subsistence farms until they were purchased by the USDA Forest Service in the 1930s.

By the 1950s, the forest cover on Puerto Rico had reached its minimum level (figure 4-5). Since then, the island's economy has shifted from a rural, agricultural-based economy to an industrialized economy that is based on manufacturing and services. Coincident with this shift has been the migration to urban centers, the abandonment of agricultural lands, and an increase in forest cover as agricultural areas naturally reforest. In 1935, approximately 46.7 percent of northeastern Puerto Rico had agricultural land cover, but by 2003, 57 percent of the island was forested (Brandeis et al. 2007). When abandoned agricultural lands are allowed to reforest naturally, they can attain mature forest biodiversity and biomass in approximately 40 years (Aide et al. 1995; Zimmerman et al. 1995a; Silver et al. 2004). Nevertheless, past land management can leave legacies in the forest composition and soil resources that can last for decades, if not centuries.

Pathogens and Insects

Pathogens and insects, like the chestnut blight of New England or the pine beetles of Central America, can result in such rapid and dramatic changes to forest structure and composition that they are often considered landscape-level disturbances (Holdenrieder et al. 2004). In Luquillo, short-term but forest-wide defoliation of *Piper* was observed following Hurricane Hugo. European bees are also considered a threat to the endangered Puerto Rican Parrot (*Amazona vittata*), and since 1970 these bees have been manually removed from cavities in important breeding areas (Snyder et al. 1987). Although other pests and pathogens are present, no large-scale pathogenic disturbance is known to have affected the LEF in recent centuries. In fact, an important unanswered question regarding the disturbance ecology of the Luquillo Mountains is why pathogenic disturbances are not more common or apparent.

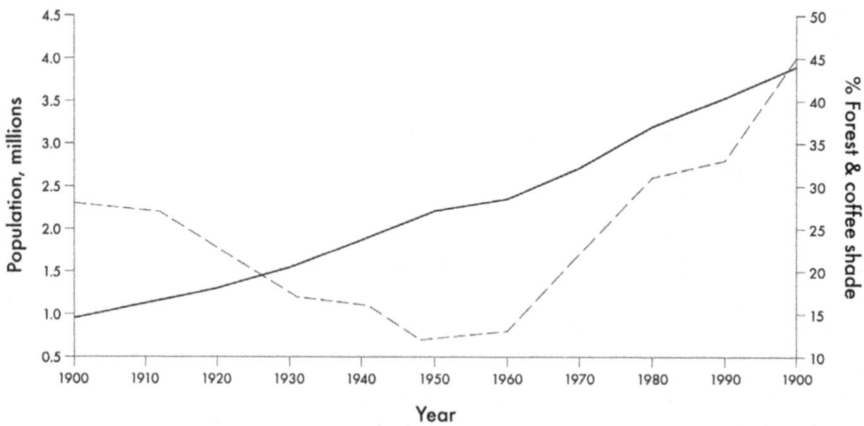

Figure 4.5 Island-wide population and forest cover (dashed line) by decade from 1900 to 2000.

Natural Disturbances in the Luquillo Mountains

Tree Mortality and Treefall Gaps

The creation of canopy gaps by individual or multiple treefalls is a common process involved in maintaining the structure and diversity of many tropical forests (Denslow 1987). The size of treefall gaps can range considerably but is typically between 50 and 100 m^2 (Hartshorn 1990). The rate of gap formation in mature tropical forests is typically around 1 gap ha^{-1} y^{-1}, and the turnover periods of forest canopy by tree fall gaps range from 50 to 165 years (table 4-6). Compared to other humid tropical forests, Luquillo has a high rate of canopy turnover due to hurricanes but a relatively low rate due to treefall gaps. The size and frequency of Luquillo gaps also varies with the topography, aspect, soil type, and forest age (Scatena and Lugo 1995). Only in riparian areas is the turnover by treefall gaps and slope failures faster than the turnover by hurricanes (table 4-7). Canopy throughfall in the center of a recent single-tree gap can be 30 to 50 percent higher than in the

Table 4.6 Canopy turnover periods by treefall gaps and hurricanes for some Neotropical forests

Location	Years	Source
Bisley, Puerto Rico: hurricane-induced defoliation; treefall, gap-induced	57–165	Scatena and Lugo 1995
Treefall gaps, Barro Colorado, Panama	62–159	Foster and Brokaw 1982
Treefall gaps, La Selva, Costa Rica	79–137	Hartshorn 1990
Treefall gaps, Tierra Firme, Amazonia	100	Uhl and Murphy 1981
Treefall gaps, Central America	62–155	Brokaw 1985
Treefall gaps, Los Tuxtlas, Mexico	61–138	Bongers et al. 1988

Table 4.7 Turnover periods in years by disturbance type and geomorphic setting for the Bisley watersheds between 1932 and 1989. (From Scatena and Lugo 1995.) Gaps = treefall gaps; Hurr = hurricanes; Slides = slope failures; Back = background, noncatastrophic mortality; dbh = diameter at breast height. Slope failures include those associated with hurricane and nonhurricane events

	Canopy area			Biomass				Stems > 10 cm dbh			
	Gaps yr	Hurr yr	Slides yr	Gaps yr	Hurr yr	Slides yr	Back yr	Gaps yr	Hurr yr	Slides yr	Back yr
Ridges	200	57	∞	250	110	∞	80	430	380	∞	75
Slopes	350	57	980	185	95	2,000	50	560	190	2,070	55
Valleys	60	57	430	40	105	680	30	110	145	400	50
Drainage	165	57	1,300	150	105	3,350	55	380	220	3,300	55

adjacent forest (Scatena 1989b). However, because the area in gaps is limited, their overall contribution to throughfall at the watershed scale is also limited (e.g., 3 percent).

The percentage of treefall gaps created by uprooted trees relative to snapped trees typically ranges between 20 and 50 percent in humid tropical forests and can be greater in steep land areas like Luquillo than in lowland tropical forests (Putz 1983; Scatena and Lugo 1995). The soil erosion associated with tree uproots contributes between 2.5 and 15 percent of the hillslope erosion in steep forested Luquillo watersheds (Larsen 1997; Larsen et al. 1999). In nearby agricultural and suburban watersheds, tree uproots account for less than 5 percent of soil erosion. The pit and mound features caused by these uproots typically occupy less than 0.1 percent of the Luquillo ground surface, whereas they can occupy as much as 60 percent in some temperate environments (Lenart et al. 2010). These differences are due in part to the dynamic surface erosion in tropical forests, which acts to remove rather than preserve the pit and mound features. Nevertheless, the pit and mound topography that results from treefalls does increase the surface storage of water and promotes the development of subsurface pipes and macropores.

The residuals of treefall gaps include an open canopy and associated microclimatic changes, coarse woody debris, and pit and mound topography. Microclimatic changes typically return to background levels within a year (Scatena 1989b). The pits and mounds created by treefalls can last for decades (Lenart et al. 2010) and are important microhabitats for certain Luquillo plants (Walker 2000).

Mass Earth Movements

Mass movements of earth are a common landform-scale disturbance in many upland humid tropical forests. In Luquillo, the velocity of downslope movement can range from the continuous downslope creep of soil profiles that occur on the order of millimeters per year (Lewis 1974) to debris flows that move tens of kilometers per hour. The frequency of Luquillo landslides and the rate of revegetation in mature forest stands have been related to bedrock geology, elevation, mean annual rainfall, and land use (Larsen and Simon 1993; Myster et al. 1997; Larsen et al. 1999). Within areas of similar geology and mean annual rainfall, mass wasting is five to eight times more frequent along roads than elsewhere and is most common on hillslopes that (1) have been anthropogenically modified, (2) have slopes greater than 12 degrees, and (3) face the prevailing trade winds.

Over the entire island of Puerto Rico, 1.2 landslide-producing storms occur each year (Larsen and Simon 1993). Storms with a total duration of 10 h or less typically require average rainfall intensities of nearly 14 mm h^{-1} in order to trigger landslides. In contrast, storms of 100 h or more can trigger landslides with an average rainfall intensity of 2 to 3 mm h^{-1}. A comparison of the Puerto Rican landslide threshold's relationship with rainfall intensities from the nearby island of Cuba indicates that all the common atmospheric systems in the Caribbean can produce landslide-generating storms (figure 4-6).

In the Luquillo Mountains, landslides are typically covered with herbaceous vegetation within 1 or 2 years, have closed canopies of woody vegetation in less

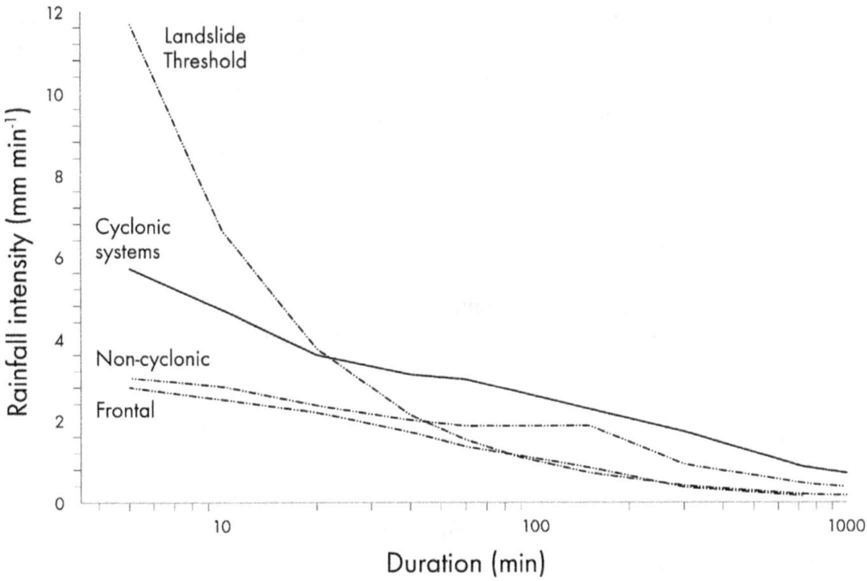

Figure 4.6 Rainfall intensity–duration curves by Caribbean storm type and a landslide rainfall intensity–duration threshold curve for Puerto Rico. (From Scatena et al. 2005.)

than 20 years, and have aboveground biomass equal to that of the adjacent forest after several decades (see chapter 5). Slope wash erosion and surface runoff from landslide scars also tend to approach adjacent forest values after a few years. After 4 years, the movement of surface soil on landslide scars can be reduced from 100 to 349 g m^{-2} y^{-1} to 3 to 4 g m^{-2} y^{-1} (Larsen et al. 1999). At the watershed scale, landslides can be major sources of stream sediment in upland humid tropical environments. They also disrupt roads and water conveyance systems and can be so chronic that certain roads in the forest need continual maintenance (Ahmad et al. 1993; Olander et al. 1998).

In summary, the changes in ecological space created by mass movements include the complete removal of above- and belowground biomass and changes in the local microclimate and soil resources. The residuals of mass movement include debris piles, unstable slopes, exposed soils, and pit and mound topography (Lenart et al. 2010). Their legacies include poor soil horizon development and the amphitheater-shaped valleys and narrow ridges that characterize much of the landscape (Scatena 1989a).

Floods and Fluvial Processes

Two general types of flood disturbances are commonly distinguished in the humid tropics: (1) seasonal inundation-type floods in which extensive areas are covered with lakelike water for extended periods (i.e., weeks to months) each year, and (2) event floods that are of relatively short duration (i.e., hours to days) and which have high-velocity stream flows (Scatena et al. 2005). In the steep Luquillo Mountains region, runoff is so rapid that only short-duration, high-intensity floods occur. However, standing floodwaters cover large parts of the surrounding coastal plain several times each century (Torres-Sierra 1996). Other areas of the tropics that have the physiographic and climatic conditions necessary to create flood disturbances and fluvial landforms similar to Luquillo and mountainous areas of the Caribbean and Central America include the following (Gupta 1988):

- river valleys of East Asia, especially Taiwan and the Philippines
- upland areas of Vietnam, Sumatra, Java, and Burma
- humid areas of the Indian subcontinent
- Madagascar and neighboring parts of coastal East Africa
- north and northeast Australia

Regression models of event-type flooding in the Luquillo Mountains indicate that both climatic and morphologic factors influence the magnitudes of peak flood discharge. Drainage area, mean annual rainfall, the 2-year 24-hour rainfall, the length of the main channel, and the total length of tributaries have been positively related to peak flood discharge and annual peak discharges (Ramos-Gines 1999; Rivera-Ramírez 1999). Likewise, the depth to bedrock and the watershed shape have been negatively correlated with peak discharge.

In the Luquillo region, the largest floods are not necessarily associated with hurricanes (table 4-5). The annual peak can occur in any month of the year but is most common in the late summer and fall (figure 4-3). In the lower Río Mameyes, flash floods (i.e., instantaneous discharge > 3.5 $m^3 s^{-1}$) occur at least once a month, and larger floods (instantaneous discharge > 18 $m^3 s^{-1}$) produced by cold fronts, tropical depressions, storms, and hurricanes occur several times per year on average (Blanco and Scatena 2005). These sudden increases in water depth and velocity increase water turbidity and flush particulate organic matter from the channels. If large enough, they can remove submerged aquatic vegetation, reduce periphyton, move bedload sediment, and rearrange aquatic habitats. Mass upstream migrations of juvenile freshwater snails can also be triggered by floods (figure 4-7).

Because of abundant bedrock and large boulders, the morphologies of the Luquillo stream channels are relatively stable and do not change dramatically following storm events. The stream hydraulic geometry is also considered relatively well developed, even in boulder-lined channels, and there are distinct longitudinal patterns in channel processes (Pike 2008; Pike et al. 2010). Floods also leave residuals, including the removal of periphyton and riparian vegetation, the addition of coarse woody debris, and the modification of aquatic habitats (Blanco and Scatena 2005, 2007). The sediments that fill the coastal plain and near-shore environments are the legacies of these processes.

Hurricanes

Hurricanes bring intense winds and rain that affect different parts of the landscape in different ways. At the scale of individual trees, winds in excess of about 100 to 130 km h^{-1} lethally damage trees within a few hours (figure 4-8). Winds in excess

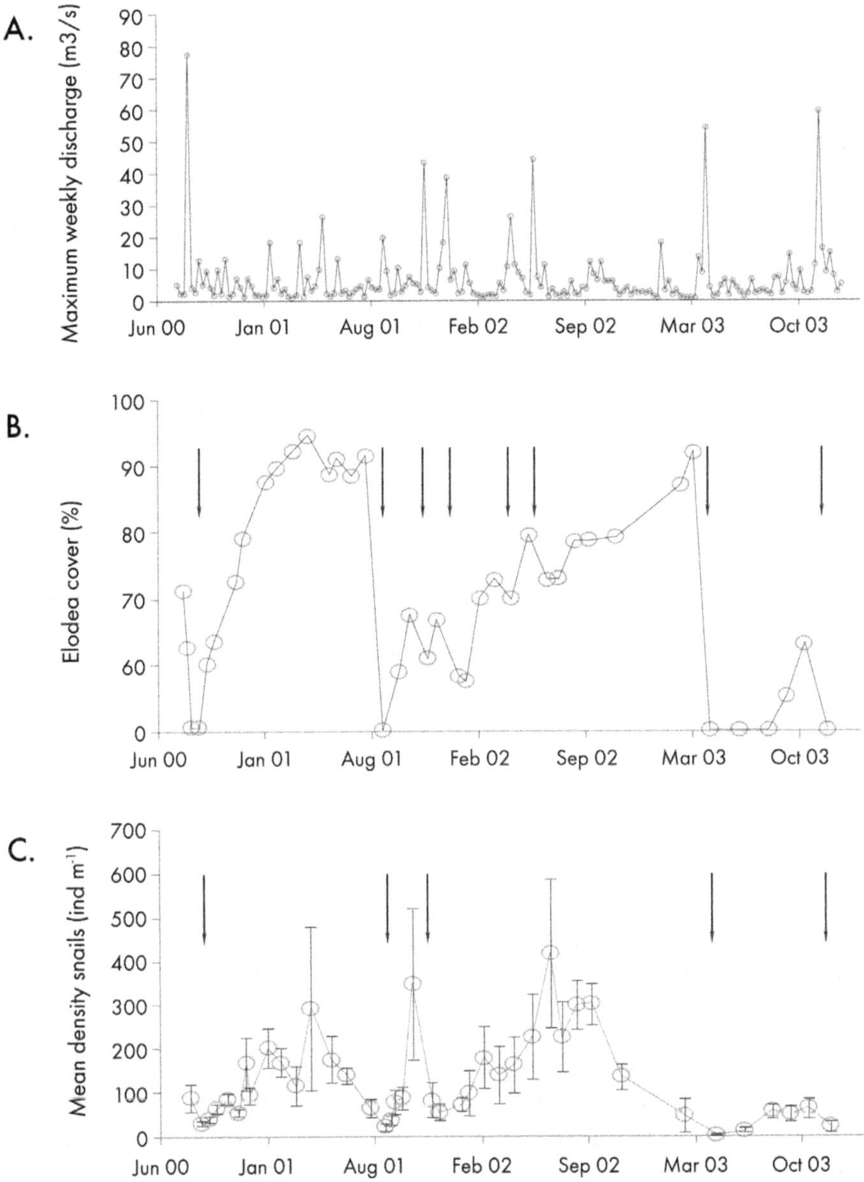

Figure 4.7 Stream discharge, cover of submerged aquatics, and density of freshwater snails in the lower Rio Mameyes between 2000 and 2002. Arrows indicate events of massive upstream snail migrations. (From Blanco and Scatena 2005.)

of 60 km h^{-1} cause large-scale defoliation and litterfall. At this scale, damage is related to the tree species, morphology, age, size, form, health, and rooting conditions. In general, fast-growing low-density woods are more susceptible to wind damage than high-density, late-successional species (Aide et al. 1995; Zimmerman et al. 1995b).

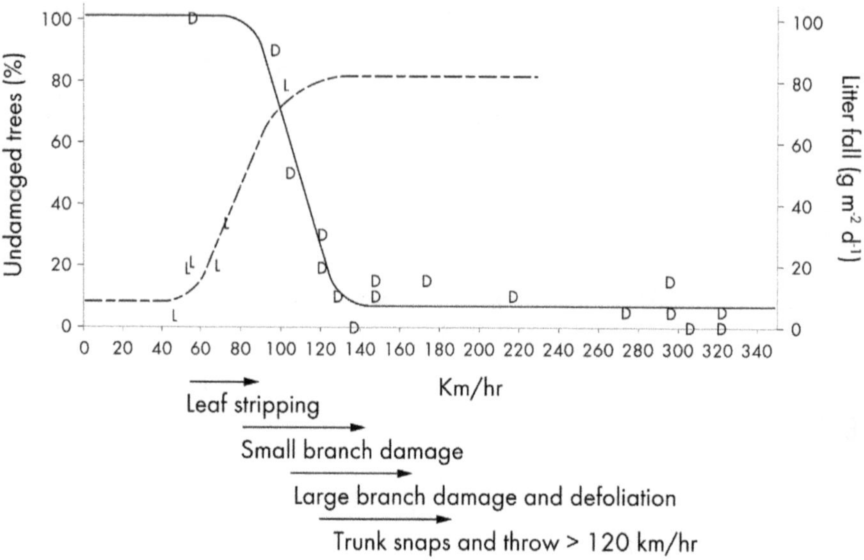

Figure 4.8 Maximum wind gusts versus percent of tree damage (D) and amount of litterfall (L) for the Luquillo Mountains. The types of canopy damage commonly observed for a given range of wind gusts are depicted by the arrows below the diagram. (After Scatena et al. 2005.)

At the landform scale, variations in wind damage result from differences in exposure and the modification of wind velocity caused by the landforms themselves. For example, valleys oriented parallel to the direction of dominant winds will receive more damage than nearby valleys that are perpendicular to the hurricane-force winds. Defoliation and the transfer of nutrients from the canopy to the forest floor can also cause major shifts in nutrient cycling pathways at the stand and landform scales (Lodge et al. 1991; Ostertag et al. 2003). Simulations of the hydrologic responses to daily rainfall following canopy defoliation suggest that significant changes in evapotranspiration, soil moisture, and stream flow occur when the forest is defoliated by 90 percent (figure 4-9). Moreover, although a 50 percent reduction in canopy leaf area does modify evapotranspiration, soil moisture, and stream flow, a 90 percent reduction in canopy leaf area can increase stream flow by over 300 percent relative to undisturbed conditions.

At the scale of the Luquillo Mountains, the spatial pattern of hurricane-induced damage can be complex and is strongly correlated with both aspect relative to the prevailing winds and forest type (Boose et al. 1994). In general, damage is spatially uniform in low-lying, uniform landscapes and more complex in dissected mountainous terrain. At the regional scale, the configuration of coastlines and mountains relative to the storm track determines how a storm will weaken when it crosses land. Factors controlling forest damage at this scale include gradients in wind velocity that are related to the size and intensity of the hurricane and large topographic features. In Puerto Rico, the Luquillo Mountains apparently influence the path of hurricanes across the island and create what is locally called the "Puerto Rican Split"—that is, the tendency for hurricanes approaching the Luquillo Mountains to be deflected to the north or south of the mountains. Apparently the 1,000 m Luquillo Mountains create enough resistance and friction to deflect the trajectory of approaching hurricanes to either the north or the south of the island.

At all scales, hurricanes create patches of survivors and new regeneration that change in structure and composition over decades (see Crow [1980] and chapter 5 for details). Nevertheless, hurricanes do not erase the signature of past land use on the species composition, and the composition of posthurricane regeneration can be directly related to the prior land use (García-Montiel and Scatena 1994; Zimmerman et al. 1995a; Thompson et al. 2002). In some stands, the forest composition still reflects the prior composition of shade coffee plantations after 100 years of abandonment and the direct impacts of several hurricanes.

Hurricane-related storm discharges can cause significant geomorphic modifications to stream channels (Scatena and Johnson 2001). Posthurricane stream-water concentrations of sediments and nutrients can also be elevated for months to years

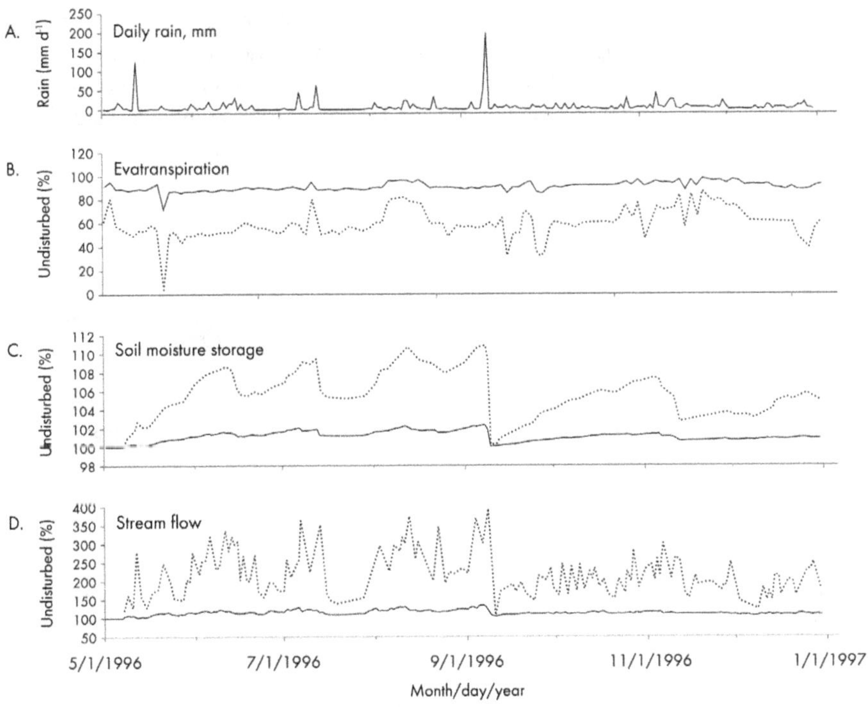

Figure 4.9 Daily rainfall and corresponding simulations of evapotranspiration, soil moisture storage, and stream flow in the Bisley Research Watersheds following simulated reductions in canopy leaf area of 50 and 90 percent. Simulations are expressed as a percent of undisturbed conditions for the same daily rainfall sequence and are represented by solid lines for a 50 percent reduction in canopy leaf area and dashed lines for a 90 percent reduction. (From Scatena et al. 2005.)

(Schaefer et al. 2000). However, suspended sediment concentrations can be lower than predicted from concentration-discharge relationships derived from nonhurricane storms of similar magnitudes (Gellis 1993). Apparently the defoliation caused by the hurricane-force winds creates residual debris dams that trap sediment and reduce suspended sediment concentrations. Nevertheless, because high stream flow can last for several days, the total sediment transported during the passage of a hurricane can be significant.

Hurricane winds also result in immediate changes in the canopy cover and aboveground biomass, the microclimate (Fernández and Fetcher 1991), and throughfall (Heartsill et al. 2007). In most tropical forest understories, the daily photosynthetic photon flux density (PPFD) is 1 to 2 percent of the value above the canopy (see, e.g., Denslow and Hartshorn 1994); at El Verde, the solar irradiance at the forest floor was 5 to 47 percent of full sunlight after Hurricane Hugo (Petty 1993; Scatena et al. 1996), a huge increase in this crucial environmental factor. For the 10 months after Hurricane Hugo, levels of understory PPFD were highly variable at a scale of 1 m, but the median was 7.7 to 10.8 mol m^{-2} d^{-1}, which is comparable to PPFD levels in a 400 m^2 treefall gap (Fernández and Fetcher 1991; Turton 1992; Bellingham et al. 1996; Fetcher et al. 1996). Values had fallen to 0.8 mol m^{-2} d^{-1} by 14 months, at which point rapid growth of *Cecropia schreberiana* overtopped the light sensors in the study. This is a clear example of how ecological space shifts rapidly over points in geographic space owing to disturbance and biotic response (see chapter 2).

In addition to these immediate changes to the forest's structure and ecological space, hurricanes also leave residuals on the landscape. These include landslides,

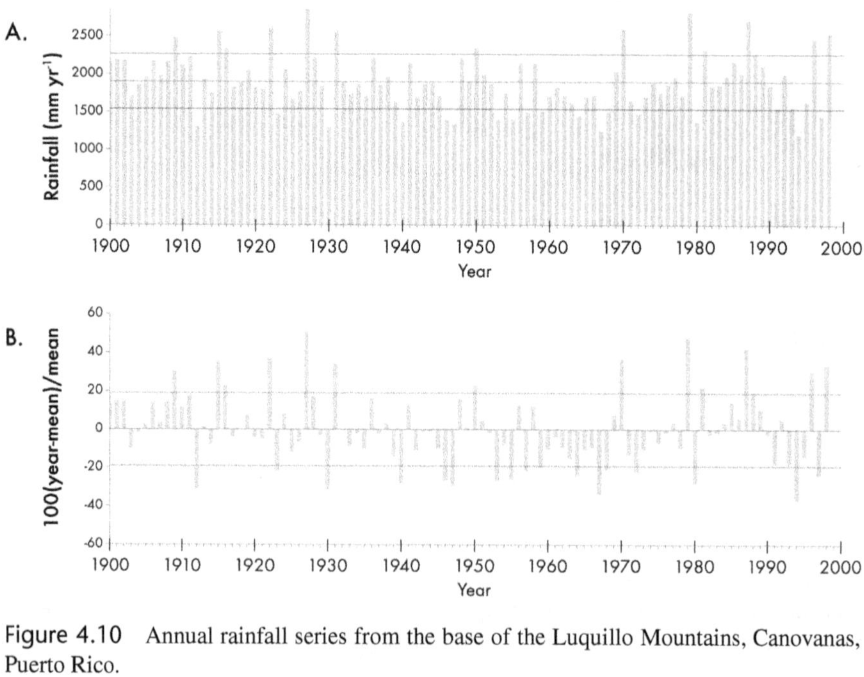

Figure 4.10 Annual rainfall series from the base of the Luquillo Mountains, Canovanas, Puerto Rico.

tip-up mounds, accumulations of coarse woody debris and litter, and large patches of defoliated or fallen trees. In turn, these residuals create opportunities for the regenerating forest and leave legacies that include cohorts of even-age stands that are distributed in patches across the mountain (see chapter 5).

Droughts

Droughts have historically been important disturbances to both the natural and human-dominated ecosystems of Puerto Rico. Rainfall records (figure 4-10) and interviews with long-term residents suggest that the 2-year drought of 1946 and 1947 was the worst drought on record. During the second year of this drought, headwater streams near El Verde were dry, crops failed, and Luquillo Mountain farmers were forced to travel to the other side of the island to find work (Alejo Estrada, University of Puerto Rico Research Technician, personal communication, 1994).

Although droughts have always been an important disturbance in the Luquillo Mountains, long-term precipitation records suggest that they might be becoming more frequent. During the 20th century, the annual precipitation had negative trends in all eight of the precipitation stations on the island, with records starting around 1900 (van der Molen 2002). The negative trends were significant in six of the eight stations and ranged from −1.59 to −4.90 mm y^{-1}. Another detailed trend analysis of 24 stations found that between 1931and 1996, 71 percent of the stations had significant decreases in monthly precipitation between May and October (Bisselink 2003). These decreases ranged between 0.6 and 2.3 mm y^{-1}. The same study also found that winter precipitation increased by 0.3 to 1.7 mm y^{-1}. Since 1987 and the initiation of the Luquillo Long-Term Ecological Research (LTER) project, the mean weekly rainfall has also decreased significantly at both the Bisley and El Verde research sites (Heartsill-Scally et al. 2007). In Bisley, the mean daily rainfall and throughfall had an average decline of 0.2 and 0.23 mm y^{-1}, respectively. Although significant, these declines are less than the average variation between years and between days. Nevertheless, islandwide, 1997, 1994, and 1991 were the second, third, and sixth driest years in the 20th century (Larsen 2002). Widespread mandatory water rationings also occurred six times on the island in the 1990s. The most severe drought, which occurred in 1994, resulted in an economic loss of $165 million (Lugo and García-Martinó 1996).

At the watershed level, the drainage density (the ratio of the length of tributaries to the length of the main channel), the percentage of the drainage basin with a northeast aspect, and the average weighted slope of the drainage basin have been used to estimate low stream flows (García-Martinó et al. 1996). At the scale of forest stands, short-term dry periods lasting weeks to months are common and have been linked to declines in the abundance of common lizards, spiders, exotic earthworms, and palaemonid river shrimp (Reagan and Waide 1996; Zou and González 1997; Covich et al. 2006). Prolonged dry periods can result in increased litterfall and decreased root biomass (Beard et al. 2005), whereas wet and drying cycles can stimulate microbial biomass growth, enhance microbial nitrogen immobilization, and impact detrital food chains (Lodge et al. 1994; Ruan et al. 2004). However,

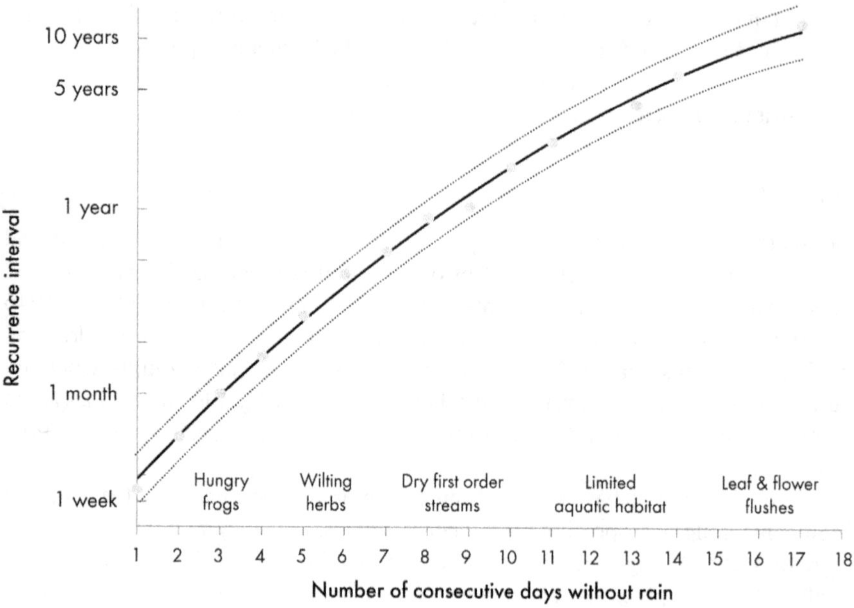

Figure 4.11 Ecological response to the number of consecutive days without rain in the tabonuco forest of the Luquillo Mountains of Puerto Rico. (Modified from Covich et al. 2006.)

these responses can be asynchronous and lag behind dry periods by weeks to months (Ruan et al. 2004). Responses to dry periods are also more apparent in well-drained ridges than in the wetter riparian valleys (Silver et al. 1999). In general, after 3 days without rain, the abundant tree frogs have empty stomachs because of the lack of insects that normally occur in wet forest litter (figure 4-11). One week without rain or canopy throughfall occurs nearly every year and causes the wilting of herbaceous vegetation in open areas such as gaps and roadways. Once every 10 to 20 years, there are enough consecutive rainless days that small headwater streams become dry and aquatic habitat becomes limiting. Unlike other disturbances, droughts cause residuals that are relatively short-lived (Beard et al. 2005), but their long-term legacies are not yet understood.

Wildfire and Lightning

Paleoclimatic evidence from the Caribbean, Amazonia, and Central America indicates that most Neotropical humid tropical forests have experienced fires and extended droughts during the past 10,000 years (Hodel et al. 1991; Servant et al. 1993). Deep ground fires have also been shown to cause massive above- and belowground biomass losses in tropical montane cloud forests in Mexico (Asbjornsen et al. 2005). Holocene charcoal stratigraphy from the north-central coast of Puerto Rico also indicates that fire frequency greatly increased at the time of human arrival to the island (Burney and Burney 1994). Nevertheless, interviews with long-time Luquillo residents indicate that there have been no extensive wildfires in the tabonuco or upper elevation forests in the past 80 years. Small patches (<0.5 ha) of roadside ferns and shrubs do burn nearly every year at lower elevations. However, the forest is generally considered too wet to sustain large wildfires.

Lightning-induced fires and tree mortality have been identified as important natural disturbances in many humid and dry tropical forests (Whitmore 1984; Horn 1991; Richards 1996; Middleton et al. 1997). In 15 years of observation in 13 ha of the Bisley watershed area, lightning strikes have killed three canopy trees (Scatena personal observation). None of these events caused fires, treefalls, or damage to multiple trees. Over a 10-year period, lightning has also damaged two of seven exposed LTER climate stations. At these rates, lightning-induced tree mortality is conservatively, and crudely, estimated at a relatively low rate of one or two trees per hectare per century. This relatively low rate of lightning might be due to the trade winds shearing the tops of developing convective clouds before they develop to the lightning-producing stage. When lightning strikes do occur, they leave isolated individual standing dead trees.

Human-Induced Disturbances in the Luquillo Mountains

The most commonly cited disturbance-generating activities that are currently operating in the Luquillo Mountains are water resource extraction, road development, and recreation (Scatena et al. 2002; Ortiz-Zayas and Scatena 2004). Long-term studies have shown that historical selective harvesting for timber and charcoal, agriculture, agroforestry, hunting, road building, water diversions, and two airplane wrecks have all caused measurable and documented changes in the ecological space of the Luquillo Mountains (see chapter 6). Unlike many of the natural disturbances, most human-induced disturbances in the Luquillo Mountains were not discrete events and instead have been cumulative and progressive in nature.

Selective Harvesting for Timber and Charcoal

The Luquillo Mountains have historically been an important source of timber and charcoal for the island. Because of the relative inaccessibility of the steeply sloping mountains, tree harvesting prior to the late 1880s was initially limited to valuable timber species, namely, ausubo (*Manilkara bidentata*) and laurel (*Magnolia splendens*) (García-Montiel and Scatena 1994). However, with increasing demand for fuelwood, tree harvesting for charcoal production became important throughout the first half of the 1900s. Today the legacy of this activity can be seen in cut tree stumps, remnant charcoal pits, and skidtrails that are scattered throughout the lower elevation areas of the forest. Comparisons of the forest structure and composition around abandoned charcoal pits and cut stumps indicate that these activities can change the local species composition. However, single-tree harvesting leaves a smaller legacy than the production of charcoal (García-Montiel and Scatena 1994). Moreover, neither disturbance creates uproot pits and mounds that facilitate regeneration, nor do they provide biomass for decomposition and nutrient recycling.

Agriculture and Agroforestry

In general, the type and magnitude of agricultural practices in the Luquillo Mountains have varied with topographic, geomorphic, and pedologic conditions. Within the Bisley watersheds, the major land use and impacts on ridges were associated with selective logging and silviculture (García-Montiel and Scatena 1994). In contrast, valleys and slopes tended to be used for agroforestry. At El Verde, areas with rocky soils were left to forest, whereas adjacent areas were cleared for pasture and cropland. Coffee cultivation, which involved liming of the soil and the cultivation of nitrogen-fixing shade trees, might have left detectable legacies in the soil pH and nitrogen following 70 years of abandonment (Beard et al. 2005). Agricultural land uses also left legacies in the species composition (see chapter 5), nutrient cycles (Silver et al. 2004), the spatial distributions of soil bacterial activity (Willig and Moorehead 1996), and the distribution of the tailless whip scorpion spider *Phrynus longipes* (Arachnida: Amblypygi) (Bloch and Weiss 2002).

Water Diversions

Water that is diverted for domestic and municipal uses is one of the major economic products of the Luquillo Mountains and accounts for about 10 percent of the total water deliveries on the island (Ortiz-Zayas and Scatena 2004). These water withdrawals have been directly linked to the reduction in the area of aquatic habitat and in the migratory routes of common aquatic species (Benstead et al. 1999; Scatena 2001; Blanco and Scatena 2005).

For decades, the standard practice has been to build small (<3 m high) dams in upland streams of the Luquillo Mountains. Water is then diverted by gravity for human uses at lower elevations. The resulting wastewater is then returned to the rivers near their estuary. This water use has been shown to alter water chemistry (Santos-Román et al. 2003) and impact the abundance and composition of aquatic life. Large dams on the island can be complete barriers to migration (Holmquist et al. 1998), and smaller diversions act as filters (Benstead et al. 1999). Since the early 1990s, the Luquillo LTER program has made significant progress in understanding the ecology and instream requirements of aquatic organisms in the Luquillo Mountains. Much of this research has been used to develop more ecologically based water management practices (see chapter 7).

Fishing and Hunting

Although poorly quantified, fishing and hunting have been, and continue to be, important community-level disturbances affecting the Luquillo Mountains. The declines of both the Puerto Rican Parrot and the plain pigeon (*Columba inornata*) have been partly attributed to hunting. Although bird hunting is not allowed within the Experimental Forest, every year birds are hunted in the region, and shell casings are commonly found near the forest boundary.

Fishing in Luquillo streams is a more common practice than hunting. Freshwater shrimp, fish, and snails are all caught on a daily basis from Luquillo Mountain streams and used for local consumption. Although historical rates of fishing are not known, interviews with long-term residents indicate that the fishing pressure on aquatic organisms is greater now than in the recent past. Moreover, during the agrarian period, local residents did not have time to fish and fished only on special occasions. When they did fish, they used the traditional hand and gig methods. The increase in free time since the late 1950s has apparently increased the fishing pressure on Luquillo Mountain streams. In addition, harvesters now use baited traps, large nets, and poison or chemical approaches to harvesting fish and shrimp. Harvest-related poisoning events cause massive mortality of shrimps and aquatic life and leave legacies that are apparent for months after an event (Greathouse et al. 2005).

Recreation

Recreational visits by island and nonisland residents are the greatest direct use of the Luquillo Mountains. In fact, the Luquillo Mountains have one of the highest visitor uses per area of any forest or grassland in the National Forest system (Scatena et al. 2002). Between 1980 and 1990, this tourism generated approximately US$5.2 million per year in economic activity. Most of the visitation occurs in the summer months and during weekdays (figure 4-12). However, most of this recreation is relatively passive and includes picnicking, swimming, and hiking in designated areas. Therefore, the direct impacts of recreation on the ecosystems of the Luquillo Mountains are considered limited and restricted to designated recreation areas and roadways. Nevertheless, where recreation is intense, it does leave residuals of trash and trampled riparian vegetation.

Road Building

All of the major roads in the Luquillo Mountains were constructed prior to 1970. Although some off-road vehicles occasionally enter the forest, this activity is currently limited to annexed lands near the community of Cubuy. The vast majority of vehicle use is on paved or maintained roads. Nevertheless, it has been shown that roads greatly increase the magnitude and frequency of landslides and promote the establishment of alien species within the forest. Mass wasting is five to eight times more frequent along roads and can affect an area that is several times the width of the road itself (Larsen and Parks 1997). The legacies left by road building include the expansion of nonnative species (Olander et al. 1998), landslides, and changes in slope morphology (Larsen and Parks 1997). Because of the aging road network, a greater frequency of landslides and road-related disturbances is expected in the future.

Future Disturbance Regimes

Although the Luquillo Mountains have had a dynamic and resilient history, past performance is not a guarantee of future behavior. In the next 100 years, the magnitude and frequency of the disturbances affecting the Luquillo Mountain forests

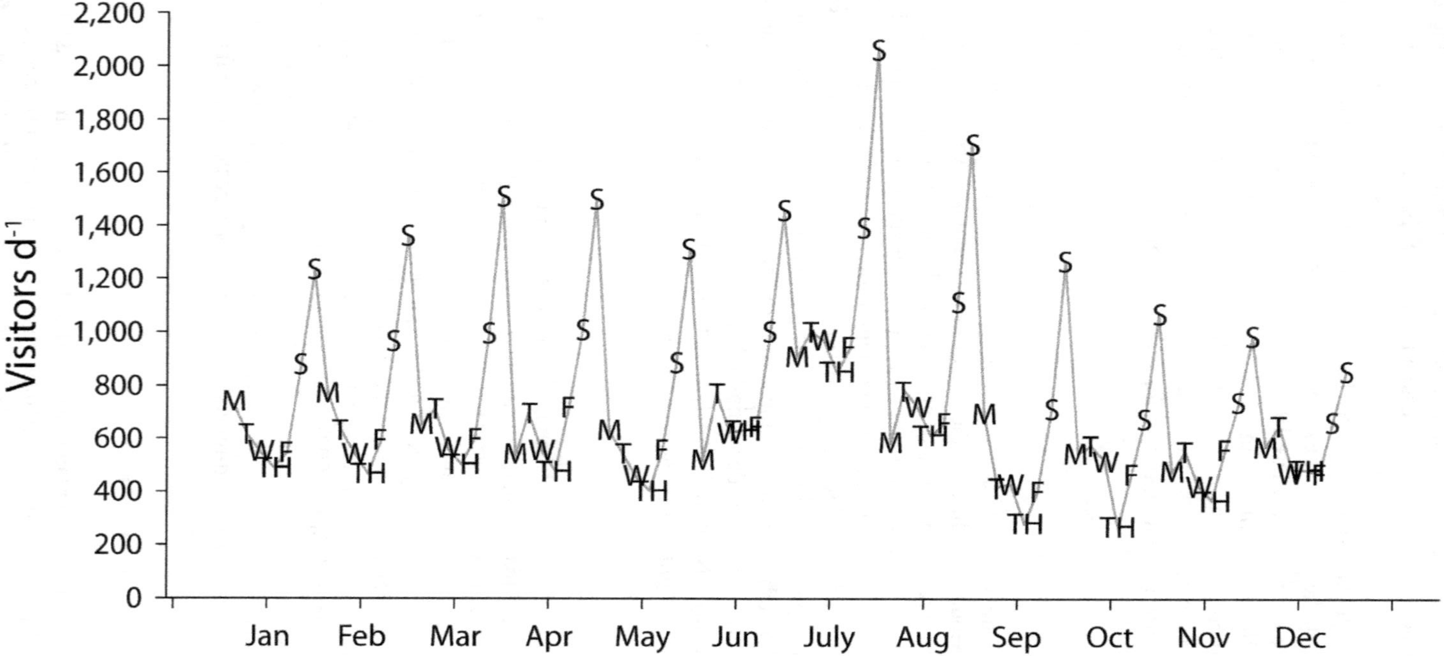

Figure 4.12 Average number of visits to the Yokahu Recreational Center in the Caribbean National Forest by day of week and month. Based on daily recreational surveys from 1980 to 1995.

can be expected to change as a result of changing local, regional, and global stressors.

Model simulation of future global-scale carbon dioxide (CO_2)-induced climate change indicates that over the next few decades, Puerto Rico might experience increases in hurricane activity (Emanuel 1987), increases in the length of the dry season, and decreases in soil moisture (Hulme and Viner 1995). Recent empirical and simulation studies also indicate that deforestation of the coastal plain reduces cloud moisture and rainfall over the island (van der Molen 2002). Increases in water diversions, land use change (Wu et al. 2007), and urban heat island effects (González et al. 2005) will also act to dry the landscape. All of these activities imply that droughts, and possible fires, will also become more common.

Although hurricane activity is expected to increase, the magnitude of the resulting hurricane-induced changes is uncertain. High-resolution computer simulations of 51 western Pacific storms under present-day and high-CO_2 conditions indicate that windspeeds will increase 3 to 7 m s^{-1} (5 to 12 percent) with a 2.2°C increase in the sea surface temperature (Knutson et al. 1998). Given the magnitude of the expected increase in windspeeds and the resilience of the Luquillo forests to wind (figures 4-8 and 4-9), increases in hurricane frequency might be more important than increases in the magnitude of hurricane winds. Simulations of the response of Luquillo Mountain forests to changes in hurricane frequency indicate that a range of forest compositions can occur with different hurricane regimes (O'Brien et al. 1992). In general, a decrease in hurricane frequency will result in mature forest with large trees, whereas an increase in frequency will result in forests that are shorter, are younger, and have a greater abundance of pioneer species and lower aboveground biomass.

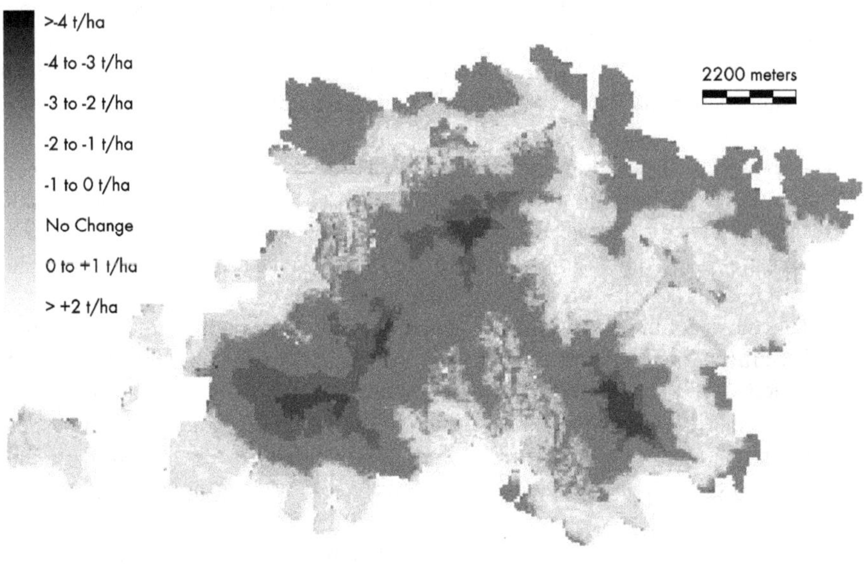

Figure 4.13 Simulated changes in soil organic carbon in response to an increase in temperature. (From Wang et al. 2002a.)

The absolute magnitudes of historical and future changes in the climate of the Luquillo Mountains are unknown. Nevertheless, the precipitation records and the presence at lower elevations of isolated large upper elevation trees does suggest that the Luquillo Mountains are drying and that the drier forest types are migrating to higher elevations (see chapter 3 for a description of forest types). Based on existing climate–elevation relationships, a change in the air temperature of 1.5°C to 2.5°C and changes in precipitation of −11 to +33 percent would drastically alter the distribution of forest types in the Luquillo Mountains (Scatena 1998). Simulations indicate that a warming of 2.0°C is likely to result in losses in soil organic carbon in the lower and higher elevations, but increased storage in the middle elevations, of the Luquillo Mountains (figure 4-13) (Wang et al. 2002a). Simulations also suggest that both gross and net primary productivity would decrease under a doubling of atmospheric CO_2 (Wang et al. 2002b).

Regardless of climate changes, changes in land use and land cover are also expected to change the hydrologic budgets of the Luquillo Mountains. Comparisons of rainfall and stream flow between the "agricultural" period of 1973–1980 and the "urbanized/reforested" period of 1988–1995 indicate that a smaller proportion of rainfall became stream flow in the urbanized/forested period because of reforestation (Wu et al. 2007). Simulations in the same study indicate that annual stream flow in northeastern Puerto Rico would decrease by 3.6 percent in a total reforestation scenario, and it would decrease by 1.1 percent if both reforestation and urbanization continue at their present rates until 2020.

Summary

- Like many humid tropical environments, the Luquillo Mountains is a dynamic ecosystem that is affected by a wide array of environmental processes and disturbances. Events that concurrently alter the environmental space of several different areas of the Luquillo Mountains occur every 2 to 5 years. Events such as hurricanes that cause widespread environmental modification occur once every 20 to 50 years.
- Although the Luquillo Mountains are the product of ancient igneous and tectonic activity, they are not as tectonically active as many tropical mountains and have been subaerial for millions of years. Nevertheless, they do receive occasional ash falls from volcanoes in the lower Caribbean, and multiple earthquakes are measured on the island each year.
- The most common disturbance-generating weather systems that affect the Luquillo Mountains are (1) cyclonic systems, (2) noncyclonic intertropical systems, (3) extratropical frontal systems, and (4) large-scale coupled ocean-atmospheric events (e.g., North Atlantic Oscillation, El Niño-Southern Oscillation). Unlike in some tropical forests, disturbances associated with the passage of the Inter-Tropical Convergence Zone or monsoonal rains are not common.
- Hurricanes are considered the most important natural disturbance affecting the structure of forests in the Luquillo Mountains. Compared to other humid tropical forests, Luquillo has a high rate of canopy turnover by hurricanes but a

relatively low rate by tree-fall gaps. Historically, pathogenic disturbances have not been uncommon.

- Hurricane-related storm discharges can cause significant geomorphic modifications to Luquillo stream channels, and stream water concentrations of sediments and nutrients can be elevated for months to years following a major hurricane. However, the largest floods are not necessarily associated with hurricanes, and the annual peak discharge can occur in any month of the year but is most common in the late summer and fall.
- Over the entire island of Puerto Rico, 1.2 landslide-producing storms occur each year. In the Luquillo Mountains, landslides are typically covered with herbaceous vegetation within 1 or 2 years, have closed canopies of woody vegetation in less than 20 years, and have an aboveground biomass equivalent to that of the adjacent forest after several decades.
- Human-induced disturbances have historically included tree harvesting for timber and charcoal, agriculture, and agroforestry. In the past few decades, water diversions, fishing and hunting, and road building have been important disturbances. Present and future human-induced disturbances are related to regional land use change, the disruption of migratory corridors, and forest drying related to coastal plain deforestation and regional climate change.

Literature Cited

Ahmad, R., F. N. Scatena, and A. Gupta. 1993. Morphology and sedimentation in Caribbean montane streams: Examples from Jamaica and Puerto Rico. *Sedimentary Geology* 85:157–169.

Aide, T. M., J. K. Zimmerman, L. Herrera, M. Rosario, and M. Serrano. 1995. Forest recovery in abandoned tropical pastures in Puerto Rico. *Forest Ecology and Management* 77:77–86.

Asbjornsen, H., N. Velázquez-Rosas, R. García-Soriano, and C. Gallardo-Hernández. 2005. Deep ground fires cause massive above-and-below ground biomass losses in tropical montane cloud forests in Oaxaca, Mexico. *Journal of Tropical Ecology* 21:427–434.

Beard, K. H., K. A. Vogt, D. J. Vogt, F. N. Scatena, A. P. Covich, R. Sigurdardottir, T. G. Siccama, and T. A. Crowl. 2005. Structural and functional responses of a subtropical forest to 10 years of hurricanes and droughts. *Ecological Monographs* 75:345–361.

Bellingham, P. J., E. V. J. Tanner, P. M. Rich, and T. C. R. Goodland. 1996. Changes in light below the canopy of a Jamaican montane rainforest after a hurricane. *Journal of Tropical Ecology* 12:699–722.

Benstead, P. J., J. G. March, C. M. Pringle, and F. N. Scatena. 1999. Effects of a low-head dam and water abstraction on migratory tropical stream biota. *Ecological Applications* 9:656–668.

Bisselink, B. 2003. Precipitation trends in Puerto Rico: Quantification and explanation of complex patterns. M.S. thesis. Vrije Universiteit, Amsterdam.

Blanco, J. F., and F. N. Scatena. 2005. Floods, habitat hydraulics and upstream migration of *Neritina virginea* (Gastropoda:Neritodae) in northeastern Puerto Rico. *Caribbean Journal of Science* 41:55–74.

Blanco, J. F., and F. N. Scatena. 2007. The spatial arrangement of *Neritina virginea* during upstream migration in a split-channel reach. *River Research and Applications* 23:235–245.

Bloch, C. P., and L. Weiss. 2002. Distribution and abundance of the whipspider *Phrynus longipes* (Arachnida: Amblypygi) in the Luquillo Experimental Forest, Puerto Rico: Response to natural and anthropogenic disturbance. *Caribbean Journal of Science* 38:260–262.

Bongers, F., J. Popma, and J. Meave del Castillo Carabias. 1988. Structure and floristic composition of a lowland rainforest of Los Tuxtlas, Mexico. *Vegetatio* 74:55–80.

Boose, E. R., D. R. Foster, and M. Fluet. 1994. Hurricane impacts to tropical and temperate forest landscapes. *Ecological Monographs* 64:396–400.

Brandeis, T. J., E. H. Helmer, and S. N. Oswalt. 2007. The status of Puerto Rico's forests, 2003. Resource Bulletin SRS-119. Asheville, NC: U.S. Department of Agriculture Forest Service, Southern Research Station.

Brokaw, N. V. L. 1985. Treefalls, regrowth and community structure in tropical forests. Pages 53–69 in S. T. A. Pickett and P. S. White, editors, *The Ecology of natural disturbance and patch dynamics*. New York: Academic Press.

Burney, D. A., and L. P. Burney. 1994. Holocene charcoal stratigraphy from Laguna Tortugero, Puerto Rico and the timing of human arrival on the island. *Archaeological Science* 21:273–281.

Chen, A. A., and M. A. Taylor. 2002. Investigating the link between early season Caribbean rainfall and the El Niño +1 year. *International Journal of Climatology* 22:87–106.

Clark, J. J., and P. R. Wilcock. 2000. Effects of land-use change on channel morphology in northeastern Puerto Rico. *Geological Society of America Bulletin* 112:1763–1777.

Connell, J. H. 1978. Diversity in tropical rain forest and coral reefs. *Science* 199:1202–1210.

Covich, A. P., T. A. Crowl, and T. Heartsill-Scalley. 2006. Effects of drought and hurricane disturbances on headwater distributions of palaemonid river shrimp (*Macrobrachium* spp.) in the Luquillo Mountains, Puerto Rico. *Journal of the North American Benthological Society*. 25:99–107.

Cox, D. P., R. F. Marvin, J. W. M'Gonigle, D. H. McIntyre, and C. L. Rogers. 1977. Potassium–argon geochronology of some metamorphic, igneous, and hydrothermal events in Puerto Rico and the Virgin Islands. U.S. Geological Survey. *Journal of Research* 5:689–703.

Crow, T. R. 1980. A rainforest chronicle a 30-year record of change in structure and composition at El Verde, Puerto Rico. *Biotropica* 12:42–55.

Denslow, J. S. 1987. Tropical treefall gaps and tree species diversity. *Annual Review of Ecology and Systematics* 18:431–451.

Denslow, J. S., and G. S. Hartshorn. 1994. Tree-fall gap environments and forest dynamic processes. Pages 120–127 in L. A. McDade, K. S. Bawa, H. A. Hespenheide, and G. S. Hartshorn, editors, *La Selva: Ecology and natural history of a Neotropical rain forest*. Chicago: University of Chicago Press.

Dillon, W., and U. T. Brink. 1999. Seismic and tsunami hazards in northeast Caribbean. *Eos* 80:309–310.

Emanuel, K. A. 1987. The dependence of hurricane intensity on tropical cyclones. *Nature* 326:483–485.

Fernández, D. S., and N. Fetcher. 1991. Changes in light availability following Hurricane Hugo in a subtropical montane forest in Puerto Rico. *Biotropica* 23:393–399.

Fetcher, N., B. L. Haines, R. A. Cordero, D. J. Lodge, L. Walker, D. S. Fernández, and W. T. Lawrence. 1996. Responses of tropical plants to nutrients and light on a landslide in Puerto Rico. *Journal of Ecology* 84:331–341.

Florentin, J. M., R. Maurrasse, and S. Gautam. 1991. Impacts, tsunamis, and the Haitian Cretaceous-Tertiary Boundary layer. *Science* 252:1690–1692.

Foster, R. B., and N. V. L. Brokaw. 1982. Structure and history of the vegetation on Barro Colorado Island. Pages 67–81 in E. G. Leigh, Jr., A. S. Rand, and D. M. Windsor, editors, *The ecology of a tropical forest: Seasonal rhythms and long-term changes*. Washington, DC: Smithsonian Institute Press.

García-Martinó, A., F. N. Scatena, G. S. Warner, and D. L. Civco. 1996. Statistical low flow estimations using GIS analysis in humid montane regions of Puerto Rico. *Journal of the American Water Resource Association* 32:1259–1271.

García-Montiel, D., and F. N. Scatena. 1994. The effect of human activity on the structure and composition of a tropical forest in Puerto Rico. *Forest Ecology and Management* 63:57–78.

Gellis, A. 1993. The effects of Hurricane Hugo on suspended sediment loads, Lago Loiza Basin, Puerto Rico. *Earth Surface Processes and Landforms* 18:505–517.

González, J., J. C. Luvall, D. Rickman, D. Comarazamy, A. J. Picón, E. W. Harmsen, H. Parsiani, N. Ramírez, R. E. Vásquez, R. Williams, R. B. Waide, and C. A. Tepley. 2005. Urban heat islands developing in coastal tropical cities. *Eos* 86:397–403.

Graham, A., and D. M. Jarzen. 1969. Studies in Neotropical paleobotany. I. The Oligocene communities of Puerto Rico. *Missouri Botanical Garden* 56:308–357.

Gray, W. M., J. D. Sheaffer, and C. W. Landsea. 1997. Climate trends associated with multidecadal variability of Atlantic hurricane activity. Pages 15–23 in H. F. Diaz and R. S. Pulwarty, editors, *Hurricanes, climate and socioeconomic impacts*. Berlin: Springer-Verlag.

Greathouse, E. A., J. G. March, and C. M. Pringle. 2005. Recovery of a tropical stream after harvest-related chlorine poisoning events. *Freshwater Biology* 50:603–615.

Gupta, A. 1975. Stream characteristics in eastern Jamaica, an environment of seasonal flow and large floods. *American Journal of Science* 275:825–847.

Gupta, A. 1988. Large floods as geomorphic events in the humid tropics. Pages 301–315 in V. R. Baker, R. C. Kochel, and P. C. Patton, editors, *Flood geomorphology*. New York: John Wiley and Sons.

Hartshorn, G. S. 1990. An overview of neotropical forest dynamics. Pages 585–599 in A. H. Gentry, editor, *Four neotropical rainforests*. New Haven, CT: Yale University Press.

Heartsill-Scalley, T., F. N. Scatena, C. Estrada, W. H. McDowell, and A. E. Lugo. 2007. Disturbance and long-term patterns of rainfall and throughfall nutrient fluxes in a subtropical wet forest in Puerto Rico. *Journal of Hydrology* 33:472–485.

Hildebrand, A. R., and W. V. Boynton. 1990. Proximal Cretaceous-Tertiary Boundary impact deposits in the Caribbean. *Science* 248:843–846.

Hodel, D. A., J. H. Curtis, G. A. Jones, A. Higuera-Gundy, M. Brenner, M. W. Binford, and K. T. Dorsey. 1991. Reconstruction of Caribbean climate change over the past 10,500 years. *Nature* 352:790–793.

Holdenrieder, O., M. Pautasso, P. J. Weisberg, and D. Lonsdale. 2004. Tree diseases and landscape processes: The challenge of landscape pathology. *Trends in Ecology and Evolution* 19(8):446–452.

Holmquist, J. G., J. M. Schmidt-Gengenbach, and B. B. Yoshioka. 1998. High dams and marine-freshwater linkages: Effects on native and introduced fauna in the Caribbean. *Conservation Biology* 12:621–630.

Horn, S. P. 1991. Fire history and fire ecology in the Costa Rican paramos. Pages 289–296 in S. C. Nodvin and T. A. Waldrop, editors, Fire and the environment: Ecological and cultural perspectives. General Technical Report SE-69. Asheville, NC: USDA Forest Service, Southern Forest Experiment Station.

Hulme, M., and D. Viner. 1995. A climate change scenario for assessing the impact of climate change on tropical rain forests. A Report for the World Wildlife Service by the

Climatic Research Unit, School of Environmental Sciences. Norwich, England: University of East Anglia.

Knutson, T. R., R. E. Tuleya, and Y. Kurihara. 1998. Simulated increase of hurricane intensities in a CO2-warmed climate. *Science* 279:1018–1020.

Larsen, M. C. 1997. Tropical geomorphology and geomorphic work: A study of geomorphic processes and sediment and water budgets in montane humid tropical forested and developed watersheds, Puerto Rico. Ph.D. dissertation. University of Colorado, Boulder, CO.

Larsen, M. C. 2002. Analysis of 20th century rainfall and streamflow to characterize drought and water resources in Puerto Rico. *Physical Geography* 22:494–521.

Larsen, M. C., and J. E. Parks. 1997. How wide is a road? The association of roads and mass wasting in a forested montane environment. *Earth Surface Processes and Landforms* 22:835–848.

Larsen, M. C., and A. Simon. 1993. A rainfall intensity-duration threshold for landslides in a humid-tropical environment, Puerto Rico. *Geografiska Annaler* 75A:13–23.

Larsen, M. C., A. J. Torres-Sánchez, and I. M. Concepción. 1999. Slopewash, surface runoff and fine-litter transport in forest and landslide scars in humid-tropical steeplands, Luquillo Experimental Forest, Puerto Rico. *Earth Surface Processes and Landforms* 2:481–502.

Lenart, M. T., D. A. Falk, F. N. Scatena, and W. R. Osterkamp. 2010. Estimating soil turnover from tree uprooting during hurricanes in Puerto Rico. *Forest Ecology and Management* 259:1076–1084.

Lewis, L. A. 1974. Slow movement of earth under tropical rain forest conditions. *Geology* 2:9–10.

Lodge, D. J., W. H. McDowell, and C. Asbury. 1994. The importance of nutrient pluses in tropical forests. *Trends in Ecology and Evolution* 9:384–387.

Lodge, D. J., F. N. Scatena, C. Asbury, and M. Sánchez. 1991. Fine litter fall and related nutrient inputs resulting from Hurricane Hugo in subtropical wet and lower montane rain forests of Puerto Rico. *Biotropica* 23:336–342.

Lugo, A. E., A. Bokkestijn, and F. N. Scatena. 1995. Structure, succession, and soil chemistry of palm forests in the Luquillo Experimental Forest. Pages 142–177 in A. E. Lugo and C. Lowe, editors, *Tropical forests: Management and ecology*. New York: Springer-Verlag.

Lugo, A. E., and F. N. Scatena. 1996. Background and catastrophic tree mortality in tropical moist, wet, and rain forest. *Biotropica* 28:585–599.

Malmgren, B. A., and A. Winter. 1998. El Niño-Southern Oscillation and North Atlantic Oscillation control of climate in Puerto Rico. *Journal of Climate* 11:2713–2717.

Masson, D. G., and K. M. Scanlon. 1991. The neotectonic setting of Puerto Rico. *Geological Society of America Bulletin* 103:144–154.

McDowell, W. H., C. Gines-Sanchez, C. E. Asbury, and C. R. Ramos-Perez. 1990. Influence of sea salt aerosols and long range transport on precipitation chemistry at El Verde, Puerto Rico. *Atmospheric Environment* 24:2813–2812.

Middleton, B. A., E. Sanchez-Rojas, B. Suedmeyer, and A Michels. 1997. Fire in a tropical dry forest of Central America: A natural part of the disturbance regime? *Biotropica* 29:515–517.

Myster, R. W., J. R. Thomlinson, and M. C. Larsen. 1997. Predicting landslide vegetation in patches on landscape gradients in Puerto Rico. *Landscape Ecology* 12:229–307.

Newsom, L. A. 1993. Native West Indian Plant Use. Ph.D. dissertation. Department of Archaeology, University of Florida, Gainesville, FL.

O'Brien, S. T., B. P. Hayden, and H. H. Shugart. 1992. Global climatic change, hurricanes, and a tropical forest. *Climate Change* 22:175–190.

Olander, L. P., F. N. Scatena, and W. L. Silver. 1998. Impacts of disturbance initiated by road construction in a subtropical cloud forest in the Luquillo Experimental Forest, Puerto Rico. *Forest Ecology and Management* 109:33–49.

Ortiz-Zayas, J. R., and F. N. Scatena. 2004. Integrated water resources management in the Luquillo Mountains, Puerto Rico: An evolving process. *International Journal of Water Resources Development* 20:387–398.

Ostertag, R., F. N. Scatena, and W. L. Silver. 2003. Forest floor decomposition following hurricane litter inputs in several Puerto Rican Forests. *Ecosystems* 6:261–273.

Petty, W. H. 1993. Seedling growth and mortality of four shade-tolerant canopy tree species in the rain forest of Puerto Rico following Hurricane Hugo. M.S. thesis. University of Tennessee, Knoxville, TN.

Pike, A. S. 2008. Longitudinal patterns in stream channel geomorphology and aquatic habitat in the Luquillo Mountains of Puerto Rico. Ph.D. dissertation. University of Pennsylvania, Philadelphia, PA.

Pike, A. S., F. N. Scatena, and E. E. Wohl. 2010. Lithological and fluvial controls on the geomorphology of tropical montane stream channels in Puerto Rico. *Earth Surface Processes and Landforms* 35(12):1402–1417.

Planos Gutiérrez, E. 1999. Análisis hidrológico de las grandes precipitaciones: Suprovincia fiscio-geográfica Cuba Occidental y Central [Hydrologic analysis of the large precipitations of the Cuban provinces]. Reporte de Investigación del Instituto de Meteorología. La Habana: Cuba.

Prospero, J. M., and P. J. Lamb. 2004. African droughts and dust transport to the Caribbean: Climate change implications. *Science* 302:1024–1027.

Putz, F. E. 1983. Treefall pits and mounds, buried seeds, and the importance of soil disturbance to pioneer trees on Barro Colorado Island, Panama. *Ecology* 64:1069–1074.

Ramos-Gines, O. 1999. Estimation of magnitude and frequency of floods for streams in Puerto Rico: New empirical models. Water Resources Investigations Report 99–4142. San Juan, PR: U.S. Geological Survey.

Reagan, D. P., and R. B. Waide. 1996. *The food web of a tropical rain forest*. Chicago: University of Chicago Press.

Reid, H. F., and S. Taber. 1919. The Puerto Rico earthquakes of October-November 1918. *Bulletin of the Seismological Society of America* 9:95–127.

Richards, P. W., editor. 1996. *The tropical rain forest: An ecological study*. Cambridge, England: Cambridge University Press.

Rivera-Ramírez, H. D. 1999. Peak discharge prediction, recession analysis, and evaluation of streamflow response to rainfall for watersheds in the Luquillo Experimental Forest, Puerto Rico. M.S. thesis. University of Connecticut, Storrs, CT.

Ruan, H. H., X. M. Zou, and F. N. Scatena. 2004. Asynchronous fluctuation of soil microbial biomass and plant litterfall in a tropical wet forest. *Plant and Soil* 260:147–154.

Santos-Román, D. M., G. S. Warner, and F. N. Scatena. 2003. Multivariate analysis of water quality and physical characteristics of selected watersheds in Puerto Rico. *Journal of the American Water Resources Association* 39: 829–839.

Scatena, F. N. 1989a. An introduction to the physiography and history of the Bisley Experimental Watersheds in the Luquillo Mountains of Puerto Rico. General Technical Report SO-72. New Orleans, LA: USDA Forest Service, Southern Research Station.

Scatena, F. N. 1989b. Watershed scale rainfall interception on two forested watersheds in the Luquillo Mountains of Puerto Rico. *Journal of Hydrology* 113:89–102.

Scatena, F. N. 1995. Relative scales of time and effectiveness of watershed processes in a tropical Montane Rain Forest of Puerto Rico. Pages 103–111 in J. E. Costa, editor,

Natural and anthropogenic influences in fluvial geomorphology. Geophysical Monograph 89. Washington, DC: American Geophysical Union.

Scatena, F. N. 1998. Climate change and the Luquillo Experimental Forest of Puerto Rico: Assessing the impacts of various climate change scenarios. *American Water Resources Association TPS* 98-92:193–198.

Scatena, F. N. 2001. Ecological rhythms and the management of humid tropical forests: Examples from the Caribbean National Forest, Puerto Rico. *Forest Ecology and Management* 5528:1–12.

Scatena, F. N., S. Doherty, H. T. Odum, and P. Karecha. 2002. An EMERGY valuation of Puerto Rico and the Luquillo Experimental Forest. USDA Forest Service General Technical Report GTR-9. San Juan, PR: U.S. Department of Agriculture Forest Service.

Scatena, F. N., and S. L. Johnson. 2001. Instream-flow analysis for the Luquillo Experimental Forest, Puerto Rico: Methods and analysis. General Technical Report IITF-GTR-11. San Juan, PR: U.S. Department of Agriculture Forest Service.

Scatena, F. N., and M. C. Larsen. 1991. Physical aspects of Hurricane Hugo in Puerto Rico. *Biotropica* 23:317–323.

Scatena, F. N., and A. E. Lugo. 1995. Geomorphology, disturbance, and the soil and vegetation of two subtropical wet steepland watersheds of Puerto Rico. *Geomorphology* 13:99–213.

Scatena, F. N., S. Moya, C. Estrada, and J. D. Chinea. 1996. The first five years in the reorganization of aboveground biomass and nutrient use following Hurricane Hugo in the Bisley Experimental Watersheds, Luquillo Experimental Forest, Puerto Rico. *Biotropica* 28:424–440.

Scatena, F. N., E. Planos Gutiérrez, and J. Schellekens. 2005. Impacts of natural disturbances on the hydrology of tropical forests. Pages 489–513 in M. Bonell and L. A. Bruijnzeel, editors, *Forest, water and people in the humid tropics*. International Hydrology Series. Cambridge, England: Cambridge University Press.

Schaefer, D. A., W. H. McDowell, F. N. Scatena, and C. E. Asbury. 2000. Effects of hurricane disturbance on stream water concentrations and fluxes in eight tropical watersheds of the Luquillo Experimental Forest, Puerto Rico. *Journal of Tropical Ecology* 16:189–207.

Schellekens, J., F. N. Scatena, L. A. Bruijnzeel, A. I. J. M. van Dijk, M. M. A. Groen, and R. J. P. van Hoogezand. 2004. Stormflow generation in a small rain-forest catchment in the Luquillo Experimental Forest, Puerto Rico. *Hydrological Processes* 18:505–530.

Seiders, V. M. 1971a. Cretaceous and lower Tertiary stratigraphy of the Gurabo and El Yunque quadrangles, Puerto Rico. Bulletin 1294-F. San Juan, PR: U.S. Geological Survey.

Seiders, V. M. 1971b. Geologic map of the El Yunque quadrangle, Puerto Rico. Miscellaneous Geologic Investigations, Map I-658. Washington, DC: U.S. Geological Survey.

Servant, M., J. Maley, B. Turcq, M. Absy, P. Bernac, M. Fournier, and M. Ledru. 1993. Tropical forest changes during the late Quaternary in African and South American lowlands. *Global and Planetary Change* 7:25–40.

Silver, W. L., L. Kueppers, and B. Ostertag. 2004. Carbon sequestration and plant community dynamics with reforestation of tropical pasture. *Ecological Applications* 14:1115–1127.

Silver, W. L., A. E. Lugo, and M. Keller. 1999. Soil oxygen availability and biogeochemistry along rainfall and topographic gradients in upland wet tropical forest soils. *Biogeochemistry* 44:301–328.

Smith, A. L., J. H. Schellekens, and A.-L. Muriel Díaz. 1998. Batholith emplacement in the northeastern Caribbean: Markers of tectonic change. Pages 99–122 in E. Lidiak and D. K. Larue, editors, *Geochemistry and tectonics of the northeastern Caribbean. Special Paper*. Vol. 322. Denver, CO: Geological Society of America.

Snyder, N. F. R., J. W. Wiley, and B. Cameron. 1987. *The parrots of Luquillo: Natural history and conservation of the Puerto Rican parrot.* Los Angeles: Western Foundation of Vertebrate Zoology.

Spence, J. E., M. A. Taylor, and A. Chin. 2004. The effect of concurrent sea-surface temperature anomalies in the tropical Pacific and Atlantic on Caribbean rainfall. *International Journal of Climatology* 24:1531–1541.

Taggart, B. E. 1992. Tectonic and eustatic correlations of radiometrically dated marine terraces in northwest Puerto Rico and the Isla de Mona. Ph.D. dissertation. Department of Geology, University of Puerto Rico-Mayagüez, Mayagüez, PR.

Thomlinson, J. R., M. I. Serrano, T. del M. López, T. M. Aide, and J. K. Zimmerman. 1996. Land-use dynamics in a post-agricultural Puerto Rican landscape (1936–1988). *Biotropica* 28:525–536.

Thompson, J. N., N. V. L. Brokaw, J. K. Zimmerman, R. B. Waide, E. M. Everham III, D. J. Lodge, C. M. Taylor, D. García-Montiel, and M. Fluet. 2002. Land use history, environment, and tree composition in a tropical forest. *Ecological Applications* 12:1344–1363.

Toon, O. B., K. Zahnle, D. Morrison, R. P. Turco, and C. Covey. 1997. Environmental perturbations caused by the impacts of asteroids and comets. *Reviews of Geophysics* 35:41–78.

Torres-Sierra, H. 1996. Flood of January 5–6, 1992 in Puerto Rico. Open-File Report 95–374. San Juan, PR: U.S. Geological Survey.

Turton, S. M. 1992. Understory light environments in a north-east Australian rain forest before and after a tropical cyclone. *Journal of Tropical Ecology* 8:241–252.

Uhl, C., and P. G. Murphy. 1981. Composition, structure and regeneration of a tierra firme forest in the Amazon Basin of Venezuela. *Tropical Ecology* 22:219–237.

van der Molen, M. K. 2002. Meteorological impacts of land use change in the maritime tropics. Ph.D. dissertation. Vrije Universiteit, Amsterdam.

Walker, L. R. 2000. Seedling and sapling dynamics of treefall pits in Puerto Rico. *Biotropica* 32:262–275.

Walker, L. R., and M. R. Willig. 1999. An introduction to terrestrial disturbance. Pages 1–16 in L. R. Walker, editor, *Ecosystems of disturbed ground.* Amsterdam: Elsevier.

Walsh, R. P. D. 1997. Climate. Pages 159–205 in P. W. Richards, editor, *The tropical rain forest*, 2nd ed. Cambridge, England: Cambridge University Press.

Wang, H., J. D. Cornell, C. A. S. Hall, and D. P. Marley. 2002a. Spatial and seasonal dynamics of surface soil carbon in the Luquillo Experimental Forest, Puerto Rico. *Ecological Modeling* 147:105–122.

Wang, H., C. A. S. Hall, F. N. Scatena, N. Fetcher, and W. Wu. 2002b. Modeling the spatial and temporal variability in climate and primary productivity across the Luquillo Mountains, Puerto Rico. *Forest Ecology and Management* 179:69–94.

White, P. S., and S. T. A. Pickett. 1985. Natural disturbances and patch dynamics: An introduction. Pages 3–13 in S. T. A. Pickett and P. S. White, editors, *The ecology of natural disturbances and patch dynamics*. San Diego, CA: Academic Press.

Whitmore, T. C. 1984. *Tropical rain forests of the Far East.* Oxford, England: Clarendon Press.

Whittaker, R. J., and J. Walden. 1992. Post-1883 ash fall on Panjang and Sertung and its ecological impact. *GeoJournal* 28:153–171.

Willig, M. R., and D. L. Moorehead. 1996. Functional diversity of soil bacteria communities in the tabonuco forest: The interaction of anthropogenic and natural disturbance. *Biotropica* 28:471–483.

Wolman, M. G., and J. P. Miller. 1960. Magnitude and frequency in geomorphic processes. *Journal of Geology* 68:54–74.

Wu, W., C. A. S. Hall, and F. N. Scatena. 2007. Modelling the impact of recent land-cover changes on the stream flows in northeastern Puerto Rico. *Hydrological Processes* 21:2944–2956.

Zimmerman, J. K., T. M. Aide, M. Rosario, M. Serrano, and L. Herrera. 1995a. Effects of land management and a recent hurricane on forest structure and composition in the Luquillo Experimental Forest, Puerto Rico. *Forest Ecology and Management* 77:65–76.

Zimmerman, J. K., E. M. Everham III, R. B. Waide, D. J. Lodge, C. M. Taylor, and N. V. L. Brokaw. 1995b. Responses of tree species to hurricane winds in subtropical wet forest in Puerto Rico: Implications for tropical tree life histories. *Journal of Ecology* 82:911–922.

Zou, X., and G. González. 1997. Changes in earthworm density and community structure during secondary succession in abandoned tropical pastures. *Soil Biology and Biochemistry* 29:627–629.

5

Response to Disturbance

Nicholas Brokaw, Jess K. Zimmerman, Michael R. Willig, Gerardo R. Camilo, Alan P. Covich, Todd A. Crowl, Ned Fetcher, Michael Gannon, Bruce L. Haines, D. Jean Lodge, Ariel E. Lugo, Randall W. Myster, Catherine M. Pringle, Joanne M. Sharpe, Frederick N. Scatena, Timothy D. Schowalter, Whendee L. Silver, Jill Thompson, Daniel J. Vogt, Kristiina A. Vogt, Robert B. Waide, Lawrence R. Walker, Lawrence L. Woolbright, Joseph M. Wunderle, Jr., and Xiaoming Zou

Key Points

- Background treefall gaps (not caused by hurricanes) are filled with plant regrowth as in other tropical forests. There is limited response by animals to treefall gaps, probably because background treefall gaps are relatively less important in these forests, which are dominated by chronic, widespread hurricane effects.
- Despite substantial effects on trees, the tree species composition changed little in the tabonuco forest after two recent hurricanes.
- Animal species show various responses to the changes in forest architecture and food resources caused by hurricanes. Bird species tend to be plastic in habitat and dietary requirements, probably due to the large changes in forest structure caused by hurricanes and regrowth, which require birds to change their foraging locations and diets.
- Although hurricane-produced debris is substantial (litterfall up to 400 times the average daily amount), decomposition, nutrient export, and trace gas emissions after hurricanes change only briefly, as rapid regrowth reasserts control over most ecosystem processes.
- In general, terrestrial ecosystem functions recover faster than structure.
- Hurricanes dump debris in streams, and floods redistribute inorganic and detrital material, as well as stream organisms, throughout the benthic environment along the stream continuum.

- Succession in landslides is slow and primarily limited by the availability of seed and by low nutrient availability, and early plant colonists, especially ferns, have a strong influence on later dynamics.
- Past land use is the most important determinant of species composition in tabonuco forest, despite repeated hurricane effects and underlying environmental variation such as in soil and topography.
- The native organisms of the Luquillo Mountains are more resilient after natural than human disturbances.

Introduction

The Luquillo Mountains are a heterogeneous landscape, produced by environmental variation (chapter 3), a varied disturbance regime (chapter 4), and varied responses to disturbances in space and time. Background treefall gaps (gaps not caused by hurricanes) open up 0.24 to 1.8 percent of tabonuco forest per year (Scatena and Lugo 1995); landslides denude a minimum of 0.08 to 0.30 percent of the Luquillo Mountains each century (Guariguata 1990); and severe hurricanes strike the Luquillo Mountains every 50 to 60 years (Scatena and Larsen 1991) and cause treefall gaps, landslides, and floods. To these disturbances we can add thousands of lesser storms, floods, and droughts over the millennia. Moreover, various kinds of human disturbance have affected nearly all forest area below 600 meters above sea level (masl) (Foster et al. 1999). The Luquillo Mountains represent many other tropical landscapes in which disturbance produces heterogeneity (Foster 1980).

How do the organisms of the Luquillo Mountains respond to this great variety, high frequency, and long history of disturbances? To describe these responses, we use the conceptual approach outlined in chapter 2. The response depends on the predisturbance conditions and the disturbance severity, which determine the conditions at the onset of response, and on the characteristics of the responding species. The initial conditions created by a disturbance can be classified as abiotic (including structure) and biotic (see figure 2-2) and are thought of as "residuals," or primary effects. Residuals are the physical manifestations of disturbance, that is, what remains of the abiotic, biotic, and structural features. These residuals shape the response to disturbance, creating secondary effects in the form of "legacies," or the long-term subsequent behavior of the ecosystem as determined by the residuals. Residuals and legacies of disturbance help explain the present condition of ecosystems (Foster et al. 2003).

The conditions at any given time can be described in terms of ecological space (see figure 2-8). Ecological space may be visualized as a multidimensional hypervolume that reflects abiotic, biotic, and structural components of a system. Disturbance changes this hypervolume by modifying these components at points located in geographical space. For instance, forest canopy changes can turn a previously shaded, cool, moist geographical point at ground level into a sunny, hot, dry point. In turn, biotic responses to disturbance, such as forest regrowth, can change conditions at that geographical point back to those of the earlier, shady, cool, moist ecological point. In such a situation, geographical space has not changed; ecological space has. The trajectory of response in ecological space depends on the intensity, duration, and extent of the disturbance, and then on the degree of resistance and resilience of biotic responses (see figure 2-9) (Lugo et al. 2002; see also chapter 4). Resistance is the degree to which a system is not affected by disturbance, as, for example, when trees are affected but not killed. Resilience is the time required for a system to return to a state that is similar to that before the disturbance, as when trees recover their predisturbance biomass.

With these concepts in mind, in this chapter we describe the responses of organisms, populations, communities, and ecosystems to the variety of disturbances in the Luquillo Mountains. We look at residuals and legacies, at resistance and resilience, and at mechanisms or processes of response. The chapter is organized according to descriptions of disturbances that act at stand to landscape scales, including background treefalls, hurricanes, floods, droughts, landslides, and various human disturbances. Most sections begin with a description of residuals—of how disturbance affects the biotic and structural environments (chapter 2)—and proceed to a discussion of legacies, or longer-term responses (abiotic effects are mainly covered in chapter 4). The chapter concludes with a discussion of the variation in responses to different disturbances and of interactions among disturbances. Variations and interactions among responses weave the tapestry of the Luquillo Mountains, encompassing landscape variation in space, and they also produce the layers of the palimpsest, encompassing persisting variations in time (chapter 1). (Chapter 6 continues the discussion of response to disturbance but empshizes the role of key species and their control of ecosystem processes.)

Response to Background Treefall Gaps

Background treefall gaps are gaps in the forest not caused by hurricanes. When a background treefall creates a forest gap (an opening through the canopy to near the ground), the gap is filled with the growth of adjacent trees, sprouts from affected trees, and seedling and sapling regeneration, and thus the gap area eventually returns to a mature phase, barring further disturbance (Hartshorn 1978; Whitmore 1978). From disturbance through recovery, this gap-phase regeneration adds diversity to the structure of the forest and to the structure of tree populations and tree communities (Brokaw and Busing 2000). Background treefalls are not a severe disturbance; there are many residuals that support response, such as mostly intact soil with nutrients and buried seeds, advance regeneration (surviving seedlings and saplings), and affected and bordering trees ready to sprout and fill the gap. Regrowth from these residuals is fast enough (cf. Fraver et al. 1998) that it reduces values of throughfall (rain reaching the forest floor) in gaps from post-treefall highs to pretreefall values in 1 year (Scatena 1990).

There have been three studies of plants in background treefall gaps in the Luquillo Mountains. The first study showed that the seedling gas exchange of the common tree species *Dacryodes excelsa* and *Sloanea berteriana* increases in gaps (Lugo 1970). The other two studies were on species composition in gaps. Both took place when the forest canopy had been developing for some 60 years without hurricane effects, and the relatively mature forest canopy had begun to open up with background gaps. In a study of 15 natural, recently formed gaps (34 to 322 m^2) in tabonuco forest (chapter 3), soil temperatures were higher in gaps than in adjacent intact forest, probably because of the higher insolation in gaps (Pérez Viera 1986). Soil humidity did not differ between gap and intact forest, whereas in some other forests it is wetter in gaps (Becker et al. 1988). There was a high density of colonizing saplings in gaps, as in other gap studies (see Brokaw 1985). The species composition of colonizers differed among gaps but typically included saplings of light-demanding species (Smith 1970), especially in some larger gaps. The species richness of saplings was higher in gaps than in intact forest because gaps have more small stems and thus a larger sample of plants (the "density effect") (Denslow 1995), and because a few gap specialists are, by definition, found mainly in gaps and are regenerating from seed. The second study in background treefall gaps was a rapid survey of all gaps in about 35 ha of tabonuco forest and aimed to record the number of seedlings and saplings of the disturbance-dependent species *Cecropia schreberiana* (Brokaw 1998). Only 34 *C. schreberiana* saplings were found, a number apparently insufficient to maintain the larger population of adults in the area, suggesting the importance of regeneration after hurricane, rather than gap, disturbance for this species (see below).

Gap-phase regeneration is less important for the tree community composition and dynamics in the tabonuco forest than in many other tropical forests because background treefall gaps are relatively few and small in this forest (Brokaw et al. 2004). Gaps are few (except in some riparian zones) (Scatena and Lugo 1995) because the periodic, simultaneous removal of many vulnerable trees by hurricanes reduces the number of treefalls between hurricanes (Lorimer 1989; Lugo and Scatena 1996; Whigham et al. 1999; Debski et al. 2000). Gaps are smaller in tabonuco forest than in some forests not affected by hurricanes because hurricanes tend to prevent trees from reaching large sizes and making large gaps when they fall (Odum 1970; Perez 1970).

Because gap creation and gap-phase regeneration are not the prevailing dynamics in the tabonuco forest, we expect little specialization by animals based on the environment of background treefall gaps. However, some species are found in higher densities in gaps than in the adjacent understory of intact forest. Community assemblages of birds differed between gaps and the understory of intact forest at El Verde, because species normally found in the canopy also frequented gaps (Wunderle 1995), but there were no species that specialized on gaps (i.e., that mainly occurred there), as found in other tropical forests (Schemske and Brokaw 1981; Wunderle et al. 2005). Coquí frogs (*Eleutherodactylus coqui*) move to gaps where debris provides the preferred humidity and shelter from predators (Stewart and Woolbright 1996; Woolbright 1996), but, as with the birds, they are not gap specialists.

Sixteen species of snails have been found in treefall gaps in tabonuco forest, but none were restricted to gaps, and community assemblages of snails did not differ between gap and nongap areas (Alvarez and Willig 1993). Five snail species were common enough for their habitat preferences to be assessed. The densities of *Austroselenites alticola*, *Megalomastoma croceum*, and *Subulina octana* did not differ between gap and intact forest; *Nenia tridens* was more abundant in gaps; and *Caracolus caracola* was more abundant in intact forest. *Nenia tridens* might gravitate toward gaps in order to eat dead plant material or the algae and fungi on dead plants. *Caracolus caracola* might avoid gaps due to its low tolerance for the heat and aridity in gaps, or because those factors reduce food quality (Alvarez and

Willig 1993). Among insects, walking sticks (Phasmodidae) are herbivores that preferentially frequent treefall gaps, presumably to eat the new plant growth there (Willig et al. 1986, 1993; Sandlin and Willig 1993; Garrison and Willig 1996).

The lack of striking differences in animal assemblages between treefall gaps and intact forest understory might be related to two factors. First, the generally small size of treefall gaps might reduce the environmental differences between gap and intact forest relative to differences found in other forests. Second, animals in the Luquillo Mountains have evolved in an environment that is strongly disturbed by hurricanes, which would favor generalist species that are adapted to both successional and mature forest stands (Waide 1991b), and so they are not especially responsive to background treefalls.

The accumulated ecosystem effects of frequent background tree mortality (not necessarily creating gaps) are comparable to the effects of less frequent but catastrophic tree mortality from hurricanes (Scatena and Lugo 1995; Lugo and Scatena 1996; see also chapter 4). For tabonuco forest, an estimated constant tree mortality of 2.0 percent y^{-1} for 100 years would release the biomass and nutrients of a forest stand two times per century, whereas two highly catastrophic events of 30 percent mortality, plus extensive effects to surviving trees (50 percent reduction of aboveground biomass) (Scatena et al. 1996), would also release nearly all tree biomass and nutrients about twice per century. Thus, although background tree mortality might not even disturb the canopy, over time it can equal some ecosystem effects of hurricanes that dramatically alter the forest structure.

Terrestrial Response to Hurricanes

Unlike background treefalls, hurricanes create a range of terrestrial disturbances, including large areas of affected and defoliated trees, individual and multiple treefall gaps, and landslides, depending on the topography and location of a site relative to the storm trajectory (chapter 4) (Brokaw and Grear 1991; Walker 1991; Larsen and Torres-Sánchez 1992). The catastrophic, sudden tree mortality (Lugo and Scatena 1996) and the extensive effects on surviving trees caused by a strong hurricane have a major influence on the distribution and quantity of biomass and nutrients, on microclimates, and on populations (figure 5-1; Walker et al. 1991). Biomass and nutrients move from the canopy to the forest floor and soil. Light floods the understory over large areas. Fine root biomass drops sharply. Many plants and some animals die. In turn, responses are manifest at all levels, from individual to ecosystem, and it is striking how resistant and resilient the organisms and ecosystem processes in the Luquillo Mountains are. In fact, numerous features of the forest return to prehurricane levels within about 5 years (Zimmerman et al. 1996).

Hurricanes, Forest Canopy Structure, and Microclimate

The upper canopy of the tabonuco forest is lower and typically smoother than that in many other tropical forests in which background treefall gaps dominate the forest dynamics (Odum 1970; Brokaw et al. 2004). For example, the compositionally

Figure 5.1 Structural effects in tabonuco forest at El Verde Research Area, Puerto Rico, resulting from Hurricane Hugo, 1989. (Photograph by Nicholas Brokaw.)

similar *Dacryodes-Sloanea* forest on the Lesser Antillean island of Dominica experiences fewer hurricanes and is much taller than tabonuco forest in the Luquillo Mountains (Perez 1970).

To account for the smooth canopy of tabonuco forest, Odum (1970) suggested that repeated hurricanes in Puerto Rican tabonuco forest have selected, evolutionarily, against the emergent habit among trees and for smaller crowns with reduced wind resistance, with both resulting in a smooth forest canopy. A more parsimonious explanation of this smooth canopy is simply that hurricanes and lesser storms repeatedly prune the extended tops and branches of trees that would otherwise grow tall and spread their canopies as in some hurricane-free forests. Thus, short trees with small crowns would be a phenotypic, not a genotypic, feature (Fetcher et al.

2000; Myster and Fetcher 2005). Persistent trade winds probably help produce a smooth canopy at higher elevations in the Luquillo Mountains (cf. Lawton 1982), but these winds might not explain the relatively smooth-canopied tabonuco forest at lower elevations, such as El Verde. At El Verde, the mean annual windspeed above the canopy (Waide and Reagan 1996) is similar to the annual windspeed on Barro Colorado Island, Panama (Brokaw et al. 2004), which has a forest of large, spreading trees and a comparatively rough canopy punctured by treefall gaps. Other tropical forests subject to cyclonic storms are typically short (de Gouvenain and Silander 2003), but the canopy structure among hurricane-disturbed forests differs widely (Brokaw et al. 2004).

In a 1.08 ha plot at El Verde, the canopy was relatively smooth before Hurricane Hugo (Brokaw and Grear 1991). After the storm, the residual canopy surface was much rougher and lower in average height than before (figure 5-2). With the sprouting of surviving trees and new regeneration, the mean height of the canopy increased, and the coefficient of variation of the height, here a measure of roughness, declined, suggesting that the canopy was redeveloping its former smoothness. Hurricanes Hortense and Georges temporarily reversed this trend toward smoothness, but 18 years after Hurricane Hugo the canopy has returned to nearly the structure it had before that hurricane, and which it had been developing since the previous major hurricane passage in 1932. The small individual tree crowns in tabonuco forest (whether genetically or phenotypically determined) confer resistance to wind effects (Everham and Brokaw 1996), and the rapid sprouting of hurricane-trimmed trees (see below) provides resilience.

Hurricanes tend to affect older forests with large trees more than they do young forests with small trees (Everham and Brokaw 1996; Grove et al. 2000; Lomascolo and Aide 2001; but see Franklin et al. 2004). Hurricane Hugo affected a relatively mature forest in the Luquillo Mountains; it was the first hurricane to cross Puerto Rico since 1956, and it passed closer to the Luquillo Mountains than any hurricane since 1867 (Scatena and Larsen 1991). Hurricane Georges, on the other hand, struck the Luquillo Mountains only 9 years after Hurricane Hugo, and therefore affected a less structurally mature forest (figure 5-2). Due to this effect and to lower storm intensity, Hurricane Georges produced smaller canopy openings and deposited less debris than did Hurricane Hugo (Lugo and Frangi 2003; Ostertag et al. 2003).

Canopy openings increase the understory light climate (chapter 4), which apparently stimulates seed germination and plant growth (see below and chapter 6) that eventually return the understory structure and light to prehurricane conditions.

Hurricanes and Terrestrial Plant Species and Communities

Tree Response

Effects and Mortality The effect on trees of Hurricanes Hugo (1989) and Georges (1998) varied across the landscape and among tree species (Walker 1991; Boose et al. 1994; Ostertag et al. 2005), but some patterns were evident (Brokaw and Walker 1991). In general, forests on slopes facing winds were more affected

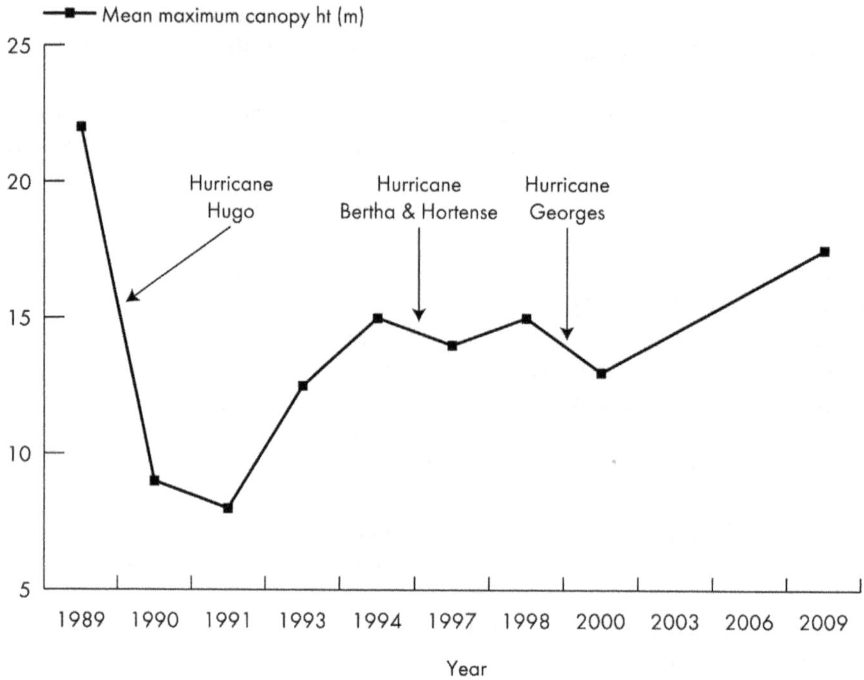

Figure 5.2 Mean height of upper surface of forest canopy before Hurricane Hugo and at points in time afterward, with effects of all hurricanes in the period indicated, at El Verde, Puerto Rico (N. Brokaw). Data from measurements at 475 points in a 1.08 ha plot.

than those on lee slopes (e.g., Walker 1991). Ridges were more affected than slopes at a colorado forest (chapter 3) site, whereas the reverse was true in a tabonuco site, probably due to the presence of stable (resistant) tabonuco trees (*Dacryodes excelsa*; see below) on ridges (Brokaw and Grear 1991; Basnet et al. 1992). Defoliation was the most common type of effect, followed by effects on small branches, the loss of large branches, and the snapping and uprooting of large stems (figure 5-3) (Brokaw and Walker 1991; Zimmerman et al. 1994). Tall trees were more likely to be defoliated, snapped, or uprooted, and tall trees with larger diameters were more likely to uproot than snap (Walker 1991; You and Petty 1991; Basnet et al. 1992; Ostertag et al. 2005). Size-specific effects varied greatly among species (Zimmerman et al. 1994). Generally, shade-tolerant species (Smith 1970) and species with dense wood lost many branches but suffered less uprooting and snapping than did light-wooded and shade-intolerant species (Zimmerman et al. 1994; Ostertag et al. 2005). Understory trees were more likely to be defoliated or snapped than uprooted (Walker et al. 1992). Direct effects on trees led to indirect effects when downed trees and limbs fell on other trees (accounting for 16 percent of all effects at one site) (Frangi and Lugo 1991) and in places where increased sunlight scalded understory juveniles and seedlings (You and Petty 1991). Effects on understory plants from falling debris are frequent in tropical forests, whether hurricane-related or not (Aide 1987; Clark and Clark 1991).

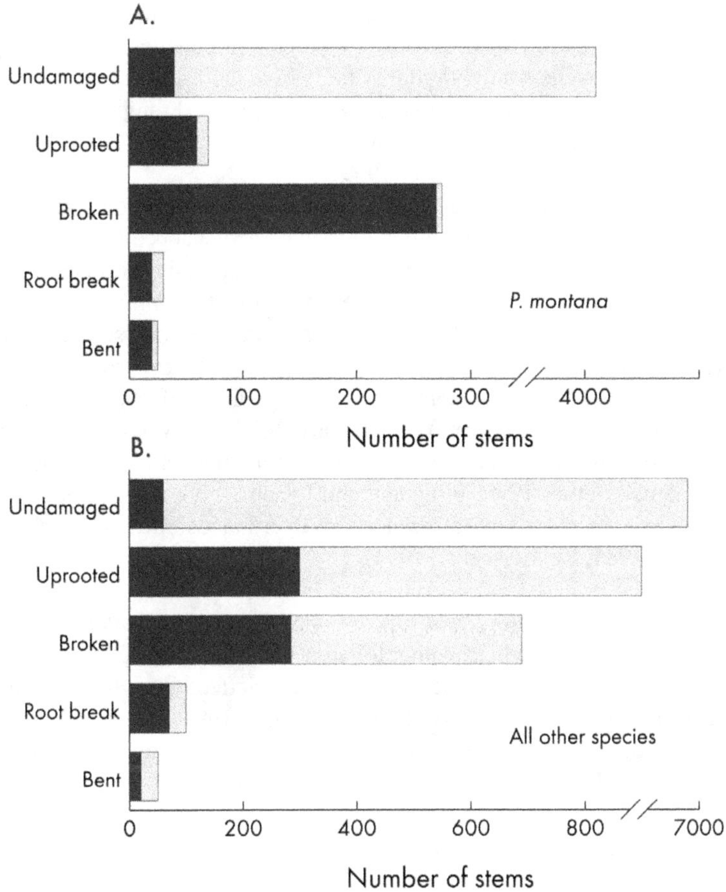

Figure 5.3 Frequency of hurricane-affected and unaffected main stems in the Luquillo Forest Dynamics Plot at El Verde, Puerto Rico (Zimmerman et al. 1994). Black portions of bars indicate tree mortality. (A) Effects on the sierra palm *Prestoea montana*. (B) Effects on all other tree species. (Used with permission from the British Ecological Society.)

Hurricanes cause catastrophic sudden tree mortality, defined as sudden mortality greater than 5 percent (Lugo and Scatena 1996). Hurricane Hugo immediately killed 9.1 percent of trees ≥10 cm in diameter at breast height (dbh) in the 16 ha Luquillo Forest Dynamics Plot (LFDP), located in the tabonuco zone at El Verde (Zimmerman et al. 1994; Thompson et al. 2004). In another study, in twenty 300 m^2 plots in the tabonuco zone, the storm had killed 7.4 percent of trees after 54 weeks, and this number rose to 13.3 percent after 171 weeks (Walker 1995). At Bisley, a site in the tabonuco forest that was especially affected by Hurricane Hugo, mortality in a 1.0 ha plot was 16.8 percent just after Hurricane Hugo and had risen to 31.6 percent 5 years later (Dallmeier et al. 1998); by that time mortality probably included some background deaths not attributable to the hurricane. In a secondary forest, mortality 21 months after Hurricane Georges was 5.2 percent y^{-1}—seven times the background mortality (Ostertag et al. 2005). However, the immediate mortality from Hurricane Hugo was only 1.0 percent in a sheltered 0.25 ha riparian forest stand, whereas the annual mortality there over the following 5 years was 2.0 percent, mainly involving dicotyledonous trees, not palms (Frangi and Lugo 1991). At an elfin forest (see chapter 3) site, mortality was 21 percent of stems in the 5 years following the hurricane (Weaver 1999).

Mortality after Hurricane Hugo was greater among the more affected trees (Dallmeier et al. 1998), especially among uprooted and snapped trees in the LFDP (figure 5-3) (Zimmerman et al. 1994). Snapping and uprooting did not necessarily kill dicotyledonous trees but did kill the palm *Prestoea montana* (table 5-1) (previously *Euterpe globosa*, and named *P. acuminata* in Henderson et al. [1995]). Mortality differed among size classes in a population of the canopy tree *Manilkara bidentata*, in which 4 percent of large trees died from direct effects and 60 percent of seedlings died from burial by litter (You and Petty 1991). Elsewhere in the Caribbean, tree mortality from hurricanes also differs greatly among sites and species (Bellingham et al. 1992; Imbert et al. 1998; Whigham and Lynch 1998), and cyclone effects in Asia and Oceania can cause higher tree mortality than that recorded in these Caribbean studies (Dittus 1985; Elmqvist et al. 1994).

Refoliation, Sprouting, and Release of Seedlings and Saplings Surviving trees respond to hurricane effects with refoliation and the sprouting of new branches; saplings (advance regeneration) respond with accelerated growth, and seedlings emerge and become established (Brokaw and Walker 1991; Everham and Brokaw 1996). In tabonuco forest, after Hurricane Hugo, leaves had regrown on some affected trees in 2 weeks and on most by 7 weeks; only 7 percent of all trees were leafless after

Table 5.1 Types of effects on trees as a percentage of trees observed in various tabonuco forest stands after Hurricane Hugo

Total trees	Defoliation	Branch effects	Crown loss	Uprooted	Snapped	Mortality	Sprouting	Source
8,579 dicots[1]	–	24.9[2]	–	9.8	8.3	9.1	64.8[3]	Zimmerman et al. 1994
4,498 palms[1]	–	–	–	1.5[4]	6.0	8.8	–	Zimmerman et al. 1994
2,278[5]	–	–	25.5	2.4	2.2	1.0	98.0[6]	Frangi and Lugo 1991
732	56[7]	13[8]	–	9	11	7.0[9] 13.1[10]	–	Walker 1991, 1995

[1]Dicots and palms ≥ 10 cm dbh.
[2]Percentage of trees with no affected stems with at least one broken branch > 10 cm dbh.
[3]Percentage of surviving trees.
[4]Stem broken above ground level.
[5]Dicots ≥ 4.0 cm dbh, palms ≥ 0.7 m tall.
[6]Palms that lost all leaves.
[7]>75 percent leaf loss, on trees not uprooted or snapped.
[8]Branches > 5 cm diameter, on trees not uprooted or snapped.
[9]Assumed dead if no leaves at 54 wk.
[10]After 171 wk.

54 weeks (figure 5-4) (Walker 1991). In a palm forest (see chapter 3) site, 98.0 percent of defoliated palms had produced an average of 4.7 new leaves by 9 months after Hurricane Hugo (Frangi and Lugo 1991). In the high-elevation elfin forest, refoliation was slower than in tabonuco forest (Walker et al. 1996b). New branches in the tabonuco forest were common; in the LFDP, 64.8 percent of surviving trees sprouted new branches, especially those suffering branch loss (Zimmerman et al. 1994). Both uprooted and snapped stems were capable of sprouting new branches from main trunks or at the top of broken stems, but shade-tolerant species sprouted more abundantly than shade intolerant species (Zimmerman et al. 1994; but see Walker 1991). The refoliation and sprouting of affected trees after Hurricane Hugo has been commonly observed in other hurricane-affected forests (Brokaw and Walker 1991; Yih et al. 1991; Bellingham et al. 1992, 1994; Everham and Brokaw 1996).

Advance regeneration is "released," that is, grows faster, after canopy disturbance provides it with more light and perhaps a larger share of soil resources (Denslow and Hartshorn 1994; Fraver et al. 1998). As mentioned above, *Manilkara bidentata* seedlings suffered much mortality from Hurricane Hugo, but surviving seedlings grew 17 times faster than before the hurricane, presumably in response to higher light (figure 5-5) (You and Petty 1991). This accelerated growth reduced the transition period from seedling to sapling from 292 to 16 weeks, which suggests how important hurricane disturbance could be for the population dynamics of this and other tree species (You and Petty 1991).

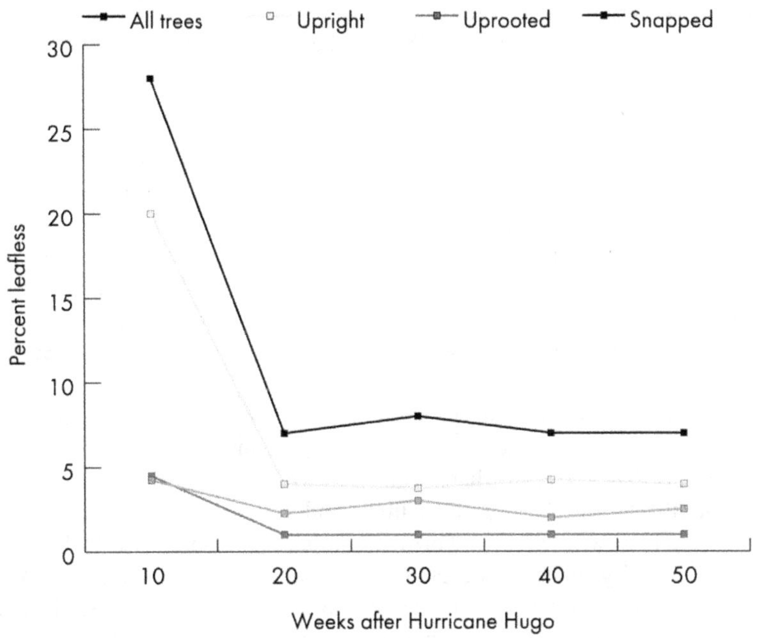

Figure 5.4 Percent of trees (≥5 cm dbh, no palms) with leaves at intervals after Hurricane Hugo, as a function of the type of effect, at El Verde, Puerto Rico (Walker 1991). (Used with permission from the Association for Tropical Biology and Conservation.)

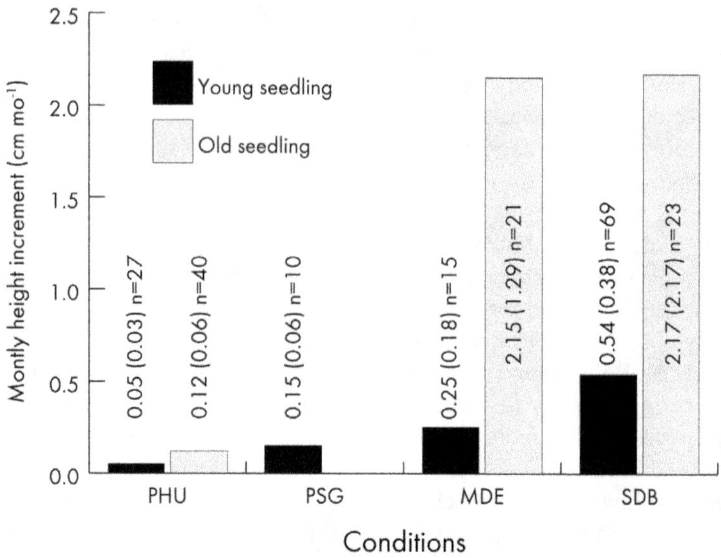

Figure 5.5 Seedling growth rates of *Manilkara bidentata* under pre- and posthurricane conditions. PHU = prehurricane understory conditions at El Verde, Puerto Rico, less than 5 percent of maximum potential photosynthetic photon flux density (MP); PSG = prehurricane small gap at El Verde, 5 to 15 percent of MP; MDE = moderately affected posthurricane site at El Verde, 45 percent of MP; SDB = severely damaged posthurricane site at the Bisley Experimental Watersheds, Puerto Rico, 64 percent of MP. Standard deviations of growth rates are in parentheses. *n* = sample size (You and Petty 1991). (Used with permission from the Association for Tropical Biology and Conservation.)

Fruit Production, Seed Dispersal, and Seedling and Sapling Dynamics After Hurricane Hugo, forest-wide fruit production declined (figure 5-6) (You and Petty 1991; Wunderle 1999), as trees presumably put energy into refoliation and sprouting. However, many seeds germinated and seedlings became established in response to altered microclimates at ground level (Guzmán-Grajales and Walker 1991; Everham et al. 1996; Scatena et al. 1996). At Bisley, seedling numbers peaked at 12 months after Hurricane Hugo, remained high until 36 months, and then declined (Scatena et al. 1996).

Posthurricane germination and establishment differed greatly among tree species in the tabonuco forest, depending on levels of light, nutrients, and litter (Guzmán-Grajales and Walker 1991; Everham et al. 1996; Walker et al. 2003). An experiment showed that the overall density of seedlings and number of seedling species were highest where litter was removed (Guzmán-Grajales and Walker 1991). However, it was mainly seedlings of early-successional species, such as *Cecropia schreberiana* and *Chionanthus domingensis*, that were denser at litter removal sites. The density of late-successional species did not increase or was reduced after litter removal. For example, *Dacryodes excelsa* seedlings declined where litter was removed, whereas *Sloanea berteriana* was not affected by litter removal. Among all species together, seedling mortality was higher and growth less where litter was removed. Given that

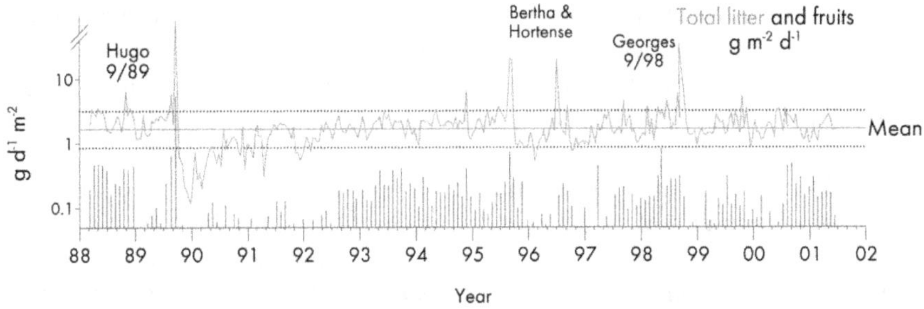

Figure 5.6 Monthly leaf litterfall (continuous line) and fruitfall (columns) in tabonuco forest in 1988–2006 at the Bisley Experimental Watersheds, Puerto Rico (F. Scatena). Named hurricanes are shown near the peaks of litterfall caused by the storms. Mean and standard deviation are shown.

hurricane litter is deposited unevenly on the forest floor, these different establishment patterns would lead to a patchy and diverse distribution of tree species.

Seeding recruitment after Hurricane Georges was modeled for nine tree species in the LFDP using maximum likelihood methods (Uriarte et al. 2005). Field data on seedlings and light were fitted to different models that included spatially explicit seedling recruitment functions. The majority of the nine species tested supported models that included at least one of several recruitment functions, as follows: (1) the estimated minimum reproductive size of parents, ranging from 9 to 48 cm dbh, influenced seedling spatial distributions; (2) bath recruitment (the presence of a uniform number of seedlings over space, regardless of the local distribution of conspecific adults) accounted for 6 to 81 percent of observed seedling recruitment; (3) light availability appeared to divide species into two groups: one that requires low light levels (<5 percent of full sunlight) for recruitment and one that performs best at high light levels (>30 percent of full sunlight); and (4) density-dependent mortality during the period between seed germination and seedling establishment shifted the mode of seedling distribution away from potential parent trees for most species. This last effect is thought to result from species-specific seed or seedling predators or pathogens. It should promote the species richness of trees by favoring the survival of rare species (Volkov et al. 2005; Wills et al. 2006), and it is noteworthy that it operates in this forest, where frequent hurricane disturbance might be expected to reduce the precision of species-specific interactions with pests.

In larger size classes (saplings through mature trees), a study of the survival and growth of 12 dominant tree species in the LFDP after Hurricane Georges (1998) revealed complex relationships among life history type, density, effects of Hurricane Hugo (1989), and size class (Uriarte et al. 2004a). However, some rough generalizations can be made. Competitive thinning of densely packed saplings that grew after the storm accounted for the majority of posthurricane mortality, particularly for secondary species. The species identity of competitors was important mainly for secondary species, whereas functional equivalence of competitors was more common among shade-tolerant species. Effects of the earlier Hurricane Hugo influenced the growth and survival of large stems of some shade tolerant species, and previously affected trees had less of a competitive effect on their neighbors (cf. Ostertag et al. 2005). Thus, the regeneration and regrowth of trees after a severe hurricane reflects a variety of influences, legacies, and species-specific patterns, all contributing to heterogeneity among stands.

Hurricane Disturbance and Tree Life Histories Two general tree life-history types appear well adapted to hurricane disturbance (Zimmerman et al. 1994). The first type is pioneers, including *Cecropia schreberiana*, *Schefflera morototoni*, and *Miconia tetrandra*. These species show little resistance to hurricanes; that is, they suffer high effects and mortality and have relatively little ability to sprout (Zimmerman et al. 1994). However, they exhibit much resilience, as they recruit abundantly from seed and grow quickly in response to conditions resulting from canopy opening (Brokaw 1998). The second type, nonpioneers, includes *Dacryodes excelsa*, *Sloanea berteriana*, *Prestoea montana*, and *Guarea guidonia*. These species lose leaves and limbs but resist fatal hurricane effects and exhibit resilience by refoliating and sprouting new branches. Other species exhibit a mix of the characteristics of these two types (Walker 1991; McCormick 1995; Lugo and Zimmerman 2002; Uriarte et al. 2004a).

Cecropia schreberiana is an example of a pioneer. It is light- and nutrient-demanding, fast growing, fecund, and short-lived (Silander and Lugo 1990; Walker et al. 1996b). Its population dynamics respond dramatically to hurricanes (Brokaw 1998). At the time of Hurricane Hugo, there were 136 *C. schreberiana* trees ≥10 cm dbh in the LFDP and fewer small stems (Brokaw 1998). More than half (52.9 percent) of these stems were killed by the hurricane (the mean mortality for other common species was 8.4 percent) (Zimmerman et al. 1994). After the hurricane, *C. schreberiana* was recruited abundantly from a soil seed bank (especially in treefall pits and mounds) (Walker 2000). Within 40 months after the hurricane, there were 10,635 *C. schreberiana* stems 1 to 10 cm dbh and 565 stems ≥10 cm dbh in the 16 ha LFDP, amounting to a 400 percent increase of trees ≥10 cm dbh) (Brokaw 1998). There was much thinning of these recruits, but some survivors grew fast; at Bisley a *C. schreberiana* grew to 27 cm dbh in the 5 years after Hurricane Hugo (Scatena et al. 1996). Posthurricane, *C. schreberiana* colonizers mature, senesce, and decline in large numbers (Crow 1980; Weaver 1989, 2002), but the species remains abundant as seeds in the soil, lying dormant and ready to form cohorts after the next hurricane disturbance (see the section "Interactions among Disturbances"). The abundance of this species seems to depend on hurricane disturbance; the regeneration of *C. schreberiana* in background treefall gaps is not sufficient to maintain the species' observed numbers in Luquillo forests (see above and Brokaw 1998).

Dacryodes excelsa is an example of a nonpioneer. During Hurricane Hugo, individuals of *D. excelsa* lost leaves and branches, but few trunks were snapped, and few stems died (Zimmerman et al. 1994). Mature *D. excelsa* are interconnected by lateral roots that form tree unions and also appear to be strongly anchored in the soil, often on rocky ridges, and thus resist uprooting (Basnet et al. 1992). The species' resilience is shown by vigorous sprouting on standing trunks (Zimmerman et al. 1994), which might be helped by its habit of root grafting to conspecifics, which could direct resources from unaffected to affected stems (Basnet et al. 1992, 1993).

Dacryodes excelsa recruits moderately from seed. Another nonpioneer, the sierra palm *Prestoea montana*, demonstrates both high resistance and resilience. It is often defoliated by hurricanes but is infrequently snapped or uprooted. It usually retains at least its youngest leaf and refoliates vigorously (Frangi and Lugo 1991; Weaver 1999). *Prestoea montana* also tolerates burial by storm debris and regrows after the debris decays (Beard et al. 2005). As with *C. schreberiana*, in places the age structure of *P. montana* exhibits clear cohorts corresponding to disturbance events (Lugo and Rivera Batlle 1987).

Not surprisingly, the tree species that seem especially resistant and/or resilient to hurricanes are among the most abundant species in this hurricane-affected forest. Other studies in the Luquillo Mountains and elsewhere also show that tree species in the tropics are resistant to hurricanes in that they generally suffer little mortality relative to effects, and they are resilient after hurricanes through sprouting, recruitment from seed, and release from suppression (Whigham et al. 1991; Bellingham et al. 1992, 1994, 1995; Boucher et al. 1994; Franklin et al. 2004). Chronic hurricanes could possibly have selected for these characteristics of trees in the Luquillo Mountains (Lugo and Zimmerman 2002); however, it is not clear that trees in the Luquillo Mountains have in fact evolved unique adaptations in response to hurricanes (but see Francis and Alemañy 2003). The responses one sees in Luquillo forests after hurricane effects (sprouting, recruitment, release) are the same responses one sees in large treefall gaps in tropical forests that lack hurricanes (e.g., Brokaw 1985; Putz and Brokaw 1989; Fraver et al. 1998) and in hurricane-affected forests where these storms are infrequent (Boucher et al. 1994). Nevertheless, though we cannot yet demonstrate any adaptation specifically to hurricanes, we can assume that hurricanes in the Luquillo Mountains have filtered out any tree species that cannot cope with these storms (Willig and Walker 1999).

Stand-Level Tree Response Early papers on large-scale, chronic storm effects emphasized how disturbance history could explain stand characteristics and tree species composition (Browne 1949; Webb 1958; Whitmore 1974; Crow 1980) and concluded that storm-prone areas might never attain equilibrium (Lugo et al. 1983). In the Luquillo Mountains, the response by trees at the stand level after major hurricane effects is first rapid and then slower but long-lasting, as in most successional sequences. The initial mortality reduces stem numbers. Stem numbers then rise with recruitment but later decline with thinning, whereas diameter class distributions shift to larger trees (Weaver 1986, 1989, 1998; Dallmeier et al. 1998; Frangi and Lugo 1998). The 16 ha LFDP was established at El Verde in 1990, the year after Hurricane Hugo, and was inventoried three times through 2002, overlapping Hurricane Georges in 1998. From a peak of recruitment after Hurricane Hugo, the overall stem numbers declined from 1993 to 2002 (table 5-2). The number of species also declined, with species losses exceeding additions at each census. Losses included some originally rare species and some uncommon and short-lived posthurricane colonizers (e.g., *Trema micrantha*).

Overall, although the numbers of a few pioneer species increased greatly (figure 5-7), the relative abundances of tree species in the LFDP changed little after Hurricane Hugo, as observed elsewhere in Puerto Rico (Fu et al. 1996; Dallmeier et al. 1998; Frangi and Lugo 1998; Pascarella et al. 2004) and in some other hurricane-affected forests worldwide (Burslem et al. 2000; Tanner and Bellingham 2006; but see Dittus 1985; Imbert et al. 1998). In fact, recurrent disturbances might tend to stabilize species composition through repeated selection for resistant and resilient species (Willig and Walker 1999). Due to this process, individual hurricanes would have a minor effect on the tree species composition (Burslem et al. 2000).

Table 5.2 Changes in numbers of individuals, stems, and species of self-supporting woody plants ≥ 1.0 cm dbh in the 16 ha Luquillo Forest Dynamics Plot, El Verde, Puerto Rico (J. Thompson, unpublished data). Negative numbers in parentheses are the numbers of species recorded in a previous census but not in the indicated census; positive numbers are the numbers of species recorded in the indicated census but not in the previous census. (Stems ≥ 10 cm dbh were censused soon after Hurricane Hugo, whereas stems ≥ 1 and < 10 cm dbh were censused in 1991–1993, after their numbers had risen due to recruitment.)

Total ≥ 1.0 cm dbh	1990	1995	2000
Individuals	90,166	71,828	68,099
Stems	108,891	89,014	85,883
Species	150	143 (−8, +1)	135 (−11, +2)

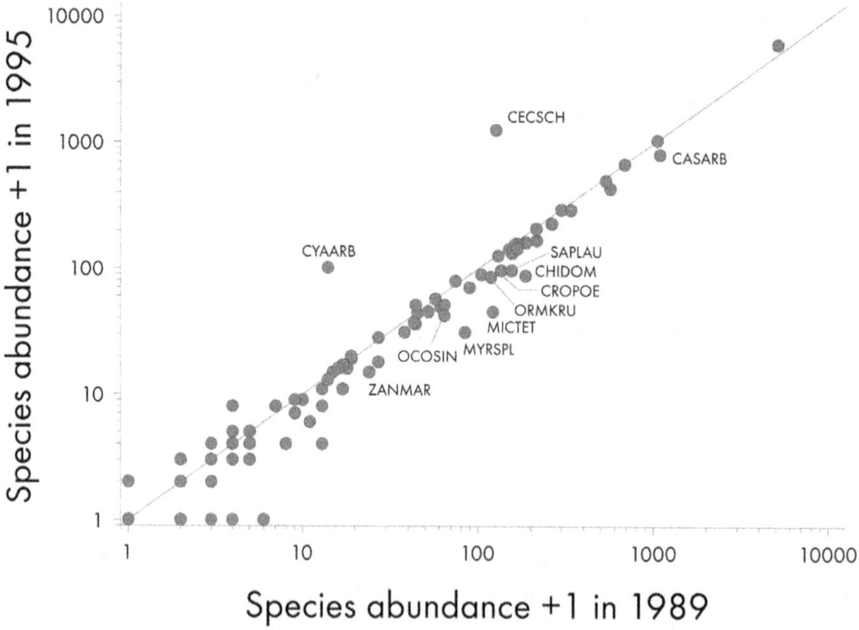

Figure 5.7 Log number of stems ≥ 10 cm dbh of tree species in 1989 (estimated) and in 2000 in the Luquillo Forest Dynamics Plot, El Verde, Puerto Rico. Equal numbers at both censuses lie on the diagonal line. Numbers below the line indicate population declines in the interval; numbers above indicate population increases (Zimmerman et al. 2010). (Used with permission from the British Ecological Society.)

The longest records of change in tabonuco forest come from a 0.72 ha plot established in 1943 (Crow 1980; Lugo 2008) and a 0.4 ha plot established in 1946 (Weaver 2002), 11 and 14 years, respectively, after the passage of Hurricane San Ciprián in 1932. Several tree inventories (stems ≥ 4.0 cm dbh) in these plots through 2005 show that the numbers of stems (Weaver 2002) and species peaked in the first 10 to 15 years after the hurricane and then decreased, as natural thinning reduced numbers and more species went locally extinct than entered the plot (Crow 1980; cf. Tanner and Bellingham 2006). In particular, secondary species (pioneers) died out after an initial pulse of recruitment (Weaver 2002). Similar patterns are evident in the colorado forest (see chapter 3), where the long-term posthurricane response includes shifts to larger tree diameters, shifts from pioneer to mature forest species, and an eventual decline in species richness over the period measured (Weaver 1986, 1989). A generalized scenario of posthurricane forest dynamics includes (1) a phase in years 0 to 10 of increasing stem density; (2) a phase in years 10 to 45 of slow but steady ingrowth, strong competition, and high and then lower mortality, especially of secondary species; and (3) a phase after about 50 years of slow ingrowth and low mortality, in which secondary species would be maintained by background treefall gap dynamics (Weaver 1998). Another suggested scenario includes a 10-year aggrading phase, a 10-year reorganization phase, a 25-year transition period, and then a 15-year period of maturity (Lugo et al. 1999). Beyond 50 to 60 years of stand development—that is, without further hurricane effects—we do not know what forests in the Luquillo Mountains would be like. Hints might come from looking at the compositionally similar *Dacryodes-Sloanea* forest on the Lesser Antillean island of Dominica. This forest experiences fewer hurricanes and is much taller than the tabonuco forest in the Luquillo Mountains (Perez 1970).

Some climate models predict an increased intensity of hurricanes (Emanuel 1987; Overpeck et al. 1990). With increased intensity, or frequency, the forest model ZELIG predicts reduced tree height and diversity in tabonuco forest (O'Brien et al. 1992), due to a reduction in the number of large, climax species. Another model, FORICO, agrees with ZELIG that the tree species richness would decline if the hurricane frequency were significantly higher, but FORICO also predicts a decline in species richness when the hurricane frequency is much less, because pioneer species would drop out (Doyle 1981; cf. Tanner and Bellingham 2006). FORICO, however, does not take into account long-term processes that might enrich forests in the absence of hurricanes. It does not take into account the possibility that absent hurricanes, the forest would grow taller and background treefall gaps would be larger, creating a more heterogeneous vertical and horizontal forest habitat that could sustain more tree species, including pioneers (as well as other plant life forms and animals) (Brokaw and Lent 1999; Brokaw et al. 2004; also see above). Also, FORICO includes only the present complement of tree species. It assumes no in situ evolution of species, which might occur more frequently in a more structurally varied forest, and it assumes no immigration of species, which might occur more frequently without the harsh filter of chronic hurricane disturbance. This filter might explain the high dominance of some tree species in the LFDP (Thompson et al. 2004) and in Puerto Rican forests generally (Lugo et al. 2002) relative to forests elsewhere.

Understory Plants and Lianas

Shrubs in the LFDP were affected by falling debris during Hurricane Hugo but then flowered abundantly (N. Brokaw, personal observation; Wunderle 1995), sprouted vigorously (Hammond 1996), and reached high densities (J. Thompson, unpublished data). An experimental study showed that the germination and establishment of a common shrub in tabonuco forest, *Palicourea riparia*, is enhanced by conditions created by disturbance (Lebrón 1979). Shrubs thus show little resistance but much resilience to hurricanes.

Ferns respond markedly to hurricane disturbance to the canopy and then to canopy closing. Depending on the species, these responses can include increased plant and leaf mortality, increased or decreased spore production and leaf production, and changes in the size of leaves produced (Sharpe 1997; Halleck et al. 2004). For example, after Hurricane Hugo, leaf production in *Nephrolepis rivularis* increased via runners sent out under the litter from existing plants, but almost all these new leaves disappeared within 5 years. Following Hurricane Georges, small plants of *Thelypteris reticulata* increased in leaf size, leaf production, and fertility, but within 5 years the same plants were again producing small, sterile leaves. In elfin forests, ferns and grasses proliferate after hurricane disturbance and can delay tree recruitment (Weaver 1986; Walker et al. 1996b). After Hurricane Hugo, herbaceous climbers and vines proliferated in some areas, but stem numbers declined rapidly with time (Walker et al. 1996b; Chinea 1999). Lianas (large, woody vines) are less abundant in the tabonuco forests studied at El Verde and Bisley than in most other tropical forests, perhaps because hurricanes strip lianas, as well as potential supporting branches, from trees (Rice et al. 2004). The common, large herb *Heliconia caribaea* was recruited where the canopy was opened by Hurricane Georges (Meléndez-Ackerman et al. 2003) but has greatly declined since (J. Thompson, unpublished data). Thus, many shrubs, herbaceous vines, herbs, and ferns capitalize on the changed ecological space in the understory after a hurricane, but some effects are short-lived.

Hurricanes and Terrestrial Consumers

Hurricanes have mixed effects on terrestrial consumers, depending on their ecologies and preferences for different ecological spaces. The increased debris promotes populations of decomposer species, but the altered three-dimensional structure and microclimate of the forest have negative effects on many other species.

Arthropods

After Hurricane Hugo, the numbers of Diptera, bark beetles, pin-hole borers, scale insects, and orb-weaving spiders all increased (Torres 1992; Schowalter 1994; Pfeiffer 1996). Herbivores increased in response to the flush of new plant growth. For example, 15 species of Lepidoptera flourished; the most common of these was *Spodoptera eridania*, which fed on 56 plant species in 31 families (Torres 1992). All these plants were early-successional species, and *S. eridania* fed exclusively on herbs and on young leaves of saplings or on sprouts of older trees; it was not found on mature leaves in tree canopies. The herbivore outbreak might have been stimulated by the abundance of palatable, young leaves in the posthurricane regeneration. Drought (which followed Hurricane Hugo) also tends to concentrate leaf nutrients and carbohydrates and reduce secondary chemicals, further increasing the leaves' palatability (Lawrence 1996). The outbreak of *S. eridania* ended when host plants were consumed and ichneumonid wasps increasingly parasitized *S. eridania* (Torres 1992). It was the first time these natural enemies of *S. eridania* had been observed.

Insect herbivores respond differently to particular tree host species after hurricanes, perhaps because tree species typically suffer different degrees of effects (Schowalter 1994; Zimmerman et al. 1994; Schowalter and Ganio 1999). The sap-sucker functional group was generally more abundant on saplings and sprouts in gaps than on trees in intact stands. This probably reflects the rapid production of shoots and foliage on which this group feeds. Not all plant species were eaten; in one study, eight tree species flushed new leaves without an increase in herbivory (Angulo-Sandoval et al. 2004). Generally, hurricanes appear to promote sap-suckers and inhibit defoliators in the forest canopy. Leaf concentrations of nitrogen, phosphorus, potassium, and calcium did not affect herbivore abundances or leaf area missing (a proxy of leaf area eaten) (Schowalter and Ganio 1999).

As mentioned, walking sticks are herbivores that can reach high densities in background treefall gaps (Willig et al. 1986), but they seem to be negatively affected by larger scale and more intense disturbances. Hurricane Hugo drastically reduced densities of the walking sticks *Lamponius portoricensis* and *Agamemnon iphimedeia* for at least 5 years (Willig and Camilo 1991; M. Willig, personal observation). *Lamponius portoricensis*, previously common, was still quite uncommon 15 years later in most areas of the Luquillo Mountains (M. Willig, unpublished data). This 15-year reduction in numbers suggests that *L. portoricensis*, and walking sticks generally, are among the least resistant and resilient of species in the tabonuco forest of Puerto Rico.

Orb-weaving spiders benefited when hurricane debris created more places for webs, as well as more sites for larval flies, adding to the spiders' food supply (Pfeiffer 1996). Debris also created diurnal refuges from predators, increasing spider survival. The big beneficiary of these changes was the orb-weaver *Leucauge regnyi*. But some species declined—for example, *Modisimus signatus*, which attaches to undersides of live leaves in the understory; many of these sites were eliminated during the hurricane.

Snails

Studies of snail response to hurricanes illustrate the complex effects of hurricanes on populations. The densities of four common snail species declined greatly after Hurricane Hugo. Six months after the hurricane, the densities of *Caracolus caracola*, *Polydontes acutangula*, *Nenia tridens*, and *Gaeotis nigrolineata* were 22, 25, <1, and <1 percent, respectively, of their prehurricane values (Willig and Camilo 1991; Secrest et al. 1996; Willig et al. 1998). But, remarkably, 5 years after the hurricane, the densities of *C. caracola* and *N. tridens* had increased to three and six times their prehurricane densities, respectively. In general, the four snail populations responded in the same fashion because the hurricane-caused changes did not disrupt patterns of correlation among environmental characteristics of the vegetation structure and plant species composition that affect snails (Secrest et al. 1996).

The steep declines followed by increases in snail populations might have been caused by strongly contrasting negative and positive effects of hurricane on snails. Hurricane effects on the forest canopy produce hot and dry microclimates (Denslow 1980; Fernández and Fetcher 1991) inhospitable to snails. Desiccation kills snails (Solem 1984), especially eggs and snails in early growth stages (Heatwole and Heatwole 1978; Riddle 1983). Thus the microclimate of canopy gaps caused by hurricanes probably restricts activity, increases mortality, and limits reproduction. However, hurricanes also produce dead plant material covered with fungi and algae, which snails eat (Alvarez and Willig 1993). After the snails had suffered the effects of a changed microclimate, canopy closure might have allowed them to take advantage of increased food and rebound strongly.

Snail response, however, is not uniform among species after every hurricane (Bloch and Willig 2006). *Oleacina glabra*, *Polydontes portoricensis*, and *Subulina octona* were more abundant after Hurricane Georges (the less intense storm) than after Hurricane Hugo, whereas *P. acutangula* exhibited the opposite pattern. This might be due to the smaller effect of Hurricane Georges in terms of gap size and debris deposition, coupled with the variable sensitivities of the snail species.

Frogs and Lizards

Hurricanes greatly affect frog and lizard populations. Numbers of adult *Eleutherodactylus coqui* frogs were not immediately affected by Hurricane Hugo but increased sharply a year later (figure 5-8), although adults were smaller than before (Woolbright 1991, 1996). In contrast, numbers of juvenile *E. coqui* at first declined but also peaked a year after the storm, and then declined and continued to vary greatly (figure 5-8). Five years after the storm, both adult and juvenile numbers had decreased to prehurricane levels. Among congeners, *E. hedricki* increased by 14 percent and *E. richmondi* decreased by 83 percent in the first 2 years after the hurricane.

Disturbance affects *E. coqui* by changing the forest floor habitat structure and microclimate. Treefalls and hurricanes add structure to the forest floor and understory by depositing debris and promoting the growth of herbs, seedlings, and saplings. All this creates moist microhabitats and refuges from predators (Reagan 1991). For example, hurricane-caused patches of *Cecropia schreberiana* and *Heliconia caribaea* (Meléndez-Ackerman et al. 2003) provide high-quality nest and retreat sites for frogs (Woolbright 1996). These favorable microsites created by hurricanes are transient; eventually, decomposition and forest maturation reduce the understory structure. Relative to background treefalls, which affect only small areas of the forest, Hurricane Hugo added understory structure at a larger scale, temporarily increasing frog survival and reproduction throughout the forest.

After Hurricane Hugo, populations of anoline lizards declined (Reagan 1991) along with the reduction in overall forest structure (figure 5-2). *Anolis* species also moved nearer to the ground, where structure increased. As the forest structure and microclimate returned to prestorm characteristics (figure 5-2), anoles responded by

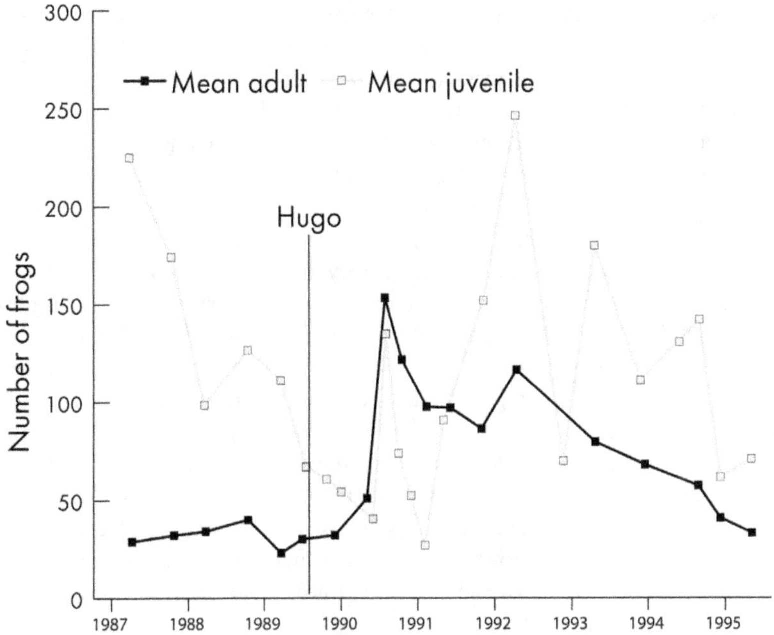

Figure 5.8 Mean population estimates for juvenile and adult *Eleutherodactylus coqui* frogs in four long-term study plots (each 400 m^2) from 1987 to 1995. Population estimates for adults for each plot were the total number of individuals marked during four nocturnal surveys. Population estimates for juveniles were the maximum count during one of three nights. Standard error for adults ranged from 1.2 to 16.1, and for juveniles from 2.5 to 51.5 (Woolbright 1996). (Used with permission from the Association for Tropical Biology and Conservation.)

reoccupying higher levels in the forest; this is a good example of organisms tracking changes in ecological space.

Bats

As with snails, hurricane effects on bat species illustrate the complex interaction between species and disturbance (Gannon and Willig 1994, 1998). Three species dominate the bat fauna of the tabonuco forest in the Luquillo Mountains. *Artibeus jamaicensis* (Jamaican fruit bat) and *Stenoderma rufum* (red fig-eating bat) are principally frugivorous, whereas *Monophyllus redmani* (Greater Antillean long-tongued bat) is nectarivorous. The effects of disturbance on bats can occur at two levels: direct effects of the hurricane (high winds, heavy rain) on the animals themselves, and indirect effects, in which changes in habitat structure or resources stimulate emigration or cause differential survivorship and reproduction (Willig and McGinley 1999).

The abundance of *Artibeus jamaicensis* quickly declined after Hurricane Hugo and remained low for about 18 months, but it was the first bat species to return to and exceed its pre–Hurricane Hugo population level. Rather than reflecting direct hurricane-caused mortality, these shifts might have reflected migration—first fleeing the

Luquillo Mountains to areas of the island that were less affected by the hurricane, and then returning to the mountains when fruiting recovered (figure 5-6). Thus, the typical demographics of *A. jamaicensis* might integrate effects over a large area, which would confer resilience after disturbance. In response to Hurricane Georges, *A. jamaicensis* declined more gradually and its numbers remained low for a longer period than after Hurricane Hugo. Because effects were more widespread from Hurricane Georges than from Hurricane Hugo throughout Puerto Rico, feeding opportunities for this frugivore might have been affected more widely by Hurricane Georges.

Stenoderma rufum was affected negatively by both hurricanes. Its numbers decreased steadily after Hurricane Hugo and were lowest at 18 months postdisturbance. An inability to disperse out of the tabonuco forest, as suggested by its normally limited foraging and home ranges (Gannon and Willig 1994), combined with increased exposure to high temperature, precipitation, and wind at roost sites (tree canopy), as well as the decreased availability of fruit, might account for its decline after Hurricane Hugo. Changes in the age structure and the scarcity of reproductive females also suggested a decline in *S. rufum* reproduction after the storm. The decline of *S. rufum* after Hurricane Georges was much faster, and even 6 years after Hurricane Georges, recovery was not obvious. Other known populations of this species are few in number and occur as isolated pockets, separated by miles of urban and deforested areas (Gannon et al. 2005). This, along with the fact that *S. rufum* is not a strong flier, suggests that immigration that restores declining populations is unlikely in this species.

Whether these changes in bat populations reflect mortality or the temporary emigration of individuals from the affected sites is not proven. For canopy-roosting species, such as *S. rufum*, mortality due to direct effects of disturbance is likely. For frugivorous species, such as *A. jamaicensis*, that roost in caves or other solid structures, direct mortality from hurricane disturbance might play a small role; instead, indirect effects (e.g., fruit crop loss) of a hurricane on these species might stimulate their dispersal to less affected areas. Consistent with these possibilities, a Puerto Rican cave population of the frugivorous bat *Eropyhlla sezekorni* showed no direct responses to disturbance immediately after Hurricane Georges but declined rapidly in the following weeks, possibly owing to a scarcity of food (Jones et al. 2001). In contrast, the abundance of nectarivorous *Monophyllus redmani* increased slightly after both hurricanes. The small increase might be due to a local increase of posthurricane flowering in gaps that predated the hurricanes (Gannon and Willig 1994; Wunderle 1995).

Reduced bat populations also have been reported after severe storms at other island sites (Willig and McGinley 1999). For example, declines after cyclones in the Pacific and Indian Oceans have been reported for populations on Guam (Wiles 1987), Samoa (Craig et al. 1984; Pierson et al. 1996), Mauritius (Cheke and Dahl 1981), and Rodrigues (Carroll 1984), and in the Caribbean on Montserrat declines have been noted after hurricanes (Pedersen et al. 1996).

Birds

Bird species were either little affected by Hurricane Hugo or resilient afterward, depending on their diet (Waide 1991a; Wunderle 1995). Insectivores (e.g., Puerto Rican Tody, *Todus mexicanus*) and omnivores (e.g., Pearly-Eyed Thrasher, *Margarops fuscatus*; Puerto Rican Tanager, *Nesospingus speculiferus*; and Puerto Rican Woodpecker, *Melanerpes portoricensis*) were little affected. For insectivores, this might be because insect prey survived in sheltered sites as pupae. As with lizards, insectivorous birds adjusted their foraging height to the posthurricane vegetation structure, occupying a reduced vertical range in their search for food (Waide 1991a). In contrast, nectarivores (e.g., Bananaquit, *Coereba flaveola*; Puerto Rican Emerald, *Chlorostilbon maugaeus*), a frugivore (Scaly-Naped Pigeon, *Columba squamosa*), and possibly one granivore (Ruddy Quail-Dove, *Geotrygon montana*) declined, either as a direct result of changes in forest structure or, in most cases, because flowers, fruit, and seeds were stripped from trees and new fruiting declined overall. The quail-dove forages while walking on the ground, looking for seeds and fruits in the litter. This movement would have been difficult in the debris- and regeneration-choked ground layer after Hurricane Hugo, and fruit and seed supplies declined in any case.

When fruiting returned to prehurricane levels, all frugivore populations (except the quail-dove) also returned to prehurricane levels, before the next breeding season, suggesting that migration rather than mortality caused the declines (Waide 1991a). Two nonforest bird species moved into the affected areas before fruiting and forest structure had recovered: the Black-Faced Grassquit (*Tiaris bicolor*), probably to eat seeds of grasses that colonized open areas, and the Red-Legged Thrush (*Turdus plumbeus*), which prefers open habitat (Waide 1991a).

Birds in the Luquillo Mountains and other hurricane-prone areas seem to have evolved plasticity in their habitat and food requirements (Waide 1991b; Wunderle et al. 1992; Whigham and Lynch 1998). In the Dominican Republic, the use of different foraging substrates and maneuvers separates bird species ecologically; they are not separated by foraging height relative to forest structure, as are some bird species in mainland forests (Latta and Wunderle 1998). This might be because hurricanes affect the forest and make it difficult for species to specialize on structure. Therefore, birds are flexible in terms of their foraging mode. Overall, the responses of the bird community are consistent with an adaptation to frequent and large-scale disturbance, which should select for flexible diet and behavior (Reagan et al. 1996; Willig and Walker 1999).

An interesting legacy of background treefall gaps is that the relatively small plants already present in these gaps at the time of Hurricane Hugo suffered relatively few effects and were oases of fruit production after the storm (Wunderle 1995; cf. Levey 1990). Both fruit production and bird abundance in these gaps peaked 93 to 156 days after the hurricane.

Hurricanes, Decomposition, and Nutrient Cycling

A hurricane transforms large quantities of live biomass to dead biomass. Massive amounts of aboveground biomass and nutrients from the live tree compartment are transferred to the forest floor in the form of leaves and coarse and small woody debris. Falling debris kills smaller plants (see above), adding to the litterfall. Dying roots add belowground detritus. These large, rapid transfers and the subsequent detrital dynamics regulate carbon and nutrient fluxes and have a profound effect on the response to hurricane disturbance (Sanford et al. 1991; Lodge et al. 1994; Scatena et al. 1996; Vogt et al. 1996; Ostertag et al. 2003).

Hurricane Debris

Normally, litterfall is fairly even through time in the forests of the Luquillo Mountains, but in just a few hours Hurricane Hugo deposited a mass of fine litter (leaves, wood < 1.0 cm in diameter) on the tabonuco forest floor at El Verde that was about 400 times (1,006 to 1,083 g m^{-2}) the average daily amount (Lodge et al. 1991; Scatena et al. 1996). Another 928 g m^{-2} fell but was suspended in the vegetation above ground. In the tabonuco forest at Bisley, the total fine litterfall during the hurricane was 1.2 times the mean annual litterfall. Altogether, the storm moved 50 percent of the prehurricane aboveground biomass to the forest floor at Bisley (figure 5-9) (Scatena et al. 1996). It moved 10 percent in a palm forest, where the fine litterfall was 2.3 g m^{-2} d^{-1} before Hurricane Hugo but 1,029 g m^{-2} during the hurricane (Frangi and Lugo 1991), or 123 percent of the prehurricane annual fine litterfall. At an elfin forest site, the storm deposited 682 times the average daily amount of fine litterfall, and another 45 g m^{-2} was suspended above ground (Lodge et al. 1991).

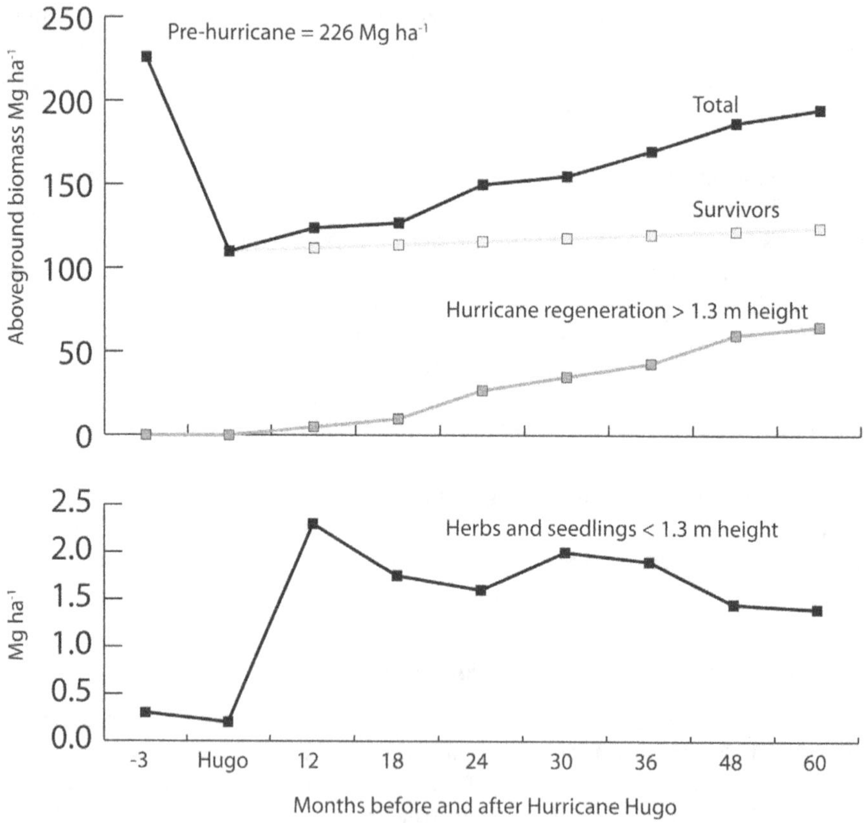

Figure 5.9 Aboveground biomass in the Bisley Experimental Watersheds, Puerto Rico, before and after Hurricane Hugo (Scatena et al. 1996). "Survivors" indicates the biomass of individuals that survived the hurricane. (Used with permission form the Association for Tropical Biology and Conservation.)

Much root biomass also was killed by the death or swaying of trees, by the drought after Hurricane Hugo, and perhaps by the depletion of nonstructural carbohydrate reserves (Parrotta and Lodge 1991; Beard et al. 2005). These strong pulses of litter and dead roots were patchy in space (Lodge et al. 1991).

Hurricane Nutrient Input, Decomposition, and Decomposers

As with the patchy litterfall during a hurricane, nutrient fluxes and pools after a hurricane are patchy in time and space due to abiotic effects such as locally variable soil, topography, and debris. Posthurricane nutrient fluxes and pools are also patchy due to biotic effects, such as variable nutrient content and decomposition time of debris, as well as variable local uptake among different plant species. The leaf litter deposited by a hurricane is green and relatively nutrient rich, unlike normal brown, senescent litter from which some nutrients have been translocated back into plants before leaf shedding (table 5-3). For example, the phosphorus content, often a limiting factor in tropical forests (Vitousek and Sanford 1986), in hurricane litter was 4.7 times (per unit volume) that in normal litter in a palm forest (Frangi and Lugo 1991). Thus, hurricanes produce an immediate pulse of nutrients from leaf litter on the forest floor, and later hurricane inputs come from the litter suspended above ground. Likewise, decomposing woody debris makes a sustained contribution to nutrient contents for years after a storm (Lodge et al. 1991; Vogt et al. 1996).

In a study imitating the decomposition of nutrient-rich, green litter (defined as having a higher nitrogen concentration and a lower lignin:nitrogen ratio), the green leaves of four common tabonuco forest tree species (*Manilkara bidentata*, *Dacryodes excelsa*, *Guarea guidonia*, *Cecropia schreberiana*) decomposed faster than brown leaves of those species (Fonte and Schowalter 2004). This faster decomposition can fuel nutrient cycling and primary productivity, which might be affected by the timing and spatial variation of decomposition. However, after Hurricane Hugo, the decomposition rates of leaf litter and fine roots of the dominant tree species tabonuco (*D. excelsa*) and sierra palm (*Prestoea montana*) did not differ between the period immediately after Hurricane Hugo and a period several years later (Bloomfield 1993; Bloomfield et al. 1993; Vogt et al. 1996). Decay constants (the time required for 99 percent material loss) of tabonuco leaf and root litter were the same in both Bisley and El Verde, as well as across different topographic positions

Table 5.3 Ratios of hurricane-caused nutrient input to total mean annual nutrient input in fine litterfall at Pico del Este (lower montane rainforest) and El Verde and Bisley (subtropical wet forest; data from Lodge et al. 1991). The site at El Verde was an especially heavily affected site

	N	P	K	Ca	Mg
Pico del Este	2.21	2.5	4.47	1.21	1.05
Bisley	1.29	1.53	2.99[a]	1.25	1.27
El Verde	1.25	2.42[b]	1.26	0.93	0.91

[a]Probably an overestimate due to leaching losses in prehurricane samples.

[b]Possibly an overestimate due to differences in classifying fine wood (see Lodge et al. 1991).

(Vogt et al. 1996), but the decay rate of fine and medium-diameter woody debris (<10 cm diameter) did vary according to the topography, possibly reflecting the effects of the hurricane on the local soil moisture (Vogt et al. 1996; Beard et al. 2005). For example, *D. excelsa* wood (3 to 6 cm in diameter) at El Verde decayed faster in riparian areas (9.9 y for 99 percent weight loss) than in upslope areas (16.1 y), where the effect of Hurricane Hugo was less. This is in contrast with Bisley, where the decay rates of wood of the same diameter were significantly faster in the drier, upslope areas (6.7 y for 99 percent weight loss) than in the riparian areas (8.4 y). Greater changes in the soil water content at Bisley than at El Verde due to Hurricane Hugo appeared to increase the decay rate of woody material, especially in the upslope areas (Vogt et al. 1996). The decomposition rates of coarse woody debris (>10 cm diameter) also differed across the Luquillo landscape and by location within each habitat (Vogt et al. 1996). For roots, the belowground decomposition of fine material took 1.5 years, but the decomposition of large roots was slower (Silver et al. 1996).

The large quantities of high-quality organic debris deposited by hurricanes stimulate the growth of microbial decomposers (Miller and Lodge 1997). For example, cord-forming fungi, such as the stinkhorns (Phallales) and *Phanerochaete flava*, were abundant after Hurricane Hugo, presumably in response to the abundant debris. On the other hand, hurricane effects allow sunlight to penetrate to the forest floor and dry the litter in some locations, which inhibits fungi (Lodge 1993).

Nutrient Export and Cycling

The massive effects of Hurricane Hugo on trees and other plants caused large losses of aboveground nutrients in vegetation (52 to 55 percent loss; see figure 5-10) and some small initial losses of nutrients in soils (Scatena et al. 1993, 1996), but these small losses were temporary, as regrowth over about 2 years restored control of the nutrient cycling. Aboveground, the largest nutrient losses were of K and N. Belowground, soils lost K and nitrate-N initially, but most exchangeable soil nutrient pools were either the same or greater than before the hurricane. Thus, most nutrients were not lost. After the hurricane, there was a temporary increase in the concentration of macronutrients in litterfall, herb, and woody seedling biomass that could be explained in part by the rise and fall in the abundance of pioneer plant species with high nutrient contents (Scatena et al. 1996). For example, aboveground N, K, and magnesium (Mg) in plants declined after Hugo due to the loss of tree stems but then accumulated rapidly in colonizing pioneers. In soils, there was increased ammonium availability and net N-mineralization and nitrification rates 4 months after Hugo, followed by a gradual decline (Steudler et al. 1991). The return of inorganic N levels in soils to prestorm values can be explained by the regrowth of roots (Parrotta and Lodge 1991), as well as by N immobilization by microbes (Zimmerman et al. 1995b). The soil organic matter content did not change after Hurricane Hugo (Silver et al. 1996).

After Hurricane Hugo, nitrous oxide (N_2O) emissions increased more than 15-fold in the first month and remained high for 7 months, at a rate three times the predisturbance value (figure 5-11) (Steudler et al. 1991). The maximum rates of this

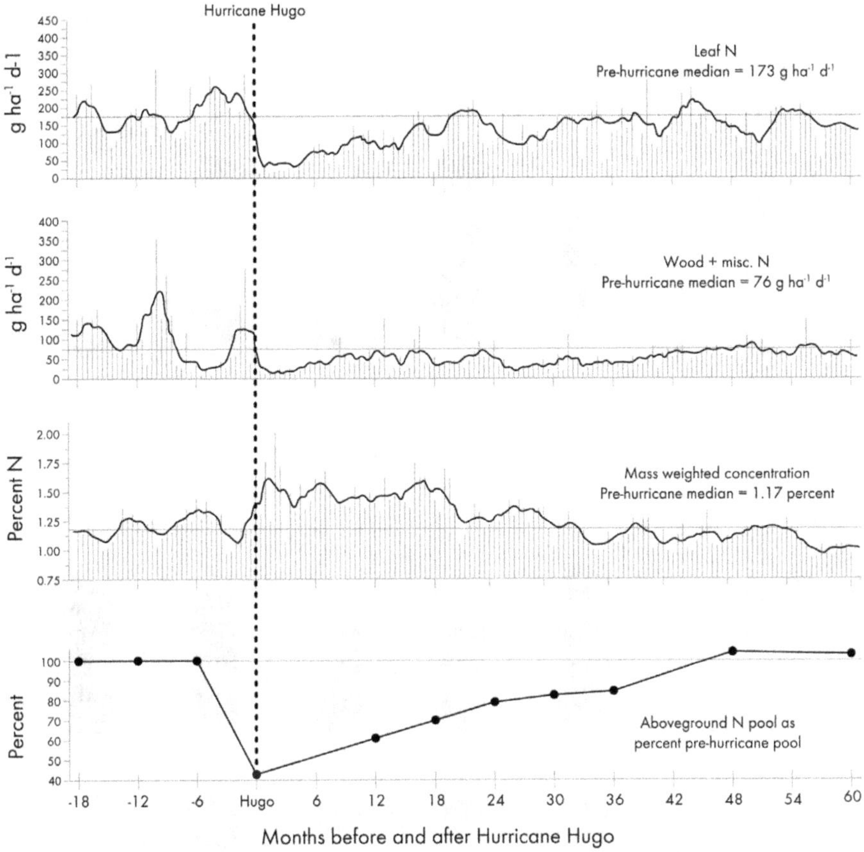

Figure 5.10 Nitrogen flux in leaf litter and wood and miscellaneous litter, the mass-weighted concentration of total litter, and the aboveground N pool as a percentage of the prehurricane pool, Bisley Experimental Watersheds, Puerto Rico, before and after Hurricane Hugo (Scatena et al. 1996). The horizontal line is the median of prehurricane values; the curve is the 2-month running average. (Used with permission from the Association for Tropical Biology and Conservation.)

flux coincided with peaks in N mineralization, nitrification, and soil nitrate pools. Carbon dioxide (CO_2) emissions were initially 64 percent those of undisturbed areas and returned to normal after 14 months. Soils were generally sinks for methane (CH_4), and its consumption decreased by half, perhaps owing to disturbance-induced changes in the nitrogen cycle. Emissions of N_2O for up to 7 months after Hurricane Georges were five times the fluxes at the same sites measured for 16 months before the storm (Erickson and Ayala 2004). During the 27 posthurricane months of this study, N_2O emissions remained at levels more than twice those of the prestorm fluxes. Soil ammonium pools decreased after Hurricane Georges and remained low during the study. Nitrate pools increased during the first year after Hurricane Georges, but not significantly (Erickson and Ayala 2004).

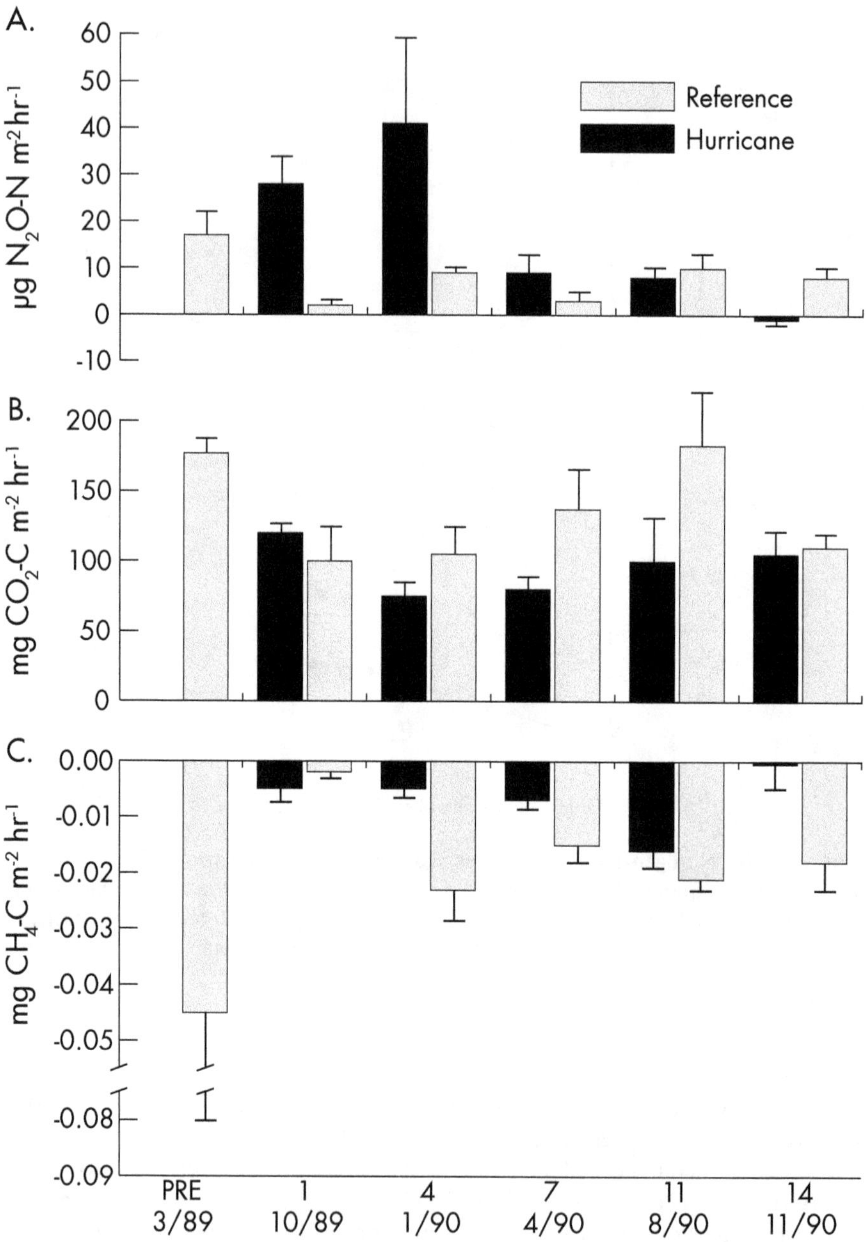

Figure 5.11 Fluxes of N_2O (A), CO_2 (B), and CH_4 (C) from reference (El Verde, Puerto Rico) and hurricane-affected (Bisley Experimental Watersheds) sites over time following disturbance (mo) and by sampling date (mo/y). Positive flux values indicate emission from the soil to the atmosphere. Negative values indicate uptake by the soil. Flux rates are the means of four chamber measurements; bars show standard errors (Steudler et al. 1991). (Used with permission from the Association for Tropical Biology and Conservation.)

Riparian Groundwater and Stream Water

Reflecting the temporary decreases of living biomass during the first 5 months after Hurricane Hugo, concentrations of all forms of nitrogen increased in riparian groundwater in a Bisley catchment, including nitrate (NO_3^-), ammonium (NH_4^+), and dissolved organic N (McDowell et al. 1996). Base cations, chloride (Cl^-), and silicon dioxide (SiO_2) also increased in groundwater over this period. The largest relative change in concentration occurred for K^+, which had increased from 0.7 mg L^{-1} to as high as 13 mg L^{-1} 5.5 years after the hurricane. At another study site, the Icacos catchment, NO_3^- concentrations peaked at 1.1 mg L^{-1} a year after the hurricane and had decreased to nearly 0.0 mg 5.5 years after the hurricane. At both sites, NO_3^- concentrations were higher in upslope sampling wells than in those closer to the stream. Most solutes had returned to background levels within 1 to 2 after the hurricane, except for K^+. Overall, riparian processes appear to reduce but not eliminate hydrologic losses of N following hurricane disturbance (McDowell et al. 1996; McDowell 2001). In the absence of riparian N retention, the total dissolved N export would be 50 percent greater at the scale of the whole Río Icacos basin (Chestnut and McDowell 2000; Madden 2004). Rapid dissimilatory nitrate reduction to NH_4^+ by microbes probably has a significant role in this process (Silver et al. 2001, 2005).

The massive defoliation caused by Hurricane Hugo produced large but short-lived increases in nutrient export in streams (figure 5-12) (Schaefer et al. 2000). Average concentrations of nitrate, potassium, and ammonium in stream water increased by 13.1, 3.6, and 0.54 kg ha^{-1} y^{-1}, respectively, for up to 2 years, representing increases of 119, 182, and 102 percent. (Nitrate, however, was not detected in streams for several weeks immediately after the hurricane, perhaps due to an increase in dead fine roots that stimulated microbial immobilization [Parrotta and Lodge 1991]). The later increase in stream water nitrate concentrations might have been caused by reduced plant uptake of nutrients or the loss of nutrients released by microbial mineralization of hurricane-derived litter. Sulphate (SO_4), chlorine (Cl), Na, Mg, and Ca showed smaller increases, and the N and K were equivalent to only 1 and 3 percent, respectively, of the N and K in the hurricane-derived plant litter (Scatena et al. 1996). After 2 years, export in streams returned to prehurricane rates, in synchrony with revegetation. Despite extensive effects on the forest, the high survival of plants, rapid revegetation, microbial immobilization of nutrients (see below), and riparian retention led to a rapid return of the stream chemistry to prestorm conditions.

Posthurricane Productivity and Biomass

Measurements of Productivity and Biomass

Hurricane Hugo reduced the aboveground forest biomass by as much as 50 percent. However, the posthurricane productivity was higher than that before the storm, and the biomass recovered quickly (Scatena et al. 1996; Weaver 2000). At Bisley, the net primary productivity (NPP) peaked within 12 to 18 months after Hurricane Hugo, and the accumulation of aboveground biomass was nearly 7 to 10 times the

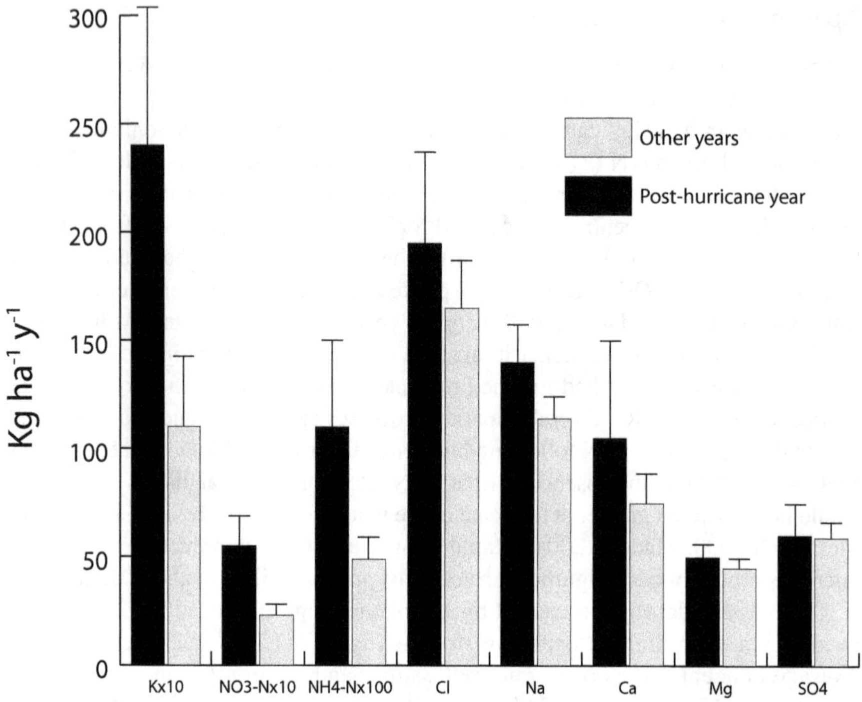

Figure 5.12 Comparison of stream chemical fluxes in the first year after Hurricane Hugo in several watersheds in the Luquillo Experimental Forest, Puerto Rico, to fluxes averaged over all other years of record (1983 to 1991–1994) (Schaefer et al. 2000). Bars show mean values across watersheds and the 95 percent confidence intervals. (Used with permission from Cambridge University Press.)

annual average, mainly due to regeneration of the pioneer *Cecropia schreberiana* (Scatena et al. 1996). Five years after Hurricane Hugo, the aboveground NPP had reached 21.5 Mg ha^{-1} y^{-1}, triple the prehurricane rate, and the aboveground biomass had reached 86 percent of the prehurricane level (Scatena et al. 1996). Of this, 35 percent was from the postdisturbance regeneration of pioneer trees, still mainly *C. schreberiana*. Nonpioneer species also responded to canopy opening, and possibly to reduced root competition, with increased growth, as in a posthurricane Jamaican forest (Tanner and Bellingham 2006). Seedlings of the dominant, nonpioneer tree *Manilkara bidentata* grew 17 times as fast after Hurricane Hugo than before, as mentioned previously (You and Petty 1991). At Bisley 1 year after Hurricane Hugo, the biomass of seedlings 0.2 to 0.5 m in height had increased to five times what it had been before the storm, and after 5 years it was three times greater than before the storm. In other parts of the forest, the aboveground net productivity of the palm *Prestoea montana* was 20 percent greater after Hurricane Hugo (Weaver 1999). This species and others respond with faster growth when coarse woody debris, the decomposition of which might supply nutrients for growth, is added to experimental plots (Beard et al. 2005; also see Zalamea-Bustillo 2005). Over the long term, in a plot established in 1943 after the 1932 hurricane, the basal area and biomass increased until the 1970s, when they appeared to reach a steady state (Crow 1980; Weaver 1986).

In the short term, Hurricane Hugo substantially reduced root biomass and aboveground biomass (Vogt et al. 1995; Beard et al. 2005). Four weeks after the storm, standing stocks of live fine roots (<3 mm diameter, to a depth of 10 cm) were 0 to 2 g m^{-2} and fluctuated greatly, probably in response to rainfall (Parrotta and Lodge 1991). In one study, it took more than a year for fine-root standing stock to return to prehurricane levels (Parrotta and Lodge 1991). In another study after Hurricane Hugo, fine roots recovered in 7 months and increased greatly at 8 months, when rainfall increased after a posthurricane drought (Beard et al. 2005). The coarse woody debris added to experimental plots increased the fine root biomass (Beard et al. 2005). This regrowth of fine roots is fast compared to regrowth in areas of tabonuco forest where all roots were experimentally removed (Kangas 1992).

One measure of productivity, litterfall, took 5 years to return to pre–Hurricane Hugo values in tabonuco forest (figure 5-6) but only 1 month to recover after minor hurricane effects (Beard et al. 2005). Elsewhere in the tabonuco forest after Hurricane Hugo, fine litterfall was at 55 to 77 percent of prestorm values at El Verde immediately after the hurricane, and at 39 to 82 percent after 5 years (Vogt et al. 1996). Variation was associated with topography; inputs of litterfall into a stream returned to prehurricane levels at a slower rate than did those into riparian and upslope areas (Vogt et al. 1996). As with fine roots and basal area increment, litter production increased with the addition of coarse woody debris (Beard et al. 2005). After Hurricane Hugo, leaf litter production was slower to recover in the high-elevation elfin forest than in tabonuco forest (Walker et al. 1996b), where tree growth is 10 times faster (figure 5-13) (Walker et al. 1996b; Waide et al. 1998).

As mentioned, coarse woody debris is potentially a long-lasting supply of nutrients, and its presence increased the basal area increment, fine root biomass, and litterfall of established trees, including the abundant palm *Prestoea montana* (Beard et al. 2005). Although coarse woody debris can boost long-term productivity, it can also depress it during the short-term pulse of nutrients after a hurricane. After Hurricane Hugo, the abundant carbon source in woody debris is thought to have stimulated the growth of microbial decomposers, which then outcompeted trees for soil N and possibly other nutrients, thereby slowing response (Lodge et al. 1994; Zimmerman et al. 1995b). Thus the removal of woody debris from experimental plots at El Verde increased the short-term rate of canopy closure and forest productivity, and fertilization without debris removal appeared to reduce competition for nutrients (Zimmerman et al. 1995b).

Tree species seem to differ in their ability to compete with decomposers for nutrients. The diameter growth of the canopy trees *Dacryodes excelsa* and *Manilkara bidentata* increased when coarse woody debris was added in their vicinity, and growth decreased when the debris was removed (Beard et al. 2005), suggesting that *Dacryodes* and perhaps *Manilkara* were able to compete effectively with decomposer microbes for nutrients. In contrast, *Cecropia schreberiana* growth declined with the addition of wood, suggesting that *Cecropia* is less well adapted for acquiring nutrients from decomposing wood and competing with microbes. Other studies have

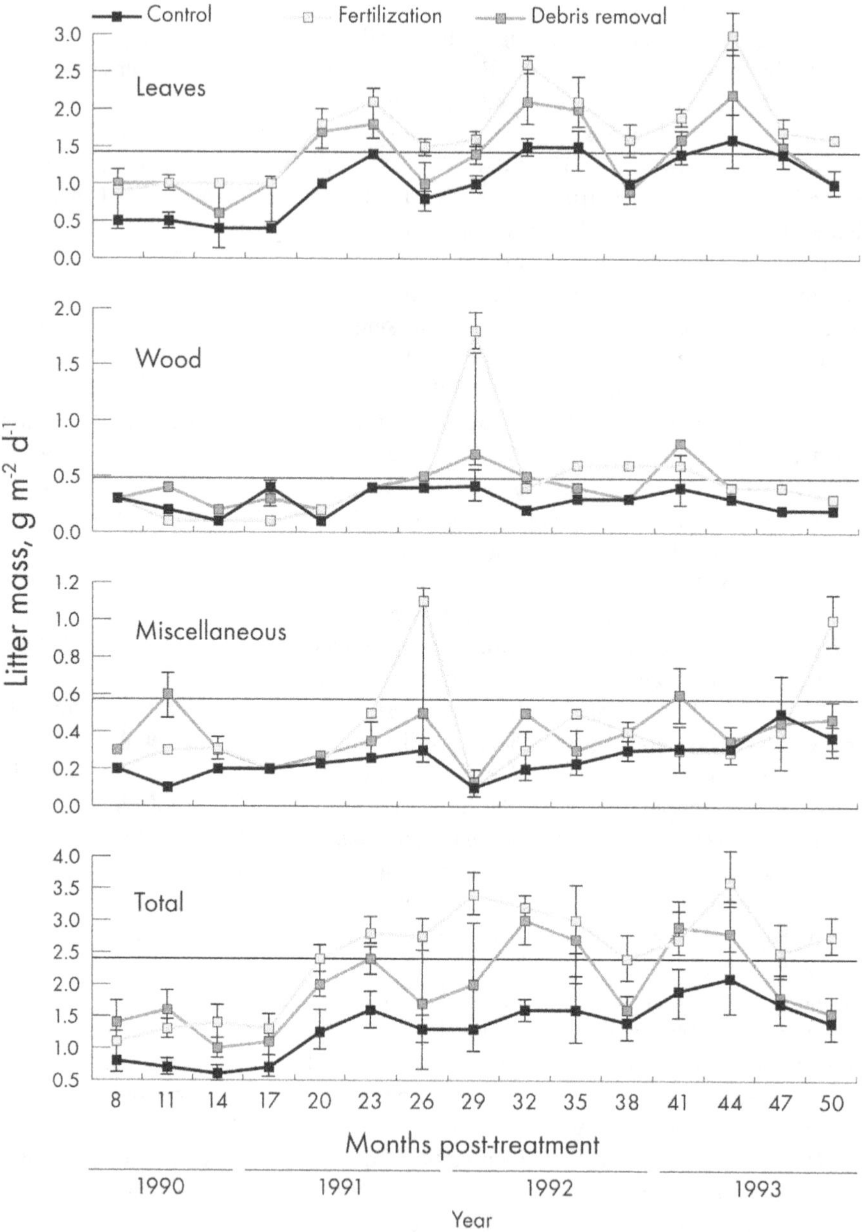

Figure 5.13 Comparison of leaf, wood, miscellaneous, and total components of litter trapped in control, fertilized, and debris removal plots in tabonuco forest following Hurricane Hugo at El Verde, Puerto Rico (Walker et al. 1996b). Horizontal lines show prehurricane annual mean litter mass (Zou et al. 1995). Mean and standard error are shown. n = 4 plots per 3-month period. (Used with permission from the British Ecological Society.)

shown *Cecropia* to be especially nutrient demanding (Walker et al. 1996b). Species effects might also depend on the relative availability of nutrients; at a nutrient-rich site, neither wood addition nor removal affected the growth of *Dacryodes* (Beard et al. 2005). Moreover, the inherent growth rates of trees, including palms, were generally maintained despite the vagaries of disturbance (Beard et al. 2005). Thus local posthurricane productivity reflects inherent site and species differences as much as, or more than, it reflects local variation in storm effects (Walker et al. 1996b; Beard et al. 2005).

The thorough study of response to Hurricane Hugo at Bisley suggested a sequence of phases in ecosystem reorganization in the first 5 years following this hurricane (Scatena et al. 1996). The first phase was a period of foliage production as hurricane survivors releafed and herbaceous vegetation and woody regeneration became established. During this phase, 75 to 92 percent of the nutrient uptake remained in aboveground vegetation. There was a relatively low rate of aboveground carbon accumulation per mole of nutrient cycled, and thus a low level of "nutrient use efficiency," measured as organic matter produced per unit of nutrient uptake (Vitousek 1982). In the second phase, there was a peak in aboveground productivity when early successional species entered the sapling and pole stages. In the third phase, the litterfall nutrient cycle was reestablished, and there was an increase in the net productivity per mole of nutrient cycled, and thus a higher nutrient use efficiency. During the 5 years following Hurricane Hugo, the Bisley forest had some of the lowest within-stand nutrient use efficiencies and some of the highest levels of aboveground productivity ever observed in the Luquillo Mountains. Thus, high productivity and rapid aboveground ecosystem reorganization can be achieved with rapid within-system cycling and inefficient within-stand nutrient use.

Modeling of Production, Biomass, and Nutrient Dynamics

The Century Soil Organic Matter Model (CENTURY) was used to synthesize knowledge of nutrient cycling and productivity and to project trends over centuries of repeated hurricanes (Sanford et al. 1991). A spatial version of CENTURY, the model TOPOECO, was used to simulate these factors over the Luquillo Mountains landscape, taking into account elevation, exposure, and effects from Hurricane Hugo (Wang et al. 2002a, 2002b, 2003; Wang and Hall 2004).

The typical biomass of the tabonuco forest is about 300 Mg ha^{-1} (Sanford et al. 1991). This is low compared to the values in many tropical lowland forests, but that is expected given that chronic hurricanes seem to limit tree size (see above). According to CENTURY simulations, biomass in the tabonuco forest would increase for up to 400 years of forest development without major disturbance, a developmental stage hurricanes never permit the forest to attain (figures 5-14 and 5-15) (Sanford et al. 1991). Although the forest biomass is low, productivity is high due to the repeated establishment of young, fast-growing trees and the repeated pulses of available nutrients. With high productivity but a low biomass because of disturbance, organic carbon ends up in the soil, and this in turn fuels productivity. Model simulations show that high soil organic carbon results in comparatively high rates of P and N mineralization. The model results are supported by observed increases in ammonium

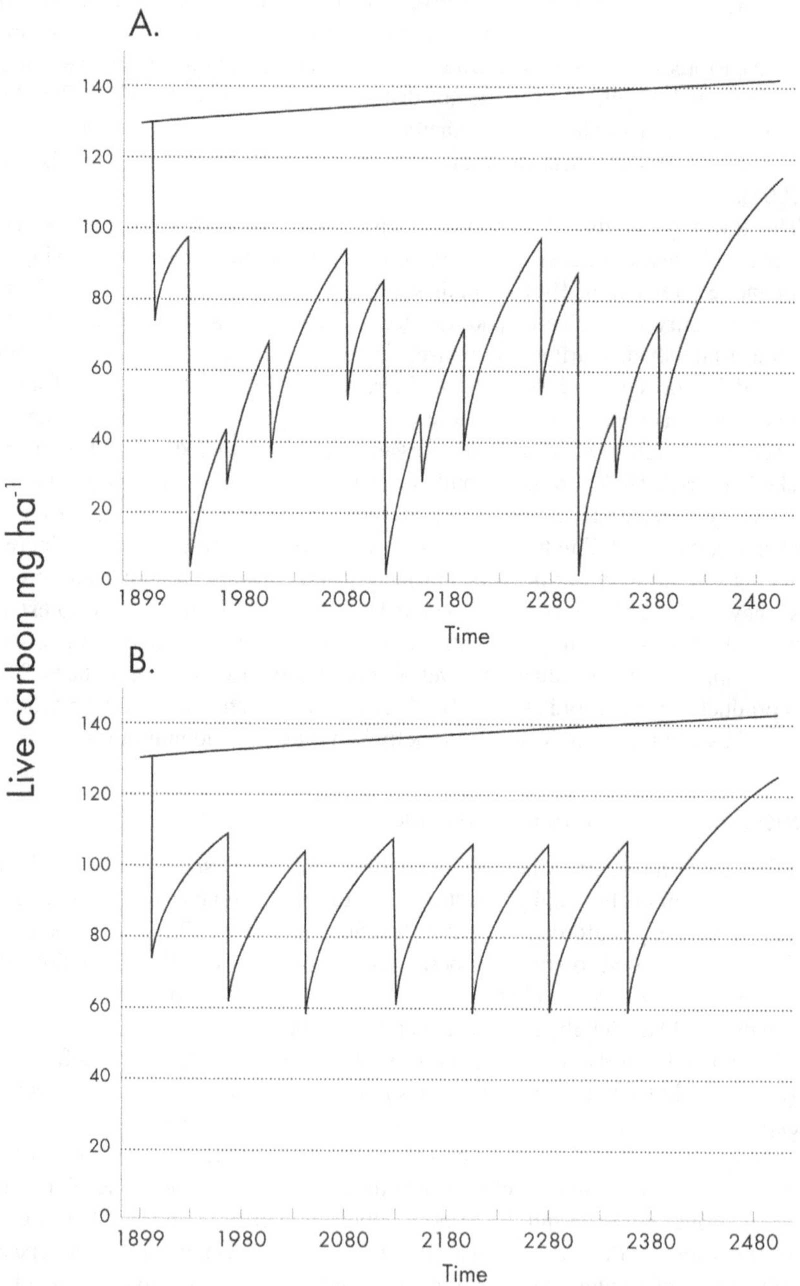

Figure 5.14 Above- and belowground carbon simulations as a function of hurricane frequency (Sanford et al. 1991). Straight line is a control (no storms). Irregular lines indicate periodic hurricane disturbance. (A) Historical hurricane disturbance projected into the future using the sequence of six hurricanes that occurred in 1899–1989. (B) Hurricane sequence of repeated Hurricane Hugo strength storms at c. 60-year intervals. (Used with permission from the Association for Tropical Biology and Conservation.)

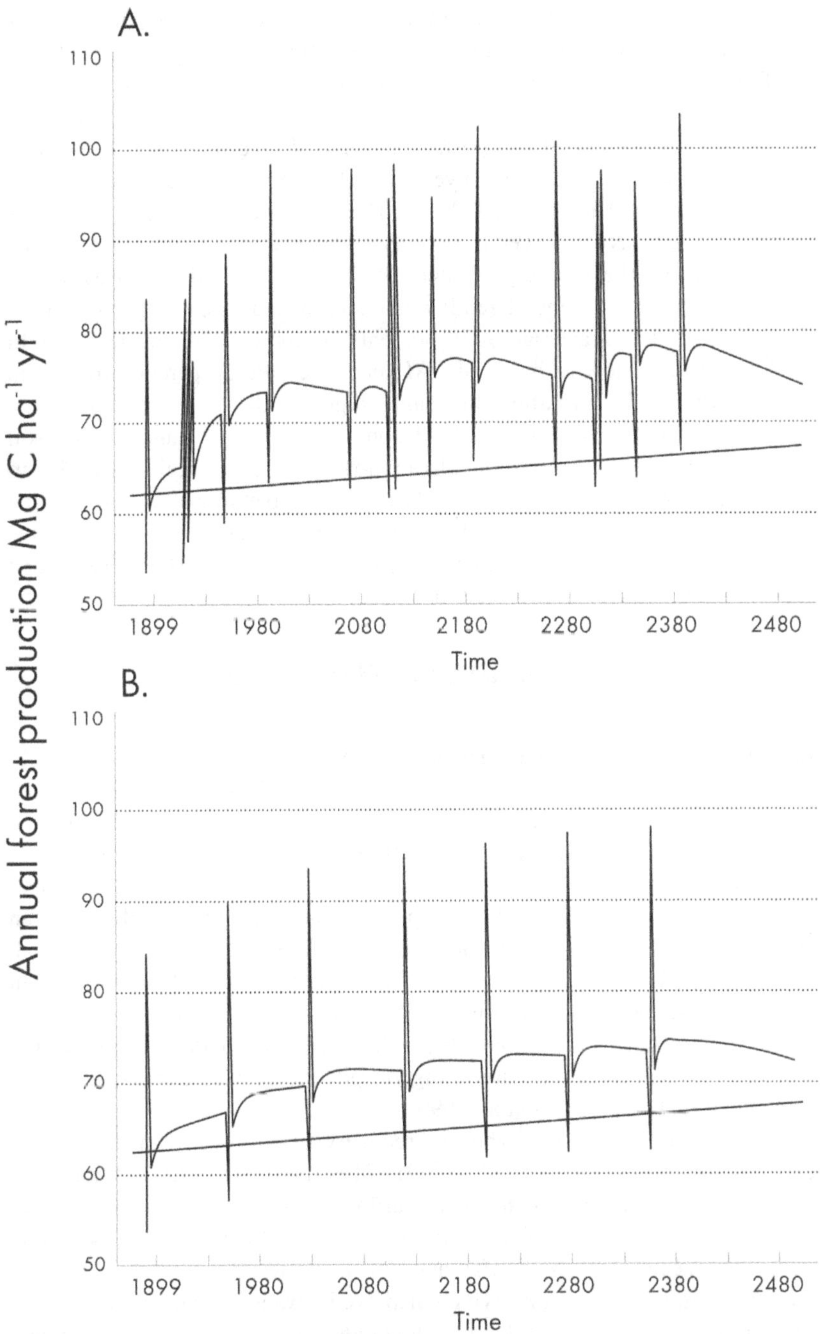

Figure 5.15 Forest production simulations in tabonuco forest as a function of hurricane frequency (Sanford et al. 1991). Straight line is a control (no storms). Irregular lines indicate periodic hurricane disturbance. (A) Historical hurricane disturbance projected into the future using the sequence of six hurricanes that occurred in 1899–1989. (B) Hurricane sequence of repeated Hurricane Hugo–strength storms at c. 60-year intervals. (Used with permission from the Association for Tropical Biology and Conservation.)

availability and net N-mineralization and nitrification rates (after an initial short-term decline owing to microbial immobilization), followed by a gradual decline 4 months after Hurricane Hugo (also matching the stream water chemistry results; see above).

The landscape model TOPECO posits that the leaf area index (LAI) recovers within 2 years in the tabonuco forest, 3 years in colorado forest, and more slowly in palm and elfin forests (Wang and Hall 2004). The elfin forest lacks pioneer species that quickly restore LAI at lower elevations (Walker et al. 1996b). The model further suggests that the recovery of tabonuco forest LAI and increases in soil organic carbon (SOC) and mineralized P would spur increases in the gross primary productivity (GPP) by an average of 30 percent 5 years after Hurricane Hugo. In palm and elfin forest, slow recovery of LAI keeps the GPP 20 percent lower than before the storm for 5 years after Hurricane Hugo. In all four vegetation types, storages of SOC, CO_2 emissions from the decomposition of SOC, and the total soil N increase slightly. However, N mineralization rates increase significantly due to the massive input of plant materials from Hurricane Hugo at low elevations and the slow decomposition at higher elevations. There is much variation in these measures because of topography as well (see above). Both CENTURY and TOPECO suggest that these responses last only a few (about 5) years.

Contrasting Recovery of Forest Function and Structure after Hurricane Effects

Within only 5 years after Hurricane Hugo severely affected forests in the Luquillo Mountains, many populations and ecosystem functions had returned to prehurricane states (Zimmerman et al. 1996; Lugo et al. 1999), but some populations and the physical structure of the forest had not. After 10 years of observation and experiment, through both hurricane and drought events, it was still evident that ecosystem processes, such as plant growth and decomposition rates, had recovered faster than elements of ecosystem structure, such as foliage and fine root biomass (Beard et al. 2005). At Harvard Forest (Massachusetts, USA), an experiment that pulled down trees in order to simulate hurricane effects revealed a similar disconnection between forest structure and function: the rapid regrowth of trees and understory vegetation quickly restored patterns of nutrient cycling, despite the slow recovery of structure (Foster et al. 1997; Cooper-Ellis et al. 1999).

A system attains steady state when its recovery time is less than the interval between disturbances (White et al. 1999). The resistance and posthurricane resilience of many populations in the Luquillo Mountains help them return to prehurricane states within the average 60-year storm interval. Examples among dominant tree species are *Dacryodes* and *Prestoea*, which resisted winds and suffered low mortality; *Manilkara*, which displayed advance regeneration that was released from suppression and helped maintain the species' abundance; and *Cecropia*, the population structure of which changed drastically but was rapidly returning to its prestorm state. Among animals, the abundances of many dominant snail, frog, lizard, bat, and bird species 5 years after Hurricane Hugo were within the range of prehurricane variation.

Overall, nutrient cycling is likewise resilient after the passage of a hurricane, for several reasons (McDowell et al. 1996; Scatena et al. 1996; Schaefer et al. 2000; Beard et al. 2005). First, the removal of aboveground biomass does not necessarily lead to a loss of soil nutrients (Silver et al. 1996). Second, many post-storm processes (microbial uptake of nutrients, root recovery, the establishment of fast-growing pioneers, high survival of dominant trees and their rapid refoliation and branch sprouting) quickly take up and store nutrients from decomposing debris. Third, coarse woody debris provides a long-term source of nutrients for continued productivity in soils that are relatively rich in any case. The result is that turnover rates of nutrients and biomass are faster than the hurricane return time, which allows ecosystem functions to achieve steady state in those intervals (Scatena 1995).

In contrast to some populations and ecosystem functions, the three-dimensional structure and biomass of the forest are slow to recover and would probably continue to change over centuries in the absence of subsequent disturbance (Sanford et al. 1991). Structure in forests of the Luquillo Mountains might always be in a state of development (cf. Lugo et al. 1983) if the time to steady state exceeds the hurricane return time. As an extreme model, it is estimated that 500 years must pass after land clearing before a recovering dipterocarp forest in Asia reaches steady state in structure and composition (Riswan et al. 1985). In the Luquillo Mountains, the time from a hurricane-affected state to a steady state of structure and composition might be faster, but it is surely longer than the average 60-year recurrence interval measured for severe hurricanes. Many tree species would continue growing large boles and spreading crowns well after 60 years, thus changing the structure and biomass of the forest, with consequences for other organisms. Two hundred years are thought to be necessary for the recovery of elfin forest in the Luquillo Mountains after effects caused by a plane wreck (Weaver 2000), and modeling suggests that, in the absence of hurricane disturbance, 400 years are required in order for biomass to level off in tabonuco forest (Sanford et al. 1991).

Some Unmet Expectations

One might expect to see certain hurricane effects that are not observed in the Luquillo Mountains. As reported above, after a short interval of response, Luquillo forests do not have highly irregular canopies (Brokaw et al. 2004) as described for the "cyclone scrub" in Australia (Webb 1958) and the "hurricane forest" in St. Vincent, which consists of low thickets with occasional vine-covered emergent trees (Beard 1945). Also, the number of true pioneer tree species, such as *Cecropia*, is not high in Luquillo forests (cf. Brokaw 1985), possibly because there are few treefall gaps to sustain pioneers between hurricanes. For the same reason, the understory of Luquillo forests is minimally cluttered with the background treefall debris and regeneration frequently encountered in some forests not struck by hurricanes (N. Brokaw, personal observation). Lastly, lianas are not as common in tabonuco forests that have been studied (Rice et al. 2004) as they are in disturbed forests elsewhere (Schnitzer and Bongers 2002).

Aquatic Response to Hurricanes and Floods

Hurricanes usually bring heavy rain, high river discharge, and fast currents, and they dump much debris into streams. They also alter the stream microclimate through effects on neighboring forest. All of this has strong effects on stream organisms and processes. Heavy rain not associated with hurricanes also produces high discharges (see chapter 4), but these are not necessarily accompanied by large inputs of debris and changed microclimates.

During Hurricane Hugo, high discharge and fast currents redistributed inorganic and detrital material, as well as stream organisms, throughout the benthic environment along the stream continuum (Vannote et al. 1980) of the Luquillo Mountains (Covich et al. 1996). Litterfall added nutrients and detritus. Debris dams formed, catching detrital food and reducing the washout of invertebrate consumers. Large debris dams persisted for months, continuously releasing microbially conditioned leaves that were carried downstream and eaten by shrimp, a key animal group in stream ecosystems (chapter 3; Crowl et al. 2001). Also after the storm, sunlight poured through the open canopy, promoting the growth of periphytic algae and increasing food for shrimp. In some areas silt covered detrital and algal food sources and refuges from predators, but it washed out within 3 months (Covich et al. 1996). Thus hurricane floods created strong residuals in Luquillo Mountain streams.

Shrimp, especially those in the family Atyidae, are abundant herbivores and detritivores in the headwater streams in the tabonuco forest and make up most of the stream biomass (Covich and McDowell 1996). Their populations were greatly affected by the immediate effects of Hurricane Hugo and by changes in the stream environment and food resources. One month after Hurricane Hugo, atyid shrimp densities were reduced on average by 50 percent in upstream pools, the shrimp apparently having been washed out, and they increased by 80 percent in downstream pools (340 to 460 masl) (Covich et al. 1991). In the next 6 months, shrimp densities increased rapidly to the highest abundances ever recorded in all sites. These high densities most likely resulted from shrimp migrating upstream from riverine pools and from the increased availability of algae and decomposing leaves as food. Shrimp populations in the middle-elevation pools then declined (Covich et al. 1996). A long-term effect appears to be that, in response to floods, shrimp favor pools where they and their food are seldom washed out (Covich et al. 1991).

Atya spp. and *Xiphocaris* shrimps respond directly to the redistribution of sedimentary material by rapidly consuming it and clearing it away via bioturbation (Pringle et al. 1993, 1999). This relationship between storms and shrimp was studied by manipulating the presence and absence of shrimp with electric fences in streams (Pringle and Blake 1994; Pringle et al. 1999). Where shrimp were excluded, there was a greater mass of fine and particulate organic material and algal biovolume than in controls with natural densities of shrimp, and there was a larger increase in the mass of sedimentary material following storms. In controls, there was no measurable accumulation of sediment under base flow conditions, and shrimp rapidly removed sediments that accumulated during storms, reducing them to near-prestorm levels within 30 hours. Thus shrimp have a significant effect on the posthurricane, postflood distribution of inorganic sediments and on fine and coarse particulate organic materials.

Benthic communities are resilient after intermediate levels of storm disturbance because debris dams catch food and reduce the washout of invertebrate consumers. Storms producing less wind and/or greater streamflow than Hurricane Hugo could cause extensive, longer-lasting decreases in populations of benthic-dwelling shrimp because there might be less input of debris and greater washout of the shrimp and their food.

Response to Droughts

Droughts affect the Luquillo Mountains, and research is beginning to reveal their effect on forests and streams. Understanding current droughts might help us foresee the consequences of predicted change toward reduced and more variable rainfall in the Luquillo Mountains.

Following Hurricane Hugo, and again in 1994–1995, there were exceptionally dry periods in the tabonuco forest (chapter 4), with measurable effects on microbes, plants, and animals. Fungal decomposers that produce mycelia on leaf surfaces appear to be especially susceptible to this drought. One such species, *Collybia johnstonii*, was a common litter decomposer in tabonuco forest before Hurricane Hugo, but during the 5 years after canopy destruction by the storm, some mycelia of *C. johnstonii* were smaller or extirpated and were replaced by more drought-tolerant species (Lodge and Cantrell 1995; Lodge 1996; see chapter 6). Fungal biovolumes in soil are closely correlated with soil moisture and decreased slowly in response to drought (Lodge 1993). For trees, hurricane effects reduced fine-root biomass, which recovered in 7 months, but the frequent droughts that followed reduced fine-root biomass such that it did not recover to prestorm levels for 10 years (Parrotta and Lodge 1991; Silver et al. 1996; Beard et al. 2005). Thus droughts can have a greater effect than hurricanes on fine roots and, consequently, on nutrient acquisition and productivity. Litterfall rates also reflected the effect of the drought. After Hurricane Hugo, aboveground litterfall inputs did not recover to prehurricane rates even after 5 years, apparently because of the posthurricane droughts (Vogt et al. 1996). For the riparian fern *Thelypteris angustifolia*, the overall leaf production did not change during the drought year of 1994; however, leaf life spans did decrease relative to earlier years (Sharpe 1997; J. Sharpe, personal observation). The possibility that drought enhanced posthurricane herbivory is discussed above.

Juvenile coquí frogs (*Eleutherodactylus coqui*) cannot survive drought (Stewart 1995), but no effects of drought on adults have been recorded that are distinguishable from background variation (L. Woolbright, personal observation). Females that retain their egg clutches during dry weather typically lay them when it starts to rain again. However, some frog species might be less hardy than *E. coqui*. Both *E. portoricensis* and *E. richmondi* disappeared from mid-elevation forests at a time roughly corresponding to the drought following Hurricane Hugo (L. Woolbright, personal observation). Posthurricane drought appears to have depressed *Anolis* lizard numbers, and a drought coincided with the lowest recorded density of spiders in one study (Pfeiffer 1996; Reagan 1996).

Droughts have many effects on stream communities and processes (Covich et al. 1998, 2000, 2003, 2006; Covich and Crowl 2002). Droughts alter the local food-web structure, detrital processing dynamics, and predator–prey dynamics. During droughts there are no or fewer flushing events, and first- and second-order streams might accumulate organic detritus and inorganic sediments that can decrease pool depth and volume. The reduced pools expose prey to predators at the top and bottom of the water column. When the pool size contracted during a drought in 1994, the density of the dominant shrimp *Atya lanipes* rose from 22 to 75 shrimp m^{-2} of pool area, and the density of another species, *Xiphocaris elongata*, increased from 5 to 14 shrimp m^{-2} of pool area. Gravid adults of both species were fewer during the drought, and the reproductive activity of *X. elongata* remained low during the year. The lowest mean abundance of the predatory shrimp *Macrobrachium* spp. occurred during the 1994 drought, the driest year of 28 years on record in the Río Espíritu Santo drainage. After that drought *Macrobrachium* increased in abundance for 6 years. Droughts increase crowding, reduce both predator and prey populations of detritivores in the short term, increase predator populations over the longer term, and depress reproduction among key detritivores. In addition, the lack of flushing during droughts reduces mortality due to physical scour and results in detrital storage that appears to provide shelter for prey in some pools.

Response to Landslides

Landslides are frequent disturbances in tropical mountains, including the Luquillo Mountains (Garwood et al. 1979; Guariguata 1990; Larsen and Torres-Sánchez 1992). Landslides provide good opportunities for research on disturbance and response because they include strong temporal and spatial gradients in light, moisture, and soil fertility and stability. These gradients permit examination of the roles of dispersal, competition, and facilitation in order to explain vegetative responses. The primary succession that occurs on landslides follows clear trajectories of response and a sequence of processes that clearly alters the ecological space (Myster and Fernández 1995; Walker et al. 1996a; Myster and Walker 1997).

Landslides consist of two or three relatively discrete zones in which soil and vegetation removal, subsequent stability, and regeneration vary. These zones include (1) an upper zone nearly devoid of vegetation that is unstable and which is colonized slowly, because it has few residuals of the previous system; (2) a lower zone in which soil and vegetation from the upper zone are deposited and which is more stable and able to support faster revegetation, as residuals of the previous vegetation are still present; and sometimes (3) a middle zone that is a "transport chute" between the upper and lower zones (Walker et al. 1996a). Soil organic matter and nutrient concentrations are generally higher in the lower zone, but light levels are typically higher in the upper zone (Fernández and Myster 1995). Temperature and soil moisture are generally higher in landslides than in adjacent forest. Low-fertility patches in landslides contrast with hurricane-affected or cleared sites where soils remain intact (Myster et al. 1997).

Succession on Landslides

The rates of change and the particular sequences of plant community composition during succession in landslides are affected by elevation, landslide size, compass orientation, surrounding vegetation, soil development, colonization dynamics, and biotic interactions (Myster and Walker 1997). Plant replacements during succession are especially evident in landslides, where there is little advance regeneration to obscure the sequence of colonization. Succession is slow, having a long plant-to-plant replacement phase, and early plant colonists have a strong influence on later dynamics (Walker et al. 1996a; see chapter 6 for details). Development is faster at lower than at higher elevations and on volcaniclastic than on other substrates. Hurricane effects can retard landslide succession. The severe disturbance of a landslide can erase land use history as an influence, but the past land use of areas of neighboring vegetation that contribute propagules is important (Myster and Walker 1997). On landslide areas in the Luquillo Mountains, the levels of soil nutrients, basal area, and plant composition start to resemble mature forest levels after about 55 years (Guariguata 1990; Zarin and Johnson 1995a, 1995b).

Landslide colonization in the Luquillo Mountains is primarily limited by the availability of dispersed seed (Walker and Neris 1993); the availability of germination microsites within the landslide itself (Myster 1997); and competition for light, water, and soil nutrients (Fetcher et al. 1996). Seed dispersal into a site is particularly important in landslide succession when the seed bank and seed producing plants were removed when soil slid downslope (Walker and Neris 1993; Myster and Fernández 1995). Seed-dispersing birds avoid barren landslide areas where there are no perches (Shiels and Walker 2003). Seed loss to predators and pathogens in landslides is small (Myster 1997). Germination varies among sites within slides, and fertilization has increased the germination of two common plant species on landslides (Shiels et al. 2006). Shrubs have the highest levels of germination among life forms (Walker and Neris 1993). Once they have germinated, the mortality of *Cecropia schreberiana* and *Inga vera* is due more to presumed competition for nutrients and less to pathogens and herbivores (Myster and Fernández 1995). However, in another study, fertilization did not increase the seedling growth of two common plant species on landslides (Shiels et al. 2006).

Succession on landslides might be slowed by the low nutrient availability in areas of soil loss. Both the base saturation and major nutrient cation concentrations are low on new landslide scars (Zarin and Johnson 1995a), but these increased in surface mineral soil (0 to 10 cm) over a 1-to-55+-year chronosequence. During this period, the recovery of N, P, K, and Mg to levels present in mature forests near landslides occurred, suggesting that, ultimately, forest recovery is not limited by a lack of those nutrients (Zarin and Johnson 1995b).

Potential sources of nutrients on landslides include atmospheric deposition, substrate weathering, and litterfall, the importance of which can change with succession. For example, there is a net increase in labile P supplied from the atmosphere and litter input, and probably from the pool of inorganic occluded P. The added P is used by the biota and returned to the soil in organic combinations. Eventually, the main source of plant-available P seems to become the labile P pool, as plants increasingly rely on the processing of readily mineralizable organic P (Frizano et al. 2002). The nutrient flux from allochthonous litterfall would differ depending on the identities of colonizing species (Shiels 2006). For example, leaves of the common colonizers *Cecropia schreberiana* and *Cyathea arborea* differ in chemical composition and decomposition rate. Similarly, nutrient flux would change as the species composition shifts during succession.

Ferns in Landslides

Ferns have a large role in succession on landslides. They stabilize soil, build organic matter, and increase soil nutrients and soil moisture, but they also inhibit the establishment of woody species (Walker 1994). Ferns are commonly found throughout the tropics on disturbed soils (Kochummen 1977; Maheswaran and Gunatilleke 1988). In fact, bare soil, such as that on landslides, is required for the germination and establishment of ferns, which can colonize abundantly in suitable conditions (Moran 2004). In the Luquillo Mountains, *Gleichenia bifida* and *Dicranopteris pectinata* can form dense thickets up to 2 m tall, in adjacent monocultures of each species or in mixed stands. Ferns spread via extensive rhizomes that grow along the soil surface. This might allow fern rhizomes to colonize nutrient-poor soils from parent plants rooted in more fertile soils on the landslide edge. The rhizomes both stabilize the soil on landslides (resistance to erosion) and are sources of invasive propagules following disturbance (resilience). Ferns have an indeterminate growth form that maximizes their use of available space, and old fronds generally remain attached while new growth rises above them. This growth habit forms a dense layer of suspended leaf litter that, together with the newer fronds, greatly reduces the light penetration below (Walker 1994). Long-term observations suggest that these fern thickets in landslides might persist at least several decades before they are eventually shaded out and replaced by trees (cf. Kochummen 1977).

Response to Human Disturbance

Human effects on tropical ecosystems are widespread in the present and pervasive in the past (Keay 1957; Barrera et al. 1977; Hartshorn 1980; Sanford et al. 1985; Gómez-Pompa and Kaus 1992; Clark 1996; chapter 7). Human disturbance can be more severe than natural disturbance, because human disturbance typically eliminates more of the previous ecosystem more uniformly and more often while leaving fewer residuals to assist in recovery (Franklin et al. 2000). In the past 500 years alone, the extensive forest cover of pre-Columbian Puerto Rico has been reduced by humans to only 6 percent of the island (Birdsey and Weaver 1987). Currently, contemporary reforestation, suburbanization, water extraction, and perhaps climate change are accelerating ecosystem change. In this section we discuss the response to past human disturbances in the Luquillo Mountains, including agriculture, forest clearcutting, road building, and radiation disturbances. (See chapters 3 and 7 for discussions of water extraction and dams and chapters 3 and 8 for discussions of introduced species.)

In 1936, about 40 percent of the area of the Luquillo Mountains was unforested or covered with secondary forest, and less than half of the overall area was continuous-canopy forest (>80 percent canopy cover) (Foster et al. 1999). Most of the unforested and secondary forest areas were below 600 masl. In fact, the tabonuco forest zone (below about 600 m) included only 8 percent "mature" forest (Wadsworth 1950; Foster et al. 1999). Many stands were never completely deforested but were heavily affected by charcoal manufacture, coffee growing, or selective tree cutting (García-Montiel and Scatena 1994; Thompson et al. 2002). Much of this area was purchased by the U.S. Forest service in the 1930s and allowed to revert to forest. By 1989, 96 percent of the Luquillo Mountains was continuous forest, and the forest area had also increased elsewhere in Puerto Rico (Birdsey and Weaver 1987; Thomlinson et al. 1996; Grau et al. 2003). Thus, in order to interpret the present structure, species composition, and ecosystem function of many Luquillo Mountains forests, we must study responses to land uses.

Secondary Forest after Agriculture

Secondary Forest Structure and Composition: Animals

Studies of a chronosequence of abandoned pastures in the vicinity of the Luquillo Mountains show that after about 40 years of regrowth, forests recover most structural and functional characteristics found in older-growth forests affected only by natural disturbance (Crow and Grigal 1979; Aide et al. 1995; Pascarella et al. 2000). At 40 years of age, these secondary forests cannot be distinguished from old-growth forest in terms of their tree density, basal area, species number, or diversity. However, the tree species composition differs greatly (Zimmerman et al. 2000). Many species composing old growth are absent, and intropduced species can dominate secondary stands at low elevations and in alluvial areas (Abelleira Martínez and Lugo 2008), especially the trees *Spathodea campanulata* and *Syzygium jambos*. This pattern obtains across Puerto Rico in lower elevation secondary forests, in which the structure and tree species richness are similar to those in less disturbed forests, but species differ from those in less disturbed forests, and endemics and very large trees are fewer (Lugo and Helmer 2004). With succession, species dominance decreases and more rare species are represented, yet the predisturbance tree species composition will take centuries to recover (Aide et al. 1996). At higher elevations, in contrast, the native species *Miconia prasina* and *Tabebuia heterophylla* dominate abandoned pastures. Dominant species in recently abandoned pastures are those good at coppicing.

Age is the key correlate of forest development in these abandoned pastures (figure 5-16) (Aide et al. 1996). The distance to old-growth forest patches had no effect on any measure of forest recovery; however, the original pastures were not so large as the old pastures studied elsewhere—for example, in the Amazon, where pasture size is important because seed input is clearly limited (Uhl 1987). The lack of *Cecropia* until late in development in these secondary forests on old pastures suggests how different human disturbance is from natural disturbances in the Luquillo Mountains, such as hurricane effects and landslides, after which *Cecropia* colonizes early in development.

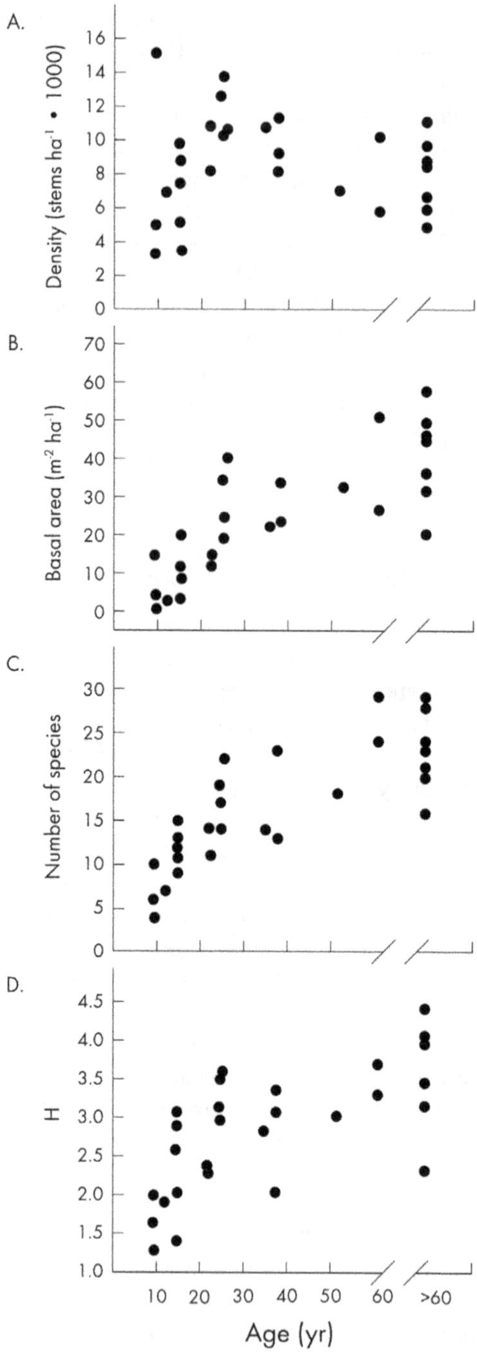

Figure 5.16 Relationship between age since pasture abandonment and (A) tree density, (B) basal area, (C) number of species, and (D) species diversity (H') in 23 abandoned cattle pastures (ages 9.5 to 690 y) and 7 sites that had been forested for ≥60 y (Aide et al. 1996). Only data from abandoned pastures were used to calculate the regression lines. (Used with permission from the Association for Tropical Biology and Conservation.)

Old pastures, abandoned coffee plantations, and relatively mature forest are dominated by different tree species. *Guarea guidonia* tends to dominate old coffee plantations, possibly in response to elevated N (Pascarella et al. 2000). In the Cayey Mountains, in central Puerto Rico, abandoned coffee plantations have a higher basal area than abandoned pastures with forest regrowth, because the coffee areas had more residual woody plants at the time of abandonment (Pascarella et al. 2000). As a result, coffee plantations do not rapidly accumulate plant species richness after abandonment. Age and elevation were also related to tree species composition in this study.

Naturally, the animal community also changes with succession in abandoned agricultural areas. The dominant frog and lizard species change as the vegetation becomes more structurally complex and the microenvironment becomes less variable during succession in abandoned pastures (Herrera Montes 2008). Earthworms, which play a key role in decomposition, shift from dominance by nonnative species to natives during succession on old agricultural lands (González et al. 2008).

Land Use and Species Composition on the Luquillo Forest Dynamics Plot

The present tree, fungal, slime mold, and bacterial species compositions of the 16 ha LFDP (see above) in tabonuco forest at El Verde all differ according to past land use on the plot (Willig et al. 1996; Huhndorf and Lodge 1997; Lodge 1997; Thompson et al. 2002). Historical records show that land use in the LFDP ranged along a gradient of severity from clearing for agriculture and clearcutting for plywood to coffee planting (probably under residual trees), timber stand improvement (thinning to improve growth), and selective logging (Thompson et al. 2002). All of these uses, except stand improvement, ended by about 1930. Aerial photographs from 1936 indicate four areas of different-percentage canopy cover in the LFDP that match historical information about previous land uses in those areas.

These historical land uses are the main determinant of present-day tree species composition among subplots in the LFDP (Thompson et al. 2002). A detailed comparison of the effects of historical land use intensity, soils, topography, elevation, and other environmental variables showed the overriding effect of historical land use intensity. In the most intensely used area, *Prestoea montana* dominated. In areas severely harvested for plywood, *Casearia arborea* was especially common. Areas of past coffee farming had few *Dacryodes excelsa* and *Manilkara bidentata*; both of these species would have been cut for timber and to decrease shade too heavy for coffee growth. Instead, a common species was *Guarea guidonia* (see above). *Dacryodes excelsa* and *M. bidentata* were relatively abundant in the area that had been selectively cut and improved for timber. The most intensely used area in the LFDP had a markedly lower stem number, richness, and diversity of tree species on both an area and a per-stem basis (Thompson et al. 2002) and fewer rare and endemic species, whereas the least affected area had the highest values for all these measures (table 5-4). Basal area was higher in the least disturbed area (Zou et al. 1995; cf. Aide et al. 1996). In the LFDP, past land use had a greater effect on forest composition and community characteristics than did either strong environmental gradients or the effects of several hurricanes after intensive land use had ceased.

Table 5.4 Forest structure, species totals, and diversity of trees with stems ≥ 10 cm D_{130} in the 16 ha Luquillo Forest Dynamics Plot at El Verde, Puerto Rico, in 1989 at the time of Hurricane Hugo. Data are presented for the whole plot and as a function of canopy cover class as determined from aerial photographs taken in 1936. Cover Classes 1, 2, and 3 had been clearcut or heavily logged and farmed or locally planted with tree crops before 1936; Class 4 was selectively logged before 1936 and from 1944 to 1953

Forest in 1989	LFDP total	Cover class (% canopy cover) in 1936			
		1 (0–20%)	2 (20–50%)	3 (50–80%)	4 (80–100%)
Area (ha)	16.00	1.16	3.96	5.64	5.24
Number of stems	13,167	866	3,401	4,572	4,328
Stem density (ha^{-1})[a]	822.9	746.6	858.8	810.6	826.0
BA (m^2 ha^{-1})[a]	36.7	36.5	35.7	35.4	40.8
Total number of species (w/o exotics)	89 (83)	32 (30)	66 (63)	62 (60)	76 (75)
Species ha^{-1}: mean (s.d.) and range[b]	44.3 (5.7) 33–52	32	44.5 (3.5) 42–47	42.7 (2.5) 40–45	48.0 (2.9) 45–51
Shannon-Wiener H' (w/o exotics)[c]	2.90 (2.86)	2.18 (2.06)	2.69 (2.68)	2.65 (2.63)	2.93 (2.92)
Rare species (w/o exotics)[d]	44 (41)	3 (2)	21 (19)	17 (17)	32 (30)
Unique species (w/o exotics)[e]	19 (16)	1 (0)	1 (0)	5 (5)	12 (11)
Endemic to Luquillo Mountains[f]	4	0	2	1	4
Endemic to Puerto Rico[f,g]	14	2	10	8	13

[a]Calculated by dividing total stems or basal area by total area for the LFDP or Cover classes.
[b]Calculated by using species totals in nonoverlapping hectares delimited within the LFDP or the cover classes (see text); includes exotics.
[c]Totals in parentheses exclude exotics.
[d]<1 stem ha^{-1} in LFDP. Totals in parentheses exclude exotics.
[e]Number found in only one cover class in LFDP; under heading "LFDP," the total of such species in the plot is given. Totals in parentheses exclude exotics.
[f]Little and Woodbury (1976).
[g]Including Luquillo Mountain endemics.

At Bisley, human land use varied with landscape position (García-Montiel and Scatena 1994). Ridges were left uncut in tabonuco forest at Bisley, slopes were planted with a coffee understory, and valleys were planted with bananas. Charcoal manufacture was controlled by the U.S. Forest Service and limited to selected non-timber trees. As in the LFDP, the local tree species composition at Bisley also reflects past local land use.

The present diversity and species composition of wood-inhabiting ascomycete and pyrenomycete fungi were compared among the areas in the LFDP with different land use histories (Lodge 1997). Only 25 to 31 percent overlap in fungal species composition occurred between areas differing in past land use. Although these areas within the LFDP also differ in tree species composition, host differences alone cannot account for the differences in the fungal communities, because only 3 to 8 of the 253 fungal species found were clearly host-specific (Lodge 1997). Both mycomycete and dictyostelid slime molds were more diverse and abundant in the more intensively used areas of the LFDP (Stephenson and Landolt 1998).

Bacteria are another group in the LFDP having a composition that differs according to land use history (Willig et al. 1996). The functional diversity of bacteria in surface soils can be assessed as the diversity of abilities to degrade different substrates (Willig et al. 1996). Bacterial functional diversity and total catabolic activity were highest in the parts of the LFDP that had the greatest human disturbance more than 60 years previously. For these bacterial communities, the higher concentrations of labile carbon in the leaf litter of secondary tree species might provide more energy with which to produce enzymes to degrade complex substrates than do the lower concentrations in the litter of primary forest trees (Willig et al. 1996).

Secondary Forest Nutrient Dynamics

During secondary forest succession, most of the important carbon fluxes associated with litter production and decomposition reestablish within a decade or two (Ostertag et al. 2008). Decomposition is affected by the tree species composition resulting from agriculture. Areas of El Verde that were farmed or clearcut during the early 1900s were colonized by secondary tree species (Thompson et al. 2002). The leaves of secondary species generally have relatively less lignin and other secondary plant compounds (Coley 1987) and should decompose relatively fast. In a comparison between a forest stand that had been a farmed area 50 years previously and a mature tabonuco forest disturbed only naturally and dominated by primary tree species, the litterfall rates were similar, but litter accumulation on the ground was less in the secondary forest (Zou et al. 1995). This suggests that decomposition was faster in the secondary forest, consistent with expectations. An experiment showed that litter from the secondary forest initially decomposed faster than litter from the less disturbed forest, a process that could be stimulated by the higher content of N and K in the secondary forest litter, as well as by the presumably lower levels of secondary compounds. However, long-term decomposition rates were the same in both forests (Zou et al. 1995). Another study of ecosystem processes along a successional sequence confirmed that litterfall rates remain similar through time (even though the basal area and tree density increase) (Marín-Spiotta et al. 2007), but litter standing stocks are lower in secondary forests (Ostertag et al. 2008).

The accumulation of soil C during secondary succession varies among sites. There was a net accumulation of soil C (at depths of 0 to 60 cm) in a 61-year-old secondary forest of 102 ± 10 Mg ha^{-1} (mean ± 1 standard error), compared to values in a nearby pasture of 69 ± 16 Mg ha^{-1} (Silver et al. 2004). This gain in soil C was due to a fast rate of soil C gain in forest soils (0.9 Mg ha^{-1} y^{-1}) and a slow rate of C loss from surface soils in the pastures (0.4 Mg ha^{-1} y^{-1}). However, a separate study revealed no net change in total soil C (0 to 1.0 m) across 80 years of reforestation (Marín-Spiotta 2006).

Soils in tabonuco forest were highly resilient to nutrient loss following a clearcutting experiment in which two 1,024 m^2 plots were stripped of all aboveground vegetation (Silver et al. 1994, 1996). Most belowground nutrient changes in the plots lasted no more than a year. Nearly all pools increased or did not change over the first 1 to 2 years following the clearcutting. During the first 5 to 6 years of succession, soil nutrient cations increased above predisturbance levels, whereas amounts of soil P and aluminum (Al) (not a nutrient) were not statistically different from predisturbance values (Silver et al. 1996). Potassium was an exception; it increased in soils shortly after disturbance, presumably due to leaching from litter, and then decreased shortly thereafter. The potassium nutrient pool was the only one to drop below the predisturbance size. Nutrient immobilization and the slow release of nutrients from the decay of dead root biomass were important means of conservation (Silver and Vogt 1993). Live root replacement was slow; it took about 10 years for fine, live root biomass to reach predisturbance levels in the clearcut plots.

The presence of certain tree species planted in order to provide shade in coffee plantations, such as *Guarea guidonia* and *Inga* spp. (Zimmerman et al. 1995a), can influence nutrient dynamics after abandonment. The experimental addition of coarse woody debris did not increase growth in old coffee plantations where there were N-fixing trees, such as *Inga* spp., but it did in areas where there had been no coffee (Beard et al. 2005) and N was perhaps limiting.

Effects of Other Disturbances

Three other types of disturbances—road building, small clearings in elfin forest, and a single radiation experiment—reveal response patterns in the Luquillo Mountains. Vegetation, environmental, and soil characteristics were compared between "roadfills" (road shoulders created by road building, 6 months and 35 y old) and the mature colorado and elfin forest nearby (Olander et al. 1998). The 6-month-old roadfills had higher light and soil temperatures, higher soil bulk densities, larger pools of exchangeable soil nutrients, and higher soil oxygen (O) than the forest sites. The roadfills also had lower soil moisture, soil organic matter, and total soil N than the forest. In the 35-year-old roadfill, the bulk density, soil pH, and P pools were statistically similar to those in mature forest, but the soil moisture, total N, and base cations were different. The biomass and plant density were much less on the 35-year-old roadfill. If roadfill areas were abandoned to revegetation, it is estimated that it would take 200 to 300 years for them to attain the biomass of mature forest. Roads also induce landslides; half the landslides in the Luquillo Mountains are associated with roads (Walker et al. 1996a).

Small clearcuts and a plane crash in the elfin forest (c. 900 masl) on Pico del Este provided information about secondary succession at this elevation (Byer and Weaver 1977; Weaver 2000). As observed after the natural disturbance of Hurricane Hugo (Walker et al. 1996b), vegetation regrowth in the elfin forest is slow compared to that in lower elevation forests. In the first 18 years of regrowth at the crash site (0.078 ha), woody sprouts, ferns, and graminoids dominated, unlike with secondary succession in the tabonuco forest, where woody plants dominate. The ferns might be favored by the very moist soil, and the graminoids by the lack of shading from taller plants. The scarcity of seedlings might be caused by rain washout on the soil surface or the lack of colonizing adaptations in a habitat that historically has had few disturbances as intensive as clearcutting or a plane crash. After 18 years in the crash site, plant heights and diameters were about half, and the biomass one-quarter, of those in surrounding undisturbed vegetation. Species compositions were similar. The radiation experiment took place in tabonuco forest, where a small area was exposed for 3 months to 10,000 curies of cesium (Odum and Drewry 1970). The radiation killed most plants and seeds within 40 m of the source and produced a forest gap with higher temperature and light and lower humidity than in surrounding forest (McCormick 1970). Because seeds and nearly all advance regeneration of primary species were killed, regenerating plants consisted almost entirely of secondary species, all of which were native. The extensive mortality also made regeneration through the first 23 years slow compared to regrowth in natural gaps and in an experimentally cleared area of similar size that was cleared at the same time as the radiation-exposure treatment (Taylor et al. 1995).

Discussion

Disturbance and response are central to the patterns and processes woven into the tapestry of the Luquillo Mountains; disturbance legacies underlie the tapestry and form the ecological palimpsest. In this section, we discuss responses to disturbance in the Luquillo Mountains in order to illustrate the concepts of ecological space, resistance and resilience, and residuals and legacies (chapter 2). We also discuss interactions among disturbances.

Ecological Space

In order to understand the biotic response to disturbance in terms of ecological space (chapter 2), we need detailed knowledge of the changes in abiotic conditions caused by disturbance (that is, the relationship of abiotic variables to geographical space), the characteristics of the multidimensional niche occupied by each species (the relationship between species abundance and abiotic variables), and the feedback of the biota on the abiotic variables. Applying the concept of ecological space emphasizes the degree to which disturbance decouples the linkage between a species' abundance and its location in geographical space, producing through time a varying ecological tapestry. In this section, we demonstrate this by contrasting the response of vegetation to hurricane versus landslide disturbance, and by looking at the response of two groups of animals (lizards and frogs) to hurricanes in the Luquillo Mountains.

Hurricanes remove forest canopy and have relatively little effect on soil (tree tip-ups by Hurricane Hugo exposed 5 percent of the soil surface area in the LFDP) (Zimmerman et al. 1994; and see Walker 2000). Removal of the forest canopy produces increased light (Fernández and Fetcher 1991), higher temperatures, and drier soil surfaces. These effects can be exacerbated by posthurricane drought (Waide 1991a). Changes to the canopy structure and the resulting debris deposition (Brokaw and Grear 1991; Lodge et al. 1991) are both patchy, so that in addition to having changed mean values, the abiotic conditions are more variable than they were before the hurricane. Following a hurricane, the hues and contrasts of the forest tapestry are more extreme.

In response to canopy opening by a severe hurricane, shrubs, herbs, seedlings, and saplings thrive in the understory (Walker et al. 1991; Scatena et al. 1996), as increased light levels reach much of the forest floor and suitable ecological conditions for these plants expand from previously isolated treefall gaps and stream edges. However, hurricanes do not set in motion changes in the vegetation that are like those of a typical secondary succession (Yih et al. 1991; Zimmerman et al. 1994). In all but the most severe cases of hurricane disturbance (Basnet et al. 1992), hurricanes do not kill many of the canopy trees, and the survivors resprout vigorously, quickly shading the understory and limiting the time for shade intolerant species in the understory to grow and reproduce (Fernández and Fetcher 1991; Walker 1991; Yih et al. 1991; Bellingham et al. 1992, 1994; Angulo-Sandoval et al. 2004). A largely undisturbed soil layer, the nutrient levels of which remain largely the same through the disturbance and beyond, supports this rapid recovery (Silver et al. 1996).

In contrast to hurricanes, landslides remove both vegetation and surface soil and expose the nutrient-poor subsoil in the zone at the top of the slide while depositing a jumbled pile of vegetation and surface soil at the bottom zone (Walker et al. 1996a). The tapestry is torn. The responses of the vegetation to these two zones contrast sharply, with the difference controlled by levels of soil organic matter and associated nutrients (Walker et al. 1996a). In the exposed mineral soil, community changes proceed slowly and include a period in which climbing ferns, grasses, and other herbaceous species dominate. In the residual forest soil in the slide, where nutrient and propagule levels are high, succession proceeds rapidly. Here, rapidly growing pioneer species are able to take advantage of high levels of light and quickly establish a canopy. Subsequent community changes follow a sequence of replacement driven by changes in ecological space (defined by nutrient and light availability) that is commonly associated with secondary succession (Walker et al. 1996a).

For animals in the forest, changes in the forest structure and in temperature and moisture regimes are the critical factors that define ecological space. We consider the hurricane responses of lizards and frogs as examples. Different *Anolis* lizard species occupy different height strata in the forest (Reagan 1996). Hurricanes disrupt forest strata (Brokaw and Grear 1991) and compress *Anolis* habitats and species within a range near the forest floor (Reagan 1991). Thus their ecological requirements leave all *Anolis* spp. in close geographical proximity after this disturbance. The understory species *Anolis gundlachi* was observed to restrict itself to the interior of debris piles after the hurricane, presumably in order to avoid high heat and desiccation (Reagan 1996), but it might have suffered increased competition from the other two canopy lizard species. In contrast, coquí frogs, whose reproduction is limited by available nesting sites on the forest floor (Stewart and Woolbright 1996), increased in abundance following Hurricane Hugo, as the animals took advantage of the increased structure at ground level (Woolbright 1996). Yet this numerical increase of coquís was delayed, possibly by negative effects of posthurricane drought on juveniles (Woolbright 1996). These examples show the degree to which disturbance decouples geographical space and the abiotic variables that constitute ecological space, which governs animal distribution and abundance.

Resistance and Resilience

As discussed above, two important components of ecosystem stability are resistance and resilience (chapter 2). Resistance is the degree to which a system is not affected by disturbance. Resilience is the time required for a system to return to a state that is indistinguishable from that before the disturbance. An ecosystem is considered resilient if the recovery time is less than the recurrence interval of disturbance. When discussing the response of forested ecosystems to disturbance, it is important to distinguish between structural (state) and functional (flux) variables (Herbert et al. 1999; Beard et al. 2005), because structural components tend to show less resilience than functional ones. For example, after Hurricane Hugo, forest biomass recovered slower (to two-thirds of the prehurricane values by 5 y posthurricane) (Scatena et al. 1996) than leaf litterfall and net primary productivity (fully recovered 3 to 5 y posthurricane) (Scatena et al. 1996; Beard et al. 2005). Some state variables recover remarkably quickly. Forest floor biomass and soil and stream nutrient pools that exhibit posthurricane change return to prehurricane levels in less than 2 years (Zimmerman et al. 1996). Population densities of many organisms that have responded positively or negatively to hurricane disturbance also return to prehurricane levels in relatively few years. Some state variables do not appear resilient to hurricane disturbance. Fine root biomass (Silver et al. 1996; Beard et al. 2005) and densities of walking sticks (insect herbivores in the Phasmodidae; M. Willig, unpublished data) have been slow to return to prehurricane levels. The community composition of canopy trees might be in eternal flux (Crow 1980; Lugo et al. 1999; Weaver 2002), changing constantly through the average interhurricane interval. Overall, however, in comparison to an average return interval of about 60 years, the Luquillo forest ecosystem seems highly resilient to hurricane disturbance, which is a surprise to those who saw the immense tangle of downwood and open canopies caused by Hurricane Hugo. Similar conclusions regarding the overall ecosystem resilience apply to drought; for example, the return intervals for many variables appeared short relative to the recurrence intervals of severe droughts (chapter 4; Beard et al. 2005). To a certain degree, the same can be said of the resilience after landslides (Walker et al. 1996a). Recovery times in the mineral soil exposed by landslides appear to be about equal to the recurrence interval (chapter 4), whereas recovery in residual forest soil is much faster than the recurrence interval of landslides.

Ecosystem resistance and resilience can be inversely related, as seen in Hawaii (Herbert et al. 1999). After being struck by a hurricane, Hawaiian forest plots that lost much leaf area (low resistance) recovered leaf area rapidly (high resilience). Plots losing less leaf area (high resistance) recovered it more slowly (low resilience). Similarly, more severe disturbance can be associated with faster recovery in the Luquillo Mountains. The Bisley area suffered more effects on and mortality of trees from Hurricane Hugo than El Verde did, but Bisley also had a faster recovery of basal area (Beard et al. 2005). A corollary to the putative trade-off in ecosystem resistance and resilience is that more resistant/less resilient ecosystems should be less responsive to supplemental nutrients in terms of growth and turnover in comparison to ecosystems that are less resistant/more resilient (Chapin et al. 1986). Indeed, changes in leaf litterfall and other community and ecosystem components of elfin forest were much less responsive to supplemental nutrients than were the same components in tabonuco forest (Walker et al. 1996b); supplemental nutrients caused leaf litterfall in tabonuco forest to return to prehurricane values only 20 months after Hurricane Hugo.

Residuals, Legacies, and Human Disturbance

The response to disturbances in the Luquillo Mountains can be interpreted in terms of residuals and legacies (chapter 2). Residuals are the immediate manifestations of disturbance, including biotic residuals, such as fallen trees, and abiotic residuals, such as the resulting increased light at the forest floor. Legacies are the subsequent behavior of the ecosystem as influenced by those residuals of the prior community. Residuals and legacies can persist for short to long terms, influencing subsequent disturbance and response and building the layers of the palimpsest covered by the current landscape tapestry. The longest-term legacies persist beyond the normal recovery time of the ecosystem and can be relatively permanent (Franklin et al. 2000). Residuals such as fallen trees, debris suspended in trees (Lodge et al. 1991), and slowly dying trees that continue to fall after a hurricane, leave legacies in the form of available nutrients and soil organic matter. Similarly, landslides leave residuals such as debris and forest soil, including buried seeds, at the base of the slide, all of which are key determinants of the ensuing legacy of successional dynamics (Walker et al. 1996a).

Human disturbance has left strong legacies in the Luquillo Mountains (García-Montiel and Scatena 1994; Zimmerman et al. 1995a; Aide et al. 1996; Erickson et al. 2001; Thompson et al. 2002; Beard et al. 2005). The residuals of charcoal production, clearcutting, coffee plantations, and pastures all leave different legacies in the ecosystem in the composition of the vegetation, the soil characteristics, or both. Some human-induced effects can be permanent, because the scale of human disturbance is large relative to the ability of species to disperse into and recolonize abandoned agricultural areas, because of permanent changes in soil characteristics, or because of both causes, evident in the characteristics of early plant regeneration and soil in abandoned pastures (Zimmerman et al. 1995a). The close correspondence of current floristic differences with past land use boundaries in the LFDP suggests that the vegetation differences are not being quickly "blurred" by seed dispersal and colonization from adjacent forest types (Thompson et al. 2002). Coffee cultivation, which required the shade of nitrogen-fixing trees (a residual of disturbance), appears to have a long-term legacy evident in the forest composition and nutrient dynamics (Erickson et al. 2001; Beard et al. 2005). Similarly, the legacy of charcoal pits is evident in local hydrology and the local persistence of palms (García-Montiel and Scatena 1994). Thus, human disturbance produces soil residuals that in turn produce long-term legacies in the vegetation composition.

Interactions among Disturbances

Having described the disturbance regime of the Luquillo Mountains (chapter 4) and begun to understand the response to disturbance events, we can begin to investigate interactions among disturbances and how they shape ecosystem dynamics in the long term. Putting the interactions between disturbances in a matrix (table 5-5) shows how the ecosystem is affected by a sequence of disturbances of either the same or a different type. At present, the interactions among hurricanes, landslides, and human disturbance are well characterized, but interactions among these disturbances and treefalls and drought are less well known (table 5-5). In the case of drought, the interaction strength is probably weak (Beard et al. 2005), except where soil drying could reduce the effects of a subsequent storm (table 5-5). Thus, these cells in the matrix might never be filled, even as we continue to study the effects of drought. It seems that the components of the ecosystem affected by interacting disturbances are largely structural or population-based, rather than functional. As noted above, many of the functional attributes of the ecosystem exhibit high resilience and might therefore be expected to be affected less by interactions among disturbances.

A matrix depiction of the interactions among disturbance types (table 5-5) suggests a way in which to conceptualize and model ecosystem dynamics in the Luquillo Mountains using Markovian or similar processes. Each position in the landscape is defined by a disturbance regime, and each disturbance causes a change in the ecosystem state and sets in motion subsequent ecosystem changes as determined by the biota (whose position in geographic space is determined by their ecological requirements or ecological space). The effects—immediate and long-term—of a particular disturbance are modified by the history of disturbance, and some (e.g., human effects) have more persistent effects than others (e.g., drought). More important, one can begin to see a way out of Margalef's (1968) difficulty whereby it is impossible to define the ecosystem state of a particular geographic space because each location has its own unique history. He wrote, "An ecosystem is a historical construction, so complex that any actual state has a negligible *a priori* probability" (Margalef 1968:30). This problem is less severe if we understand the effects of disturbances on the ecosystem, even if the interactions among disturbances are common. The situation is further resolved by the fact that ecosystem resilience erases many effects of previous disturbances. Finally, this approach emphasizes the value of long-term observations of particular ecosystems. It is only through long-term measurements of disturbance and response that we can begin to fully understand how disturbances interact and determine ecosystem dynamics.

Summary

The organisms of the Luquillo Mountains respond to background treefalls, hurricanes, landslides, floods, droughts, and human disturbances. Background treefalls (not caused by hurricanes) are filled with plant regrowth as in other tropical forests. There is limited response by animals to treefall gaps, probably because background treefall gaps are relatively less important in these forests dominated by chronic, widespread hurricane effects. Hurricanes in the Luquillo Mountains appear to create a low, smooth-canopied forest (after regrowth), which is in contrast to the growth in some other forests that are not disturbed by these storms. Regrowth occurs via sprouting, the growth of advance regeneration, and recruitment from seed, and tree species exhibit both resistance to and resilience after hurricanes. Despite

Table 5.5 Summary of some interactions among disturbances noted in the Luquillo Mountains

	Initial disturbance				
Subsequent disturbance	*Treefall*	*Hurricane*	*Landslide*	*Drought*	*Human*
Treefall		Rate of isolated treefalls following hurricane disturbance determined by the death of affected trees (Walker 1995; Uriarte et al. 2004a; Ogle et al. 2006)		Might reduce susceptibility of some trees to uprooting in a rainstorm	Effects on species composition might have indirect effects on treefall rate and recovery dynamics; see below
Hurricane	Low stature of recovering vegetation reduces disturbance; fruiting shrubs become important for frugivorous birds (Wunderle 1995)	Depends on interhurricane interval and time for which forest canopy has recovered, particularly woody biomass; shorter intervals reduce distur-bance effects (Lugo et al. 1999; Canham et al. 2010). Many popula tions (e.g., trees [Canham et al. 2010], snails [Bloch and Willig 2006]) exhibit individual-istic responses based on life histories, habitat affinities, and other factors.	Existing landslides might suffer additional slides (Walker et al. 1996a). Elsewhere, there is little effect because of the low stature of vegetation; important exceptions have been described in which a hurricane has altered the successional trajectory of a landslide (Myster and Walker 1997).		Important in secondary forests because the low stature of recovering vegetation reduces disturbance (Pascarella et al. 2004; Uriarte et al. 2004b) while secondary species dominating older forest lead to increased damage (Everham and Brokaw 1996; Ogle et al. 2006)
Landslide		High rainfall associated with hurricanes causes many slides in susceptible areas (e.g., Scatena and Larsen 1991)	Additional sliding is common in many landslides; this instability is important in vegetation dynamics (Walker et al. 1996a)	Will increase rainfall amounts necessary in order to cause landslides	Human modification of topography might promote landslides, e.g., roads (Guar-iguata 1990; Walker et al. 1996a)

Table 5.5 *(continued)*

	Initial disturbance				
Subsequent disturbance	*Treefall*	*Hurricane*	*Landslide*	*Drought*	*Human*
Drought	Drying of soil surface might be increased, but reduced root biomass might leave higher levels of moisture at depth (Becker et al. 1988).	Drought following Hurricane Hugo in Luquillo had manifest effects on the ecosystem (Walker et al. 1991), but drought does not always follow hurricanes. Debris from hurricanes reduces drying at the forest floor and regulates stream habitat changes at low flows.	Additional sliding is suppressed due to drying; promotes drought-resistant vegetation in exposed mineral soil (Walker et al. 1996a)		At large scale, deforestation might promote drought frequency; this is untested below the scale of the island of Puerto Rico (van der Molen et al. 2010). Higher productivity secondary forests might be resistant to drought effects (Beard et al. 2005).
Human		Lower elevations of Luquillo mountains were abandoned in the 1930s due to hurricane effects (Scatena 1989).	Along roads, a landslide will result in stabilization efforts associated with road rebuilding or, alternatively, road abandonment (e.g., southern portion of Rt. 191).		

the effects on trees, the tree species composition changed little in the tabonuco forest after two recent hurricanes. Density-dependent mortality partly controls the species composition of regrowth. Understory plants grow and flower vigorously after hurricanes, but lianas apparently do not proliferate. Animal species show various responses to the changes in forest architecture and food resources caused by hurricanes. The populations of most herbivorous arthropods increase in response to vigorous plant regrowth. Snails capitalize on hurricane detritus while suffering from exposure to hot and dry conditions where canopy is removed. Lizards change their foraging locations, and the population of the abundant frog *Eleutherodactylus coqui* increases because hurricane litter provides juveniles with refuges from predators. Bat populations decline or emigrate after hurricanes, as fruiting declines, but they return as fruiting recovers, with variations among bat species. Bird species tend to be plastic in habitat and dietary requirements, probably due to the large changes in forest structure caused by hurricanes and regrowth, which make it necessary for birds to change their foraging locations and diets. Although hurricane-produced debris is substantial (litterfall up to 400 times the average daily amount), decomposition, nutrient export, and trace gas emissions after hurricanes change only briefly, as rapid regrowth reasserts control over most ecosystem processes. For example, concentrations of nitrogen increased in riparian groundwater after a hurricane, but within 2 years the export in streams returned to prehurricane rates. Hurricanes reduce aboveground forest biomass by as much as 50 percent, but productivity is stimulated, and biomass accumulates rapidly. Woody debris boosts productivity, but it also stimulates microbial decomposers, which can outcompete trees for soil N and possibly other nutrients, thereby slowing tree response. In general, terrestrial ecosystem functions recover faster than structure. Hurricanes dump debris in streams, and floods redistribute inorganic and detrital material, as well as stream organisms, throughout the benthic environment along the stream continuum. Hurricanes create debris dams that catch detrital food and reduce the washout of invertebrate consumers. A hurricane flood apparently washed shrimp downstream, but in the next 6 months shrimp densities increased rapidly to the highest abundances ever recorded in all sites, probably owing to migration upstream and the increased availability of algae and decomposing leaves as food. Droughts concentrate inorganic and detrital material and make stream organisms more susceptible to predation. In terrestrial habitats, droughts limit juvenile frog survival and fungi and limit fine root and litterfall recovery after hurricanes. Landslides consist of relatively discrete zones in which soil and vegetation removal, subsequent stability, and regeneration vary. Succession in landslides is slow, with a long plant-to-plant replacement phase, and early plant colonists, especially ferns, have a strong influence on later dynamics. Landslide colonization is primarily limited by the availability of dispersed seed and by low nutrient availability. The natural reforestation of pastures in the Luquillo Mountains area has produced forests that resemble older growth in most measures of structure and function; for example, most of the important C fluxes associated with litter production and decomposition reestablish within a decade or two. However, these secondary forests are dominated by introduced tree species, and some old growth species are missing. Past land use is the most important determinant of species composition in secondary tabonuco forest, despite repeated hurricane effects and underlying environmental variation, such as in soil and topography. The organisms of the Luquillo Mountains are more resilient after natural than human disturbances.

Literature Cited

Abelleira Martínez, O., and A. E. Lugo. 2008. Post sugar cane succession in moist alluvial sites in Puerto Rico. Pages 73–92 in R. W. Myster, editor, *Post-agricultural succession in the Neotropics*. New York: Springer.

Aide, T. M. 1987. Limbfalls: A major cause of sapling mortality for tropical forest plants. *Biotropica* 19:284–285.

Aide, T. M., J. K. Zimmerman, L. Herrera, M. Rosario, and M. Serrano. 1995. Forest recovery in abandoned tropical pastures in Puerto Rico. *Forest Ecology and Management* 77:77–86.

Aide, T. M., J. K. Zimmerman, M. Rosario, and H. Marcano. 1996. Forest recovery in abandoned cattle pastures along an elevational gradient in northeastern Puerto Rico. *Biotropica* 28:537–548.

Alvarez, J., and M. R. Willig. 1993. Effects of treefall gaps on the density of land snails in the Luquillo Experimental Forest of Puerto Rico. *Biotropica* 25:100–110.

Angulo-Sandoval, P., H. Fernández-Marín, J. K. Zimmerman, and T. M. Aide. 2004. Changes in patterns of understory leaf phenology and herbivory following hurricane damage. *Biotropica* 36:60–67.

Barrera, A., A. Gómez-Pompa, and C. Vásquez-Yanes. 1977. El manejo de las selvas por los Mayas: Sus implicaciones silvícolas y agrícolas [The management of forests by the Mayans: Implications for forestry and agriculture]. *Biotica* 2:47–61.

Basnet, K., F. N. Scatena, G. E. Likens, and A. E. Lugo. 1992. Hurricane Hugo: Damage to a tropical rain forest in Puerto Rico. *Journal of Tropical Ecology* 8:47–55.

Basnet, K., F. N. Scatena, G. E. Likens, and A. E. Lugo. 1993. Ecological consequences of root grafting in tabonuco (*Dacryodes excelsa*) trees in the Luquillo Experimental Forest, Puerto Rico. *Biotropica* 25:28–35.

Beard, J. S. 1945. The progress of plant succession on the Soufriere of St Vincent. *Journal of Ecology* 33:1–9.

Beard, K. H., K. A. Vogt, D. J. Vogt, F. N. Scatena, A. P. Covich, R. Sigurdardottir, T. C. Siccama, and T. A. Crowl. 2005. Structural and functional responses of a subtropical forest to 10 years of hurricanes and droughts. *Ecological Monographs* 75:345–361.

Becker, P., P. E. Rabenold, J. R. Idol, and A. P. Smith. 1988. Water potential gradient for gaps and slopes in a Panamanian tropical moist forest's dry season. *Journal of Tropical Ecology* 4:173–184.

Bellingham, P. J., V. Kapos, N. Varty, J. R. Healey, E. V. J. Tanner, D. L. Kelly, J. W. Dalling, L. S. Burns, D. Lee, and G. Sidrak. 1992. Hurricanes need not cause high mortality: The effects of Hurricane Gilbert on forests in Jamaica. *Journal of Tropical Ecology* 8:217–223.

Bellingham, P. J., E. V. J. Tanner, and J. R. Healey. 1994. Sprouting of Jamaican montane forests, after a hurricane. *Journal of Ecology* 82:747–758.

Bellingham, P. J., E. V. J. Tanner, and J. R. Healey. 1995. Damage and responsiveness of Jamaican montane tree species after disturbance by a hurricane. *Ecology* 76:2562–2580.

Birdsey, R. A., and P. L. Weaver. 1987. Forest area trends in Puerto Rico. Research note SO-331. New Orleans, LA: USDA Forest Service, Southern Experiment Station.

Bloch, C. P., and M. R. Willig. 2006. Context-dependence of long-term responses of terrestrial gastropod populations to large-scale disturbance. *Journal of Tropical Ecology* 22:111–122.

Bloomfield, J. 1993. Nutrient dynamics and the influence of substrate quality on the decomposition of leaves and fine roots of selected tree species in a lower montane tropical rain forest in Puerto Rico. Ph.D. dissertation. Yale University, New Haven, CT.

Bloomfield, J., K. A. Vogt, and D. J. Vogt. 1993. Decay rate and substrate quality of fine roots and foliage of two tropical tree species in the Luquillo Experimental Forest, Puerto Rico. *Plant and Soil* 150:233–245.

Boose, E. R., D. R. Foster, and M. Fluet. 1994. Hurricane impacts to tropical and temperate forest landscapes. *Ecological Monographs* 64:369–400.

Boucher, D. H., J. H. Vandermeer, M. A. Mallona, N. Zamora, and I. Perfecto. 1994. Resistance and resilience in a directly regenerating rainforest: Nicaraguan trees of the Vochysiaceae after Hurricane Joan. *Forest Ecology and Management* 68:127–136.

Brokaw, N. V. L. 1985. Gap-phase regeneration in a tropical forest. *Ecology* 66:682–687.

Brokaw, N. V. L. 1998. *Cecropia schreberiana* in the Luquillo Mountains of Puerto Rico. *The Botanical Review* 64:91–120.

Brokaw, N., and R. T. Busing. 2000. Niche versus chance and tree diversity in forest gaps. *Trends in Ecology and Evolution* 15:183–188.

Brokaw, N., S. Fraver, J. S. Grear, J. Thompson, J. K. Zimmerman, R. B. Waide, E. M. Everham III, S. P. Hubbell, and R. B. Foster. 2004. Disturbance and canopy structure in two tropical forests. Pages 177–194 in E. Losos and E. G. Leigh, Jr., editors, *Tropical forest diversity and dynamism: Results from a long-term tropical forest network.* Chicago: University of Chicago Press.

Brokaw, N. V. L., and J. S. Grear. 1991. Forest structure before and after Hurricane Hugo at three elevations in the Luquillo Mountains, Puerto Rico. *Biotropica* 23:386–392.

Brokaw, N. V. L., and R. A. Lent. 1999. Vertical structure. Pages 373–399 in M. L. Hunter, Jr., editor, *Maintaining biodiversity in forest ecosystems.* Cambridge, England: Cambridge University Press.

Brokaw, N. V. L., and L. R. Walker. 1991. Summary of the effects of Caribbean hurricanes on vegetation. *Biotropica* 23:442–447.

Browne, F. G. 1949. Storm forest in Kelantan. *Malayan Forester* 12:28–33.

Burslem, D. F. R. P., T. C. Whitmore, and G. C. Brown. 2000. Short-term effects of cyclone impact and long-term recovery of tropical rain forest on Kolombangara, Solomon Islands. *Journal of Ecology* 88:1063–1078.

Byer, M. D., and P. L. Weaver. 1977. Early secondary succession in an elfin woodland in the Luquillo Mountains of Puerto Rico. *Biotropica* 9:35–47.

Canham, C. D., J. Thompson, J. K. Zimmerman, and M. Uriarte. 2010. Variation in susceptibility to hurricane damage as a function of storm severity in Puerto Rican tree species. *Biotropica* 42:87–94.

Carroll, J. B. 1984. The conservation and wild status of the Rodrigues fruit bat, *Pteropus rodricensis. Myotis* 21–22:148–154.

Chapin, F. S. III, P. M. Vitousek, and K. Van Cleve. 1986. The nature of nutrient limitation in plant communities. *American Naturalist* 127:48–58.

Cheke, A. S., and J. F. Dahl. 1981. The status of bats on western Indian Ocean islands with special reference to *Pteropus. Mammalia* 45:205–238.

Chestnut, T. J., and W. H. McDowell. 2000. Nitrogen and carbon dynamics in the riparian and hyporheic zones of a tropical stream: Luquillo Mountains, Puerto Rico. *Journal of the North American Benthological Society* 19:199–214.

Chinea, J. D. 1999. Changes in the herbaceous and vine communities at the Bisley experimental watersheds, Puerto Rico, following Hurricane Hugo. *Canadian Journal of Forest Research* 29:1433–1437.

Clark, D. B. 1996. Abolishing virginity. *Journal of Tropical Ecology* 12:735–739.

Clark, D. B., and D. A. Clark. 1991. The impact of physical damage on canopy tree regeneration in tropical rain forest. *Journal of Ecology* 79:447–457.

Coley, P. D. 1987. Interspecific variation in plant anti-herbivore properties: The role of habitat quality and rate of disturbance. *New Phytologist* 106:251–263.

Cooper-Ellis, S., D. R. Foster, G. Carlton, and A. Lezberg. 1999. Forest response to catastrophic wind: Results from an experimental hurricane. *Ecology* 80:2683–2696.

Covich, A. P., and T. A. Crowl. 2002. Reorganization of benthic stream food webs in response to drought-altered population densities: Effects on rainforest streams. *Verhandlungen der Internationale Vereinigung für Theorestische und Angewandte Limnologie* 28:1172–1175.

Covich, A. P., T. A. Crowl, and T. Heartsill-Scalley. 2006. Effects of drought on palaemonid river shrimp (*Macrobrachium* spp.) in headwaters of the Luquillo Mountains, Puerto Rico. *Journal of the North American Benthological Society* 25:99–107.

Covich, A. P., T. A. Crowl, S. L. Johnson, and M. Pyron. 1996. Distribution and abundance of tropical freshwater shrimp along a stream corridor: Response to disturbance. *Biotropica* 28:484–492.

Covich, A. P., T. A. Crowl, S. L. Johnson, and F. N. Scatena. 1998. Drought effects on pool morphology and benthic macroinvertebrates in headwater streams of the Luquillo Experimental Forest, Puerto Rico. *American Water Resources Association TPS* 98:91–96.

Covich, A. P., T. A. Crowl, S. L. Johnson, D. Varza, and D. L. Certain. 1991. Post-Hurricane Hugo increases in Atyid shrimp abundances in a Puerto Rican montane stream. *Biotropica* 23:448–454.

Covich, A. P., T. A. Crowl, and F. N. Scatena. 2000. Linking habitat stability to floods and droughts: Effects on shrimp in montane streams, Puerto Rico. *Verhandlungen der Internationale Vereinigung für Theorestische und Angewandte Limnologie* 27:2430–434.

Covich, A., T. A. Crowl, and F. N. Scatena. 2003. Effects of extreme low flows on freshwater shrimps in a perennial tropical stream. *Freshwater Biology* 48:1199–1206.

Covich, A. P., and W. H. McDowell. 1996. The stream community. Pages 433–459 in D. P. Reagan and R. B. Waide, editors, *The food web of a tropical rain forest.* Chicago: University of Chicago Press.

Craig, P., P. Trail, and T. E. Morrell. 1984. The decline of fruit bats in American Samoa due to hurricanes and overhunting. *Biological Conservation* 69:261–266.

Crow, T. R. 1980. A rainforest chronicle: A 30 year record of change in structure and composition at El Verde, Puerto Rico. *Biotropica* 12:42–55.

Crow, T. R., and D. F. Grigal. 1979. Numerical analysis of arborescent communities in the rain forest of the Luquillo Mountains, Puerto Rico. *Vegetatio* 40:135–146.

Crowl, T. A., W. H. McDowell, A. P. Covich, and S. L. Johnson. 2001. Freshwater shrimp effects on detrital processing and nutrients in a tropical headwater stream. *Ecology* 82:775–783.

Dallmeier, F., J. A. Comiskey, and F. N. Scatena. 1998. Five years of forest dynamics following Hurricane Hugo in Puerto Rico's Luquillo Experimental Forest. Pages 231–248 in F. Dallmeier and J. A. Comiskey, editors, *Forest biodiversity in North, Central and South America, and the Caribbean.* UNESCO Man and the Biosphere Series, Vol. 21. Pearl River, NY: Parthenon Publishing Group.

Debski, I., D. F. R. P. Burlsem, and D. Lamb. 2000. Ecological processes maintaining differential tree species distributions in an Australian subtropical rain forest: Implications for models of species coexistence. *Journal of Tropical Ecology* 16:387–415.

de Gouvenain, R. C., and J. A. Silander, Jr. 2003. Do tropical storm regimes influence the structure of tropical lowland rainforests? *Biotropica* 35:166–180.

Denslow, J. S. 1980. Patterns of plant species diversity during succession under different disturbance regimes. *Oecologia* 46:18–21.

Denslow, J. S. 1995. Disturbance and diversity in tropical rain forests: The density effect. *Ecological Applications* 5:962–968.

Denslow, J. S., and G. S. Hartshorn. 1994. Tree-fall gap environments and forest dynamic processes. Pages 120–127 in L. A. McDade, K. S. Bawa, H. A. Hespenheide, and G. S. Hartshorn, editors, *La Selva: Ecology and natural history of a Neotropical rain forest.* Chicago: University of Chicago Press.

Dittus, W. P. J. 1985. The influence of cyclones on the dry evergreen forest of Sri Lanka. *Biotropica* 17:1–14.

Doyle, T. W. 1981. The role of disturbance in the gap dynamics of a montane rain forest: An application of a tropical forest succession model. Pages 56–73 in D. C. West, H. H. Shugart, and D. B. Botkin, editors, *Forest succession: Concepts and application.* New York: Springer-Verlag.

Elmqvist, T., W. E. Rainey, E. D. Pierson, and P. A. Cox. 1994. Effects of tropical cyclones Ofa and Val on the structure of a Samoan lowland rain forest. *Biotropica* 26:384–391.

Emanuel, K. A. 1987. The dependence of hurricane intensity on climate. *Nature* 362:483–485.

Erickson, H. E., and G. Ayala. 2004. Hurricane-induced nitrous oxide fluxes from a wet tropical forest. *Global Change Biology* 10:1155–1162.

Erickson, H. E., M. Keller, and E. Davidson. 2001. Nitrogen oxide fluxes and nitrogen cycling during secondary succession and forest fertilization in the humid tropics. *Ecosystems* 4:67–84.

Everham, E. M. III, and N. V. L. Brokaw. 1996. Forest damage and recovery from catastrophic wind. *The Botanical Review* 62:113–185.

Everham, E. M. III, R. W. Myster, and E. VanDeGenachte. 1996. Effects of light, moisture, temperature, and litter on the regeneration of five tree species in the tropical montane wet forest of Puerto Rico. *American Journal of Botany* 83:1063–1068.

Fernández, D. S., and N. Fetcher. 1991. Changes in light availability following Hurricane Hugo in a subtropical montane forest in Puerto Rico. *Biotropica* 23:393–399.

Fernández, D. S., and R. W. Myster. 1995. Temporal variation and frequency distribution of photosynthetic photon flux densities on landslides in Puerto Rico. *Tropical Ecology* 36:73–87.

Fetcher, N., R. A. Cordero, and J. Voltzow. 2000. Lack of ecotypic differentiation: Plant response to elevation, population origin, and wind in the Luquillo Mountains, Puerto Rico. *Biotropica* 32:225–234.

Fetcher, N., B. L. Haines, R. A. Cordero, D. J. Lodge, L. Walker, D. S. Fernández, and W. T. Lawrence. 1996. Responses of tropical plants to nutrients and light on a landslide in Puerto Rico. *Journal of Ecology* 84:331–341.

Fonte, S. J., and T. D. Schowalter. 2004. Decomposition of greenfall vs. senescent foliage in a tropical forest ecosystem in Puerto Rico. *Biotropica* 36:474–482.

Foster, D. R., J. D. Aber, J. M. Melillo, R. D. Bowden, and F. A. Bazzaz. 1997. Forest response to disturbance and anthropogenic stress. *BioScience* 47:437–445.

Foster, D. R., M. Fluet, and E. R. Boose. 1999. Human or natural disturbance: Landscape-scale dynamics of the tropical forests of Puerto Rico. *Ecological Applications* 9:555–572.

Foster, D., F. Swanson, J. Aber, I. Burke, N. Brokaw, D. Tilman, and A. Knapp. 2003. The importance of land-use legacies to ecology and conservation. *BioScience* 53:77–88.

Foster, R. B. 1980. Heterogeneity and disturbance in tropical vegetation. Pages 75–92 in M. E. Soulé and B. A. Wilcox, editors, *Conservation Biology: An Evolutionary–Ecological Perspective*. Sunderland, MA: Sinauer, Inc.

Francis, J. K., and S. E. Alemañy. 2003. Hurricane damage to mahogany crowns associated with seed source. Pages 94–102 in A. E. Lugo, J. C. Figueroa Colón, and M. Alayón, editors, *Big-leaf mahogany: Genetics, ecology, and management*. New York: Springer.

Frangi, J., and A. E. Lugo. 1991. Hurricane damage to a flood plain forest in the Luquillo Mountains of Puerto Rico. *Biotropica* 23:324–335.

Frangi, J., and A. E. Lugo. 1998. A flood plain palm forest in the Luquillo Mountains of Puerto Rico five years after Hurricane Hugo. *Biotropica* 30:339–348.

Franklin, J., D. R. Drake, K. R. McConkey, F. Tonga, and L. B. Smith. 2004. The effects of Cyclone Waka on the structure of lowland tropical rain forest in Vava'u, Tonga. *Journal of Tropical Ecology* 20:409–420.

Franklin, J. F., D. Lindenmayer, J. A. MacMahon, A. McKee, J. Magnuson, D. A. Perry, R. Waide, and D. Foster. 2000. Threads of continuity: Ecosystem disturbance, recovery, and the theory of biological legacies. *Conservation Biology in Practice* 1:9–16.

Fraver, S., N. V. L. Brokaw, and A. P. Smith. 1998. Delimiting the gap phase in the growth cycle of a Panamanian forest. *Journal of Tropical Ecology* 14:673–681.

Frizano, J., A. H. Johnson, D. R. Vann, and F. N. Scatena. 2002. Soil phosphorus fractionation during forest development on landslide scars in the Luquillo Mountains, Puerto Rico. *Biotropica* 34:17–26.

Fu, S., C. Rodríguez Pedraza, and A. E. Lugo. 1996. A twelve year comparison of stand changes in a mahogany plantation and a paired natural forest of similar age. *Biotropica* 26:515–524.

Gannon, M. R., M. R. Duran, A. Kurta, and M. R. Willig. 2005. *Bats of Puerto Rico: An island focus and Caribbean perspective*. Lubbock, TX: Texas Tech University Press.

Gannon, M. R., and M. R. Willig. 1994. The effects of Hurricane Hugo on bats of the Luquillo Experimental Forest of Puerto Rico. *Biotropica* 26:320–321.

Gannon, M. R., and M. R. Willig. 1998. Long-term monitoring protocol for bats: Lessons from the Luquillo Experimental Forest of Puerto Rico. Pages 271–291 in F. Dallmeier and J. Comiskey, editors, *Forest biodiversity in North, Central and South America, and the Caribbean: Research and monitoring*. Man and the Biosphere Series, Vol. 21. Pearl River, NY: Parthenon Publishing Group.

García-Montiel, D. C., and F. N. Scatena. 1994. The effect of human activity on the structure and composition of a tropical forest in Puerto Rico. *Forest Ecology and Management* 63:57–78.

Garrison, R. W., and M. R. Willig. 1996. Arboreal invertebrates. Pages 183–245 in R. B. Waide and D. P. Reagan, editors, *The food web of a tropical rain forest*. Chicago: University of Chicago Press.

Garwood, N. C., D. P. Janos, and N. Brokaw. 1979. Earthquake-caused landslides: A major disturbance to tropical forests. *Science* 140:997–999.

Gómez-Pompa, A., and A. Kaus. 1992. Taming the wilderness myth. *BioScience* 42:271–279.

González, G., C.-Y. Huang, and S.-S. Chuang. 2008. Earthworms and post-agricultural succession. Pages 115–138 in R. W. Myster, editor, *Post-agricultural succession in the Neotropics*. New York: Springer.

Grau, H. R., T. M. Aide, J. K. Zimmerman, J. R. Thomlinson, E. Helmer, and X. Zou. 2003. The ecological consequences of socioeconomic and land-use changes in postagricultural Puerto Rico. *BioScience* 53:1159–1168.

Grove, S. J., S. M. Turton, and D. T. Siegenthaler. 2000. Mosaics of canopy openness induced by tropical cyclones in lowland rain forests with contrasting management histories in northeastern Australia. *Journal of Tropical Ecology* 16:883–894.

Guariguata, M. R. 1990. Landslide disturbance and forest regeneration in the upper Luquillo Mountains of Puerto Rico. *Journal of Ecology* 78:814–832.

Guzmán-Grajales, S. M., and L. R. Walker. 1991. Differential seedling responses to litter after Hurricane Hugo in the Luquillo Experimental Forest, Puerto Rico. *Biotropica* 23:407–413.

Halleck, L. F., J. M. Sharpe, and X. Zou. 2004. Understory fern responses to post-hurricane fertilization and debris removal. *Journal of Tropical Ecology* 20:173–183.

Hammond, E. P. 1996. Regeneration of understory shrubs in the Luquillo Experimental Forest, Puerto Rico. M.S. thesis. University of Georgia, Athens, GA.

Hartshorn, G. S. 1978. Treefalls and tropical forest dynamics. Pages 617–638 in P. B. Tomlinson and M. H. Zimmerman, editors, *Tropical trees as living systems*. Cambridge, England: Cambridge University Press.

Hartshorn, G. S. 1980. Neotropical forest dynamics. *Biotropica* 12:23–30.

Heatwole, H., and A. Heatwole. 1978. Ecology of the Puerto Rican camaenid tree-snails. *Malacologia* 17:241–315.

Henderson, A., G. Galeano, and B. Bernal. 1995. *Field guide to the palms of the Americas*. Princeton, NJ: Princeton University Press.

Herbert, D. A., J. H. Fownes, and P. M. Vitousek. 1999. Hurricane damage to Hawaiian forest: Nutrient supply rate affects resistance and resilience. *Ecology* 80:908–920.

Herrera Montes, A. 2008. The relation between forest structure and herpetofaunal community structure during forest succession. M.S. thesis. University of Puerto Rico-Río Piedras, Río Piedras, Puerto Rico.

Huhndorf, S. M., and D. J. Lodge. 1997. Host specificity among wood inhabiting pyrenomycetes (Fungi, Ascomycetes) in a wet tropical forest in Puerto Rico. *Tropical Ecology* 38:307–315.

Imbert, D., A. Rousteau, and P. Labbé. 1998. Ouragans et diversité biologique dans les forêts tropicales. L'exemple de la Guadeloupe [Hurricanes and biological diversity in tropical forests. The example of Guadeloupe]. *Acta Oecologica* 19:251–262.

Jones, K. E., K. E. Barlow, N. Vaughan, A. Rodríguez-Durán, and M. R. Gannon. 2001. Short-term impact of extreme environmental disturbance on the bats of Puerto Rico. *Animal Conservation* 4:59–66.

Kangas, P. 1992. Root regrowth in a subtropical wet forest in Puerto Rico. *Biotropica* 24:463–465.

Keay, R. W. J. 1957. Wind-dispersed species in a Nigerian forest. *Journal of Ecology* 45:471–478.

Kochummen, K. M. 1977. Natural plant succession after farming in Kepong. *Malaysian Forester* 40:61–78.

Larsen, M. C., and A. J. Torres-Sánchez. 1992. Landslides triggered by Hurricane Hugo in eastern Puerto Rico, September 1989. *Caribbean Journal of Science* 28:13–120.

Latta, S. C., and J. M. Wunderle, Jr. 1998. The assemblage of birds foraging in native West Indian pine (*Pinus occidentalis*) forests of the Dominican Republic during the non-breeding season. *Biotropica* 30:645–656.

Lawrence, W. T. 1996. Plants: The food base. Pages 17–51 in D. P. Reagan and R. B. Waide, editors, *The food web of a tropical rain forest*. Chicago: University of Chicago Press.

Lawton, R. O. 1982. Wind stress and elfin stature in a montane rain forest tree: An adaptive explanation. *American Journal of Botany* 69:1224–1230.

Lebrón, M. L. 1979. An autecological study of *Palicourea riparia* Bentham as related to rain forest disturbance in Puerto Rico. *Oecologia* 42:31–46.

Levey, D. J. 1990. Habitat-dependent fruiting behaviour of an understorey tree, *Miconia centrodesma*, and tropical treefall gaps as keystone habitats for frugivores in Costa Rica. *Journal of Tropical Ecology* 6:409–420.

Little, E. L., Jr., and R. O. Woodbury. 1976. Trees of the Caribbean National Forest, Puerto Rico. Forest Service Research Paper ITF-20. Río Piedras, PR: International Institute of Tropical Forestry.

Lodge, D. J. 1993. Nutrient cycling by fungi in wet tropical forests. Pages 37–57 in S. Isaac, J. C. Frankland, R. Watling, and A. J. S. Whalley, editors, *Aspects of tropical mycology*. British Mycological Society Symposium Series 19. Cambridge, England: Cambridge University Press.

Lodge, D. J. 1996. Microorganisms. Pages 53–108 in D. P. Reagan and R. B. Waide, editors, *The food web of a tropical rain forest*. Chicago: University of Chicago Press.

Lodge, D. J. 1997. Factors related to diversity of decomposer fungi in tropical forests. *Biodiversity and Conservation* 6:681–688.

Lodge, D. J., and S. Cantrell. 1995. Fungal communities in wet tropical forests: Variation in time and space. *Canadian Journal of Botany* 73(Suppl. 1):S1391–1398.

Lodge, D. J., W. H. McDowell, and C. P. McSwiney. 1994. The importance of nutrient pulses in tropical forests. *Trends in Ecology and Evolution* 9:384–387.

Lodge, D. J., F. N. Scatena, C. E. Asbury, and M. J. Sánchez. 1991. Fine litterfall and related nutrient inputs resulting from Hurricane Hugo in subtropical wet and lower montane rain forests of Puerto Rico. *Biotropica* 23:336–342.

Lomascolo, T., and T. M. Aide. 2001. Seed and seedling bank dynamics in secondary forests following Hurricane Georges in Puerto Rico. *Caribbean Journal of Science* 37:259–270.

Lorimer, C. G. 1989. Relative effects of small and large disturbances on temperate hardwood forest structure. *Ecology* 70:565–567.

Lugo, A. 1970. Photosynthetic studies on four species of rain forest seedlings. Pages I81–I102 in H. T. Odum and R. F. Pigeon, editors, *A tropical rain forest: A study of irradiation and ecology at El Verde, Puerto Rico*. Oak Ridge, TN: U.S. Atomic Energy Commission.

Lugo, A. E. 2008. Visible and invisible effects of hurricanes on forest ecosystems: An international review. *Austral Ecology* 33:368–398.

Lugo, A. E., M. Applefield, D. J. Pool, and R. B. McDonald. 1983. The impact of Hurricane David on the forests of Dominica. *Canadian Journal of Forest Research* 13:201–211.

Lugo, A. E., J. Figueroa Colón, and F. N. Scatena. 1999. The Caribbean. Pages 593–622 in M. G. Barbour and W. D. Billings, editors, *North American terrestrial vegetation*. Cambridge, England: Cambridge University Press.

Lugo, A. E., and J. L. Frangi. 2003. Changes in necromass and nutrients on the forest floor of a palm floodplain forest in the Luquillo Mountains of Puerto Rico. *Caribbean Journal of Science* 39:265–272.

Lugo, A. E., and E. Helmer. 2004. Emerging forests on abandoned land: Puerto Rico's new forests. *Forest Ecology and Management* 190:145–161.

Lugo, A. E., and C. T. Rivera Batlle. 1987. Leaf production, growth rate, and age of the palm *Prestoea montana* in the Luquillo Experimental Forest, Puerto Rico. *Journal of Tropical Ecology* 3:151–161.

Lugo, A. E., and F. N. Scatena. 1996. Background and catastrophic tree mortality in tropical moist, wet, and rain forests. *Biotropica* 28:585–599.

Lugo, A. E., F. N. Scatena, W. Silver, S. Molina Colón, and P. G. Murphy. 2002. Resilience of tropical wet and dry forests in Puerto Rico. Pages 195–225 in L. H. Gunderson and L. Pritchard, Jr., editors, *Resilience and the behavior of large-scale systems*. Washington, DC: Island Press.

Lugo, A. E., and J. K. Zimmerman. 2002. Ecological life histories. Pages 191–213 in J. A. Vozzo, editor, *Tropical tree seed manual*. Agriculture Handbook 721. Washington, DC: U.S. Department of Agriculture Forest Service.

Madden, N. T. 2004. Scaling up: Nitrogen flux at the stream-soil interface in the Río Icacos, PR. M.S. thesis. University of New Hampshire.

Maheswaran, J., and I. A. U. N. Gunatilleke. 1988. Litter decomposition in a lowland rain forest and a deforested area in Sri Lanka. *Biotropica* 20:90–99.

Margalef, R. 1968. *Perspectives in ecological theory*. Chicago: University of Chicago Press.

Marín-Spiotta, E. 2006. Controls on above and belowground carbon storage during tropical reforestation. Ph.D. dissertation. Berkeley, CA: University of California.

Marín-Spiotta, E., R. Ostertag, and W. L. Silver. 2007. Long term patterns in reforestation of tropical pastures: Plant community composition and aboveground biomass accumulation. *Ecological Applications* 17:828–839.

McCormick, J. F. 1970. Direct and indirect effects of gamma radiation on seedling diversity and abundance in a tropical forest. Pages D201–D211 in H. T. Odum and R. F. Pigeon, editors, *A tropical rain forest: A study of irradiation and ecology at El Verde, Puerto Rico*. Oak Ridge, TN: U.S. Atomic Energy Commission.

McCormick, J. F. 1995. A review of the population dynamics of selected tree species in the Luquillo Experimental Forest, Puerto Rico. Pages 224–257 in A. E. Lugo and C. Lowe, editors, *Tropical forests: Management and ecology*. New York: Springer-Verlag.

McDowell, W. H. 2001. Hurricanes, people, and riparian zones: Controls on nutrient losses from forested Caribbean watersheds. *Forest Ecology and Management* 154:443–451.

McDowell, W. H., C. P. McSwiney, and W. B. Bowden. 1996. Effects of hurricane disturbance on groundwater chemistry and riparian function in a tropical forest. *Biotropica* 28:577–584.

Meléndez-Ackerman, E., C. Calisto-Pérez, M. Morales-Vargas, and J. Fumero-Cabán. 2003. Post-hurricane recovery of a herbaceous understorey plant in tropical rain forest in Puerto Rico. *Journal of Tropical Ecology* 19:677–684.

Miller, R. M., and D. J. Lodge. 1997. Fungal responses to disturbance-agriculture and forestry. Pages 65–84 in K. Esser, P. A. Lemke, and D. T. Wicklow, editors, *Environmental and microbial relationships*. Vol. V. New York: Springer-Verlag.

Moran, R. C. 2004. *A natural history of ferns*. Portland, OR: Timber Press.

Myster, R. W. 1997. Seed predation, disease and germination on landslides in Neotropical lower montane wet forest. *Journal of Vegetation Science* 8:55–64.

Myster, R. W., and D. S. Fernández. 1995. Spatial gradients and patch structure on two Puerto Rican landslides. *Biotropica* 27:149–159.

Myster, R. W., and N. Fetcher. 2005. Ecotypic differentiation and plant growth in the Luquillo Mountains of Puerto Rico. *Journal of Tropical Forest Science* 16:413–419.

Myster, R. W., J. R. Thomlinson, and M. C. Larsen. 1997. Predicting landslide vegetation in patches on landscape gradients in Puerto Rico. *Landscape Ecology* 12:299–307.

Myster, R. W., and L. R. Walker. 1997. Plant successional pathways on landslides in Puerto Rican landslides. *Journal of Tropical Ecology* 13:165–173.

O'Brien, S. T., B. P. Hayden, and H. H. Shugart. 1992. Global climatic change, hurricanes, and a tropical forest. *Climatic Change* 22:175–190.

Odum, H. T. 1970. Rain forest structure and mineral cycling homeostasis. Pages H3–H52 in H. T. Odum and R. F. Pigeon, editors, *A tropical rain forest: A study of irradiation and ecology at El Verde, Puerto Rico*. Oak Ridge, TN: U.S. Atomic Energy Commission.

Odum, H. T., and G. Drewry. 1970. The cesium source at El Verde. Pages C23–C36 in H. T. Odum and R. F. Pigeon, editors, *A tropical rain forest: A study of irradiation and ecology at El Verde, Puerto Rico*. Oak Ridge, TN: U.S. Atomic Energy Commission.

Ogle, K., M. Uriarte, J. Thompson, J. Johnstone, A. Jones, Y. Lin, E. McIntire, and J. Zimmerman. 2006. Implications of vulnerability to hurricane damage for long-term survival of tropical tree species: A Bayesian hierarchical analysis. Pages 98–123 in J. S. Clark and A. E. Gelfand, editors, *Applications of computational statistics in the environmental sciences: Hierarchical bayes and MCMC Methods*. Oxford, England: Oxford University Press.

Olander, L. P., F. N. Scatena, and W. L. Silver. 1998. Impacts of disturbance initiated by road construction in the Luquillo Experimental Forest, Puerto Rico. *Forest Ecology and Management* 109:33–49.

Ostertag, R., E. Marin-Spiotta, W. L. Silver, and J. Shulten. 2008. Litterfall and decomposition in relation to soil carbon pools along a secondary forest chronosequence in Puerto Rico. *Ecosystems* 11:701–714.

Ostertag, R., F. N. Scatena, and W. L. Silver. 2003. Forest floor decomposition following hurricane litter inputs in several Puerto Rican forests. *Ecosystems* 6:261–273.

Ostertag, R., W. L. Silver, and A. E. Lugo. 2005. Factors affecting mortality and resistance to damage following hurricanes in a subtropical moist forest. *Biotropica* 37:16–24.

Overpeck, J. T., D. Rind, and R. Goldberg. 1990. Climate-induced changes in forest disturbance and vegetation. *Nature* 343:51–53.

Parrotta, J., and D. J. Lodge. 1991. Fine root dynamics in a subtropical wet forest following hurricane disturbance in Puerto Rico. *Biotropica* 23:343–347.

Pascarella, J. B., T. M. Aide, M. I. Serrano, and J. K. Zimmerman. 2000. Land-use history and forest regeneration in the Cayey Mountains, Puerto Rico. *Ecosystems* 3:217–228.

Pascarella, J. B., T. M. Aide, and J. K. Zimmerman. 2004. Short-term response of secondary forests to hurricane disturbance in Puerto Rico, USA. *Forest Ecology and Management* 199:379–393.

Pedersen, S. C., H. H. Genoways, and P. W. Freeman. 1996. Notes on bats from Montserrat (Lesser Antilles) with comments concerning the effects of Hurricane Hugo. *Caribbean Journal of Science* 32:206–213.

Perez, J. W. 1970. Relation of crown diameter to stem diameter in forests of Puerto Rico, Dominica, and Thailand. Pages B105–B122 in H. T. Odum and R. F. Pigeon, editors, *A tropical rain forest: A study of irradiation and ecology at El Verde, Puerto Rico*. Oak Ridge, TN: U.S. Atomic Energy Commission.

Pérez Viera, I. E. 1986. Tree regeneration in two tropical rain forests. M.S. thesis. University of Puerto Rico-Río Piedras, Río Piedras, Puerto Rico.

Pfeiffer, W. J. 1996. Arboreal arachnids. Pages 247–271 in D. P. Reagan and R. B. Waide, editors, *The food web of a tropical rain forest*. Chicago: University of Chicago Press.

Pierson, E. D., T. Elmqvist, W. E. Rainey, and P. A. Cox. 1996. Effects of tropical cyclonic storms on flying fox populations on the South Pacific Islands of Samoa. *Conservation Biology* 10:438–451.

Pringle, C. M., and G. A. Blake. 1994. Quantitative effects of atyid shrimps (Decapoda: Atyidae) on the depositional environment in a tropical stream: Use of electricity for experimental exclusion. *Canadian Journal of Fisheries and Aquatic Sciences* 51:1443–1450.

Pringle, C. M., G. A. Blake, A. P. Covich, K. M. Buzby, and A. Finley. 1993. Effects of omnivorous shrimp in a montane tropical stream: Sediment removal, disturbance of sessile invertebrates and enhancement of understory algal biomass. *Oecologia* 93:1–11.

Pringle, C. M., N. H. Hemphill, W. McDowell, A. Bednarek, and J. March. 1999. Linking species and ecosystems: Effects of different macrobiotic assemblages on interstream differences in organic matter. *Ecology* 80:1860–1872.

Putz, F. E., and N. V. L. Brokaw. 1989. Sprouting of broken trees on Barro Colorado Island, Panama. *Ecology* 70:508–512.

Reagan, D. P. 1991. The response of *Anolis* lizards to hurricane-induced habitat changes in a Puerto Rican rain forest. *Biotropica* 23:468–474.

Reagan, D. P. 1996. Anoline lizards. Pages 321–345 in D. P. Reagan and R. B. Waide, editors, *The food web of a tropical rain forest*. Chicago: University of Chicago Press.

Reagan, D. P., G. R. Camilo, and R. B. Waide. 1996. The community food web: Major properties and patterns of organization. Pages 462–510 in D. P. Reagan and R. B. Waide, editors, *The food web of a tropical rain forest*. Chicago: University of Chicago Press.

Rice, K., N. Brokaw, and J. Thompson. 2004. Liana abundance in a Puerto Rican forest. *Forest Ecology and Management* 190:33–41.

Riddle, W. A. 1983. Physiological ecology of land snails and slugs. Pages 431–461 in W. D. Russell-Hunter, editor, *The Mollusca*. Vol. 6. Orlando, FL: Academic Press.

Riswan, S., J. B. Kenworthy, and K. Kartawinata. 1985. The estimation of temporal processes in tropical rain forest: A study of primary mixed dipterocarp forest in Indonesia. *Journal of Tropical Ecology* 1:171–182.

Sandlin, E. A., and M. R. Willig. 1993. The effects of age, sex, prior experience, and intraspecific food variation on diet composition of a tropical folivore (Phasmatodea: Phasmatidae). *Environmental Entomology* 22:625–633.

Sanford, R. L., W. J. Parton, D. S. Ojima, and D. J. Lodge. 1991. Hurricane effects on soil organic matter dynamics and forest production in the Luquillo Experimental Forest, Puerto Rico: Results of simulation modeling. *Biotropica* 23:364–372.

Sanford, R. L., J. Saldarriaga, K. E. Clark, C. Uhl, and R. Herrera. 1985. Amazon rainforest fires. *Science* 227:53–55.

Scatena, F. N. 1989. An introduction to the physiography and history of the Bisley Experimental Watersheds in the Luquillo Mountains of Puerto Rico. U.S. Forest Service General Technical Report SO-72. Washington, DC: USDA Forest Service.

Scatena, F. N. 1990. Watershed scale rainfall interception on two forested watersheds in the Luquillo Mountains of Puerto Rico. *Journal of Hydrology* 113:89–102.

Scatena, F. N. 1995. Relative scales of time and effectiveness of watershed processes in a tropical montane rain forest of Puerto Rico. Pages 103–111 in *Natural and anthropogenic influences in fluvial geomorphology*. Geophysical Monograph 89. Washington, DC: American Geophysical Union.

Scatena, F. N., and M. C. Larsen. 1991. Physical aspects of Hurricane Hugo in Puerto Rico. *Biotropica* 23:317–323.

Scatena, F. N., and A. E. Lugo. 1995. Geomorphology, disturbance, and the soil and vegetation of two subtropical wet steepland watersheds of Puerto Rico. *Geomorphology* 13:199–213.

Scatena, F. N., S. Moya, C. Estrada, and J. D. Chinea. 1996. The first five years in the reorganization of aboveground biomass and nutrient use following Hurricane Hugo in the Bisley Experimental Watersheds, Luquillo Experimental Forest, Puerto Rico. *Biotropica* 28:424–440.

Scatena, F. N., W. Silver, T. Siccama, A. Johnson, and M. J. Sánchez. 1993. Biomass and nutrient content of the Bisley Experimental Watersheds, Luquillo Experimental Forest, Puerto Rico, before and after Hurricane Hugo. *Biotropica* 25:15–27.

Schaefer, D. A., W. H. McDowell, F. N. Scatena, and C. E. Asbury. 2000. Effects of hurricane disturbance on stream water concentrations and fluxes in eight tropical forest watersheds of the Luquillo Experimental Forest, Puerto Rico. *Journal of Tropical Ecology* 16:189–207.

Schemske, D. W., and N. Brokaw. 1981. Treefalls and the distribution of understory birds in a tropical forest. *Ecology* 62:938–945.

Schnitzer, S. A., and F. Bongers. 2002. The ecology of lianas and their role in forests. *Trends in Ecology and Evolution* 17:223–230.

Schowalter, T. D. 1994. Invertebrate community structure and herbivory in a tropical rain forest canopy in Puerto Rico following Hurricane Hugo. *Biotropica* 26:312–319.

Schowalter, T. D., and L. M. Ganio. 1999. Invertebrate communities in a tropical rain forest canopy in Puerto Rico following Hurricane Hugo. *Ecological Entomology* 24:191–201.

Secrest, M. F., M. R. Willig, and L. L. Peppers. 1996. The legacy of disturbance on habitat associations of terrestrial snails in the Luquillo Experimental Forest, Puerto Rico. *Biotropica* 28:502–514.

Sharpe, J. M. 1997. Leaf growth and demography of the rheophytic fern *Thelypteris angustifolia* (Willdenow) Proctor in a Puerto Rican rainforest. *Plant Ecology* 130:203–212.

Shiels, A. B. 2006. Leaf litter decomposition and substrate chemistry of early successional species on landslides in Puerto Rico. *Biotropica* 38:348–353.

Shiels, A. B., and L. R. Walker. 2003. Bird perches increase forest seeds on Puerto Rican landslides. *Restoration Ecology* 11:457–465.

Shiels, A. B., L. R. Walker, and D. B. Thompson. 2006. Organic matter inputs create variable resource patches on Puerto Rican landslides. *Plant Ecology* 184:223–236.

Silander, S. R., and A. E. Lugo. 1990. *Cecropia peltata* L.: Yagrumo hembra, trumpet tree. Pages 244–249 in *Silvics of North America: Hardwoods*. Vol. 2. Agricultural Handbook 654. Washington, DC: U.S. Department of Agriculture.

Silver, W. L., D. J. Herman, and M. K. Firestone. 2001. Dissimilatory nitrate reduction to ammonium in tropical forest soils. *Ecology* 82:2410–2416.

Silver, W. L., L. M. Kueppers, A. E. Lugo, R. Ostertag, and V. Matzek. 2004. Carbon sequestration and plant community dynamics following reforestation of tropical pasture. *Ecological Applications* 14:1115–1127.

Silver, W. L., F. N. Scatena, A. H. Johnson, T. G. Siccama, and M. J. Sánchez. 1994. Nutrient availability in a montane wet tropical forest in Puerto Rico: Spatial patterns and methodological considerations. *Plant and Soil* 164:129–145.

Silver, W. L., F. N. Scatena, A .H. Johnson, T. G. Siccama, and F. Watt. 1996. At what temporal scales does disturbance affect belowground nutrient pools? *Biotropica* 28:441–457.

Silver, W. L., A. W. Thompson, A. Reich, J. J. Ewel, and M. K. Firestone. 2005. Nitrogen cycling in tropical plantation forests: Potential controls on nitrogen retention. *Ecological Applications* 15:1604–1614.

Silver, W. L., and K. A. Vogt. 1993. Fine root dynamics following single and multiple disturbances in a subtropical wet forest ecosystem. *Journal of Ecology* 81:729–738.

Smith, R. F. 1970. The vegetation structure of a Puerto Rican rain forest before and after short-term gamma radiation. Pages D103–D140 in H. T. Odum and R. F. Pigeon, editors, *A tropical rain forest: A study of irradiation and ecology at El Verde, Puerto Rico*. Oak Ridge, TN: U.S. Atomic Energy Commission.

Solem, A. 1984. A world model of land snail diversity and abundance. Pages 6–22 in A. Solem and A. C. van Bruggen, editors, *World-wide snails: Biogeographical studies on non-marine Mollusca*. Leiden, The Netherlands: E. J. Brill.

Stephenson, S. L., and J. C. Landolt. 1998. Dictyostelid cellular slime molds in canopy soils of tropical forests. *Biotropica* 30:657–661.

Steudler, P. A., J. M. Melillo, R. D. Bowden, M. S. Castro, and A. E. Lugo. 1991. The effects of natural and human disturbances on soil nitrogen dynamics and trace gas fluxes in a Puerto Rican wet forest. *Biotropica* 23:356–363.

Stewart, M. M. 1995. Climate driven population fluctuations in rain forest frogs. *Journal of Herpetology* 29:437–446.

Stewart, M. M., and L. L. Woolbright. 1996. Amphibians. Pages 273–320 in D. P. Reagan and R. B. Waide, editors, *The food web of a tropical rain forest*. Chicago: University of Chicago Press.

Tanner, E. V. J., and P. J. Bellingham. 2006. Less diverse forest is more resistant to hurricane disturbance: Evidence from montane forests in Jamaica. *Journal of Ecology* 94:1003–1010.

Taylor, C. M., S. Silander, R. B. Waide, and W. J. Pfeiffer. 1995. Recovery of a tropical forest after gamma irradiation: A 23-year chronicle. Pages 258–285 in A. E. Lugo and C. Lowe, editors, *Tropical forests: Management and ecology*. Berlin: Springer-Verlag.

Thomlinson, J. R., M. I. Serrano, T. del M. López, T. M. Aide, and J. K. Zimmerman. 1996. Land-use dynamics in a post-agricultural Puerto Rican landscape (1936–1988). *Biotropica* 28:525–536.

Thompson, J., N. Brokaw, J. K. Zimmerman, R. B. Waide, E. M. Everham III, D. J. Lodge, C. M. Taylor, D. García-Montiel, and M. Fluet. 2002. Land use history, environment, and tree composition in a tropical forest. *Ecological Applications* 12:1344–1363.

Thompson, J., N. Brokaw, J. K. Zimmerman, R. B. Waide, E. M. Everham III, and D. A. Schaefer. 2004. Luquillo Forest Dynamics Plot. Pages 540–550 in E. Losos and E. G. Leigh, Jr., editors, *Tropical forest diversity and dynamism: Results from a long-term tropical forest network*. Chicago: University of Chicago Press.

Torres, J. A. 1992. Lepidoptera outbreaks in response to successional changes after the passage of Hurricane Hugo in Puerto Rico. *Journal of Tropical Ecology* 8:285–298.

Uhl, C. 1987. Factors controlling succession following slash-and-burn agriculture in Amazonia. *Journal of Ecology* 75:377–407.

Uriarte, M., C. D. Canham, J. Thompson, and J. K. Zimmerman. 2004a. A neighborhood analysis of tree growth and survival in a hurricane-driven tropical forest. *Ecological Monographs* 74:591–614.

Uriarte, M., C. D. Canham, J. Thompson, J. K. Zimmerman, and N. Brokaw. 2005. Seedling recruitment in a hurricane-driven forest: Light limitation, density-dependence, and the spatial distribution of parent trees. *Journal of Ecology* 93:291–304.

Uriarte, M., L. W. Rivera, J. K. Zimmerman, T. M. Aide, A. G. Power, and A. S. Flecker. 2004b. Effects of land use history on hurricane damage and recovery in a neotropical forest. *Plant Ecology* 174:49–58.

van der Molen, M. K., H. F. Vugts, L. A. Bruijnzeel, F. N. Scatena, R. A. Pielke, Sr., and L. J. M. Kroon. 2010. Meso-scale climate change due to lowland deforestation in the maritime tropics. Pages 527–537 in L. A. Bruijnzeel, F. N. Scatena, and L. S. Hamilton, editors, *Tropical montane cloud forests: Science for conservation and management*. Cambridge, England: Cambridge University Press.

Vannote, R. L., G. W. Minshall, K. W. Cummins, J. R. Sedell, and C. E. Cushing. 1980. The river continuum concept. *Canadian Journal Fisheries and Aquatic Sciences* 37:130–137.

Vitousek, P. M. 1982. Nutrient cycling and nutrient use efficiency. *American Naturalist* 119:553–572.

Vitousek, P. M., and R. L. Sanford, Jr. 1986. Nutrient cycling in tropical forest. *Annual Review of Ecology and Systematics* 17:137–167.

Vogt, K. A., D. J. Vogt, H. Asbjornsen, and R. A. Dahlgren. 1995. Roots, nutrients and their relationship to spatial patterns. *Plant and Soil* 168–169:113–123.

Vogt, K. A., D. J. Vogt, P. Boon, A. Covich, F. N. Scatena, H. Asbjornsen, J. L. O'Hara, J. Pérez, T. G. Siccama, J. Bloomfield, and J. F. Ranciato. 1996. Litter dynamics along stream, riparian and upslope areas following Hurricane Hugo, Luquillo Experimental forest, Puerto Rico. *Biotropica* 28:458–470.

Volkov, I., J. R. Banavar, F. He, S. P. Hubbell, and A. Maritan. 2005. Density dependence explains tree species abundance and diversity in tropical forests. *Nature* 438:658–661.

Wadsworth, F. H. 1950. Notes on the climax forests of Puerto Rico and their destruction and conservation prior to 1900. *Caribbean Forester* 11:38–47.

Waide, R. B. 1991a. The effect of Hurricane Hugo on bird populations in the Luquillo Experimental Forest, Puerto Rico. *Biotropica* 23:475–480.

Waide, R. B. 1991b. Summary of the response of animal populations to hurricanes in the Caribbean. *Biotropica* 23:508–512.

Waide, R. B., and D. P. Reagan. 1996. The rain forest setting. Pages 1–16 in D. P. Reagan and R. B. Waide, editors, *The food web of a tropical rain forest*. Chicago: University of Chicago Press.

Waide, R. B., J. K. Zimmerman, and F. N. Scatena. 1998. Controls of primary productivity: Lessons from the Luquillo Mountains in Puerto Rico. *Ecology* 79:31–37.

Walker, L. R. 1991. Tree damage and recovery from Hurricane Hugo in Luquillo Experimental Forest, Puerto Rico. *Biotropica* 23:379–385.

Walker, L. R. 1994. Effects of fern thickets on woodland development on landslides in Puerto Rico. *Journal of Vegetation Science* 5:525–532.

Walker, L. R. 1995. Timing of post-hurricane tree mortality in Puerto Rico. *Journal of Tropical Ecology* 11:315–320.

Walker, L. R. 2000. Seedling and sapling dynamics of treefall pits in Puerto Rico. *Biotropica* 32:262–275.

Walker, L. R., N. V. L. Brokaw, D. J. Lodge, and R. B. Waide, editors. 1991. Special issue: Ecosystem, plant, and animal responses to hurricanes in the Caribbean. *Biotropica* 23:313–521.

Walker, L. R., D. J. Lodge, S. M. Guzmán-Grajales, and N. Fetcher. 2003. Species-specific seedling responses to hurricane disturbance in a Puerto Rican forest. *Biotropica* 35:472–485.

Walker, L. R., and L. E. Neris. 1993. Posthurricane seed rain dynamics in Puerto Rico. *Biotropica* 25:408–418.

Walker, L. R., J. Voltzow, J. Ackerman, D. S. Fernández, and N. Fetcher. 1992. Immediate impact of Hurricane Hugo on a Puerto Rican rain forest. *Ecology* 73:691–694.

Walker, L. R., D. J. Zarin, N. Fetcher, R. W. Myster, and A. H. Johnson. 1996a. Ecosystem development and plant succession on landslides in the Caribbean. *Biotropica* 28:566–576.

Walker, L. R., J. K. Zimmerman, D. J. Lodge, and S. Guzmán-Grajales. 1996b. An altitudinal comparison of growth and species composition in hurricane-damaged forests in Puerto Rico. *Journal of Ecology* 84:877–889.

Wang, H., J. D. Cornell, C. A. S. Hall, and D. P. Marley. 2002a. Spatial and seasonal dynamics of surface soil carbon in the Luquillo Experimental Forest, Puerto Rico. *Ecological Modelling* 147:105–122.

Wang, H., and C. A. S. Hall. 2004. Modeling the effects of Hurricane Hugo on spatial and temporal variation in primary productivity and soil carbon and nitrogen in the Luquillo Experimental Forest, Puerto Rico. *Plant and Soil* 262:69–84.

Wang, H., C. A. S. Hall, J. D. Cornell, and M. H. P. Hall. 2002b. Spatial dependence and the relationship of soil organic carbon and soil moisture in the Luquillo Experimental Forest, Puerto Rico. *Landscape Ecology* 17:671–684.

Wang, H., C. A. S. Hall, F. N. Scatena, N. Fetcher, and W. Wu. 2003. Modeling the spatial and temporal variability in climate and primary productivity across the Luquillo Mountains, Puerto Rico. *Forest Ecology and Management* 179:69–94.

Weaver, P. L. 1986. Hurricane damage and recovery in the montane forests of the Luquillo Mountains of Puerto Rico. *Caribbean Journal of Science* 22:53–70.

Weaver, P. L. 1989. Forest changes after hurricanes in Puerto Rico's Luquillo Mountains. *Interciencia* 14:181–192.

Weaver, P. L. 1998. Hurricane effects and long-term recovery in a subtropical rain forest. Pages 249–270 in F. Dallmeier and J. A. Comiskey, editors, *Forest biodiversity in North, Central and South America, and the Caribbean: Research and monitoring*. UNESCO Man and the Biosphere Series, Vol. 21. Pearl River, NY: Parthenon Publishing Group.

Weaver, P. L. 1999. Impacts of Hurricane Hugo on the dwarf cloud forest of Puerto Rico's Luquillo Mountains. *Caribbean Journal of Science* 35:101–111.

Weaver, P. L. 2000. Elfin woodland recovery 30 years after a plane wreck in Puerto Rico's Luquillo Mountains. *Caribbean Journal of Science* 36:1–9.

Weaver, P. L. 2002. A chronology of hurricane-induced changes in Puerto Rico's lower montane forest. *Interciencia* 27:252–258.

Webb, L. J. 1958. Cyclones as an ecological factor in tropical lowland rain forest, north Queensland. *Australian Journal of Botany* 6:220–228.

Whigham, D. F., M. B. Dickinson, and N. V. L. Brokaw. 1999. Background canopy gap and catastrophic wind disturbances in tropical forests. Pages 223–252 in L. R. Walker, editor, *Ecosystems of the world: 16 Ecosystems of disturbed ground*. Amsterdam: Elsevier.

Whigham, D. F., and J. F. Lynch. 1998. Responses of plants and birds to hurricane disturbances in a dry tropical forest of Quintana Roo, Mexico. Pages 165–186 in F. Dallmeier and J. A. Comiskey, editors, *Forest biodiversity in North, Central and South America, and the Caribbean: Research and monitoring*. UNESCO Man and the Biosphere Series, Vol. 21. Pearl River, NY: Parthenon Publishing Group.

Whigham, D. F., I. Olmsted, E. C. Cano, and M. E. Harmon. 1991. Impacts of Hurricane Gilbert on trees, litterfall, and woody debris in a dry tropical forest in the northeastern Yucatan Peninsula. *Biotropica* 23:434–441.

White, P. S., J. Harrod, W. H. Romme, and J. Betancourt. 1999. Disturbance and temporal dynamics. Pages 281–312 in W. T. Sexton, R. C. Szaro, N. C. Johnson, and A. J. Malk, editors, *Ecological stewardship: A common reference for ecosystem management*. New York: Elsevier.

Whitmore, T. C. 1974. Change with time and the role of cyclones in tropical rain forest on Kolombangara, Solomon Islands. Commonwealth Forestry Institute Paper 46. Oxford, England: Commonwealth Forestry Institute.

Whitmore, T. C. 1978. Gaps in the forest canopy. Pages 639–655 in P. B. Tomlinson and M. H. Zimmerman, editors, *Tropical trees as living systems*. Cambridge, England: Cambridge University Press.

Wiles, G. J. 1987. Current research and future management of Marianas fruit bats (Chiroptera: Pteropodidae) on Guam. *Australian Mammalogy* 10:93–95.

Willig, M. R., and G. R. Camilo. 1991. The effect of Hurricane Hugo on six invertebrate species in the Luquillo Experimental Forest of Puerto Rico. *Biotropica* 23:455–461.

Willig, M. R., R. W. Garrison, and A. Bauman. 1986. Population dynamics and natural history of a neotropical walking stick, *Lamponius portoricensis* Rehn (Phasmatodea, Phasmatidae). *Texas Journal of Science* 38:121–137.

Willig, M. R., and M. A. McGinley. 1999. Animal responses to natural disturbance and roles as patch generating phenomena. Pages 633–657 in L. R. Walker, editor, *Ecosystems of the world: 16 Ecosystems of disturbed ground*. Amsterdam: Elsevier.

Willig, M. R., D. L. Moorhead, S. B. Cox, and J. C. Zak. 1996. Functional diversity of soil bacterial communities in the tabonuco forest: Interaction of anthropogenic and natural disturbance. *Biotropica* 28:471–483.

Willig, M. R., E. A. Sandlin, and M. R. Gannon. 1993. Structural and taxonomic components of habitat selection in the Neotropical folivore, *Lamponius portoricensis* (Phasmatodea: Phasmatidae). *Environmental Entomology* 22:634–641.

Willig, M. R., M. F. Secrest, S. B. Cox, G. R. Camilo, J. F. Cary, J. Alvarez, and M. R. Gannon. 1998. Long-term monitoring of snails in the Luquillo Experimental Forest of Puerto Rico: Heterogeneity, scale, disturbance, and recovery. Pages 293–322 in F. Dallmeier and J. Comiskey, editors, *Forest biodiversity in North, Central and South American, and the Caribbean: Research and monitoring*. Pearl River, NY: Parthenon Publishing Group.

Willig, M. R., and L. R. Walker. 1999. Disturbance in terrestrial ecosystems: Salient themes, synthesis, and future directions. Pages 747–767 in L. R. Walker, editor, *Ecosystems of the world: 16 Ecosystems of disturbed ground*. Amsterdam: Elsevier.

Wills, C., K. E. Harms, R. Condit, D. King, J. Thompson, F. He, H. C. Muller-Landau, P. Ashton, E. Losos, L. Comita, S. Hubbell, J. LaFrankie, S. Bunyavejchewin, H. S. Dattaraja, S. Davies, S. Esufali, R. Foster, N. Gunatilleke, S. Gunatilleke, P. Hall, A. Itoh, R. John, S. Kiratiprayoon, S. Loo de Lao, M. Massa, C. Nath, M. N. S. Noor, A. R. Kassim, R. Sukumar, H. S. Suresh, I.-F. Sun, S. Tan, T. Yamakura, and J. Zimmerman. 2006. Non-random processes contribute to the maintenance of diversity in tropical forests. *Science* 311:527–531.

Woolbright, L. L. 1991. The impact of Hurricane Hugo on forest frogs in Puerto Rico. *Biotropica* 23:462–467.

Woolbright, L. L. 1996. Disturbance influences long-term population patterns in the Puerto Rican frog, *Eleutherodactylus coqui* (Anura: Leptodactylidae). *Biotropica* 28:493–501.

Wunderle, J. M., Jr. 1995. Responses of bird populations in a Puerto Rican forest to Hurricane Hugo: The first 18 months. *The Condor* 97:879–896.

Wunderle, J. M., Jr. 1999. Pre- and post-hurricane fruit availability: Implications for Puerto Rican Parrots in the Luquillo Mountains. *Caribbean Journal of Science* 35:249–264.

Wunderle, J. M., D. J. Lodge, and R. B. Waide. 1992. Short-term effects of Hurricane Gilbert on terrestrial bird populations on Jamaica. *Auk* 109:148–166.

Wunderle, J. M., Jr., M. R. Willig, and L. M. P. Henriques. 2005. Avian distribution in treefall gaps and understorey of *terra firme* forest in the lowland Amazon. *Ibis* 147:109–129.

Yih, K., D. H. Boucher, J. H. Vandermeer, and N. Zamora. 1991. Recovery of the rain forest of southeastern Nicaragua after destruction by Hurricane Joan. *Biotropica* 23:106–113.

You, C., and W. H. Petty. 1991. Effects of Hurricane Hugo on *Manilkara bidentata*, a primary tree species in the Luquillo Experimental Forest of Puerto Rico. *Biotropica* 23:400–406.

Zalamea-Bustillo, M. 2005. Soil biota, nutrients, and organic matter dynamics under decomposing wood. M.S. thesis. University of Puerto Rico-Río Piedras, Río Piedras, Puerto Rico.

Zarin, D. J., and A. H. Johnson. 1995a. Base saturation, nutrient cation, and organic matter increases during early pedogenesis on landslide scars in the Luquillo Experimental Forest, Puerto Rico. *Geoderma* 65:317–330.

Zarin, D. J., and A. H. Johnson. 1995b. Nutrient accumulation during primary succession in a montane tropical forest, Puerto Rico. *Soil Science Society of America Journal* 59:1444–1452.

Zimmerman, J. K., T. M. Aide, M. Rosario, M. Serrano, and L. Herrera. 1995a. Effects of land management and a recent hurricane on forest structure and composition in the Luquillo Experimental Forest, Puerto Rico. *Forest Ecology and Management* 77:65–76.

Zimmerman, J. K., L. S. Comita, J. Thompson, M. Uriarte, and N. Brokaw. 2010. Patch dynamics and community metastability of a subtropical forest: Compound effects of natural disturbance and human land use. *Landscape Ecology* 25:1099–1111.

Zimmerman, J. K., E. M. Everham III, R. B. Waide, D. J. Lodge, C. M. Taylor, and N. V. L. Brokaw. 1994. Responses of tree species to hurricane winds in subtropical wet forest in Puerto Rico: Implications for tropical tree life histories. *Journal of Ecology* 82:911–922.

Zimmerman, J. K., J. B. Pascarella, and T. M. Aide. 2000. Barriers to forest regeneration in abandoned pastures in Puerto Rico. *Restoration Ecology* 8:350–360.

Zimmerman, J. K., W. M. Pulliam, D. J. Lodge, V. Quiñones, N. Fetcher, S. Guzmán-Grajales, J. A. Parrotta, C. E. Asbury, L. R. Walker, and R. B. Waide. 1995b. Nitrogen immobilization by decomposing woody debris and the recovery from tropical wet forest from hurricane damage. *Oikos* 73:23–45.

Zimmerman, J. K., M. R. Willig, L. R. Walker, and W. L. Silver. 1996. Introduction: Disturbance and Caribbean ecosystems. *Biotropica* 28:414–423.

Zou, X., C. P. Zucca, R. B. Waide, and W. H. McDowell. 1995. Long-term influence of deforestation on tree species composition and litter dynamics of a tropical rain forest in Puerto Rico. *Forest Ecology and Management* 78:147–157.

6

When and Where Biota Matter

Linking Disturbance Regimes, Species Characteristics, and Dynamics of Communities and Ecosystems

Todd A. Crowl, Nicholas Brokaw, Robert B. Waide, Grizelle González, Karen H. Beard, Effie A. Greathouse, Ariel E. Lugo, Alan P. Covich, D. Jean Lodge, Catherine M. Pringle, Jill Thompson, and Gary E. Belovsky

Key Points

- Individual biota or taxa sometimes have a disproportionate effect on food web or ecosystem dynamics.
- The differences in the architecture of tree species (e.g., *Dacryodes excelsa*) alter wind disturbance magnitude and effects through the dissipation of wind energy.
- Ferns and earthworms can enhance the recolonization rate on bare soils following disturbances through modification of the physical microenvironment and nutrient availability.
- Freshwater shrimp and earthworms alter nutrient availability in the streams and soils, altering processing rates through effects on detrital processing.
- Small vertebrate species such as anolis lizards and tree frogs (*Coquis*) significantly alter food web dynamics through direct consumption of herbivorous insects and their cycling of important, limiting nutrients.

Introduction

Organismal ecologists traditionally have been interested in the distribution and abundance of organisms, whereas ecosystem ecologists have been interested in the biological and chemical controls of the pools and fluxes of nutrients and materials.

Over the past 2 decades, however, the importance of species, species traits, and populations affecting ecosystem processes has emerged. As critical components of biodiversity are lost, so too might be a number of critical ecosystem services (Loreau et al. 2001, 2002; Raffaelli 2004; Solan et al. 2004; Zavaleta and Hulvey 2004), although uncertainty and dissent characterize the generality of the association. Thus, understanding the way in which variation in biodiversity is connected to variation in ecosystem processes is a grand, but elusive, challenge in ecology (Naeem et al. 1995; Chapin et al. 1997; Naeem 1998) that has generated controversy in recent years (Hodgson et al. 1998; Lawton et al. 1998; Emmerson and Raffaelli 2000). From the beginning of the Rain Forest Project (Odum and Pigeon 1970; see chapter 1) and continuing into the Luquillo Long-Term Ecological Research (LTER), research in the Luquillo Mountains has addressed the entire ecological continuum from individual ecophysiology and behavior to populations, communities, and ecosystems. In this chapter, we explore how particular species or groups of similar species affect the disturbance sequence and ultimately affect community and ecosystem function in our highly disturbed forest ecosystem.

Species Diversity and Ecosystem Function

Several reviewers have examined the links among levels of the biological hierarchy (Schultze and Mooney 1994; Jones and Lawton 1995; Johnson et al. 1996; Chapin et al. 1997, 1998; Grime 1997; Loreau 2000; Kinzig et al. 2002; Duffy et al. 2007). Three major hypotheses regarding the role of species diversity in ecosystem function have emerged. The suggestion that every species matters (the rivet hypothesis) was the first hypothesis (Ehrlich and Ehrlich 1981). Other hypotheses were posed that suggested a more holistic view. The second and third hypotheses both come under the rubric of the "redundancy hypothesis" (Walker 1992; Lawton and Brown 1993; Frost et al. 1995). These hypotheses are similar in suggesting that, rather than individual species dictating ecosystem function, it is the presence of functional groups within communities that is critical (Covich et al. 2004; Boulton et al. 2008). For example, as long as all trophic levels of a food web exist, the overall ecosystem properties will be maintained. Similarly, with regard to plant communities, ecosystem function is thought to depend on functional plant groups defined by phenology, physiology, and morphology (Vitousek and Hooper 1993; Hooper et al. 2005).

These ideas have coalesced into suggested relationships between species diversity and the maintenance of ecosystem function (Ricklefs and Schluter 1993; Schultze and Mooney 1994; Jones and Lawton 1995; Rosenzweig 1995; Walker and Steffan 1996; Kinzig et al. 2002). A number of investigators have found that high species diversity yields high and stable levels of primary productivity (Tilman and Downing 1994; Naeem et al. 1995; Naeem 1998), although debate continues regarding patterns and mechanisms (Waide et al. 1999; Mittelbach et al. 2001). Studies in aquatic ecosystems have focused on species richness and are just beginning to consider other components of diversity such as the relative abundances of species, functional dominance relationships, and trophic structure (Covich et al. 2004; Boyero et al. 2006; Boulton et al. 2008; Duffy et al. 2007).

To date, several studies have identified linkages between species composition and ecosystem properties. Perhaps the most common pattern is the relationship between biogeochemical cycles and species composition (Carpenter et al. 1987; Pastor et al. 1987; Vitousek and Walker 1989; Zak et al. 1990). For example, carbon storage and fluctuations have been tied to species properties. Ewel et al. (1991) found that soil organic matter increased with increasing species richness to an intermediate level but did not respond to further species additions. In most cases, the nutrient availability and exchange rates are higher and more available when the primary consumer biomass and species diversity are high (Tilman 1982; McNaughton 1985; Carpenter 1988; Power 1990), although some biomes such as mangroves or pine forests show high productivity and nutrient use efficiency with relatively few species (Lugo et al. 1990).

Ecosystem Engineers

Another hypothesis relating species to ecosystem function holds that particular species or groups of species have such disproportionate effects on energy flux (keystone species *sensu* Paine [1966]) that they drive or control ecosystems. It has also been stated that the ability of an organism to regulate a system might not be related to its abundance, biomass, or rate of energy use, but rather to the ability of the organism to affect the organisms with which it interacts (Chew 1974; Paine 1980; Moore and Walter 1988). Therefore, the resolution of food web interactions requires the consideration of both taxa and trophic levels, in which functional groups then serve as a link between species interactions and energy flow (Moore and Walter 1988). The disproportionate effects of a particular species on ecosystem functions or processes could occur through biotic interactions (Paine 1966) or through alterations to the physicochemical environment. Species affecting the localized physical environment have been termed "ecosystem engineers" (Lawton and Jones 1993; Willig and McGinley 1999).

The activities of animal species, especially primary consumers, are linked to the rate and quantity of resource availability (Huntley 1991). Animals affect the movement of energy through the ecosystem by feeding directly on living tissue (herbivory), which affects the rate of primary production and alters plant community composition (MacLean 1974). Animals can also affect decay processes and nutrient cycling; thus if key taxa are excluded from or added to litter, the decomposition of plant material might be altered (Butcher et al. 1971; Seastedt 1984; Moore and Walter 1988; Wall and Moore 1999; González and Seastedt 2001). This is the evidence for most arguments regarding species effects on ecosystem function, in which the predicted response depends on changes to the system following species loss (regarding species gained, see Lugo and Helmer 2003; Helmer 2004; Lugo and Brandeis 2005). Most of the empirical data summarized to date have generally documented changes in a critical ecosystem process such as productivity or decomposition rates as a function of the number of species present (Cuevas et al. 1991; Pimm 1991; Lugo 1992; Tilman and Downing 1994; Cuevas and Lugo 1998; Doak et al. 1998; Kinzig et al. 2002). Fewer reviews look at how the existing species affect and modify local environmental conditions in the face of both external (exogenous) and internal (endogenous) factors.

Studying Species Effects on Ecosystems in the Luquillo Mountains

This focus on understanding the relationships among species, communities, and ecosystem functions and dynamics is now at the forefront of ecology. It is a subject that is effectively addressed by multidisciplinary, long-term ecological research teams such as those in LTER. The LTER programs are designed to collect and synthesize information linking species and population dynamics with key ecosystem properties (e.g., Hobbie et al. 2003), and the Luquillo LTER, in particular, is well suited to address the question of the importance of species to ecosystems for several reasons. First, as a tropical wet forest biome, Luquillo has the highest diversity of plant species of any LTER site (see chapter 3). Second, it has a long history of population monitoring and experimentation. Indeed, the site has had a large number of organismal ecologists involved with it since its conception. These strengths, as well as the legacy provided by the efforts of H. T. Odum to understand material and energy flux at this site, solidify the Luquillo LTER's background and approach to tying all levels of ecological organization together.

In the Luquillo Mountains, disturbance followed by species and ecosystem response provides an opportunity to look at the relationships discussed above. Thus far in this book, the authors have argued that disturbances, both natural and anthropogenic, are critical organizing and determining events in the Luquillo Mountains. They have shown that, through localized changes in environmental conditions, disturbances dictate the potential ecological space and the potential niches available to the species pool. Moreover, they have suggested that in order to understand the spatial and temporal dynamics of a community and its ecosystem processes, one must understand the immediate local effects of single disturbance events, as well as the cumulative effects of the disturbance regime (chapter 2). In this chapter, we show how this conceptual approach can be applied to relations between various species or functional groups and ecosystem responses to disturbance in the Luquillo Mountains. First, we review the events during and after a disturbance that form the context of these relationships. Then we look at examples, including detailed case studies, of the interface between biota and the disturbance sequence that illustrate the effect of particular species and functional groups on ecosystem function in the Luquillo Mountains. Finally, we summarize our findings and suggest further work on these questions.

The Disturbance Sequence

A single disturbance in a localized area can be represented as a sequence of related events (figure 6-1). First, the agent of the disturbance (the physical force) impinges on some set of the existing community (the interface between the force and the biota). That physical force then affects some subset of the biota through direct mortality, changes in reproductive success, altered metabolism, the redistribution of nutrients, or changes in environmental gradients. The amount of change in biomass redistribution and environmental parameters determines the physical severity of the disturbance, sets the stage for the remaining community response, and provides an opportunity for species invasions. Finally, the remaining species and any invaders in the affected area respond to the physical effects through changes in resource acquisition, food web linkages, and colonization dynamics.

Organisms both interact with and respond to the sequence of events associated with a disturbance (Walker 1999). First, some species might dampen or ameliorate the immediate effects of a disturbance through structural attributes (for example, tall height or deep roots of trees) or other life-form or life history adaptations that allow organisms to physically buffer the direct effects of the disturbance agent. If a tree species has a tall, strong stem with deep roots, its presence might deflect or absorb much of the energy associated with high winds (Zimmerman et al. 1994). The presence of that species will then decrease the overall force, decreasing the severity for the rest of the community. This can result in lowered biomass redistribution and mortality, as well as decreased changes in primary environmental drivers such as light and soil moisture.

Following disturbances, the residual and newly arriving species (chapter 2) might affect the response of the local community through physical modification (e.g., soil aeration, habitat structure, or nutrient cycling), environmental gradient change (shading, increased moisture, energy availability), or biotic interactions (seed banks, dispersal, decomposition, competition). If the remaining species do

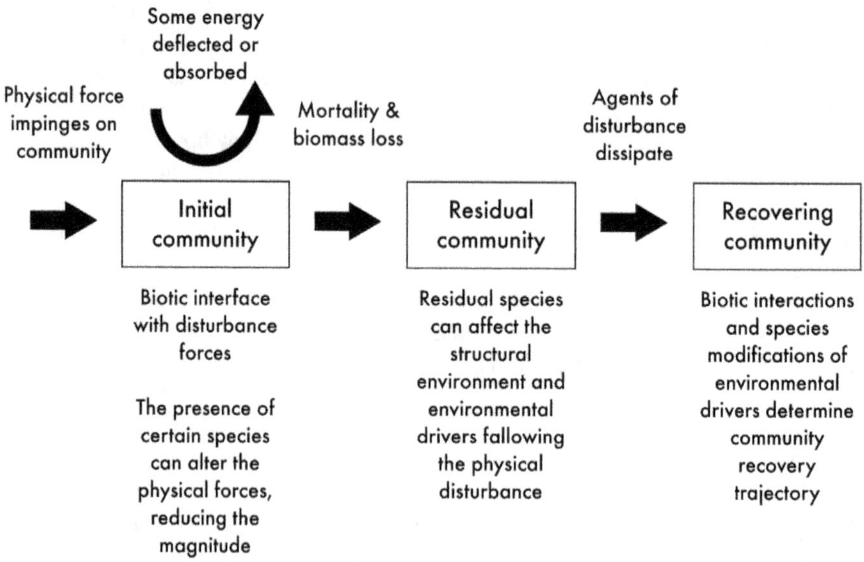

Figure 6.1 The disturbance sequence. The first event is some physical event impinging on an existing community. Certain key species might be able to diminish the physical force (and effects) due to a structural adaptation. The physical force then manifests itself through direct mortality effects, as well as the loss and redistribution of biomass.

not significantly alter the physical environment, then response to disturbance will simply follow the trajectory determined by the resultant change in the physical attributes (figure 6-1). However, if some of the new species combinations do alter the localized environment, the rate of response might be faster, with the previous species composition being reconstituted rapidly or replaced by a new combination.

The Biotic Interface with the Disturbance Sequence

The common natural disturbance events in the Luquillo Mountains are treefalls, landslides, and hurricanes, all of which are related to high wind and rainfall (chapters 3, and 4). Hurricanes and treefalls open the canopy, increase light availability, change soil moisture and humidity, and redistribute carbon and nutrients from living biomass in standing vegetation to dead biomass on or near the ground or into streams (Lodge and McDowell 1991; Walker et al. 1991; Fernández and Myster 1995; Lugo and Lowe 1995; Fetcher et al. 1996; see chapter 4). Landslides include similar changes, as well as the mass movement of soils and the nutrients therein (Walker et al. 1996). The examples that follow suggest that the rate of succession and the resultant community depend heavily on biotic feedbacks during and after disturbances (see figure 2-7, chapter 2).

Biotic Interface with Disturbance Forces

The amount of alteration to the forest canopy following hurricanes is highly variable (Walker et al. 1992). This variation is explained in part by the topographic location, especially the aspect and proximity to ridgetops (Boose et al. 1994), but also by the tree species composition (Walker 1991). For example, the abundant sierra palm (*Prestoea montana*) resists wind effects (Frangi and Lugo 1991; Reed 1998; Zimmerman and Covich 2007). These trees often lose their leaves, but the stems are left unaffected and quickly produce new leaves (Brokaw and Walker 1991). The groups of deeply rooted and root-grafted tabonuco trees (*Dacryodes excelsa*) might contribute to this species' evident resistance to wind, as it is common on ridge tops (Basnet et al. 1992, 1993). Presumably, it might also shelter other trees from wind (Lugo et al. 2000). The two architectural characteristics of having flexible stems and being short relative to the surrounding canopy are the best predictors of which tree species will maintain their importance values following disturbances (Frangi and Lugo 1991; Brokaw and Walker 1991). At a larger scale, smooth canopies (no emergent trees) reduce wind impacts and effects (Lugo et al. 2000).

Biotic Effects on the Postdisturbance Environment

After a disturbance event, the remaining species (residuals) affect the environmental drivers (e.g., light, moisture, nutrients) of the local environment. Some modifications are instantaneous, owing to structural attributes such as residual tree canopies or leaf resprouting from surviving trees. The rate and amount of these ameliorating factors depend on the presence of species that are resistant to the physical disturbance forces or which have adaptations for quick recovery. Above, we discuss the effects of resistant trees. Here we discuss the resilience of colonizing trees and the formation of debris dams, with biotic and abiotic components, that affect the post-disturbance environment.

Biotic Feedbacks on Environmental Drivers after Disturbance

Many tree species in the Luquillo Mountains have adapted to quickly releaf following disturbances or to germinate from a predisturbance seed bank (seeds dormant in the soil). *Casearia arborea*, *Tabebuia heterophylla*, *Myrica deflexa*, *Cecropia schreberiana*, and *Prestoea montana* (sierra palm) all began releafing within 1 to 2 weeks following Hurricane Hugo (Fernández and Fetcher 1991; Walker 1991). As a result, within 10 months of Hurricane Hugo, most light hitting the forest floor was diffuse light with photosynthetic photon flux densities (PPFD) below 400 μmol m^{-2} s^{-1} (Fernández and Fetcher 1991). For the 10 months after Hurricane Hugo, levels of understory PPFD were highly variable at a scale of 1 m, but the median was 7.7 to 10.8 mol m^{-2} d^{-1}, which is comparable to PPFD levels in a 400 m^2 treefall gap (Fernández and Fetcher 1991; Turton 1992; Bellingham et al. 1996; Fetcher et al. 1996). Values had fallen to 0.8 mol m^{-2} d^{-1} by 14 months, when the rapid growth of *Cecropia schreberiana* and other species overtopped the light sensors in this study. This is a clear example of how ecological space shifts rapidly over points in geographic space, owing to disturbance and resilient biotic response (chapter 2).

Species make a difference in the cycling of nutrients in the Luquillo Mountains, and these differences are important for succession after a natural disturbance or forest restoration after land abandonment following agricultural use (Lugo et al. 2004). Table 6-1 contains 12 parameters that influence nutrient cycling and which are themselves affected by the tree species. Each of these parameters differs with species, giving us an understanding of how ecosystem function is influenced as species change through succession, and providing us with an opportunity to manage stand characteristics by manipulating the species composition of the stand. Scatena et al. (1996) found that following Hurricane Hugo, early successional species such as *Cecropia schreberiana* exhibited rapid rates of biomass accumulation and nutrient immobilization while returning nutrient-rich leaves to the forest floor (box 6-1). Nutrient use efficiency was low during this period of early secondary succession after a hurricane. Brown and Lugo (1990) reported that, in general, successional species and young forests are characterized by high rates of nutrient uptake and high rates of nutrient circulation through litterfall. Retranslocation rates for these species are usually low. In contrast, mature forest species, such as *Dacryodes excelsa*, have low rates of nutrient uptake and high rates of nutrient retranslocation (Lugo 1992). Their litterfall is low in nutrients, and their nutrient use efficiency is high. These species tend to conserve and reuse nutrients.

Table 6.1 Examples of how tree species influence nutrient cycling attributes of stands (from Lugo et al. 2004)

Nutrient cycling attribute	Implications for restoration
Uptake rate	Capacity to grow in the site
Retranslocation rate	Regulates the quality of litterfall, reduces the uptake requirement
Return to the forest floor	Opportunity for recycling and improvement of site fertility
Accumulation in biomass	Sink function and retention of nutrients on site
Distribution between above- and belowground compartments	Determines opportunity for building soil fertility belowground vs. circulating nutrients aboveground
Quality of tissue	Influence on decomposition and consumption rates by fungi, bacteria, and soil organisms
Efficiency of recycling	High efficiency favors living plants (reuse); low efficiency makes more nutrients available for the rest of the system
Efficiency of storage	High efficiency favors the sink function
Episodic return	Introduces pulses of nutrient availability
Episodic retranslocation	Causes periodic changes in the quality of litterfall
Episodic mast production	Can dominate the nutrient return pathway and favor particular nutrient cycling pathways
Episodic change in use efficiency	Causes periodic changes in the quality of plant tissue

Box 6.1. Case Study 1—*Cecropia schreberiana* recruitment affects forest structure and nutrient dynamics.

The biology of *Cecropia schreberiana* both reflects the hurricane-driven dynamics of forest in the Luquillo Mountains and helps drive those dynamics (Brokaw 1998). *Cecropia schreberiana* is a light-demanding, fast-growing pioneer tree, and its population responds dramatically to hurricanes (chapter 5). After Hurricane Hugo, *C. schreberiana* was recruited abundantly from a soil seed bank; for example, there were about 11,200 *C. schreberiana* stems ≥ 1 cm in diameter at breast height (dbh) in the 16 ha Luquillo Forest Dynamics Plot after Hurricane Hugo, whereas there had been no more than 200 before (J. Thompson, unpublished data). This abundant colonization helped reestablish the forest canopy and modified the microclimate of the understory (see above). In many places in the Luquillo Mountains, *C. schreberiana* was the only tree forming a canopy after the hurricane.

As a rapidly and abundantly colonizing species, *C. schreberiana* plays key roles in ecosystem function and in the development of forest structure and composition after disturbance. Silander (1979) hypothesized that colonizing stands of *C. schreberiana* conserve nutrients in the recovering forest ecosystem by efficiently acquiring nutrients. Colonizing *C. schreberiana* did quickly concentrate nutrients after Hurricane Hugo. In heavily damaged tabonuco stands in the Bisley watersheds, the highest aboveground net primary productivity in the 5 years after the hurricane occurred in the second year as a result of massive recruitment of *C. schreberiana* saplings to the ≥1.3 m tall size class (Scatena et al. 1996). This productivity was achieved as large amounts of nutrient-poor necromass in the watersheds were replaced by nutrient-rich tissue in fast-growing colonizers. *Cecropia schreberiana* was prominent among these pioneers and had particularly high concentrations of potassium (K) and magnesium (Mg) in its foliage. In addition, leaf litter from *C. schreberiana* decays relatively slowly due to high lignin concentrations (La Caro and Rudd 1985; González and Seastedt 2001), releasing nutrients gradually. *Cecropia schreberiana* might similarly store nutrients in landslides (Walker et al. 1996) and treefall gaps (Walker 2000). Thus it appears that *C. schreberiana* performs a "key function" at the plant - soil interface (Silver et al. 1996; cf. Silander 1979) by having a disproportionately large role in capturing and storing nutrients from decomposing plants after disturbances.

The posthurricane dynamics of *C. schreberiana* are also directly important to some animals. The coquí frog (*Eleutherodactylus coqui*) uses the large fallen leaves of *C. schreberiana* as nest sites, and in the 5 years after Hurricane Hugo this frog was especially abundant where *C. schreberiana* was abundant (Woolbright 1996). Similarly, the lizard *Anolis gundlachi* was more abundant where it could use *C. schreberiana* saplings as understory perches (Reagan 1991).

Debris Dams and Root Mats after Disturbance

In steep forested ecosystems such as the Luquillo Mountains, terrestrial debris dams, created by boulders and tree roots, are important for the retention of leaf litter and the nutrients therein on slopes. Dams also promote the creation of litter mats by linked fungi that protect the soil surface and reduce losses of soil nutrients through erosion while at the same time reducing siltation in streams and reservoirs. The effects of debris dams in reducing erosional losses are quite evident in the Luquillo Mountains. Evidence of erosion on bare, steep slopes has been observed as "erosion columns," small columns of soil protected from the impact of raindrops by pebbles. Furthermore, approximately one-quarter to one-third of the forest floor at El Verde shows evidence of overland flow during extreme rainfall events.

Terrestrial debris dams created by boulders are relatively permanent. Long-lived debris dams include large surface structural roots of certain tree species (*Dacryodes excelsa*, *Pterocarpus officinalis*, and *Pisonia subcordata*). Moderately long-lived debris mats are bound together by rootlike structures (rhizomorphs) consisting of cords and the hyphae of white rot basidiomycete fungi. Some individual mycelia of *Collybia johnstonii* persist in more or less the same place for at least 20 years (D. J. Lodge, personal observation).

Collybia johnstonii is one of the dominant litter-binding basidiomycete species on the forest floor of the tabonuco forest during long interhurricane periods (Lodge and Asbury 1988). The rootlike structures of *C. johnstonii* and various species of *Marasmius* and *Marasmiellus* significantly reduced the rate of leaf litter export on slopes exceeding 30 percent (Lodge and Asbury 1988). Unlike most of the other litter-mat forming basidiomycetes, however, *C. johnstonii* produces a superficial mycelium on leaf surfaces and is very sensitive to drying of the litter layer (Miller and Lodge 1997). Of the 20 mycelial mats of *C. johnstonii* that had been monitored at El Verde beginning before Hurricane Hugo, 8 individuals died, and 9 others were so reduced that they went undetected and were nonfunctional for 5 years after the hurricane opened the canopy (Lodge and Cantrell 1995). Following Hurricane Hugo, more stress-tolerant species of *Marasmiellus* and *Marasmius* partially replaced the diminished function of *C. johnstonii* in binding litter together into mats, and fallen branches and trunks took on a greater role in creating terrestrial debris dams.

Debris dams in streams also form via the accumulation of leaf litter by roots and boulders. Erosional processes along stream channels expose roots, which entrap palm fronds and other leaves. Woody lianas are also associated with these accumulations whenever they hang into the stream channel. These living roots and lianas can remain in place for many years. Following Hurricane Hugo, large amounts of palm fronds and branches were held in place for months by debris dams. The dams retained fine sediments and organic matter. Dam formation in numerous pools slowed down the high stream flow during the intense rainfall associated with the hurricane and reduced the washout of benthic invertebrates. Large accumulations of leaf litter in debris dams provided detrital food resources and protective cover from predators for numerous decapod crustaceans following the hurricane (Covich et al. 1991). Stream flows slowly undercut the organic debris, and within 12 months the sediments and leaf litter had been washed downstream. By then, additional riparian leaf production had resulted in a relatively continuous supply of leaf litter to the stream detritivores, and debris dams were much smaller and transitory.

Biotic Feedbacks and Successional Community Dynamics

There has been much debate about the mechanisms and trajectories of succession and the predictability of ecosystem states following disturbances (see Walker and del Moral [2003] for a review). One paradigm emerging from this debate is that biotic factors (e.g., the effect of species on biogeochemical cycling, shifts in trophic interactions, the loss of native seed sources) are crucial elements that influence the rate and trajectory of succession (Suding et al. 2004). Individual species that either remain or colonize following disturbances often have distinctive traits that can change ecosystem characteristics such as rates of resource turnover, nutrient distribution, and competitive balances (D'Antonio and Meyerson 2002). These observations have resulted in a number of autogenic models of succession in which species' interactions with each other and the local environment drive the rates and trajectories of succession

(Connell and Slatyer 1977; Grime 1979; Noble and Slatyer 1980). Below, we describe species interactions and feedbacks that have strong influences on how communities respond to various disturbances in the Luquillo Mountains.

Animal – Soil – Ecosystem Interactions after Disturbance

Soil fauna modify the soil environment by mixing organic and mineral particles and by changing the water infiltration and aeration regimes (figure 6-2). Tillering by soil fauna directly alters the soil's physical, chemical, and biological properties, and the effects of substrate modification by soil fauna on decomposition are diverse. The breakdown of litter by soil fauna increases the leaching of nutrients and expands the surface area for microbial use. Soil fauna can also augment the nutrient pool in soil solution by adding nitrogenous compounds present in their excreta and dead tissue (González and Zou 1999a, 1999b; Hendrix et al. 1999; González 2002).

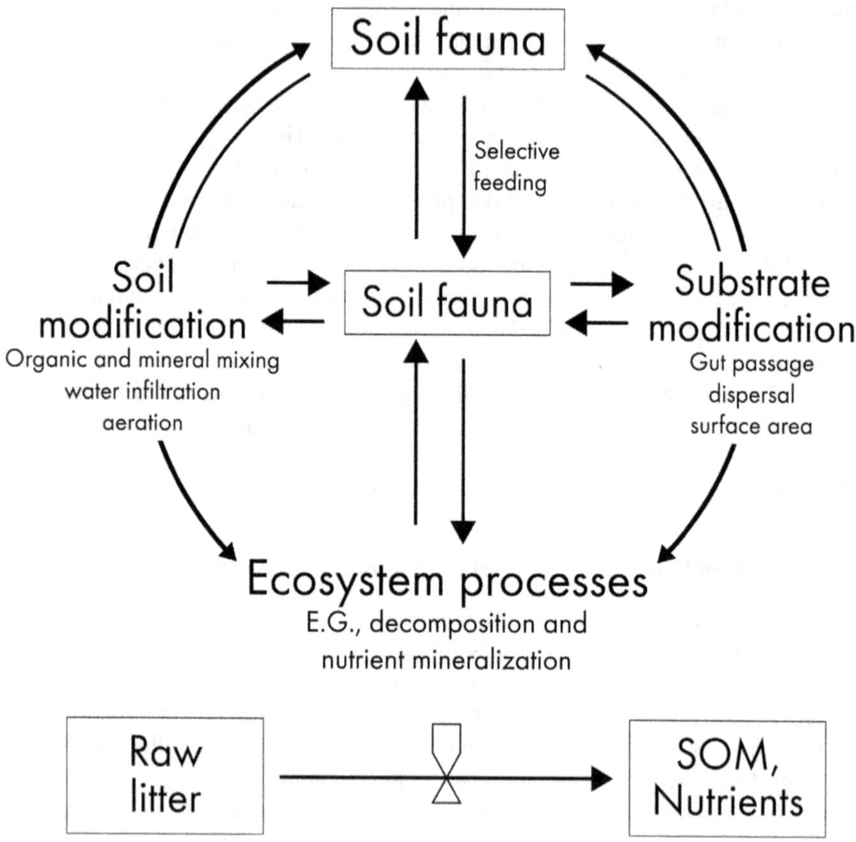

Figure 6.2 Conceptual model indicating direct and indirect paths by which soil fauna affect ecosystem processes (e.g., decomposition and mineralization) and the interaction with microorganisms. SOM = soil organic matter. (Modified from González et al. 2001.)

The majority of the energy and nutrients obtained by plants eventually becomes incorporated in dead organic matter or detritus (Wiegert and Evans 1967; Seastedt 1984). The fragmentation of the detritus, the transfer of organic matter and nutrients into the soils, and the release of carbon dioxide to the atmosphere are essential for sustaining the productivity of ecosystems (Vitousek 1982). In the Luquillo Mountains, soil fauna greatly influence these processes (González 2002); faunal effects on litter decomposition can account for up to 66 percent of the decay rate (González and Seastedt 2001) (figure 6-3). The Luquillo Mountains are a site of high abundance of soil micro- and macrofauna and diversity of functional groups (González and Seastedt 2000, 2001).

Earthworms make up the highest biomass among the soil fauna in the tabonuco forest (Odum and Pigeon 1970), and their abundance and community composition can be greatly altered by disturbance (González et al. 1996; Zou and González 1997) (box 6-2). Earthworms appear to be a significant factor in postdisturbance soil nutrient dynamics (Liu and Zou 2002). Following Hurricane Hugo, Liu and Zou (2002) experimentally removed earthworms via electro-shocking. In areas that had earthworms removed, litter decay rates decreased by 20 to 50 percent. In addition, the soil respiration decreased from 4.7 to 9.4 g m^{-2} d^{-1} in control plots to 3.8 to 6.6 g m^{-2} d^{-1} in earthworm removal plots, for a 20 to 36 percent increase in carbon dioxide (CO_2) evolution with earthworms. Liu and Zou (2002) concluded that the change in soil respiration was due to a decrease in microbial activity when earthworms were absent. This conclusion was based on the lack of response of any other physical changes to the soils, such as pH or moisture or oxygen content.

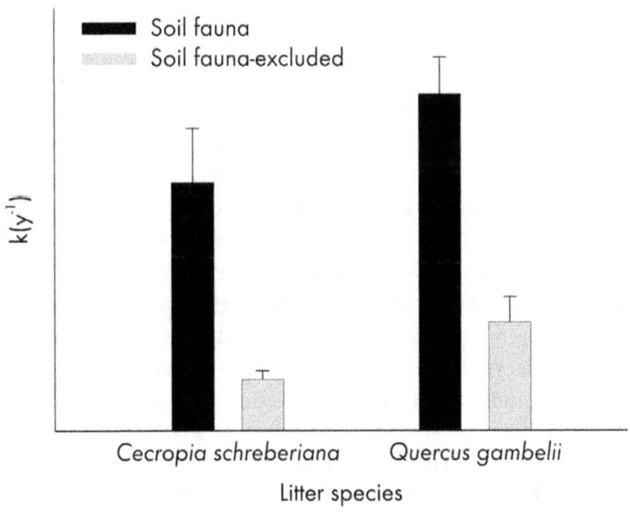

Figure 6.3 Mean decay rates (*k*) for *Cecropia schreberiana* and *Quercus gambelii* litter in soil fauna and soil fauna-excluded treatments in the tabonuco forest. (Redrawn from González and Seastedt 2001.)

Box 6.2. Case Study 2—Earthworms affect soil processes and nutrient cycling.

About 29 earthworm species have been described in Puerto Rico. Twelve of these have been recorded in the tabonuco forest of the Luquillo Mountains (González et al. 1996; Zou and González 1997; González et al. 1999a, 1999b). Among those 12 earthworm species, *Pontoscolex corethrurus*, *Amynthas rodericensis*, and *Ocnerodrilus occidentalis* are not native to Puerto Rico. *Pontoscolex corethrurus* and *A. rodericensis* are found in anthropogenically disturbed sites. The other species are native to Puerto Rico and include *P. spiralis*, *Estherella gatesi*, *E. montana*, *Borgesia sedecimsetae*, *B. montana*, *Onychochaeta borincana*, *Neotrigaster rufa*, *Trigaster longissimus*, and *T. yukiyu*. Earthworms are classified into endogeic, anecic, and epigeic species, representing soil, soil and litter, and litter feeders, respectively. Endogeic earthworms in the tabonuco forest include *P. corethrurus*, *P. spiralis*, *B. sedecimsetae*, and *O. borincana*. *Trigaster longisimus* and *N. rufa* are considered anecic and epigeic species, respectively. *Amynthas rodericensis*, *E. gatesi*, and *E. montana* are epianecic species.

Earthworm abundance and community structure differ between upland areas and riparian areas in the mature tabonuco forest. Earthworm density and fresh weight in the upland area average 118 individuals m^{-2} and 43.4 g m^{-2}, respectively (González et al. 1999a). These values are 68 individuals m^{-2} and 23.5 g m^{-2} in the riparian area. The distribution pattern of earthworms in both upland and riparian areas is clumped, but it is more aggregated in the riparian areas (González et al. 1999a, 1999b).

Disturbances play an important role in altering earthworm abundance and community structure. Although the overall earthworm abundance might recover quickly, anecic earthworms often disappear in newly exposed soils or where root mats were lifted during a treefall (Camilo and Zou 2001). Human activities can drastically change both earthworm abundance and community structure in the Luquillo Mountains. The conversion of tabonuco forest to tropical pastures increased the earthworm density from less than 100 to over 1,000 earthworms m^{-2} (Zou and González 1997; Sánchez et al. 2003). This was largely due to an increase in the nonnative endogeic *P. corethrurus*; native earthworms and earthworm diversity decreased. However, the natural regeneration of secondary forests on abandoned pastures promotes the recovery of both anecic earthworms and native species (González et al. 1996; Sánchez et al. 2003). Earthworm density and fresh weight in secondary forests are twice those in pine (*Pinus caribaea*) and mahogany (*Swietenia macrophylla*) plantations and do not differ between plantations (González et al. 1996).

Pontoscolex corethrurus dominated both secondary forests and plantations, but native earthworms occurred only in secondary forests, suggesting that naturally regenerated secondary forests are preferable to plantations for maintaining high levels of earthworm density, biomass, and native species (González et al. 1996).

Earthworms are closely associated with ecosystem processes in the tabonuco forest. An exclusion experiment demonstrated that a reduction in earthworm density could reduce the decomposition rate of plant leaf litter and soil respiration by 20 to 30 percent (Liu and Zou 2002). Earthworm exclusion also increased surface runoff of water, soil erosion, and downslope exports of organic materials. Furthermore, the presence of earthworms increased soil nitrogen (N) availability and the growth of *Cecropia* seedlings (González and Zou 1999a, 1999b).

Animal - Plant - Ecosystem Interactions after Disturbance

Walking sticks (Phasmidae) preferentially frequent treefall gaps, and lepidoptera populations increased after Hurricane Hugo in response to the flush of new leaves (chapter 5). These herbivores can have a significant effect on hydrology and nutrient turnover after a hurricane. Loss of foliage can increase throughfall and canopy turnover of N, K, and calcium (Ca.). As mentioned in chapter 5, hurricanes appear to generally promote sap-suckers and inhibit defoliators in the forest canopy (Schowalter 1994; Schowalter and Ganio 1999). Sap-sucking insects excrete honeydew, thereby creating a flow of water and sugars from plants to soil that can affect soil processes. Contributions of labile carbon to soils by sap-suckers during recovery might contribute to nutrient retention in microbial biomass. Canopy opening and the increased flux of water and nutrients by defoliators during later successional stages might contribute to greater nutrient cycling via their excretions and to reduced moisture stress owing to leaf consumption during dry periods. Schowalter (1995) reported that although the overall densities of herbivorous insects (primarily heteropterans) did not change following Hurricane Hugo, their spatial distribution was altered, resulting in high-density patches. In these areas, herbivory increased significantly, especially on early-successional plant species. Schowalter (1995) speculated that concentrating herbivores in localized areas could alter the production and survivorship of some species.

Some of the most dramatic examples of animals affecting primary producer biomass, plant decomposition, and nutrient cycling come from the streams draining the Luquillo Mountains (box 6-3). Numerous studies on the role of freshwater shrimp have found significant linkages between shrimp, the algal community, leaf decomposition, and fine particulate organic matter (FPOM) dynamics (Covich 1988a, 1988b; Covich et al. 1991, 1996, 1999; Pringle et al. 1993; Pringle and Blake 1994; Pringle et al. 1999; Crowl et al. 2000, 2001, 2002, 2006; see also chapter 8).

Box 6.3. Case Study 3—Freshwater shrimp and crabs affect primary producers, detrital processing, and nutrient cycling.

The streams of the Luquillo Mountains are dominated both numerically and in terms of biomass by two species of freshwater shrimp (Crowl and Covich 1994; Covich and McDowell 1996; Covich et al. 1996, 2009; Cross et al. 2008; Kikkert et al. 2009). *Atya lanipes* is primarily a filter feeder/collector/scraper (Fryer 1977; Hobbs and Hart 1982; Covich 1988a), and *Xiphocaris elongata* is a shredder/predator/particle feeder (Fryer 1977; Covich 1988a). Over the past 10 years, field and laboratory experiments, as well as large-scale monitoring, have shown how each of the two major decapod guilds (shredders and collectors) affect the composition, rate, and transport of the detrital pool derived from baseline litter inputs (e.g., Pringle and Blake 1994; Pringle et al. 1999; March et al. 2001) and from pulses associated with disturbances such as hurricanes and droughts (Covich et al. 2000; Crowl et al. 2001). Shrimp species composition has a significant effect on the retention of organic carbon and nitrogen within the small tributary streams draining the Luquillo Mountains. Moreover, species-specific (or functional group) processing significantly alters the size fraction and nutrient concentrations available to the remaining community. Microbial or physical processing of detrital material is also important, but these shrimp species are important in fundamentally different ways, in terms of both transport and retention of detritus and nutrient cycling. Furthermore, unique effects of these shrimp species operate during typical flow conditions (pulsed flows often interrupting base flows), during periods of low stream flow (droughts), and following large litter inputs (e.g., hurricanes).

Xiphocaris elongata and *A. lanipes* have dramatic effects on organic matter accumulation, decomposition, and nutrient composition during base flows in the headwaters of the Luquillo Mountains (Pringle et al. 1993; Crowl et al. 2001; Cross et al. 2008). In electric exclusion experiments within pools, scraping and brushing by *A. lanipes* maintained low standing stocks of epilithic biofilms and fine particulate organic matter (FPOM), carbon, and nitrogen in control treatments. In contrast, exclusion treatments had high and variable levels of FPOM and nutrients occurring on rocks (Pringle and Blake 1994; Pringle et al. 1999; March et al. 2002). Although *A. lanipes* feeding decreases the quantity of epilithic FPOM, the remaining FPOM is of a higher food quality (i.e., lower carbon-to-nitrogen ratio) (Pringle et al. 1999). Similarly, shredding by *X. elongata* causes higher leaf decomposition rates in controls than in electric exclusion treatments (March et al. 2001). These base-flow effects of shrimp species act to obscure the effects of high flows, which scour and redistribute organic materials throughout the stream channel. Pulsed flows caused high variability in FPOM and nutrient levels in exclusion treatments, but not in controls. *Atya lanipes* rapidly removes deposited FPOM, restoring low levels of epilithic organic matter within 1 to 2 days following storm-flow events (Pringle and Blake 1994; Pringle et al. 1999).

The patch-scale experiments discussed above have shown that *X. elongata* and *A. lanipes* play an important role in detrital processing during base flows and following the frequent pulsed flows characteristic of Luquillo Mountain streams. Experimental simulation of hurricane-level leaf fall shows that these decapods are particularly critical to detrital processing following hurricane disturbance. The presence of *X. elongata* results in high rates of direct leaf breakdown and downstream export of suspended fragments within 7 days, and these rates continued until the end of the experiment (figure 6-4). For coarse particulate organic matter (CPOM) and medium particulate organic matter (MPOM), pools with *X. elongata* continued to have the highest amounts of export throughout the experiment, with up to 90 percent of the original leaf material being broken down and exported as smaller size fractions.

Atya lanipes did not appear to significantly alter concentrations of medium or coarse particulates until the end of the experiment (figure 6-4). More CPOM was exported from pools containing *A. lanipes* than from the pools without shrimp. This difference suggests that *A. lanipes* enhanced the amount of leaf breakdown once microbial conditioning occurred. The FPOM export was significantly decreased in pools with *A. lanipes* relative to pools with *X. elongata* or without shrimp. These results are expected, given that *A. lanipes* is known to be an effective filter feeder (Covich 1988b). The transport of all size fractions of leaf material was greatly increased during the first 17 days of the experiment. These results suggest that both species of shrimp are extremely important in both the breakdown and the retention of leaf-litter-derived organic particles, especially of disturbance-level inputs.

The results show increased concentrations of dissolved organic matter (DOC) as high as 6.5 mg L^{-1} resulting from shrimp-mediated leaf decomposition; pools without shrimp never exceeded 3.8 mg L^{-1} (Crowl et al. 2001). The increase in stream DOC concentrations that we observed in the presence of *X. elongata* is large relative to typical conditions. Before Hurricane Hugo, low-flow DOC concentrations for streams in the Luquillo Mountains were 1 to 1.5 mg L^{-1}, with peak concentrations of 4 to 5 mg L^{-1} found during storms (McDowell and Asbury 1994). *Xiphocaris elongata* produced considerably more DOC (328 μg mg^{-1}) (DOC production was estimated as the difference in concentrations between pools with and without shrimp). This difference suggests that the mechanism of DOC

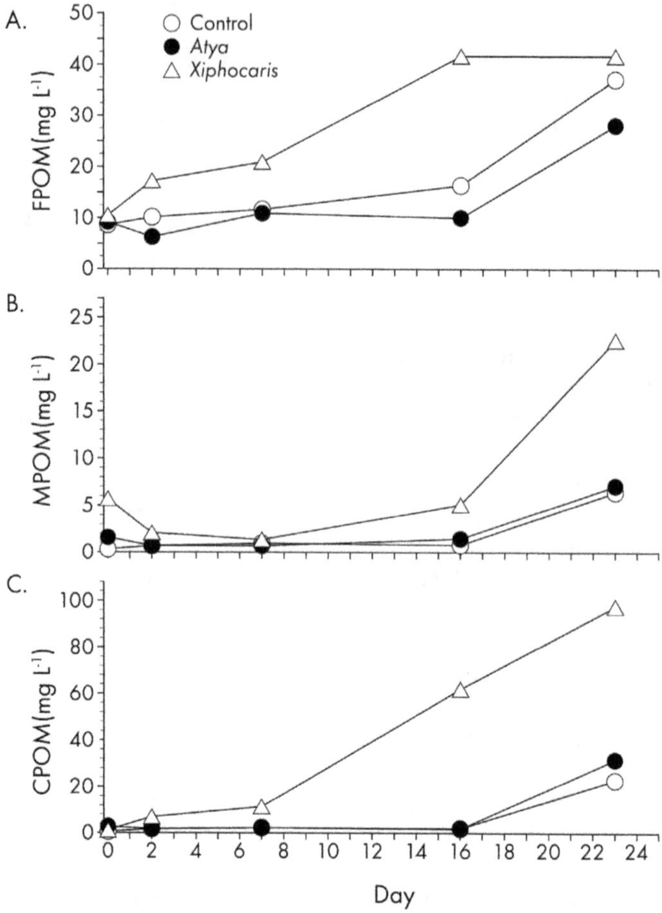

Figure 6.4 Particulate organic matter production as a function of shrimp species. *Xiphocaris* increase all three size fractions ([A]–[C]) of coarse particulate matter. The presence of the filter-feeding *Atya* results in a decrease in the smallest particles (A), presumably due to consumption. (Modified from Crowl et al 2001.)

production differs between the genera, perhaps because of differences in feeding techniques.

Although the effects of shrimp on detrital processing have been examined under a variety of disturbance conditions (i.e., base flows, flash floods, droughts, and hurricane detrital pulses), the effects of shrimp on in-stream primary producers have been studied only in the context of typical Luquillo Mountain stream flow conditions (base flows periodically interrupted by flash floods).

The conversion of leaf litter and primary production into shrimp biomass and the breakdown of CPOM into smaller particles that are more readily available to insect larvae provide a mechanism for retaining carbon and other nutrients following disturbance events in these steeply sloped headwater streams. In the absence of these consumers, nutrients would be flushed downstream and out of the drainage basin. Juvenile crabs and shrimps are confined to the stream, and their grazing and detritivore functions enhance the release of nutrients available to primary producers and bacteria. These detritivorous and algivorous shrimp and juvenile crabs in turn are prey for larger shrimp, crabs, fish, and birds that might further serve to keep the nutrients from being washed downstream. Some of these consumers (e.g., amphibious crabs such as *Epilobocera sinuatifrons*) move the nutrients from the stream back to the surrounding forest and enhance nutrient cycling (Covich and McDowell 1996; Zimmerman and Covich 2003; Fraiola 2006) and act to conserve nutrients locally. These connections potentially have significant effects on the food web structure and overall productivity of these streams.

At a larger scale, organic matter transport and storage appear to be highly dependent on the species composition and abundance of the shrimp assemblages among streams. In a one-time sampling survey of six streams varying in decapod abundance, Pringle et al. (1999) found that standing stocks of FPOM were highly correlated with shrimp abundance. When we combined all information on discharge, physical characteristics, and decapod assemblage and densities across four streams over 8 years, we were able to explain between 32 and 62 percent of the variation in organic matter storage and transport (table 6-2). In almost all cases, the density and species composition of the shrimp explained the highest amount of variation, with physical variables rarely being important. This suggests that biotic processing by decapods is the most important driving variable for organic matter processing, at least in these small headwater streams.

Previous work reported important biotic effects of decapods on the overall community dynamics (e.g., Crowl and Covich 1994; Pringle et al. 1999; Crowl et al. 2001; March et al. 2001) and organic matter transport (Crowl et al. 2002) in small pool experiments. These analyses suggest that biotic interactions occur at a larger scale and over a wider range of hydrologic conditions than previously noted. The species-specific roles of benthic macroinvertebrates that shred leaf litter in tropical streams are being studied over a wide biogeographical range (Meyer and O'Hop 1983; Dobson et al. 2002; Covich et al. 2004; Boyero et al. 2006; Boulton et al. 2008). Our results indicate a major role for macroinvertebrates such as decapod crustaceans, while also demonstrating that microbial processing is especially important for some types of leaf litter, with distinct effects of secondary chemicals in leaves (Wright and Covich 2005a, 2005b).

Table 6.2 Stepwise regression results for the relationship between particulate organic matter storage and transport and decapod and hydrologic parameters

Response variable	Predictor	*R*-square	F	*p*
CPOM (drift)	*Xiphocaris* density	61.4	24.2	<0.0001
FPOM (drift)	*Atya* density	20.2	5.3	0.0313
	Xiphocaris density	10.1	3.9	0.0587
CPOM (benthic)	*Xiphocaris* density	22.5	8.0	0.0181
	Discharge	9.3	3.3	0.0951
FPOM (benthic)	*Atya* density	22.8	8.7	0.0098
	Pool depth	6.1	2.8	0.1013

Secondary Consumers and Ecosystems following Disturbance

Although most experiments involving food web dynamics and succession have just recently begun, a number of studies provide anecdotal evidence that animals (largely frogs and lizards) might be important in terms of altering herbivorous insect populations, herbivory rates, and nutrient cycling in regrowing forest patches. Dial and Roughgarden (1995) reported that reductions in the numbers of anolis lizards resulted in a twofold increase in herbivory on plants. This occurred via two distinct pathways. First, lizards directly consume herbivorous insects (especially orthopterans), thereby decreasing herbivore pressure. Lizards also consume spiders that consume predacious insects. When lizard densities were decreased, spider densities were increased. This resulted in a decrease in insect predators and an ensuing increase in insect herbivores.

Perhaps the most conspicuous species in the Luquillo Mountains are the endemic terrestrial frogs (box 6-4). Although 16 *Eleutherodactylus* species are recognized in Puerto Rico, of the species found in the Luquillo Mountains, *Eleutherodactylus coqui* is the most widespread and abundant (Rivero 1978). *Eleutherodactylus coqui* attains extremely high densities (20,570 individuals ha^{-1} on average) and has the greatest biomass of any vertebrate in the tabonuco forests (Reagan and Waide 1996; Stewart and Woolbright 1996). At these densities, frog predation on insects and frog excretion have important effects on food web dynamics and nutrient cycling (Beard et al. 2002), and these should be even more important after hurricane disturbances, when debris on the forest floor increases frog reproduction and abundance (chapter 5).

Biodiversity: Structure and Function

Most of the research on the relationship between biodiversity and ecosystem processes has focused on patterns with respect to species richness and productivity (Kinzig et al. 2002; Loreau et al. 2002; Wilsey et al. 2005), frequently in ecosystems dominated by low-stature vascular plants (e.g., Gross et al. 2000; Chalcraft et al. 2004). Theory concerning richness and productivity predicts positive monotonic,

Box 6.4. Case Study 4—*Eleutherodactylus coqui* influence invertebrate communities, nutrient cycling, and plant growth rates.

Eleutherodactylus coqui are generalist predators, consuming an estimated 114,000 prey items (mostly invertebrates) ha^{-1} $night^{-1}$ (Stewart and Woolbright 1996). In order to study *E. coqui* effects on invertebrate communities, experiments were conducted in the Bisley Experimental Forest at both large (20 m × 20 m plots) and small scales (1 m × 1 m) using exclosures and enclosures, respectively (Beard et al. 2003a). The effects of *E. coqui* on herbivorous invertebrates was reflected in reduced herbivory rates on two potted plant species, *Piper glabrescens* and *Manilkara bidentata*, at both spatial scales (Beard et al. 2003a) (figure 6-5). *Eleutherodactylus coqui* were also found to reduce flying invertebrates (mostly Dipterans) at both scales (figure 6-6), although there was a positive relationship between *E. coqui* and flying invertebrate abundances in control plots at the larger scale (Beard et al. 2003a) (figure 6-7). Despite the fact that stomach content

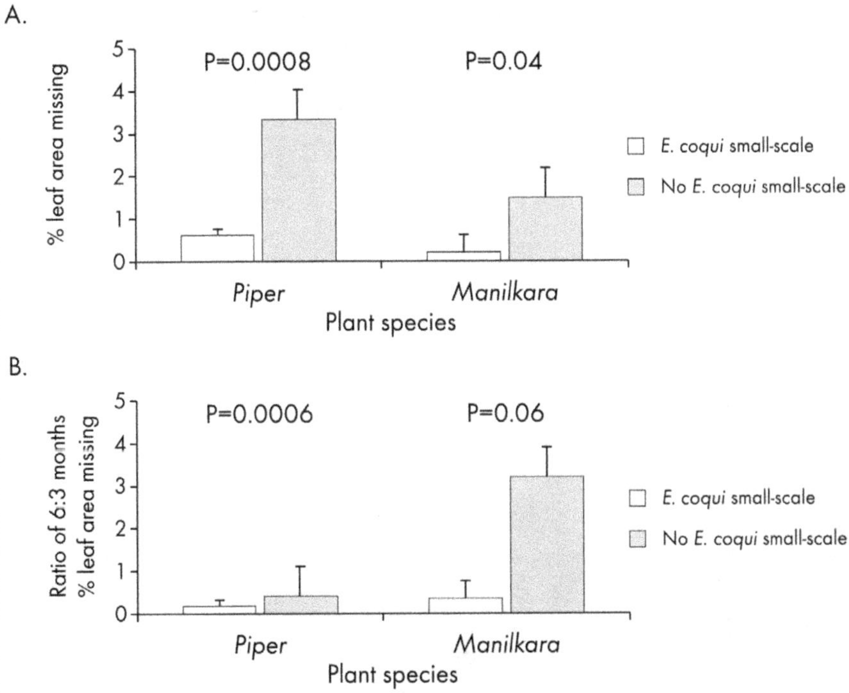

Figure 6.5 Herbivory measurements for plants (+ SE) grown in enclosures and plots with and without *Eleutherodactylus coqui* in the Bisley Watersheds, Luquillo Experimental Forest, Puerto Rico. (A) Mean percent leaf area missing after 4 months for both *Piper* and *Manilkara* in the small-scale experiment. (B) Mean ratio of percent leaf area missing from new leaves at 6 months compared to that missing at 3 months for both *Piper* and *Manilkara* in the large-scale experiment.

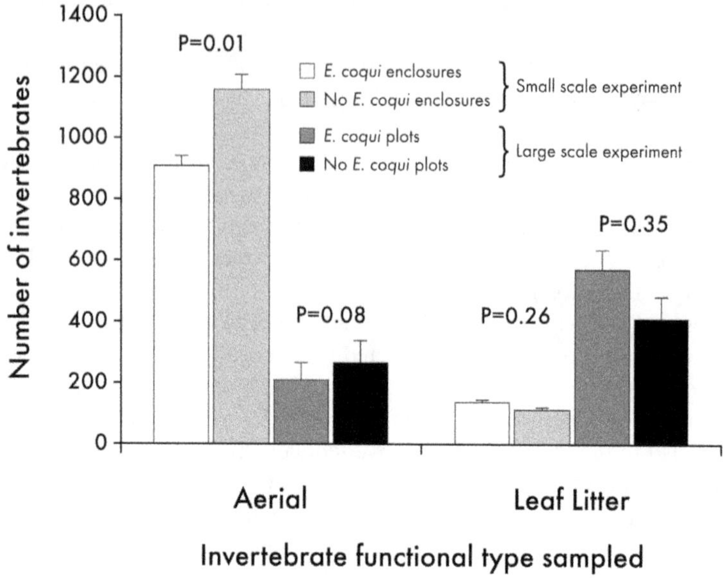

Figure 6.6 Mean total number of aerial and leaf litter invertebrates (+ SE) in enclosures and plots with and without *Eleutherodactylus coqui* in the Bisley Watersheds, Luquillo Experimental Forest, Puerto Rico. $N = 10$ for the small-scale experiment, and $N = 3$ for the large-scale experiment.

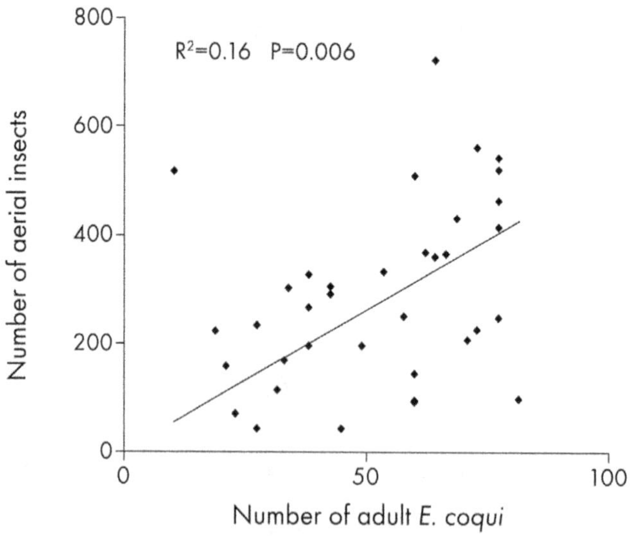

Figure 6.7 Number of aerial insects as a function of adult *Eleutherodactylus* densities.

analyses revealed that *E. coqui* consumes leaf litter invertebrates (Stewart and Woolbright 1996), populations of litter invertebrates did not change significantly with treatment in either experiment (Beard et al. 2003a) (figure 6-6). As would be expected of a predator consuming large numbers of prey items, *E. coqui* increased nutrient availability in the small-scale experiment (the increase was measurable as changes in the throughfall chemistry). More specifically, *E. coqui* increased concentrations of DOC, ammonia (NH_4^+), nitrate (NO_3^-), dissolved organic nitrogen, Ca, iron (Fe), Mg, Mn, phosphorus (P), K, and zinc (Zn) in leachate coming off foliage by 60 to 100 percent (Beard et al. 2002). *Eleutherodactylus coqui* also increased leaf litter decomposition rates and nutrient concentrations of K and P in decomposed litter by 14 percent and 16 percent, repsectively (Beard et al. 2002). The leaf area of the two potted plant species, *P. glabrescens* and *M. bidentata*, also increased with *E. coqui* (figure 6-8). For *P. glabrescens*, other plant growth variables increased, including stem height growth and the number of new leaves and stems produced (Beard et al. 2003a). Both the higher rate of leaf litter decomposition and the increase in leaf production suggest that *E. coqui* might significantly contribute to the rates at which limiting nutrients cycle in this forest, especially at a microsite scale.

The increases in leaf litter decomposition rates and plant growth rates with *E. coqui* present occur through a nutrient cycling effect, as opposed to a trophic cascade (Sin et al. 2008). Although top-down effects on ecosystem productivity through nutrient cycling by a vertebrate predator have been demonstrated in aquatic ecosystems, this is one of the first examples demonstrating the importance of this mechanism for a vertebrate predator in a terrestrial ecosystem. Nutrient cycling effects might be important in this system because frog nitrogenous waste products are in the form of urea, whereas invertebrate waste products are often the least soluble form of nitrogenous waste, uric acid (Beard et al. 2002). Similarly, frog carcasses are more likely to decompose faster and thus release nutrients faster into the substrate than invertebrate remains would. In contrast, inverterbrate remains could act as a nutrient sink owing to the slow decomposition of chitinous exoskeletons (Seastedt and Tate 1981).

The effects of *E. coqui* are likely to be greatest following disturbance events because increases in population abundance occur when breeding habitat increases near the forest floor (Woolbright 1996). The type, frequency, and severity of the disturbance (i.e., the amount of habitat structure added to the forest floor) will determine the extent of the increase in abundance (Woolbright 1991). For example, Hurricane Georges was not as severe as Hurricane Hugo, and whereas adult numbers roughly doubled after Hurricane Georges, adult numbers increased sixfold following Hurricane Hugo (Woolbright 1996).

In order for the ecosystem to recover after a hurricane, and for net primary production to return to the predisturbance level, plants must regain the foliage area lost (Scatena et al. 1993). Heavy grazing of postdisturbance invertebrates could slow this recovery (Torres 1992), but postdisturbance

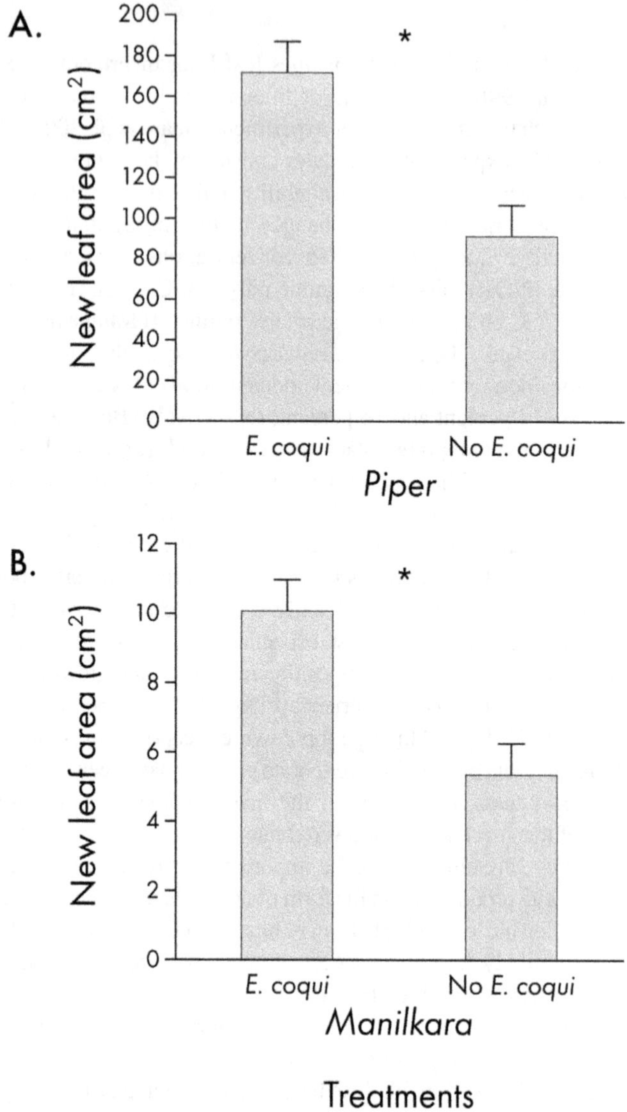

Figure 6.8 Leaf production (+ SE) with and without coquís for *Piper* (A) and *Manilkara* (B).

increases of *E. coqui* could reduce this effect. *Eleutherodactylus coqui* abundance might also increase the rate of recovery by increasing the supply of limiting nutrients to microbes that decompose increased necromass. In addition, greater *E. coqui* densities might aid recovery by increasing nutrient availability to plants because of higher nutrient content in throughput and soils (Vogt et al. 1996). These effects might be especially manifested at the scale of individual plant species and especially relevant for early-successional plant species (Beard et al. 2003b).

negative monotonic, or modal relationships (Rosenzweig 1995), with empirical support for all three (Waide et al. 1999; Mittelbach et al. 2001). Moreover, a growing consensus is that the form and parameterization of such relationships are scale dependent (Moore and Keddy 1989; Pastor et al. 1996; Weiher 1999; Chase and Leibold 2002; Scheiner et al. 2008).

Despite the spectacular diversity of species in tropical forests, little is known about their relationship or that of species traits with productivity in tropical forests, much less the extent to which other aspects of community structure (e.g., species evenness, dominance, rarity, or diversity) alter ecosystem function (Wilsey et al. 2005). Equally true, it is unclear how biota and their linkage with ecosystem processes will depend on the manner in which the importance of species is weighted in measures of evenness, dominance, or diversity (i.e., weighting by proportional abundance or proportional mass). Although to date we have not designed studies to specifically test the role of biodiversity (species richness and abundances) in the disturbance sequence in the Luquillo Mountains, we have initiated analyses across our existing long-term plots and along our various gradients toward this end. We have also provided a number of examples of cases in which individual species and their traits are important in affecting the disturbance sequence through physical and biological pathways.

Summary

We have documented a number of species or species groups that have considerable impacts on the disturbance regime and the ensuing dynamics following disturbances. Measurable effects include the dissipation of energy during wind events (*Dacryodes exclesa*), the enhanced recolonization of bare soils (ferns), and alterations of nutrient availability through food web dynamics (frogs and lizards) and detrital processing (earthworms and freshwater shrimp). Although we have not directly addressed the importance of biodiversity itself, it is clear that the loss of species from the aforementioned taxa would certainly result in a significant alteration of pattern and process in this forest ecosystem.

Literature Cited

Basnet, K., F. N. Scatena, G. E. Likens, and A. E. Lugo. 1992. Hurricane Hugo: Damage to a tropical rain forest in Puerto Rico. *Journal of Tropical Ecology* 8:47–55.

Basnet, K., F. N. Scatena, G. E. Likens, and A. E. Lugo. 1993. Ecological consequences of root grafting in tabonuco (*Dacryodes excelsa*) trees in the Luquillo Experimental Forest, Puerto Rico. *Biotropica* 25:28–35.

Beard, K. H., A. Eschtruth, K. A. Vogt, D. J. Vogt, and F. N. Scatena. 2003a. Amphibian effects on its prey and ecosystem: Evidence from two spatial scales. *Journal of Tropical Ecology* 19:607–617.

Beard, K. H., S. McCullough, and A. Eschtruth. 2003b. A quantitative assessment of habitat preferences for the Puerto Rican terrestrial frog, Eleutherodactylus coqui. *Journal of Herpetology* 37:10–17.

Beard, K. H., K. A. Vogt, and A. Kulmatiski. 2002. Top-down effects of a terrestrial frog on nutrient concentrations in a subtropical forest. *Oecologia* 133:583–559.

Bellingham, P. J., E. V. J. Tanner, P. M. Rich, and T. C. R. Goodland. 1996. Changes in light below the canopy of a Jamaican montane rainforest after a hurricane. *Journal of Tropical Ecology* 12:699–722.

Boose, E. R., D. R. Foster, and M. Fluet. 1994. Hurricane impacts to tropical and temperate forest landscapes. *Ecological Monographs* 64:369–400.

Boulton, A. J., L. Boyero, A. P. Covich, M. Dobson, S. Lake, and R. Pearson. 2008. Are tropical streams ecologically different from temperate streams? Pages 257–284 in D. Dudgeon and C. Cressa, editors, *Tropical stream ecology*. San Diego, CA: Elsevier Academic Press.

Boyero, L., R. G. Pearson, and R. Camacho. 2006. Leaf breakdown in tropical streams: The role of different species in ecosystem functioning. *Archiv für Hydrobiologia* 166:453–466.

Brokaw, N. V. L. 1998. *Cecropia schreberiana* in the Luquillo Mountains of Puerto Rico. *The Botanical Review* 64:91–120.

Brokaw, N. V. L., and L. R. Walker. 1991. Summary of the effects of Caribbean hurricanes on vegetation. *Biotropica* 23:442–447.

Brown, S., and A. E. Lugo. 1990. Tropical secondary forests. *Journal of Tropical Ecology* 6:1–32.

Butcher, J. W., R. Snider, and R. J. Snider. 1971. Bioecology of edaphic Collembola and Acarina. *Annual Reviews in Entomology* 9:147–178.

Camilo, G. R., and X. Zou. 2001. Soil fauna in managed forests: Lessons from the Luquillo Experimental Forest, Puerto Rico. Pages 289–303 in R. A. Fimble, A. Grajal, and J. G. Robinson, editors, *The cutting edge: Conserving wildlife in logged tropical forest*. New York: Columbia University Press.

Carpenter, S. R., editor. 1988. *Complex interactions in lake communities*. New York: Springer-Verlag.

Carpenter, S. R., J. F. Kitchell, J. R. Hodgson, P. A. Cochran, J. J. Elser, M. M. Elser, D. M. Lodge, D. Kretchmer, X. He, and C. N. von Ende. 1987. Regulation of lake primary productivity by food web structure. *Ecology* 68:1863–1876.

Chalcraft, D. R., J. W. Williams, M. D. Smith, and M. R. Willig. 2004. Scale dependence in the species-richness-productivity relationship: The role of species turnover. *Ecology* 85:2701–2708.

Chapin, F. S., O. E. Sala, I. C. Burke, J. P. Grime, D. U. Hooper, W. K. Lauernroth, A. Lombard, H. A. Mooney, A. R. Mosier, S. Naeem, S. W. Pacala, J. Roy, W. L. Steffen, and D. Tillman. 1998. Ecosystem consequences of changing biodiversity. *Bioscience* 48:45–52.

Chapin, F. S., B. H. Walker, R. J. Hobbs, D. U. Hooper, J. H. Lawton, O. E. Sala, and D. Tilman. 1997. Biotic control over functioning of ecosystems. *Science* 277:500–503.

Chase, J. M., and M. A. Leibold. 2002. Spatial scale dictates the productivity-biodiversity relationship. *Nature* 416:427–430.

Chew, R. M. 1974. Consumers as regulators of ecosystems: An alternative to energetics. *Ohio Journal of Science* 72:359–370.

Connell, J. H., and R. O. Slatyer. 1977. Mechanisms of succession in natural communities and their role in community stability and organization. *American Naturalist* 111:1119–1144.

Covich, A. P. 1988a. Atyid shrimp in the headwaters of the Luquillo Mountains, Puerto Rico: Filter feeding in natural and artificial streams. *Verhandlungen der Internationalen Vereinigung für Theoretische und Angewandte Limnologie* 23:2108–2113.

Covich, A. P. 1988b. Geographical and historical comparisons of neotropical streams: Biotic diversity and detrital processing in highly variable habitats. *Journal of the North American Benthological Society* 7:361–386.

Covich, A. P., M. C. Austen, F. Barlocher, E. Chauvet, B. J. Cardinale, C. L. Biles, P. Inchausti, O. Dangles, M. Solan, M. O. Gessner, B. Statzner, and B. Moss. 2004. The role of biodiversity in the functioning of freshwater and marine benthic ecosystems. *BioScience* 54:767–775.

Covich, A. P., T. A. Crowl, C. L. Hein, M. J. Townsend, and W. H. McDowell. 2009. Predator-prey interactions in river networks: Comparing shrimp spatial refugia in two drainage basins. *Freshwater Biology* 54:450–465.

Covich, A. P., T. A. Crowl, S. L. Johnson, and M. Pyron. 1996. Distribution and abundance of tropical freshwater shrimp along a stream corridor: Response to disturbance. *Biotropica* 28:484–492.

Covich, A. P., T. A. Crowl, S. L. Johnson, D. Varza, and D. L. Certain. 1991. Post-hurricane Hugo increases in atyid shrimp abundances in a Puerto Rican montane stream. *Biotropica* 23:448–454.

Covich, A. P., T.A. Crowl, and F. N. Scatena. 2000. Linking habitat stability to floods and droughts: Effects on decapods in montane streams. *Verhandlungen Internationale vereinigung fur Theoretische und Angewandte Limnologie*. 27:1–5.

Covich, A. P., and W. H. McDowell. 1996. The stream community. Pages 433–460 in D. P. Reagan and R. B. Waide, editors, *The food web of a tropical rain forest*. Chicago: University of Chicago Press.

Covich, A. P., M. A. Palmer, and T. A. Crowl. 1999. Role of benthic invertebrate species in freshwater ecosystems. *BioScience* 49:119–127.

Cross, W. F., A. P. Covich, T. A. Crowl, J. P. Benstead, and A. Ramírez. 2008. Secondary production, longevity and resource consumption rates of freshwater shrimps in two tropical streams with contrasting geomorphology and food web structure. *Freshwater Biology* 53:2504–2519.

Crowl, T. A., N. Bouwes, M. J. Townsend, A. P. Covich, and F. N. Scatena. 2000. Estimating the potential role of freshwater shrimp on an aquatic insect assemblage in a tropical headwater stream: A bioenergetics approach. *Verhandlungen der Internationalen Vereinigung für Theoretische und Angewandte Limnologie* 27:2403–2407.

Crowl, T. A., and A. P. Covich. 1994. Responses of freshwater shrimp to chemical and tactile stimuli from large decapod predators: Implications for habitat selection. *Journal of the North American Benthological Society* 13:291–298.

Crowl, T. A., A. P. Covich, F. N. Scatena, R. Phillips, M. J. Townsend, and D. K. Vinson. 2002. Particulate organic matter dynamics in tropical headwater streams: A comparison of biotic and abiotic factors. *Verhandlungen der Internationalen Vereinigung fur theoretische und angewandte Limnologie* 28:923–927.

Crowl, T. A., W. H. McDowell, A. P. Covich, and S. L. Johnson. 2001. Freshwater shrimp effects on detrital processing and nutrients in a tropical headwater stream. *Ecology* 82:775–783.

Crowl, T. A., V. Welsh, T. Heartsill-Scalley, and A. P. Covich. 2006. Effects of different types of conditioning on rates of leaf-litter shredding by *Xiphocaris elongata*, a Neotropical freshwater shrimp. *Journal of the North American Benthological Society* 25:198–208.

Cuevas, E., S. Brown, and A. E. Lugo. 1991. Above and belowground organic matter storage and production in a tropical pine plantation and a paired broadleaf secondary forest. *Plant and Soil* 135:257–268.

Cuevas, E., and A. E. Lugo. 1998. Dynamics of organic matter and nutrient return from litterfall in stands of ten tropical tree plantation species. *Forest Ecology and Management* 112:263–279.

D'Antonio, C., and L. A. Meyerson. 2002. Exotic plant species as problems and solutions in ecological restoration: A synthesis. *Restoration Ecology* 10:703–713.

Dial, R., and J. Roughgarden. 1995. Experimental removal of insectivores from rain forest canopy: Direct and indirect effects. *Ecology* 76:1821–1834.

Doak, D. F., D. Bigger, E. K. Harding, M. A. Marvier, R. E. O'Malley, and D. Thomson. 1998. The statistical inevitability of stability-diversity relationships in community ecology. *American Naturalist* 151:264–276.

Dobson, M., A. Magana, J. M. Mathooko and F. K. Ndegwa. 2002. Detritivores in Kenyan highland streams: More evidence for the paucity of shredders in the tropics? *Freshwater Biology* 47:909–919.

Duffy, J. E., B. J. Cardinale, K. E. France, P. B. McIntyre, E. Thebault, and M. Loreau. 2007. The functional role of biodiversity in ecosystems: Incorporating trophic complexity. *Ecology Letters* 10:522–538.

Ehrlich, P. R., and A. H. Ehrlich. 1981. *Extinction: The causes and consequences of the disappearance of species*. New York: Random House.

Emmerson, M. C., and D. G. Raffaelli. 2000. Detecting the effects of diversity on measures of ecosystem function: Experimental design, null models and empirical observations. *Oikos* 91:195–203.

Ewel, J. J., M. J. Mazzarino, and C. W. Berish. 1991. Tropical soil fertility changes under monocultures and successional communities of different structure. *Ecological Applications* 1:289–302.

Fernández, D. S., and N. Fetcher. 1991. Changes in light availability following Hurricane Hugo in a subtropical montane forest in Puerto Rico. *Biotropica* 23:393–399.

Fernández, D. S., and R. W. Myster. 1995. Temporal variation and frequency distribution of photosynthetic photon flux densities on landslides in Puerto Rico. *Tropical Ecology* 36:73–87.

Fetcher, N., B. Haines, R. Cordero, D. J. Lodge, L. R.Walker, D. Fernandez, and W. T. Lawrence. 1996. Responses of tropical plants to nutrients and light on a landslide in Puerto Rico. *Journal of Ecology* 84:331–341.

Fraiola, K. M. S. 2006. Juvenile *Epilobocera sinuatifrons* growth rates and ontheogenetic shifts in feeding in wild populations. Master's thesis. University of Georgia, Athens.

Frangi, J. L., and A. E. Lugo. 1991. Hurricane damage to a flood plain forest in the Luquillo Mountains of Puerto Rico. *Biotropica* 23:324–335.

Frost, T. M., S. R. Carpenter, A. R. Ives, and T. K. Kratz. 1995. Species compensation and complementarity in ecosystem function. In C. G. Jones and J. H. Lawton, editors, *Linking species and ecosystems*. New York: Chapman and Hall.

Fryer, G. 1977. Studies on the functional morphology and ecology of the atyid prawns of Dominica. *Philosophical Transactions of The Royal Society of London* 277:57–129.

González, G. 2002. Soil organisms and litter decomposition. Pages 315–329 in R. S. Ambasht and N. K. Ambasht, editors, *Modern trends in applied terrestrial ecology*. New York: Kluwer Academic/Plenum Publishers.

González, G., R. Ley, S. K. Schmidt, X. Zou, and T. R. Seastedt. 2001. Soil ecological interactions: Comparisons between tropical and subalpine forests. *Oecologia* 128:549–556.

González, G., and T. R. Seastedt. 2000. Comparison of the abundance and composition of litter fauna in tropical and subalpine forests. *Pedobiologia* 44:545–555.

González, G., and T. R. Seastedt. 2001. Soil fauna and plant litter decomposition in tropical and subalpine forests. *Ecology* 82:955–964.

González, G., and X. M. Zou. 1999a. Earthworm influence on N availability and the growth of *Cecropia schreberiana* in tropical pasture and forest soils. *Pedobiologia* 43:824–829.

González, G., and X. M. Zou. 1999b. Plant and litter influences on earthworm abundance and community structure in a tropical wet forest. *Biotropica* 31:486–493.

González, G., X. M. Zou, and S. Borges. 1996. Earthworm abundance and species richness in abandoned tropical croplands: Comparisons between tree plantations and secondary forests. *Pedobiologia* 40:385–391.

Grime, J. P. 1979. *Plant strategies and vegetation processes*. Chichester, England: John Wiley and Sons.

Grime, J. P. 1997. Biodiversity and ecosystem function. *Science* 277:1260–1261.

Gross, K. L., M. R. Willig, L. Gough, R. Inouye, and S. B. Cox. 2000. Patterns of species density and productivity at different spatial scales in herbaceous plant communities. *Oikos* 89:417–427.

Helmer, E. H. 2004. Forest conservation and land development in Puerto Rico. *Landscape Ecology* 19:29–40.

Hendrix, P. F., S. L. Lachnicht, M. A. Callaham, Jr., and X. M. Zou. 1999. Stable isotopic studies of earthworm feeding ecology in tropical ecosystems of Puerto Rico. *Rapid Communications in Mass Spectrometry* 13:1295–1299.

Hobbie, J. E., S. R. Carpenter, N. B. Grimm, J. R. Gosz, and T. R. Seastedt. 2003. The US Long-Term Ecological Research Program. *BioScience* 53:21–32.

Hobbs, H. H., Jr., and C. W. Hart, Jr. 1982. The shrimp genus *Atya* (Decapoda: Atyidae). *Smithsonian Contributions to Zoology* 364:1–143.

Hodgson, J. G., K. Thompson, P. J. Wilson, and A. Bogaard. 1998. Does biodiversity determine ecosystem function? The Ecotron experiment reconsidered. *Functional Ecology* 12:843–856.

Hooper, D. U., F. S. Chapin, J. J. Ewel, A. Hector, P. Inchausti, S. Lavorel, J. J. Lawton, D. M. Lodge, M. Loreau, S. Naeem, B. Schmid, H. Setela, A. J. Symstad, J. Vandermeer, and D. A. Wardle. 2005. Effects of biodiversity on ecosystem functioning: A consensus of current knowledge. *Ecological Monographs* 75:3–35.

Huntley, N. 1991. Herbivores and the dynamics of communities and ecosystems. *Annual Review of Ecology and Systematics* 22:477–503.

Johnson, K. H., K. A. Vogt, H. J. Clark, O. J. Schmidt, and D. J. Vogt. 1996. Biodiversity and productivity and stability of ecosystems. *Tends in Ecology and Evolution* 11:372–377.

Jones, C. G., and J. H. Lawton, editors. 1995. *Linking species and ecosystems*. London: Chapman and Hall.

Kikkert, D. A., T. A. Crowl, and A. P. Covich. 2009. Upstream migration of amphidromous shrimps in the Luquillo Experimental Forest, Puerto Rico: Temporal patterns and environmental cues. *Journal of the North American Benthological Society* 28:233–246.

Kinzig, A., D. Tilman, and S. W. Pacala, editors. 2002. *Functional consequences of biodiversity: Experimental progress and theoretical extensions*. Princeton, NJ: Princeton University Press.

La Caro, F., and R. L. Rudd. 1985. Leaf litter disappearance rates in Puerto Rican montane rain forest. *Biotropica* 17:269–276.

Lawton, J. H., and V. K. Brown. 1993. Redundancy in ecosystems. Pages 255–270 in E. D. Schulze and H. A. Mooney, editors, *Biodiversity and ecosystem function*. New York: Springer-Verlag.

Lawton, J. H., and C. G. Jones. 1993. Linking species and ecosystem perspectives. *Trends in Ecology and Evolution* 8:311–313.

Lawton, J. H., S. Naeem, L. J. Thompson, A. Hector, and M. J. Crawley. 1998. Biodiversity and ecosystem function: Getting the Ecotron experiment in its correct context. *Functional Ecology* 12:843–856.

Liu, Z. G., and X. M. Zou. 2002. Exotic earthworms accelerate plant litter decomposition in a Puerto Rican pasture and a wet forest. *Ecological Applications* 12:1406–1417.

Lodge, D. J., and C. E. Asbury. 1988. Basidiomycetes reduce export of organic matter from forest slopes. *Mycologia* 80:888–890.

Lodge, D. J., and S. Cantrell. 1995. Fungal communities in wet tropical forests: Variation in time and space. *Canadian Journal of Botany* 73(Suppl. 1)H5.1:S1391–S1398.

Lodge, D. J., and W. H. McDowell. 1991. Summary of ecosystem-level effects of Caribbean hurricanes. *Biotropica* 23:373–378.

Loreau, M. 2000. Biodiversity and ecosystem functioning: Recent theoretical advances. *Oikos* 91:3–17.

Loreau, M., S. Naeem, and P. Inchausti. 2002. *Biodiversity and ecosystem functioning: Synthesis and perspectives.* Oxford, England: Oxford University Press.

Loreau, M., S. Naeem, P. Inchausti, J. Bengtsson, J. P. Grime, A. Hector, D. U. Hooper, M. A. Huston, D. Rafffaelli, B. Schmid, D. Tilman, and D. A. Wardle. 2001. Biodiversity and ecosystem functioning: Current knowledge and future challenges. *Science* 294:804–808.

Lugo, A. E. 1992. Comparison of tropical tree plantations with secondary forests of similar age. *Ecological Monographs* 62:1–41.

Lugo, A. E., and T. J. Brandeis. 2005. A new mix of alien and native species coexist in Puerto Rico's landscapes. Pages 484–509 in D. F. R. P. Burslem, M. A. Pinard, and S. E. Hartley, editors, *Biotic interactions in the tropics: Their role in the maintenance of species diversity.* Cambridge, England: Cambridge University Press.

Lugo, A. E., M. M. Brinson, and S. Brown. 1990. Synthesis and search for paradigms in wetland ecology. Pages 447–460 in A. E. Lugo, M. M. Brinson, and S. Brown, editors, *Ecosystems of the world.* Vol. 15. Amsterdam: Elsevier.

Lugo, A. E., and E. Helmer. 2003. Emerging forests on abandoned land: Puerto Rico's new forests. *Forest Ecology and Management* 190:145–161.

Lugo, A. E., and C. Lowe, editors. 1995. *Tropical forests management and ecology.* New York: Springer-Verlag.

Lugo, A. E., C. S. Rogers, and S. W. Nixon. 2000. Hurricanes, coral reefs and rainforests: Resistance, ruin and recovery in the Caribbean. *Ambio* 29:106–114.

Lugo, A. E., W. L. Silver, and S. Molina Colón. 2004. Biomass and nutrient dynamics of restored Neotropical forests. *Water, Air and Soil Pollution: Focus* 4:731–764.

Maclean, S. F. 1974. Primary production, decomposition, and the activity of soil invertebrates in tundra ecosystems: A hypothesis. Pages 197–206 in A. J. Halding, D. W. Heal, S. F. MacLean, and P. W. Flanagan, editors, *Soil organisms and decomposition in tundra.* Stockholm: Tundra Biome Steering Committee.

March, J. G., J. P. Benstead, C. M. Pringle, and M. W. Ruebel. 2001. Linking shrimp assemblages with rates of detrital processing along an elevational gradient in a tropical stream. *Canadian Journal of Fisheries and Aquatic Sciences* 58:470–478.

March, J. G., C. M. Pringle, M. J. Townsend, and A. I. Wilson. 2002. Effects of freshwater shrimp assemblages on benthic communities along an altitudinal gradient of a tropical island stream. *Freshwater Biology* 47:377–390.

McDowell, W. H., and C. E. Asbury. 1994. Export of carbon, nitrogen, and major ions from three tropical montane watersheds. *Limnology and Oceanography* 39:111–125.

McNaughton, S. J. 1985. Ecology of a grazing system: The Serengeti. *Ecological Monographs* 55:259–294.

Meyer, J. L., and J. O'Hop. 1983. Leaf-shredding insects as a source of dissolved organic carbon in headwater streams. *The American Midland Naturalist* 109:175–183.

Miller, R. M., and D. J. Lodge. 1997. Fungal responses to disturbance—Agriculture and forestry. Pages 65–84 in K. Esser, P. A. Lemke, and D. T. Wicklow, editors, *Environmental and microbial relationships.* Amsterdam: Springer-Verlag.

Mittelbach, G. G., C. F. Steiner, S. M. Scheiner, K. L. Gross, H. L. Reynolds, R. B. Waide, M. R. Willig, S. I. Dodson, and L. Gough. 2001. What is the observed relationship between species richness and productivity. *Ecology* 82:2381–2396.

Moore, D. R. J., and P. A. Keddy. 1989. The relationship between species richness and standing crop in wetlands: The importance of scale. *Vegetatio* 79:99–106.

Moore, J. C., and D. E. Walter. 1988. Arthropod regulation of micro- and mesobiota in below-ground detrital food webs. *Annual Reviews in Entomology* 33:419–439.

Naeem, S. 1998. Species redundancy and ecosystem reliability. *Conservation Biology* 12:39–45.

Naeem, S. L., J. Thompson, S. P. Lawler, J. H. Lawton, and R. M. Woodfin. 1995. Empirical evidence that declining species diversity may alter the performance of terrestrial ecosystems. *Philosophical Transactions of the Royal Society of London B* 347:249–262.

Noble, I. R., and R. O. Slatyer. 1980. The use of vital attributes to predict successional changes in plant communities subject to recurrent disturbances. *Plant Ecology* 43:5–21.

Odum, H. T., and R. F. Pigeon, editors. 1970. *A tropical rain forest: A study of irradiation and ecology at El Verde, Puerto Rico*. Oak Ridge, TN: U.S. Atomic Energy Commission.

Paine, R. T. 1966. Foodweb complexity and community stability. *American Naturalist* 100:65–75.

Paine, R. T. 1980. Food webs: Linkage, interaction strength and community infrastructure. *Journal of Animal Ecology* 49:667–686.

Pastor, J., A. Downing, and H. E. Erickson. 1996. Species-area curves and diversity-productivity relationships in beaver meadows of Voyageurs National Park, Minnesota, USA. *Oikos* 77:399–406.

Pastor, J., R. H. Gardner, V. H. Dale, and W. M. Post. 1987. Successional changes in nitrogen availability as a potential factor contributing to spruce declines in boreal North America. *Canadian Journal of Forest Research* 17:1394–1400.

Pimm, S. L. 1991. *The balance of nature? Ecological issues in the conservation of species and communities*. Chicago: University of Chicago Press.

Power, M. E. 1990. Effects of fish in river food webs. *Science* 250:81–814.

Pringle, C. M., and G. A. Blake. 1994. Quantitative effects of atyid shrimp (Decapoda: Atyidae) on the depositional environment in a tropical stream: Use of electricity for experimental exclusion. *Canadian Journal of Fisheries and Aquatic Sciences* 51:1443–1450.

Pringle, C. M., G. A. Blake, A. P. Covich, K. M. Buzby, and A. M. Finley. 1993. Effects of omnivorous shrimp in a montane tropical stream: Sediment removal, disturbance of sessile invertebrates, and enhancement of understory algal biomass. *Oecologia* 93:1–11.

Pringle, C. M., N. Hemphill, W. H. McDowell, A. Bednarek, and J. G. March. 1999. Linking species and ecosystems: Different biotic assemblages cause inter-stream differences in organic matter. *Ecology* 80:1860–1872.

Raffaelli, D. 2004. How extinction patterns affect ecosystems. *Science* 306:1141–1142.

Reagan, D. P. 1991. The response of *Anolis* lizards to hurricane-induced habitat changes in a Puerto Rican rain forest. *Biotropica* 23:468–474.

Reagan, D. P., and R. B. Waide, editors. 1996. *The food web of a tropical rain forest*. Chicago: University of Chicago Press.

Reed, A. M. 1998. Scale-dependent influences of riparian processes on the dominant detritivore of a headwater stream in Puerto Rico. Master's thesis. Colorado State University, Ft. Collins.

Ricklefs, R. W., and D. Schluter, editors. 1993. *Species diversity in ecological communities*. Chicago: University of Chicago Press.

Rivero, J. A. 1978. *Los anfibios y reptiles de Puerto Rico* [*The amphibians and reptiles of Puerto Rico*]. San Juan, Puerto Rico: Universidad de Puerto Rico, San Juan.

Rosenzweig, M. L. 1995. *Species diversity in space and time*. Cambridge, England: Cambridge University Press.

Sánchez, Y., X. M. Zou, S. Borges, and H. H. Ruan. 2003. Recovery of native earthworms in abandoned tropical pastures. *Conservation Biology* 17:999–1006.

Scatena, F. N., S. Moya, C. Estrada, and J. D. Chinea. 1996. The first five years in the reorganization of aboveground biomass and nutrient use following Hurricane Hugo in the Bisley Experimental Watersheds, Luquillo Experimental Forest, Puerto Rico. *Biotropica* 28:424–440.

Scatena, F. N., W. Silver, T. Siccama, A. Johnson, and M. J. Sánchez. 1993. Biomass and nutrient content of the Bisley Experimental Watershed, Luquillo Experimental Forest, Puerto Rico, before and after Hurricane Hugo, 1989. *Biotropica* 25:15–27.

Scheiner, S. M., S. B. Cox, M. R. Willig, G. G. Mittelbach, C. Osenberg, and M. Kaspari. 2008. Species richness, species-area curves and Simpson's paradox. *Evolution Ecology Research* 2:791–802.

Schowalter, T. D. 1994. Invertebrate community structure and herbivory in a tropical rainforest canopy in Puerto Rico following Hurricane Hugo. *Biotropica* 26:312–319.

Schowalter, T. D. 1995. Canopy invertebrate community response to disturbance and consequences of herbivory in temperate and tropical forests. *Selbyana* 16:41–48.

Schowalter, T. D., and L. M. Ganio. 1999. Invertebrate communities in a tropical rain forest canopy in Puerto Rico following Hurricane Hugo. *Ecological Entomology* 24:191–201.

Schultze, E.-D., and H. A. Mooney, editors. 1994. *Biodiversity and ecosystem function*. Berlin: Springer-Verlag.

Seastedt, T. R. 1984. The role of microarthropods in decomposition and mineralization processes. *Annual Review of Entomology* 29:25–46.

Seastedt, T. R., and C. M. Tate. 1981. Decomposition rates and nutrient contents of arthropod remains in forest litter. *Ecology* 62:13–19.

Silander, S. R. 1979. A study of the ecological life history of *Cecropia peltata* L., an early secondary successional species in the rain forest of Puerto Rico. Master's thesis. University of Tennessee, Knoxville, TN.

Silver, W. L., S. R. Brown, and A. E. Lugo. 1996. Effects of changes in biodiversity on ecosystem function in tropical forests. *Conservation Biology* 10:17–24.

Sin, H., K. H. Beard, and W. C. Pitt. 2008. An invasive frog, *Eletherodactylus coqui*, increases new leaf production and leaf litter decomposition rates through nutrient cycling in Hawaii. *Biological Invasions* 10:335–345.

Solan, M., B. J. Cardinale, A. L. Downing, K. A. M. Engelhardt, J. L. Ruesink, and D. S. Srivastava. 2004. Extinction and ecosystem function in the marine benthos. *Science* 306:1177–1180.

Stewart, M. M., and L. L. Woolbright. 1996. Amphibians. Pages 363–398 in D. P. Reagan and R. B. Waide, editors. *The food web of a tropical rain forest*. Chicago: University of Chicago Press.

Suding, K. N., K. L. Gross, and G. R. Houseman. 2004. Alternative states and positive feedbacks in restoration ecology. *Trends in Ecology and Evolution* 19:46–53.

Tilman, D. 1982. *Resource competition and community structure*. Princeton, NJ: Princeton University Press.

Tilman, D., and J. A. Downing. 1994. Biodiversity and stability in grasslands. *Nature* 367:363–365.

Torres, J. A. 1992. Lepidoptera outbreaks in response to successional changes after the passage of Hurricane Hugo in Puerto Rico. *Journal of Tropical Ecology* 8:285–298.

Turton, S. M. 1992. Understory light environments in a north-east Australian rain forest before and after a tropical cyclone. *Journal of Tropical Ecology* 8:241–252.

Vitousek, P. 1982. Nutrient cycling and nutrient use efficiency. *American Naturalist* 119:553–572.

Vitousek, P. M., and D. U. Hooper. 1993. Biological diversity and terrestrial ecosystem biogeochemistry. Pages 3–14 in E.-D. Schulze and H. A. Mooney, editors. *Biodiversity and ecosystem function.* Berlin: Springer-Verlag.

Vitousek, P. M., and L. R. Walker. 1989. Biological invasion by *Myrica faya* in Hawaii: Plant demography, nitrogen fixation, and ecosystem effects. *Ecological Monographs* 59:247–265.

Vogt, K. A., D. J. Vogt, P. Boon, A. P. Covich, F. N. Scatena, H. Asbjornsen, J. L. O'Hara, T. G. Siccama, J. Bloomfield, and J. F. Ranciato. 1996. Litter dynamics along stream, riparian, and upslope areas following Hurricane Hugo, Luquillo Experimental Forest, Puerto Rico. *Biotropica* 28:458–470.

Waide, R. B., M. R. Willig, C. F. Steiner, G. G. Mittelbach, L. Gough, S. I. Dodson, G. P. Juday, and R. Parmenter. 1999. The relationship between primary productivity and species richness. *Annual Review of Ecology and Systematics* 30:257–300.

Walker, B. H. 1992. Biodiversity and ecological redundancy. *Conservation Biology* 6:18–23.

Walker, B., and W. Steffan, editors. 1996. *Global change in terrestrial ecosystems.* Cambridge, England: Cambridge University Press.

Walker, L. R. 1991. Tree damage and recovery from Hurricane Hugo in Luquillo Experimental Forest, Puerto Rico. *Biotropica* 23:379–385.

Walker, L. R., editor. 1999. *Ecosystems of disturbed ground.* Ecosystems of the world, Vol. 16. Amsterdam: Elsevier.

Walker, L. R. 2000. Seedling and sapling dynamics in treefall pits in Puerto Rico. *Biotropica* 32:262–275.

Walker, L. R., and R. del Moral. 2003. *Primary succession and ecosystem rehabilitation.* Cambridge, England: Cambridge University Press.

Walker, L. R., D. J. Lodge, N. V. L. Brokaw, and R. B. Waide. 1991. An introduction to hurricanes in the Caribbean. *Biotropica* 23:313–316.

Walker, L. R., J. Voltzow, J. D. Ackerman, D. S. Fernández, and N. Fetcher. 1992. Immediate impact of Hurricane Hugo on a Puerto Rican rain forest. *Ecology* 73:691–694.

Walker, L. R., D. J. Zarin, N. Fetcher, R. W. Myster, and A. H. Johnson. 1996. Ecosystem development and plant succession on landslides in the Caribbean. *Biotropica* 28:566–576.

Wall, D. H., and J. C. Moore. 1999. Interactions underground Soil biodiversity, mutualism, and ecosystem processes. *BioScience* 49:109–117.

Weiher, E. 1999. The combined effects of scale and productivity on species richness. *Journal of Ecology* 87:1005–1011.

Wiegert, R. G., and F. C. Evans. 1967. *Investigations of secondary productivity in grasslands.* Warsaw: Polish Academy of Science.

Willig, M. R., and M. A. McGinley. 1999. The response of animals to disturbance and their roles in patch generation. Pages 633–658 in L. R. Walker, editor, *Ecosystems of disturbed ground.* Ecosystems of the world, Vol. 16. Amsterdam: Elsevier.

Wilsey, B. J., D. R. Chalcraft, C. M. Bowles, and M. R. Willig. 2005. Relationships among indices suggest that richness is an incomplete surrogate for grassland biodiversity. *Ecology* 86:1178–1184.

Woolbright, L. 1996. Disturbance influences long-term population patterns in the Puerto Rican frog, *Eleutherodactylus coqui* (Anura: Leptodactylidae). *Biotropica* 28:493–501.

Woolbright, L. L. 1991. The impact of Hurricane Hugo on forest frogs in Puerto Rico. *Biotropica* 23:462–467.

Wright, M. S., and A. P. Covich. 2005a. The effect of macroinvertebrate exclusion on leaf breakdown rates in a tropical headwater stream. *Biotropica* 37:403–408.

Wright, M. S., and A. P. Covich. 2005b. Relative importance of bacteria and fungi in a tropical headwater stream: Leaf decomposition and invertebrate feeding preference. *Microbial Ecology* 20:1–11.

Zak, D. R., P. M. Grofman, K. S. Pregitzer, S. Christensen, and J. M. Tieje. 1990. The vernal dam: Plant-microbe competition for nitrogen in northern hardwood forest. *Ecology* 71:654–656.

Zavaleta, E. S., and K. B. Hulvey. 2004. Realistic species losses disproportionately reduce grassland resistance to biological invaders. *Science* 306:1175–1177.

Zimmerman, J. K., E. M. Everham III, R. B. Waide, D. J. Lodge, C. M. Taylor, and N. V. L. Brokaw. 1994. Responses of tree species to hurricane winds in subtropical wet forest in Puerto Rico: Implications for tropical tree life histories. *Journal of Ecology* 82:911–922.

Zimmerman, J. K. H., and A. P. Covich. 2003. Distribution of juvenile crabs (*Epilobocera sinuatifrons*) in two Puerto Rican headwater streams: Effects of pool morphology and past land-use legacies. *Archiv für Hydrobiologie* 158:343–357.

Zimmerman, J. K. H., and A. P. Covich. 2007. Damage and recovery of riparian sierra palms (*Prestoea acuminata* var. *montana*) after Hurricane Georges: Influence of topography, land use, and biotic characteristics. *Biotropica* 39:43–49.

Zou, X. M., and G. González. 1997. Changes in earthworm density and community structure during secondary succession in abandoned tropical pastures. *Soil Biology and Biochemistry* 29:627–629.

7

Management Implications and Applications of Long-Term Ecological Research

Ariel E. Lugo, Frederick N. Scatena, Robert B. Waide, Effie A. Greathouse, Catherine M. Pringle, Michael R. Willig, Kristiina A. Vogt, Lawrence R. Walker, Grizelle González, William H. McDowell, and Jill Thompson

Key Points

- Uses and conservation of tropical forests reflect the economic and social circumstances of their associated human populations.
- Conservation efforts in the Luquillo Mountains have benefited from research activity since the 1920s.
- Early research in Puerto Rico focused on descriptions of flora and fauna, tree nurseries and plantation establishment, tree growth, and forest products, whereas recent research focuses on ecosystem functioning and services, climate change, landscape scale patterns, disturbances, and land use legacies.
- Ecological information from both aquatic and terrestrial ecosystems facilitates the sustainable use of natural resources while informing methods for conserving ecosystems and their services.
- Results from research also help in the interpretation of environmental change and in the design of resource conservation strategies in the face of uncertainty.
- A new era of conservation based on ecological knowledge is emerging. Conservation is increasingly based on sustainable development goals and implemented in collaboration with citizens. Management in this era will be more flexible in outlook and adaptable to a continuously changing environment.
- We give examples of surprise events for which we have no explanation, and which we did not have the means to anticipate. These examples collectively demonstrate that the management of complex ecosystems requires continuous long-term research.

Introduction

In this chapter, we highlight the contributions of long-term ecological research (LTER) activity to tropical forest conservation issues. We begin by reviewing the context for management activities in the Luquillo Experimental Forest (LEF), including both the historical changes in forest cover and the research activity that resulted from these changes. We then review what management is and its relationship to disturbances and conservation, and we follow with a discussion of the implications of long-term research in current conservation issues in the tropics. To do this, we use four examples from the Luquillo LTER. Next we describe applications of LTER results for tropical conservation and provide six examples. Finally, we address the question of how the research conducted in the LEF has informed and challenged past paradigms developed in the tropics, and we finish the chapter with what we consider to be our future research needs and priorities. Our approach in this chapter is to be illustrative and synthetic, rather than to provide a comprehensive review of the implications and applications of our research.

Management in the Luquillo Experimental Forest

The LEF, also known as the El Yunque National Forest (previously the Caribbean National Forest), constitutes the core of forests in the Luquillo Mountains and is a site where management and research are concentrated. Currently, between 38 and 58 percent of the LEF is considered primary forest (Lugo 1994). Primary forests are areas where forest cover has existed continuously for centuries. The rest of the LEF experienced changes in forest cover owing to human activities (Scatena 1989; García Montiel and Scatena 1994; Foster et al. 1999; Thompson et al. 2002; Lugo et al. 2004). Aerial photography, available since 1936, allows quantification of changes in the forest cover of the LEF. In 1936, 34 percent of the current LEF was deforested or secondary forest, and 49 percent had >80 percent cover (Foster et al. 1999). In 1989, more than 97 percent of the LEF had continuous forest cover. By 2002, the LEF was almost 100 percent forested. Outside of this forest boundary, however, land cover had changed significantly, mostly transitioning from agriculture to forest and urban landscapes (Lugo et al. 2004).

Between 1936 and 1995, the landscape outside the periphery of the LEF experienced a cycle of fragmentation and consolidation (Lugo 2002), while the economy changed from an agricultural, solar-based economy to one based on fossil fuels. During a period of intense agricultural use (early 20th century), most of the land surrounding the LEF was being managed for agriculture. During this time, there were small urban fragments and patches of forest cover. In satellite images of the area, fragments were defined as groups of pixels comprising homogeneous land cover such as forest, agriculture, or urban. Landscape fragmentation reached a peak in the 1980s following the abandonment of agricultural lands (Lugo 2002). This peak coincided with the increasing dominance of small patches of regenerating forests throughout the region, as well as patches of urban cover. Increases in forest and urban land cover types were generally at the expense of low-lying agricultural lands, a process that contributed most significantly to the higher rates of fragmentation recorded during this period. More recently, these urban and forest patches have coalesced, resulting in a smaller number of larger patches of both forest and urban areas (i.e., less fragmentation).

The products and services required by the socioeconomic system in the communities surrounding the LEF changed dramatically from the time of small-scale agricultural activities of indigenous peoples to the extractive period of the Spanish government, to the agricultural period of the 20th century that was dominated by coastal sugar cane plantations, to the present with the current high-energy urban system. For example, indigenous peoples introduced fire as an ecological factor in Puerto Rico and locally modified valleys and flood plains for agriculture (Scatena 1989; Burney and Burney 1994). The Spanish mined rivers and exploited existing natural capital by inventorying and cutting valuable trees (chapter 1). Their main interest was using, rather than conserving, resources, but they also established guidelines to protect riparian zones and control erosion (Wadsworth 1949, 1970; Scatena 1989). During the early 1900s, land managers in Puerto Rico were influenced by a "the world is my garden" mentality and were keen to restore significantly altered landscapes to their previous condition. To improve conditions and even improve upon nature, they planted what they thought were the most desirable trees, as well as those that offered economic returns. Forest managers also seeded streams with fish species (i.e., trout that were highly valued in temperate zones but which did not do well in Puerto Rico [Erdman 1984]), built dams to harness stream power, and sought to improve tree growth. The ecological impacts of natural disturbances were not considered because the focus was on improving the dire livelihood of humans in Puerto Rico (Murphy 1916; Zon and Sparhawk 1923; Roberts 1942). Recreation was a luxury for most people at this time and involved making trips to the city, not to the forest.

In response to the need to improve the living conditions of the rural population on the island, research related to resource uses and improving the harvesting of products became a prominent activity in the LEF after the 1930s. Over the next 65 years, research in the Luquillo Mountains evolved from its initial focus on reforestation, forest products, and increasing land productivity to an emphasis on the maintenance of ecosystem functions and services (Wadsworth 1970, 1995; Lugo and Mastroantonio 1999). The research history in the LEF shows that adding new lines of research did not preclude the continuation of older lines of research, emphasizing the building of fundamental knowledge supporting forest conservation. For example, the first decade of U.S. Department of Agriculture (USDA) Forest Service research (1939 to 1949) identified 17 lines of scientific inquiry for its research agenda. At present, there are 69 active lines of research. They include all but one of the lines of inquiry identified during the first decade (figure 7-1). From the outset, Forest Service research was mechanistic and empirical, as was management, which used a top-down approach in which technical people provided the research agenda to those implementing the research on the ground. However, by the 1950s, Forest Service research had begun to consider input from farmers and adapt project implementation based on local knowledge. It employed local people in reforestation projects, road maintenance, and other public works. Analysis of the research productivity of Forest Service scientists and collaborators shows that all lines of research ultimately contribute to forest conservation (table 7-1).

Until the 1980s, much of the research in the Luquillo Mountains was based on repeated field observations and typically involved little more than measuring tapes, field books, and physically strenuous work. Scientists established plots in areas selected to represent the region and recorded changes in the structure and species

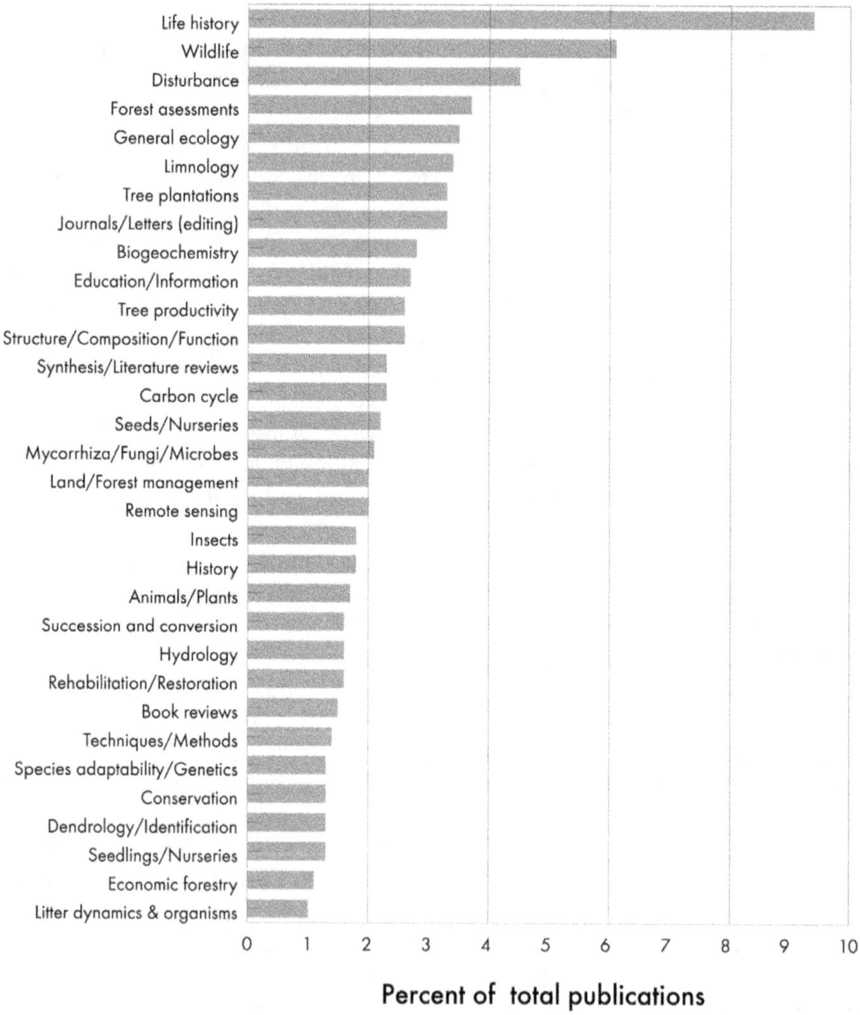

Figure 7.1 Number of publications of the International Institute of Tropical Forestry according to topic of research. The total number of publications represented is 2,000 over a period of 65 years. Data by decade of research are available from the senior author. For each publication, only the dominant topic or topics were recorded.

composition over time. Since the 1980s, there have been major advancements in the tools for research, but the goal is similar: to understand changes in ecological space over time. Although researchers continue to study many of the research plots and plantations that were established in the 1940s and 1950s, technology now allows for a greater scope of field research, with an increased capacity for data collection, storage, and analysis.

Table 7.1 Twenty topics of research in the Luquillo Mountains and their relevance to conservation

Topic	Relevance
Tree and vine identification	Increases fundamental knowledge for conservation activities
Parrot recovery	Prevents the extinction of a species
Tree species selection for different sites	Increases the effectiveness of reforestation
Reforestation techniques	Ensures the success of tree planting programs
Tree nursery techniques	Ensures the effectiveness of developing trees for reforestation and planting programs
Urban tree plantings	Allows the establishment of green areas in urban settings
Silvicultural treatment for cutover and volunteer forests	Allows the management of secondary forests for multiple uses
Properties of Caribbean woods, drying, and preservative treatments	Provides knowledge in support of wood-using industries
Rehabilitation of landslides	Allows the reestablishment of forests and the stabilization of hazardous slopes
Wood production via plantations	Increases the productivity of the land and reduces pressure on native forests
Techniques for the long-term monitoring of tree growth, tree turnover, and wildlife abundance	Saves time and increases the effectiveness of biodiversity-monitoring programs
Restoration of biodiversity on degraded lands	Returns degraded lands to productive use
Vegetation surveys and forest inventories	Assesses biodiversity and contributes criteria for the sustainability of development
Understanding tropical forests	Increases knowledge of forests (the main land cover of these regions and the principal providers of ecological services to people)
Safe water yields from watersheds	Assesses the amount of clean and abundant water for society
Chemical content and composition of plant, soil, water, and air in the tropics	Increases understanding of the functioning of forests and helps estimate their role in cleaning air and water
Understanding tropical forest function	Increases knowledge of the services provided by forests to people
Understanding forest disturbances such as land cover change, hurricanes, global change, and ionizing radiation	Increases knowledge about strategies for managing change
Understanding how silvicultural practices influence wildlife populations, soil fertility, greenhouse gas emissions, and water yield	Increases knowledge needed in order to maintain biodiversity while managing stands
Site effects on tree growth	Provides strategies for maintaining tree growth in spite of land cover changes

Management, Disturbances, and Conservation

Management is a regime of human-motivated interventions that can be understood within the context of disturbance ecology. Many management practices associated with agriculture and forestry alter ecological space in order to create a state that enhances the production of valued, often introduced or domesticated species such as ornamentals, crops, timber, or livestock. In these schemes, management retards natural secondary succession via the selective application of system-specific activities that ideally mimic natural disturbances (e.g., burning, canopy opening) or their effects. Alternatively, management might represent intervention by humans in an existing anthropogenically disturbed site (e.g., mines, clearcuts, and abandoned pastures). In this case, the intent is to increase resiliency; hasten succession toward a particular goal; or restore biodiversity, the productive capacity of the ecosystem, and ecosystem services. If the goal is to manage those sites so that they achieve a state with desired functional characteristics such as forest productivity or the production of clean water, without regard for the biotic composition or structure, then rehabilitation (*sensu* Brown and Lugo 1994) is the focus of activities. If the goal is to manage the system such that its state is within defined boundaries of the predisturbance condition with respect to the biotic composition, structure, and function, then restoration is the focus. The pervasive nature of anthropogenic disturbance often means that natural ecosystems become reduced in their extent and highly fragmented, with altered climatic and environmental conditions (Wiens 1976; Sala et al. 2000). This approach often increases the likelihood of localized species extinction, as well as decoupled ecosystem processes and services. Ecosystem-based management techniques attempt to design local environments in order to reduce the likelihood of species extinction and ensure the continued provision of ecosystem services.

Anthropogenic disturbances have been occurring for millennia in Puerto Rico and elsewhere (Crosby 1986; Perlin 1989). Signatures of these disturbances remain in forest landscapes and modify the current ecosystem structure and function at varying spatial scales across continents (Diamond 2005; Mann 2005). Past human disturbances have modified ecosystems at equally broad temporal scales, so that it has become difficult to identify what is a "natural" ecosystem (Cronon 1996). Thus, recent management activities often occur in ecosystems that have already been modified by humans. The changes that have occurred in the forests of Puerto Rico are a testament to the dynamic nature of ecosystems in the face of human and natural disturbances (see chapters 1 and 4).

Given the objectives of management activities discussed above, it is critical to link management with conservation. We believe that management and conservation must be synonymous. Considering them as conflicting activities is a false dichotomy that hinders the protection of biodiversity and ecosystem services. The importance of considering management and conservation as facets of each other became apparent early in the management of tropical forests because many tropical economies are extraction-based. Because many hot spots of biodiversity in the world are also located in tropical areas that have high poverty rates and political instability, significant challenges in implementing conservation projects exist in these regions (Wilshusen et al. 2002).

However, not until the early 1990s did the scientific community and resource managers outside the tropics recognize these challenges and acknowledge that the successful implementation of conservation projects requires the simultaneous linking of management activities with conservation efforts and the inclusion of the constraints imposed by society (Wells and Brandon 1992; Vogt et al. 2000, 2002b).

A new discipline, Conservation Biology, developed in the 1980s with three guiding principles (Meffe and Carroll 1997) that conservation is based on: (1) evolutionary change, (2) dynamic ecology, and (3) the human presence. These principles orient conservation activities toward the stewardship of natural biodiversity through sustainable development. Thus, the aims of conservation are the same as those of resource management. Aldo Leopold said it best (Meine 1987:148) in an observation about the use of land by farmers:

> This paper proceeds on two assumptions. The first is that there is only one soil, one flora, one fauna, and hence only one conservation problem. Each acre should produce what it is good for, and no two are alike. Hence a certain acre may serve one, or several, or all of the conservation groups. The second [assumption] is that economic and aesthetic land uses can and must be integrated, usually on the same acre. The ultimate issue is whether good taste and technical skill can both exist in the same land owner.

The importance of linking conservation and sustainable development is widely accepted by the scientific community and conservation practitioners. However, formally linking these concepts in order to produce a mechanistically based tool with which to assess the sustainability of resource uses and conservation has been difficult in complex human landscapes (Wells and Brandon 1992; Vogt et al. 1997, 2002b). In this book we treat "conservation" as synonymous and interchangeable with "management," because in principle their goals are the same: to "save all the parts" (*sensu* Leopold 1953) and to satisfy human needs sustainably, using the best science available, within the social and cultural context of the people that depend upon the products and services of an ecosystem. Both management and conservation include the goal of the preservation of wilderness by excluding humans. But conservation or management actions are more likely to succeed if they incorporate human needs and activities (Salwasser 1997). A preservation-only agenda usually fails, as would an agenda that excluded preservation.

Implications of Luquillo Long-Term Ecological Research

This book's synthesis of LTER at the LEF has led to a series of broad generalizations that are useful guidelines for forest managers dealing with management issues at various scales of organization from populations to watersheds and life zones. In this section, we present four examples that demonstrate particular conservation actions and strategies. These four examples include an examination of the implications of ecological space for choosing management units, species life histories in relation to disturbances, the limits of forest resilience, and the notion of ecosystem self-organization. These four examples have helped us develop an approach to and understanding of particular management situations that are common in the tropics.

Ecological Space and Management Units

Shifts in ecological space (chapter 2), whether from local or global change, influence the species composition and rates of ecological processes in affected sites. Depending upon the causal force of the change, the movement in ecological space can be cyclical or directional. Hurricane Hugo, for example, caused a large but temporary change in ecological space across the Luquillo Mountains; after a decade or so, the species composition at a scale of hectares returned to prehurricane conditions (chapter 5). In sites where human-induced change of ecological space altered soil structure and chemistry, species shifts have lasted decades and appear irreversible (Thompson et al. 2002; Lugo 2004; Lugo and Helmer 2004), challenging the concept of "recovery to the original state."

Shifts in ecological space have significant implications for conservation and for the ability of managers to achieve the desired outcomes or management objectives. In this context, the selection of boundaries for conservation units becomes critical to the effectiveness of the conservation activity, and therefore we begin our discussion with the issue of management units.

The initial step for forest management is to identify management units in geographical space. Traditionally, foresters subdivide forests into compartments, and compartments into stands. These are spatial units with similar objectives and management tools, and the long-term objective is to produce similar ecological conditions across the compartment. The criteria for compartment identification usually include geography (delimited by roads or other geographic features), history (past treatments or uses of the compartment), and the purpose for which the compartment is to be managed. Stands are smaller spatial units within compartments with similar species composition, tree age, or structure. The criteria for compartment and stand identification are subjective, and these designations have practical value to foresters because of their flexibility.

Management and conservation activities occur in geographic space. The success of these activities depends upon our understanding of ecological phenomena and our ability to manipulate ecological space constrained by past and future land uses and disturbances. Thus, managers or conservationists who do not understand ecological space and focus their attention solely on geographic space are in danger of failing to achieve their objectives, or they might create long-term conflicts. For example, managers might select the boundaries of their management units based on logistical requirements (e.g., road access, land ownership, etc.). However, such spatial mapping tends to produce ecologically heterogeneous compartments. Soil conditions, drainage, and topography vary within a stand or compartment, and this environmental heterogeneity translates into variable responses to management or natural disturbances. Moreover, the boundaries of a particular stand or compartment might overlap different watersheds and/or parts of watersheds, which would limit their usefulness for managing aquatic resources.

Catena and Watershed Management Units

Catenas and watersheds are geographical or spatial units that coincide with ecological units of function and, as such, provide a logical means to integrate spatial and ecological approaches to management. In a watershed, for example, it is possible to simultaneously manage terrestrial and aquatic ecosystems while also maximizing the effects on water resources. Given the importance of water resources to humans and the need to conserve the biodiversity of aquatic systems, watersheds are sound units of forest management for conservation purposes. However, variability in the response to management within a watershed is a confounding factor. In large or complex watersheds, it is necessary to stratify the watershed by life zone (*sensu* Holdridge 1967), geology, soil type, and land cover in order to identify environmental conditions that are as similar as possible and which are likely to respond uniformly to natural or management disturbances. Such an approach will identify legacies of land uses and vulnerability to future disturbances that affect the function and structure of the management unit. In short, by identifying homogeneous units of ecological function, the heterogeneous and less predictable responses of complex watersheds to natural or anthropogenic disturbances are minimized (Lugo et al. 1999a, 1999b).

Within any hillslope in the watershed, the catena is a practical and ecologically valid spatial unit for stratifying the watershed (Weaver 1987; Scatena and Lugo 1995). A catena is a topographic continuum from ridge to slope to valley (figure 7-2) that is interconnected by the mass transfer and exchange of water and material, resulting in distinct edaphic and geomorphic conditions at each level. In general, soil properties (e.g., moisture, grain size, nutrient and carbon content) and forest attributes (e.g., tree density and basal area, rates of tree mortality and primary productivity, species richness) differ in a predictable manner among levels of the catena in any given watershed. For example, stands located on ridges behave differently from those found growing on slopes or in valleys. In the LEF, stands located in valleys within the Luquillo Mountains have more access to nutrients and water and exhibit faster turnover of biomass, whereas those found on ridges have slower biomass turnover rates and accumulate more biomass (Scatena and Lugo 1995). When ridge areas are modified by management or disturbance, the effect will be the transport of materials to adjacent slopes and valleys by downslope movement. The convergence of ecological function, environmental conditions, and biotic responses to these conditions according to the level of the catena—that is, ridges, slopes, or valleys—translates into a higher likelihood that the responses of stands to management interventions will be similar and have low variability within a topographic sector of the catena.

Although a catena is useful for categorizing different ecological conditions between ridges, slopes, and valleys, catenas with different elevations, aspects, or underlying geology can behave differently. In addition, the patterns of variation along catenas in the Luquillo Mountains (discussed above) do not necessarily occur in catenas in other locations. The important management guideline is the need to recognize and group similar ecological conditions within watersheds so that management actions will be as effective as possible by targeting the treatments to the capabilities of the sites. Hillslope catenas can provide one framework for defining ecological conditions.

A watershed contains many catenas, and a watershed approach to conservation allows a focus on the interaction between land and water resources, rather than the

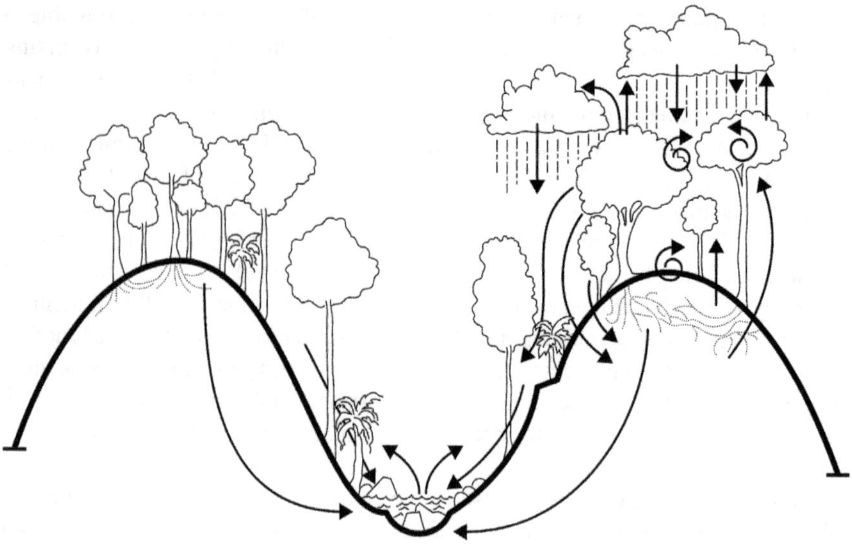

Figure 7.2 A catena in the Luquillo Experimental Forest illustrating variation in vegetation physiognomy and the fluxes of mass and nutrients (from Lugo and Scatena 1995).

consideration of each as a separate ecosystem. The importance of a watershed approach to understanding land-water relations has been well established in other LTER sites, such as Hubbard Brook (Likens and Bormann 1995). The land-water connectivity is observed clearly when rainfall produces a connective water surface layer, rivers flood over flood plains, materials are leached from terrestrial to aquatic systems, or organisms move between the land and the water. The movement of waters from terrestrial to aquatic systems influences the food supply and water quality for stream organisms. The movement of stream waters over flood plains influences the nutrient status and productivity of terrestrial ecosystems. Understanding this connectivity and its effects on the biota leads to insights into watershed management. In addition to watershed processes that feature water moving downstream from headwaters to coastal waters, the movement of organisms uphill during annual migrations contributes to downstream-upstream connectivity (Pringle 2000b).

A terrestrial example that illustrates one of the threats to upland forests from events in the lowlands is the movement of introduced species and terrestrial wildlife from adjacent urban systems into the Luquillo Mountains. This is the case with *Syzygium jambos*, an introduced tree species that is spreading upstream into the riparian areas of montane streams, and even into the mature forest (Brown et al. 2006). The movement of the Pearly-eyed Thrasher (*Margarops fuscatus*) from the lowlands to the uplands of the Luquillo Mountains (Arendt 2006) and that of the roof or black rat *Rattus rattus* (Odum et al. 1970a; Weinbren et al. 1970) are examples for animal populations.

The movement of organisms is a pathway of influence to upland systems through their surrounding interfaces (land-water, land-land, and land-air). For example, wildlife from the Luquillo Mountains supplement their diets in the lowlands. Puerto Rican parrots fly to agricultural fields in search of food (Snyder et al. 1987), as secondary forests within and outside the Luquillo Mountains have higher fruit production than do mature forests within the mountains (Lugo and Frangi 1993). Perhaps a landscape mosaic of forest types and ages is important for the support of native wildlife in the Luquillo Mountains.

Turbulence at the Interfaces within Management Units

Management unit connections with adjacent ecosystems and the atmosphere must be taken into account when designing conservation activities. It is helpful to visualize a management unit as a multidimensional volume having multiple interfaces with adjacent ecosystems and the atmosphere. The interfaces are surfaces where two ecosystems or sectors of ecosystems meet and interact through the exchange of materials, energy, and information. A sharp environmental gradient occurs at an interface, or, in some cases (as in the aerobic-anaerobic interface), there is a discontinuity in ecological space. As we show below, high flux rates and, at times, high turbulence characterize many of the processes at the interfaces. The dynamics of interfaces are best exemplified by atmospheric gas and wind interactions with forest canopies at the atmosphere-canopy interface. Gaseous diffusion between leaf surfaces and the atmosphere controls ecosystem productivity, and strong winds dissipating energy against the canopy transfer biomass to the forest floor and represent a major disturbance of Caribbean forests. Tropical forest management must attend to the processes at the interfaces of management units, including their coupling to climatic and atmospheric circulation patterns (Lugo and Scatena 1992).

In chapter 3, we document three of many possible interfaces within individual ecosystems where ecological processes occur at particularly rapid rates. These are the terrestrial-aquatic, aerobic-anaerobic, and canopy-atmosphere interfaces. The terrestrial-aquatic and the canopy-atmosphere interfaces are spatial interfaces (as is the interface between a stream and the groundwater), but the aerobic-anaerobic interface is a functional interface that can occur at any place (including spatial interfaces) and any time depending on environmental conditions (McClain et al. 2003). Thus, aerobic-anaerobic interfaces can occur within the canopy, within the soil, in wetlands, or at the edges of streams and other aquatic ecosystems. Managing ecosystems requires that managers be alert to the shifting positions of interfaces both in time and in space. The aerobic-anaerobic interface gains importance as well, given its significance to the production of methane and other greenhouse gases.

At the terrestrial-aquatic interface, or riparian zone, at the lower end of catenas, McDowell et al. (1992, 1996) documented dramatic changes in both the form and the total concentration of inorganic nitrogen and dissolved organic matter across redox gradients. Decades of research on temperate watersheds have shown that the maintenance of a vegetated buffer zone is critical to maintaining water quality in forested and agricultural landscapes (Peterjohn and Correll 1984; Simmons et al. 1992; Triska et al. 1993; Lowrance et al. 1997; Naiman et al. 2005). More recent work in an urban setting shows that even in heavily managed landscapes, the role of the riparian zone in reducing nutrient loading to surface waters can be critical (Groffman et al. 2002).

Data from the LEF and the Amazon basin suggest that in humid tropical climates, riparian processes are especially important in maintaining water quality (McDowell et al. 1992, 1996; McClain et al. 1994). In both sites, remarkable transformations of nitrogen (N) occur in the riparian flood plain (nitrate [NO_3-D] loss and ammonium [NH_4^+] accumulation), and a significant N loss (40 percent or more) occurs at the narrow groundwater-stream interface. These plot-scale studies have been expanded to a reach-scale analysis in the Río Icacos basin of the Luquillo Mountains. Chestnut and McDowell (2000) found, by comparing upslope groundwater and stream chemistry, that the stream export of N would be six to eight times as high in the absence of N losses at the stream-groundwater interface. The export of dissolved organic carbon was reduced fourfold by passage through the riparian zone.

The management implications of these results are striking. Riparian zones (the terrestrial-aquatic interface) represent biogeochemical "hot spots" that have a disproportionate effect on watershed nutrient losses relative to their spatial area (McClain et al. 2003). They are also critical for maintaining aquatic habitat (Heartsill-Scalley and Aide 2003) and aquatic food webs (Covich and McDowell 1996). When maintained in a vegetated state, riparian zones maintain water quality and terrestrial and aquatic species diversity in humid tropical environments. They are critical and priority areas of tropical landscape management (McDowell 2001). Luquillo LTER research on the structure and composition of riparian zones has provided insight into how land managers can define the appropriate width for riparian zones based on local ecological conditions (Scatena 1990).

Silver et al. (1999) demonstrated the importance of the aerobic-anaerobic interface in soils when they discovered periodic anaerobic conditions in all forest types of the Luquillo Mountains. They found high rates of methane production during oxygen-free periods. This finding in turn led to the discovery of a new pathway of the N cycle associated with the aerobic-anaerobic interface (Silver et al. 2001). Because the dissimilatory reduction of NO_3^- to NH_4^+ without oxygen is a N-conservation pathway that favors N immobilization by plants, these aerobic-anaerobic interfaces will feed back to regulate plant growth rates by affecting the nutrient supply and root respiration rates. The aerobic-anaerobic interface will also influence the vegetative community composition, as only plants adapted to periodic anaerobic conditions will survive. Survival requires specialized structures such as aerenchymatous tissue, pneumatophores, lenticels, stilt roots, etc. (Benzing 1991). Management of the aerobic-anaerobic interface is possible using a variety of approaches. One is the manipulation of water levels and water turnover. Slowing down water turnover or increasing water levels favors anaerobic conditions, and the opposite actions favor aerobic conditions. Other management mechanisms involve manipulating vegetation in riparian zones and wetlands or the alteration of river channels or the topography.

Recognition of the high rate of fluxes at ecosystem interfaces is of paramount importance to forest conservation (Hunter 1990; Saunders et al. 1991; Silver et al. 1996b). In Puerto Rico, Silver et al. (1996a) focused attention on the capture, retention, transfer, and recapture of nutrients at the atmospheric-terrestrial, biotic (organism-organism), plant-soil, and terrestrial-hydrologic interfaces. Organisms or attributes of forests at these interfaces included epiphytes and nitrogen-fixing organisms in the atmospheric-terrestrial interface; live plant tissue and dead plant tissue in biotic interfaces; root mats, fine roots, mycorrhizae, and symbiotic nitrogen fixers in the plant-soil interface; and coarse woody debris, roots, and soil microbes in the terrestrial-hydrologic interface. Many disturbance forces dissipate energy at these interfaces, thereby affecting critical sectors of the ecosystem. For example, hurricane winds directly interact with the canopy-atmosphere interface, and floods affect riparian and aerobic-anaerobic interfaces (chapter 6). Interactions between disturbance forces and the inherent high rates of fluxes at ecosystem interfaces mean that these are places where the biota are especially active and continually adjusting to turbulent environmental conditions (Silver et al. 1996b).

Another aspect of the functioning of interfaces within particular ecosystems is their spatial extent in terms of area or volume. For example, cloud penetration in forests can either be limited to the upper canopy or encompass the whole volume of the ecosystem down to the soil surface. This affects the distribution of epiphytes, which in colorado forests can extend their habitat from the canopy to the forest floor. The riparian zone, as well as the area involving aerobic-anaerobic interfaces, can expand significantly after heavy and prolonged rainfall (or contract with drought), with concomitant changes in aeration, nutrient distribution, and mechanical effects on the forest floor.

Disturbances and Species Life Histories

An understanding of the natural history of tree species is fundamental to species conservation. Information about natural history traits such as time to first reproduction, light tolerance, growth rates, regeneration potential, fecundity, and germination rates is required in order to propagate species, reforest lands, establish tree plantations, and successfully grow a tree crop. Research on the subject of tree life history in the Luquillo Mountains has been a priority, as evidenced by the 189 publications on this subject (figure 7-1). A manual with life history information for 101 tree species that grow in Puerto Rico (Francis and Lowe 2000) has been instrumental in supporting land management activities in Puerto Rico and throughout the Caribbean. Specialized manuals on the life history of urban trees (Schubert 1979) or best management practices (Wadsworth 1997; Ruiz 2002) have also been widely used. In spite of the importance and applicability of past life history research for plantation management and degraded land rehabilitation, such applications have traditionally not considered large and infrequent disturbances (*sensu* Dale et al. 2001).

Traditionally, the selection of tree species for plantations and reforestation has been based on their growth and yield potential rather than on their resistance to disturbances such as wind events. However, Liegel (1982, 1984) found that there were differences among species in their resistance to high winds. When planted together in mixed-species plantations, individuals of *Pinus oocarpa* exhibited six times the mean blowdown and twice the survival and structural effects of individuals of *P. caribaea*. Not surprisingly, after Hurricane Hugo, pine plantations in the Luquillo Mountains were destroyed beyond repair, demonstrating the importance of understanding the role of disturbances when collecting life history information. Life history attributes also proved relevant to the response to hurricane winds of pioneer and nonpioneer tree species in the LEF (Zimmerman et al. 1994) and urban trees in San Juan and Florida (Duryea et al. 2006). In short, life history traits help us understand the mechanisms for species-specific responses to disturbances and allow managers to anticipate which species will cope better with conditions before and after the disturbance.

The life history traits of trees and other organisms exposed to large and infrequent disturbances include common characteristics that make organisms adaptive to surviving under a particular disturbance regime. For example, some tree species from hurricane-prone regions have relatively short life spans, an early age of first reproduction, leaf heterophylly (ability to grow both shade-tolerant and shade-intolerant foliage and to change quickly between the two types), advanced regeneration, conspicuous sprouting, tree unions, and low ratios of crown to stem area (Lugo and Zimmerman 2002). These adaptations by species to environmental conditions might or might not be special adaptations to hurricanes, but they underscore the importance of understanding ecological rhythms (discussed later) and the life history characteristics of species that are targeted by managers.

Understanding the Limits of Forest Resilience

The forests of the Luquillo Mountains have been disturbed experimentally with ionizing radiation, chemical defoliants, clearcutting, selective cutting and planting, canopy removal, the manipulation of wood input, and fertilization (chapter 4). Scientists have also documented the effects of wind, rainfall, landslides, the conversion to different types of agriculture, and road construction (chapter 5). After each of these events, much of the forest cover and structure has returned to predisturbance conditions, but at different rates, and with different species composition (Lugo et al. 2000; Lugo 2004; Lugo and Helmer 2004). Despite the differences, however, generalizations about the resilience (*sensu* Holling 1973; Carpenter et al. 2001) of these ecosystems, even if tentative, are possible.

At least three aspects of resilience need attention from a conservation perspective. The first is the level of resilience, which differs among the components of a particular ecosystem (Zimmerman et al. 1996). For example, after Hurricane Hugo and experimental harvests, tabonuco forests at the watershed scale exhibited high resilience in leaf area index, litterfall, and root production, but lower resilience in species composition (Ewel 1977; Devoe 1989; Silver 1992; Scatena et al. 1996; Lugo et al. 2002; Beard et al. 2005). At the catena level within a watershed, however, litterfall recovered at a slower rate in the riparian zone than in the ridge and mid-slope areas (Vogt et al. 1996). At the watershed scale, biomass and basal area resilience were intermediate in response to hurricane disturbance. The resilience of biomass after a hurricane differed depending on the legacies or residuals existing at each site, with biomass recovering rapidly in areas with legacies of higher soil N from past land uses (Beard et al. 2005). Second, the state of the ecosystem relative to legacies and the level of maturity influences its resilience. For example, the resilience of forest biomass differs with the type of prior land use (Silver et al. 2000) or nutrient legacies from prior land uses (Beard et al. 2005). Biomass accumulated slowest after pasture abandonment, at a medium rate after abandoned agricultural crops, and fastest for prior coffee plantations (Silver et al. 2000). The resilience of forests on sites previously occupied by coffee plantations was attributed to high levels of soil N from nitrogen-fixing trees (*Inga* spp.) planted to provide shade to coffee trees (Beard et al. 2005). Finally, the level of resilience is influenced by the nature of the disturbance in relation to a particular disturbance event. For example, given the same conditions and timing, a forest might exhibit high resilience to wind disturbances but low resilience to landslides, because landslides remove soil and significantly delay succession (Walker 1999), whereas wind mostly knocks down biomass. Similarly, different places within a landslide are differentially resilient because of the distribution of organic residuals (chapter 5).

Each ecosystem structural component has a particular time scale at which it cycles (Scatena 1995). For example, leaf biomass cycles within a year, whereas woody biomass turns over at a scale of decades, and certain soil carbon fractions cycle much more slowly. The evaluation of resilience involves both slow and fast response variables (Carpenter et al. 2001), which have distinct effects on ecosystem resilience. Fast variables allow for a rapid response to disturbance and quick readjustment. Slow variables exhibit less of a short-term response but are critical for the long-term persistence of ecosystems. The rate of turnover of ecosystem compartments is a function of ecological space. In general, for a given compartment, tropical ecosystems have faster rates than temperate ecosystems, and within the tropics moist forests have faster rates than dry forests (Brown and Lugo 1982).

Soils in the tabonuco forest, for example, are not permanently affected by the formation of a single experimental canopy gap, as the regrowth of biomass is extremely rapid (Silver 1992). After chronic use of the same soil in agriculture and pastures, however, land degradation produces arrested succession and a shift in the ecosystem state (Silver et al. 2000). In the Luquillo Mountains, it took 6 decades of forest development to convert pastures into mature species-rich closed-canopy forests (Silver et al. 2000, 2004). Returning forest cover to sites such as pastures with arrested succession requires the repair of the soil structure and fertility (slow variables), which can be accomplished through the planting of selected tree species effective at ameliorating soil chemical characteristics (Parrotta 1995, 1999).

Rates of ecosystem processes are useful measures of resilience, but they can be deceiving when used to compare systems. Based on differences in net primary productivity and response to fertilization, Waide et al. (1998) concluded that low-elevation forests had greater resilience to hurricanes than high-elevation forests (table 7-2). However, because the biomass of high-elevation forests is lower than that in low-elevation forests, the ratio of primary productivity to biomass is similar in the two forests (0.053 and 0.044), suggesting similar rates of biomass turnover (19 to 22 y) and therefore resilience (table 7-2). If a forest failed to recover a sufficient level of biomass between hurricanes, it would not persist in the disturbance regime of the Luquillo Mountains over time.

Table 7.2 Biomass, productivity, and potential resilience of low- and high-elevation forests in the Luquillo Experimental Forest. Time for aboveground biomass turnover is the ratio of aboveground biomass to net aboveground biomass production. Values are estimated at approximately steady-state conditions. After disturbances, biomass decreases and rates are assumed to increase at both elevations, leading to roughly equal resilience (rate of return to predisturbance biomass). All data are from Weaver and Murphy (1990)

Parameter	Low-elevation forest	High-elevation forest
Net aboveground biomass production (Mg ha^{-1} y^{-1})	10.5	3.7
Litterfall (Mg ha^{-1} y^{-1})	8.6	3.1
Aboveground biomass (Mg ha^{-1})	198.0	83.0
Time for aboveground biomass turnover (y)	19	22

The estimate of biomass recovery in table 7-2 supports the notion that these forests are capable of restoring the biomass of mature states in about 20 to 25 years, or less than the 60 years available between successive hurricanes. Sixty years is the average return rate for direct hurricane hits in the Luquillo Mountains (Scatena and Larsen 1991). Therefore, at the current rate of primary productivity, these forests have sufficient time to reestablish the biomass expected of mature forests in the hurricane belt. This also implies that although the biomass and primary productivity of the lowlands and uplands of the Luquillo Mountains are different, both forest types are able to bounce back between disturbance events. This is to be expected, as both forest types have occurred under the same disturbance regime for millennia, and are therefore likely to contain species that have evolved to reach some level of biomass maturity in the average time interval between natural disturbance events (Lugo et al. 2002). Those species incapable of reproducing between disturbance events are unlikely to persist.

After a large and infrequent disturbance, what one sees is the devastation caused by the hurricane, landslide, fire, or volcanic explosion. The inherent tendency of forest managers is to restore the damaged forest or somehow accelerate succession in order to avoid further site degradation. The tangles of weeds that naturally invade affected sites immediately after the disturbance event do not appear sufficient in relation to a restoration goal. If one understands and considers the limits to resilience as discussed in this section, or nature's self-organization capacity (next section), a different management strategy emerges. When restoring landslides in the Luquillo Mountains, the Forest Service learned that those slides that were seeded with introduced herbaceous plants recovered at a slower rate than those allowed to regenerate through natural succession. Vegetation recovery after the eruption of Mount St. Helens (Dale et al. 2005) and littoral recovery after the *Exxon Valdez* oil spill (Parker and Wiens 2005) followed the same principle. A critical management strategy involves recognizing when to intervene and when not to. This requires some understanding of the ecological condition of sites after a disturbance and evaluating whether natural resilience mechanisms will restore the original ecosystem, or whether resilience capacity somehow has been lost and, therefore, management interventions are required. As with all management interventions, the management objectives and underlying ecological assumptions play a critical part in deciding what to do (Parker and Wiens 2005).

Self-Organization

Self-organization is the sorting of species through a variety of replacement mechanisms after a disturbance event. It occurs at all scales of size and complexity, from the small fragmented patch within an urban environment to large landscapes responding to a variety of natural and anthropogenic disturbances. Self-organization is a component of ecosystem resilience (Gunderson 2000) because it allows the system to reform and continue to function in a particular state (e.g., forest or pasture). After a disturbance or large shift in ecological space, many plant species will germinate, and organisms of all types will arrive at a site via either natural or artificial vectors. These species will compete for space and resources with species that survived the disturbance event. In the absence of human management, environmental conditions result in some species establishing, growing, and reproducing while competition eliminates less well-adapted species. Positive species interactions (facilitation) can also influence species composition (Callaway and Walker 1997). The resulting mix of species at the site emerges from interspecific interactions in the context of new environmental and structural conditions. The process is known as self-organization because the systems that prevail are self-reinforcing and might be different from those of the past (Odum 1988, 1989).

There are numerous examples of self-organization in the Luquillo Mountains. For example, after Hurricane Hugo devastated whole forest stands, thickets of weeds and vines covered hectares of land, and the expectation was that it would be difficult for the forest to regenerate through such a tangle of vegetation (Chinea 1999). But trees did grow through the weeds (many of which were introduced species), and a closed-canopy forest emerged with a species composition similar to that of the prehurricane forest (Scatena and Lugo 1995; Scatena et al. 1996).

Self-organization also occurs after human intervention with succession. In the 1930s to 1950s, the Forest Service planted introduced and native tree species in many degraded sites, and these plantations were allowed to develop with minimal management after the first decade (Marrero 1950). Today, mature forests occur in these former degraded lands, but their species composition is different from the original and from that of adjacent native forests of similar age (Lugo 1992; Silver et al. 2004).

When allowed to proceed unhindered, self-organization is an effective conservation tool because it is powered by natural forces rather than by the costly subsidies required for human intervention (Odum et al. 2000). Self-organization allows natural forces to determine the trajectory of ecosystem change at no cost to managers other than time. The challenge is to recognize when to allow self-organization to continue unabated ("rolling with the punches") and when to nudge it one way or another in order to meet conservation or economic goals (Lugo 1988).

Applications of Long-Term Ecological Research

Research in the Luquillo Mountains has been instrumental in developing management systems for tropical forests within and outside Puerto Rico. Long-term ecological research adds new information and insight to the body of knowledge concerning tropical forest dynamics. Hereafter, we use all available information (LTER and non-LTER) to illustrate applications of research to six tropical forest conservation issues: ecosystem services, restoring land to productivity, restoring biodiversity to deforested sites, water abstraction and conservation of biodiversity, the management of roads and landslides, and living with environmental change. This information and its application are of universal value and are not limited to the insular conditions of the Luquillo Mountains. Sites in particular systems might differ in terms of landscape types, disturbance forces, ecological cycles, or species being managed. However, regardless of the geographic location, the scientific principles and their applications to management or conservation do not change.

Ecosystem Services

Ecosystems have always provided people with products and services, but the market system focused on the value of products (e.g., wood, meat, fruit) while assuming services (e.g., clean water and air) were free externalities to economies. Ecosystem services include those that are species dependent and those that are whole-ecosystem dependent. The cleaning of rocks in streams by shrimp (discussed below) or the aeration of soils by earthworms are ecosystem services that are species dependent. In contrast, clean water at the bottom of a watershed is a service provided by the whole ecosystem, and the species composition within the watershed is not as critical. Today, ecosystem services are increasingly recognized as important to the quality of life and the functioning of economies (Daily and Ellison 2002; Scherr et al. 2004). Moreover, many believe that the ability of an ecosystem to deliver services is limited or that it might change when ecosystem states change, making the "free externality" an uncertainty for many economies. As a result, the importance of nonmarket economics is gaining importance in the analysis of economic development (Costanza and Daly 1987). Unfortunately, no agreement exists concerning ways to quantify the value of nonmarket services of ecosystems to the economy.

EMERGY Evaluation

Managing a complex system with multiple objectives and constraints, such as a national forest, requires an integrated assessment and decision-making tools. A diversity of metrics is required in order to assess ecosystem processes, impacts, and services that occur over a wide range of time scales. EMERGY analysis (Odum 1996) has been developed as a tool for quantifying and clarifying the relationships between environmental services and their effects on ecological space over different time scales. EMERGY, an energy-based measure of resource contribution and influence, is defined as the solar energy required in order to produce a flow or storage of another type of energy. EMERGY evaluation evolved from the need to develop a system of economic analysis in which market and nonmarket products and services are measured with the same units (Odum 1996). It uses a systems approach to the analysis of energy and dollar flow through natural and economic systems (figure 7-3).

Using figure 7-3, one can estimate the relative importance of natural energy inputs to the watershed (left side of the diagram) and the energy inputs to a watershed that cost money (i.e., purchased energy) (right side of the diagram). In this example (units of sej ha^{-1} y^{-1} × 10^{12}), the ratio of natural (2,765) to purchased (2,447) energy input is 1:1. In addition, one can compare the sum of all the outputs of watershed services (bottom right; 8,336) and compare that with the sum of all inputs (natural plus purchased = 5,212). This results in a ratio of 1.6. The ratio would be 3.4 if one compared outputs in services (8,336) to human investment in purchased energy inputs (2,447). In summary, figure 7-3 shows a watershed in which the contribution of natural energy to its functioning and value to society is higher than that of the purchased energy. It also shows that the return of the investment for management is positive (between 1.6 and 3.4 times).

This systems-based approach, which was developed by H. T. Odum (1971), partly from his research on energy and mass flows in the Luquillo Mountains, can be used to evaluate the flows and storages within a defined ecosystem boundary. The synthesis requires an inventory of all forms of energy and all types of materials in every flow within the system and their expression in units of solar EMERGY or EMdollars (EM$). "EMdollars" refers to the proportion of the system's buying

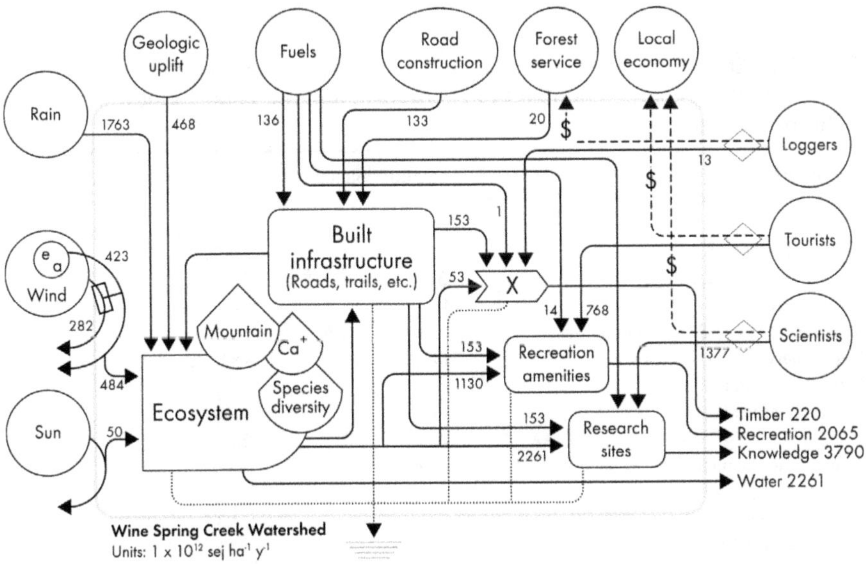

Figure 7.3 Systems diagram of the ecological–economic interface of the Wine Spring Creek watershed, North Carolina (from Tilley and Swank 2003). Circles represent outside inputs to the watershed. Symbols are according to Odum (1996). More details can be found in the text.

power that is supported by the system's solar EMERGY and is a standard way of expressing a monetary value of services and storages not traditionally accounted for in standard economics. Examples of natural values or services are carbon sequestration, the provision of water quality, transpired moisture, photosynthetic production, and forest biomass. Any transaction involving market or nonmarket economics is expressed in EMERGY units in order to ensure that the human and natural economies are measured with common units.

Two examples of the application of EMERGY analysis to forestry issues are those of Odum et al. (2000) and Tilley and Swank (2003). Odum et al. (2000) showed that natural reforestation in Puerto Rico returned a net benefit with an accumulation of national wealth 15 to 25 times the money invested over the 10 to 20 years required for canopy closure. Tilley and Swank used EMERGY analysis to evaluate alternative uses of forested watersheds in the southern Appalachian Mountains and found that ecosystem services were worth 40 times the amount of money invested in their management. The public value of annual forest production in southern Appalachia is compared to that in the LEF in table 7-3. Odum (1996) offers many more examples of these values.

In the Luquillo Mountains, EMERGY analysis has been used to analyze the forest as a system coupled to the larger ecologic-economic system of Puerto Rico (Scatena et al. 2002). This analysis indicates that rainfall and tectonic uplifting are the largest environmental inputs to the forest. The interaction between these inputs produces an erosional landscape in which the EMERGY of biological processes are less than the EMERGY associated with the physical and chemical sculpturing of the landscape. This erosional landscape undergoes a systematic shift from physically

Table 7.3 EMERGY values associated with ecological processes and human activities in the Luquillo Mountains of Puerto Rico and the Wine Spring Creek watershed in North Carolina.

Parameters	Luquillo Mountains	Wine Spring Creek
Ecological processes		
Precipitation, chemical	2,128	1,603
Precipitation, geopotential	1,372	525
Transpiration	744	440
Stream discharge, chemical potential	2,128	2,055
Net primary productivity, live biomass	744	982
Human activities		
Research information	685	3,445
Recreation	3,451	2,065
Annual road maintenance (EM$ km^{-1} y^{-1})	22,323	4,136
EMERGY indices		
Social flows/environmental flows	3.5	1.03
EM$ ratio, 1992 (sej $dollar^{-1}$)	1.64×10^{12}	1.12×10^{12}

Based on Tilley and Swank (2003) and Scatena et al. (2002). Except where indicated, units are in EM$ ha^{-1} y^{-1}. EMERGY is the available energy of one kind, previously used—either indirectly or directly—to make a product or service. EMERGY dollars (EM$) is the total amount of dollar flow generated in the entire economy supported by a given amount of solar EMERGY input. The unit sej is solar emjoules, an energy flow unit corrected for its solar energy equivalence (Odum 1995).

to biologically dominated EMERGY flows with decreasing elevation (Scatena et al. 2002). The lower elevations are relatively efficient at accumulating biomass, and the upper elevations are relatively efficient at accumulating soil organic matter.

A similar analysis done for a watershed in North Carolina allows a comparison with the Luquillo Mountains (table 7-3). The comparison shows that the wet Luquillo Mountains have greater physical and chemical erosion rates and support a considerable amount of recreation. One result of the differences in the ecological characteristics of the two locations is that the cost of maintaining roads in the highly erosive Luquillo Mountains is over five times that in the Wine Spring Watershed in North Carolina.

In the Luquillo Mountains, about 59 percent of the total annual EMERGY flows are from human activities. Most of this is from tourism. This relatively high level of human activity reflects the high value that society places on the national forest. However, the amount of environmental work that was required over millennia in order to build the natural capital of the forest was 9 to 50 times the current market value of property adjacent to the Luquillo Mountains (Scatena et al. 2002). This underestimated value of the land surrounding the Luquillo Mountains is one reason for the tremendous urban pressure on them (Lugo et al. 2004). The analysis also indicated that the environmental effect of extracting water is almost 300 times that of building roads, in large part because the effects of downstream releases of treated sewage are almost nine times those associated with water removal. Natural and artificial wetlands and the reduced use of artificial channels and other structures that promote runoff at the expense of recycling water might conserve water resources and offset these effects of treated sewage (Kent et al. 2000).

Species-Dependent Ecosystem Services

Tropical forest managers face the challenge of dealing with species-rich ecosystems, usually with little information about the ecological role that each species plays in the functioning of the ecological system. The traditional solution has been to focus attention only on those species with known commercial value (usually a few tree species). As research reveals the ecological importance of individual species or groups of species, managers must pay attention to a greater fraction of the species components of forests. One approach is to use life history information to help focus management activities (see above). As we show below, life history information for species other than timber trees is rapidly accumulating for the Luquillo Mountains. Some groups of species draw the attention of managers based on their native/nonnative status. Here we discuss the role of species in providing desired ecological services as another criterion for identifying species that might require the attention of managers. We end this section with a few guidelines for managing species for ecological services.

Freshwater shrimp species provide many ecosystem services for recreational users within the Luquillo Mountains and for residents and users of surrounding ecosystems. Recreational shrimp harvesting is one example that has been studied by Luquillo LTER scientists (Kartchner and Crowl 2002). Palaemonids (i.e., shrimp such as *Macrobrachium* spp.), which reach lengths > 230 mm, are a particularly prized catch in recreational freshwater shrimp fisheries. Atyid and xiphocaridid shrimps, with small maximum lengths (<100 mm), are also caught but are viewed as less challenging and exciting to capture and eat. Although shrimping provides an ecosystem product, the primary purpose of most shrimping in the Luquillo Mountains is recreation. Interviews with fisherman who harvest Luquillo shrimps reveal that shrimp harvesting is a relaxing hobby, a family tradition, or a means of interacting with nature.

Providing an opportunity for recreational shrimping might be the service that is most obvious to visitors of the Luquillo Mountains. However, shrimp also provide an additional service relevant to recreation, as revealed by basic research on stream ecology. The feeding activities (scraping and brushing) of atyid shrimp rapidly remove sediment, organic matter, and algae that accumulate on rock surfaces between storms and provide the service of water and habitat cleansing (Pringle and Blake 1994; Pringle et al. 1999; March et al. 2002; see chapter 6) (figure 7-4). After one storm, 440 to 620 g m^{-2} of dry mass accumulated on rocks where shrimp had been excluded, whereas shrimps in control treatments removed the sediment within 30 hours. To those visiting the Luquillo Mountains streams, this service contributes to the aesthetics of montane rivers (i.e., clean boulders and clear water). This service might also contribute to safety. If the stream bottom is visible and less slick after algae and organic matter are removed from rock surfaces, hiking in the streams is less treacherous.

In mountain streams outside the Luquillo Mountains, shrimp provide similar cleaning services, but over a wider range of conditions resulting from human effects. Survey work combined with *in situ* shrimp exclosure and enclosure experiments have shown that atyid shrimp are able to clean rock surfaces even in streams with relatively high percentages of agricultural land cover in the catchments (15 to 45 percent) and, consequently, high levels of dissolved nutrients (up to 1500 μg nitrate-N L^{-1}) (Greathouse 2006b). This indicates that shrimp foraging can reduce or prevent algal blooms in mountain streams that drain mixed land uses.

In mid- and low-elevation stream ecosystems immediately surrounding the Luquillo Mountains, grazing by goby fish (E. Greathouse, personal observation) and snails (March et al. 2002; Blanco and Scatena 2005) can provide rock-cleaning services, replacing the function of shrimp foraging observed at high elevation. However, in mid- and low-elevation rivers and estuaries, freshwater shrimps still provide important food resources for predatory fishes (e.g., mountain mullet [*Agonostomus monticola*], sleepers [e.g., *Gobiomorus dormitor*], and American eel [*Anguilla rostrata*]), which are important for both recreational and commercial fisheries (Covich and McDowell 1996; Nieves 1998; March and Pringle 2003). The "freshwater" larval shrimps pass through the estuaries and river as they migrate to saltwater, and they return to freshwater as juveniles (March et al. 1998). Larval shrimps are an important food source for estuarine fishes (Freeman et al. 2003), judging from the prevalence of shrimp larvae in the guts of fish collected from an estuary prior to the construction of a low-head dam (Corujo Flores 1980). The low-head dam now removes an estimated 34 to 62 percent of drifting larval shrimp (Benstead et al. 1999). The results of recent Luquillo LTER studies in the same estuary indicate that despite the reduction in shrimp availability, juvenile

Figure 7.4 Shrimp exclusion in the Quebrada Sonadora at El Verde, Luquillo Experimental Forest, Puerto Rico. The electric treatment (rear hoop) excludes shrimps and fishes and contains high levels of benthic particulate material, as evidenced by the dark brown depositional layer. In contrast, the unelectrified control treatment (the hoop in the foreground) has no visible particulate material because of shrimp foraging activities. (Photograph by Catherine M. Pringle.)

freshwater shrimps are still a key food resource for estuarine fishes, including fishes of commercial importance such as *Bairdiella* spp. and *Centropomus* spp. (Smith 2008).

Long-term ecological research has also shown that the ecological services provided by freshwater shrimps have significant implications for stream ecosystem function (chapter 6). Humans benefit from these functions because they are part of the regulating services of ecosystems (Millennium Ecosystem Assessment 2005). When shrimp are present, levels of benthic inorganic sediments, organic material, carbon, and nitrogen in streams are lower and less variable than when shrimp are absent. Moreover, higher rates of leaf decomposition and export of fine particulate organic matter at base flow occur in the presence of xiphocaridid shrimps. Through these effects, shrimps influence the availability of nutrients to other trophic levels. We are still elucidating the linkages between shrimps and ecosystem components and services of high practical and aesthetic value. For example, what are the effects of shrimp foraging on aquatic insect emergence or on terrestrial predators, including Puerto Rico's flagship species, coquí frogs (*Eleutherodactylus portoricensis*)?

Research in the Luquillo Mountains has discovered other examples of species-dependent services—for example, the roles of (1) ferns in the stabilization of soil and the reduction of downstream sedimentation (Walker 1994; Shiels and Walker 2003), (2) coquís in accelerating plant growth and decomposer activity after hurricanes (Beard et al. 2002, 2003), (3) earthworms and other soil fauna in accelerating litter decay or aerating soils (González and Seastedt 2000, 2001), and (4) *Cecropia* in the restoration of forest conditions after disturbances (Brokaw 1998).

The examples presented above illustrate that individual species or functional groups of species perform important ecological services. People benefit from these services by obtaining food or fiber products, clean water and air, aesthetics, etc. Thus, managers must attend to the sustainability of these species and species groups and not focus only on trees with economic value. In the Luquillo Mountains, life history information on shrimp species has suggested practices that help sustain these populations. For example, as we discuss below, water abstraction schedules and the location of intakes are designed to sustain shrimp populations. Proposals for stream channel modification, which affects habitats and migration routes, must take into consideration and mitigate any effects on migrations of aquatic fauna. Similarly, the reproductive success of coquís could be enhanced with properly located artificial nesting sites (Stewart and Woolbright 1996). Finally, managers now have justification for the protection of specialized habitats such as roosting sites for bats, wet locations for earthworms, or riparian zones that harbor a disproportionate concentration of species that deliver ecological services.

Managing Time: The Relevance of Ecological Cycles and the Time Tax

Ecological processes, and thus their associated ecosystem services and products, are time dependent. The reliable functioning of complex ecosystems such as those of the Luquillo Mountains depends on numerous cyclic and noncyclic processes that occur at different time steps ranging from nanoseconds to millennia (Scatena 1995). Those processes that occur rapidly are said to involve rapid variables because the variables have a fast turnover rate; this is the case with photosynthetic or respiration rates or leaf turnover. There are processes that take decades or centuries to unfold, and they are said to involve slow variables; the recovery of soil nutrient and organic matter supplies after degradation is an example of this. There is a need to be aware of the types of variables manipulated by managers because there is less risk when fast variables are manipulated than when the manipulation involves slow variables (Crépin 2007). A mistake takes longer to rectify when dealing with a slow variable than when manipulating a fast one. In essence, time management is a critical component of ecosystem management because conservation actions are time dependent, as they require the manipulation of ecological processes with different time steps. Human activity influences ecological processes that are either too fast (less than days) or too slow (over centuries) for effective management. For example, the lunar-controlled changes in plant secondary chemistry that affect herbivory rates on palm leaves occur at scales of weeks (Vogt et al. 2002a). The harvesting of palms (as nontimber forest products) can therefore affect herbivore population dynamics. Opportunities for management differ with the time step involved, with particular challenges when dealing with time intervals that are too short or too long.

EMERGY analysis, which uses time as a source of embodied value, and concepts of ecological space are useful in defining and evaluating natural resource management problems. However, managers need operating rules to explicitly address the problems. One method, developed directly from the time-sequence approach suggested by Scatena (2001), is to coordinate management activities with ecological cycles, which are defined broadly as biological or environmental processes that repeatedly occur at definable intervals. These cycles are commonly related to climate (e.g., phenological patterns), daily processes (e.g., diel or circadian cycles), life histories (e.g., reproduction or feeding behavior), disturbance (e.g., successional cycles), and physical and biogeochemical cycles (e.g., tides).

Knowledge of ecological cycles in the Luquillo Mountains has been applied in order to enhance the conservation of endangered species such as the Puerto Rican Parrot, allocate water supply to municipal watersheds, satisfy recreational demands, and boost productivity of natural forests and plantations (table 7-4). Although the timing of management activities with the ecological cycles has definable benefits, this type of dynamic management is not without costs or tradeoffs. It requires knowledge of the response of ecosystems and organisms to environmental

Table 7.4 Examples of ecological rhythms used in natural resource management in the Luquillo Mountains. (From Scatena 2001.)

Ecological cycle or life history trait	Management objective	Management guidelines	Source
Annual reproductive cycle and daily foraging behavior	Protect endangered Puerto Rican Parrot	Limit management activity by season and time of day	USDA, FS Southern Region 1997, Snyder et al. (1987)
Diurnal habitat preference	Define instream flow requirements for resident biota	Develop nighttime and daytime instream flow requirements	Johnson and Covich (2000)
Diurnal and seasonal larval release	Maintain migratory aquatic biota	Restrict water withdrawals by night and season	Benstead et al. (1999)
Weekly and seasonal recreational-use patterns	Maintain aquatic recreation downstream of water intakes	Restrict water withdrawals during summer weekends	Scatena (2001)
Diurnal and seasonal dissolved oxygen cycles	Minimize eutrophication by sewage plant effluent	Reduce releases during nighttime and low-flow periods	Scatena (2001)
Annual growth rates and light responses	Improve timber yields	Selective thinning by density and species	Wadsworth (1997)
Regeneration in natural tree fall gaps	Sustain timber resources	Harvesting that mimics natural gaps	Odum (1996)
Annual phenology and response to canopy opening	Sustainable mahogany plantations	Limited harvesting during seed set, thinning after canopy disturbances	Wang and Scatena (2003)

conditions; institutional memory; and an administrative commitment to long-term environmental monitoring, data analysis, and synthesis.

In some instances, time constrains the achievement of conservation goals. This might happen if the conservation goal requires the completion of ecological processes that take longer than the time that is available or desirable for the manager. When this happens, managers attempt to accelerate the process, such as with the establishment of plantations to increase the rate of wood production, or when artificial reforestation accelerates natural forest reestablishment. Sometimes it is impossible to accelerate processes because of specific conditions of natural or anthropogenic origins. The "time tax" is a symbolic way of expressing the idea that when we affront nature—eroding the soil, for example—a certain amount of time is required in order to rehabilitate productive conditions or at least overcome degradation (Lugo 1988). Often, if the disturbance is sufficiently intense, the system flips (or bifurcates, *sensu* Ludwig et al. [2002]) to a different state further removed from the familiar or desirable state (Crépin 2007). For example, the vegetation on a site that was deforested and farmed does not return to forest after abandonment but becomes permanent pasture or grassland. When the forest fails to redevelop, succession is arrested, and the pasture system might persist for decades with less structural development than the original forest ecosystem.

Restoring Degraded Lands to Productivity

Deforestation is an acute anthropogenic disturbance that, when followed by crop production or livestock grazing, results in a chronic condition that lasts as long as farmers or ranchers perceive a net benefit from their efforts. In many tropical areas of high rainfall and heavily leached or thin soils, farming is not sustainable without significant inputs of fossil fuels (Hall et al. 2000). Consequently, farmlands in these regions are abandoned relatively quickly (years to decades) when there is no access to external subsidies. Abandoned lands are usually degraded to some degree; they are eroded, and remaining soils are compacted or nutrient depleted. Without human intervention, they might slowly return to forest cover or remain as pastures in a state of arrested succession. The establishment of secondary forests following the abandonment of agricultural lands is a pantropical phenomenon responsible for the current era of secondary vegetation that characterizes the tropics (Brown and Lugo 1990). In many places, where human-induced fires or cattle grazing prevent forest reestablishment, the burning and grazing activities exacerbate land degradation, and the vegetation does not return to a forest physiognomy after abandonment (Goldammer 1992; Laurance et al. 1997). Therefore, the restoration of forest productivity requires an understanding of the types and intensities of past and present disturbances on the sites of interest.

The first step in restoring degraded lands to productive forest in the tropics is to limit chronic disturbances such as fire and cattle grazing that hinder forest succession. Many sites will gain forest cover in the absence of cattle and fire, whereas others will not recover because soils might be eroded, compacted, or nutrient depleted, or there might be less soil water available. Some organisms also might be affected by conditions such as high temperatures and low humidity in disturbed sites. Seed predators might consume the few available tree propagules, or distances might preclude effective seed rain to the site. In short, ecological space will have shifted to nonforest conditions, weakening or destroying the resilience of the original system.

Tree planting might be required in order to reestablish forests on sites with ecological conditions that have caused the persistence of pastures. The success of tree planting as a management strategy has been pantropical and includes the recovery of forests on pastures dominated by *Imperata*, a grass thought to prevent forest reestablishment once it gains hold of a site (Kosonen et al. 1997). Determining which tree species to plant and where to plant them requires scientific knowledge. In Puerto Rico, foresters planted both native and introduced tree species for timber and restoration purposes (Marrero 1950). During decades of study, nearly 500 native and introduced tree species were tested before it was determined that 32 species (mostly introduced species) were best for timber production (Wadsworth 1995). Today in the Neotropics, the use of tree plantations of native tree species for the sole purpose of establishing forest conditions has gained popularity and success (Butterfield and Fisher 1994).

In the LEF, forest cover was restored through a combination of tree planting on pastures (Silver et al. 2000, 2004), line plantings in degraded forests (Weaver and Bauer 1986), tree planting in agricultural fields (taungya system) (Weaver 1989), and natural succession (Lugo 1992). Initial plantings of tree species were designed to increase wood production by establishing tree species with high timber yields (e.g., pines, mahogany, *Eucalyptus*) (Francis 1995). This approach was successful in establishing high-yielding forest plantations (Lugo 1992) and productive secondary forests (Silver et al. 2000, 2004) that improved the soil organic matter content and nutrient accumulation (Lugo et al. 2004). Many of these positive results occurred over 6 decades (Silver et al. 2004) at the La Condesa site within the LEF (table 7-5). This site, which was planted with introduced and native species, was a carbon source for several decades before becoming a carbon sink (Silver et al. 2004). The loss of soil carbon originating from pasture soils was initially faster than carbon gain by plantation trees. These secondary forests and others in the vicinity support wildlife, improve soil conditions, and protect water supplies and watershed values. The data in table 7.5 also show that the forest stand at La Condesa had more aboveground biomass and productivity but less root biomass than a nearby native forest of similar age.

Rates of succession after pasture abandonment are variable in sites where secondary succession advances without human intervention (Zimmermann et al. 1995; Aide et al. 1996; Silver et al. 2000). The land cover at the time of abandonment influences the speed of succession. More rapid succession and accumulation of biomass characterized sites that were less disturbed at the time of abandonment, as opposed to sites that were pastures or bare soil at the time of abandonment. Succession in heavily disturbed sites was slower than in native forests after a natural disturbance (Aide et al. 1995; Silver et al. 2000). Such differential rates of succession and biomass accumulation in relation to past land use represent the time tax mentioned above. Learning when to enter the successional cycle in order

Table 7.5 Characterization of a 55-year-old subtropical moist forest restored through the planting of 13 tree species on a degraded pasture at La Condesa, Luquillo Experimental Forest, Puerto Rico.

Parameter	La Condesa	Mature tabonuco
State variable: Mg ha^{-1}		
Soil organic matter	204	161
Aboveground biomass	160	80
Fine root biomass	2.5	9.0
Annual rates		
Accumulation of tree species	1 species y^{-1}	1 species y^{-1}[a]
Accumulation of aboveground biomass	2.8	2.6
Litterfall	10.6 to 12.9	9.7 to 11.3
Aboveground net primary productivity	14.9	12.3
Fine root productivity	0.3	—[b]
Net primary productivity	15.4	—
Soil organic matter accumulation	1.8	—
Net soil organic matter sink	1.1	—

Data are from Silver et al. (2004). Data are based on trees > 9.1 cm in diameter at breast height and a forest area of 4.64 ha. Roots were sampled to a depth of 10 cm, and soil to 1 m. Mature tabonuco data are from Lugo (1992) for a native stand > 50 y at a similar elevation.

[a] Mean for native secondary forest succession (Lugo et al. 1993).

[b] No data.

to promote recovery is an example of managing with ecological cycles, and it requires an understanding of the rates of succession under different ecological conditions.

Of social interest is the role that *parceleros* (farmers who lived on LEF lands at the time of land acquisition by the Forest Service) played in the restoration of lands in the Luquillo Mountains (Wadsworth 1995). The Forest Service allowed *parceleros* to continue planting crops in their fields as long as they also planted and cared for trees selected by the Forest Service. This tree-planting method is known as the taungya system, and it is effective in plantation establishment (Weaver 1989). When planted trees reached heights that prevented farming, the Forest Service acquired lands for the *parceleros* outside the boundaries of the LEF, and the *parceleros* moved. This partnership allowed the agency to reforest larger areas at a faster rate than would have been possible otherwise. It also gave local communities the opportunity to adjust to the new ecological space, and it provided *parceleros* with continuous access to areas of ecological space suitable for farming via geographic relocation.

Restoring Biodiversity to Deforested Sites

The first critical step in restoring biodiversity to deforested sites is to reestablish forest cover. The expectation is that with the reestablishment of forest cover, other components of biodiversity (understory plants, soil organisms, and fauna) will follow. The reestablishment of forest cover can be accomplished by allowing tree regeneration to proceed naturally or by planting native or introduced trees (Lugo

1997; Lugo and Helmer 2004). Once trees are established, the next critical step is canopy closure. A fundamental change in the microclimate of the forest occurs following canopy closure. A closed-canopy forest is effective in attracting seed vectors and other organisms that require shade and cooler temperatures for reproduction and growth. The species richness of the understory in a young forest is higher than in the canopy, and many understory species will enter the stand's canopy in the future (Wadsworth and Birdsey 1983; Lugo 1988, 1992). In Puerto Rico, the number of tree species per hectare increases at a rate of one tree species per year of succession (Lugo et al. 1993), so that within 5 decades or so the number of tree species per unit area approaches the level for undisturbed stands (50 to 60 species per hectare; Lugo et al. 2002). Similar rates have been observed in landslides (Guariguata 1990; Myster and Walker 1997).

In the La Condesa site, the Forest Service planted 13 tree species on an abandoned pasture. Fifty-five years after the original planting, this site supported 70 tree species with diameter at breast height (dbh) ≥ 9.1 cm in an area of 4.64 ha; the tree species consisted of a new mixture of native and introduced taxa (Silver et al. 2004). The processes of tree species enrichment—through planting followed by natural secondary succession, or by secondary succession following abandonment of sites—is common in Puerto Rico, where the abandonment of agricultural lands is widespread (Lugo and Helmer 2004; Lugo and Brandeis 2005).

Ways in which reforestation restores ecosystem functions are shown in table 7-5, in which La Condesa is compared with a nearby native forest of similar age. At La Condesa the forest ecosystem is operating at rates comparable to those of mature native forests. In fact, most measures of ecosystem function in restored sites at the Luquillo Mountains show little if any differences compared to native forests of similar age (Lugo 1992).

The establishment of tree cover and the succession of species that follow alter site conditions such as the microclimate or the quantity and chemistry of organic inputs to soils (Zou and González 1997). These alterations accelerate the establishment of other types of organisms at the site. As a result, the overall site biodiversity is enhanced. The change through succession in species composition, biomass, and the function of earthworms is particularly well documented in the Luquillo Mountains (González et al. 1996; Zou and González 1997; González and Zou 1999; Liu and Zou 2002; Sánchez de León et al. 2003). With the conversion of pastures to forests, the richness of earthworm species increases owing to the presence of both introduced and native earthworms. The restoration of earthworm species after tree establishment is an example of how restoring tree cover affects other components of forest biodiversity. A similar pattern is known for the restoration of understory plant species (Lugo 1992) and birds (Cruz 1987, 1988).

Water Supply and Conservation of Freshwater Biota

In the early 1990s, approximately half of the water draining the Luquillo Mountains was diverted on a daily basis for human consumption (Naumann 1994). At the time, this amount of water supplied the needs of about 22 percent of the island's population. By 2005, daily withdrawals accounted for 70 percent of median daily water flow (Crook et al. 2007). However, owing to poor maintenance of the delivery infrastructure, nearly half of the water is lost in transmission or stolen before it is delivered to customers, and the actual population served is only about 11 percent (Ortiz Zayas and Scatena 2004).

The method most commonly used for harvesting water from rivers and streams is to dam the main channel and extract water for as long as is necessary or possible (figure 7-5). Many streams in the Luquillo Mountains are pumped dry by this procedure, particularly during dry periods and extended droughts (Naumann 1994; Crook et al. 2007). Moreover, streams and rivers in Puerto Rico and the rest of the Caribbean have been modified heavily with structures such as concrete channels, dams, or water intakes (figure 7-6) (Pringle and Scatena 1999a, 1999b; March et al. 2003; Greathouse et al. 2006a). Dam construction is an increasing conservation problem of global proportions, particularly in developing countries (figure 7-7) (Pringle et al. 2000a). Damming and overharvesting of river water are fundamentally incompatible with the conservation of stream biota because dewatering streams or converting them to ponds and reservoirs changes the nature of the stream ecosystem, with consequent negative effects on the survival of native freshwater biota. However, the application of a combination of science-based measures and technology offers hope for the conservation of stream biota.

Important animals (shrimps, snails, fishes) within river ecosystems of the Luquillo Mountains are migratory, with most of the migrations occurring at night (March et al. 1998; Benstead et al. 1999; Johnson and Covich 2000). Adults reproduce in freshwater, and larvae drift downstream to estuaries. Juveniles later return

Figure 7.5 The low-head dam on the Río Fajardo. Low-head dams can cause mortality of drifting larval freshwater shrimps. (Photograph by Kelly Crook.)

to freshwater (Benstead et al. 2000). Dams obstruct these migrations, facilitate predation on sensitive life stages, reduce biodiversity above the dam, and favor the establishment of introduced species. High dams without releases of water over their spillways are impenetrable barriers that prevent migration and eliminate native fish and shrimp from upstream reaches (Holmquist et al. 1998), with cascading effects on food web dynamics and ecosystem-level processes (Greathouse et al. 2006b). Furthermore, as dams and water abstraction reduce stream flow, saltwater intrusion from the ocean reaches 2 to 3 km upstream, with concomitant predation on freshwater species by saltwater fish (Pringle 1997). Because of the migratory life cycles of many stream organisms, such impacts in the lowlands are transmitted upstream, affecting not only lowland river and estuarine reaches but also the biota of headwater systems such as those in the Luquillo Forest (Pringle 1997).

Issues of water quality compound the conservation problem associated with water supplies. Industrialization often produces new compounds (such as pesticides and fertilizers) that are introduced into tropical aquatic systems, where they are novel to the biota and might be toxic (Meybeck et al. 1989). In Puerto Rico and

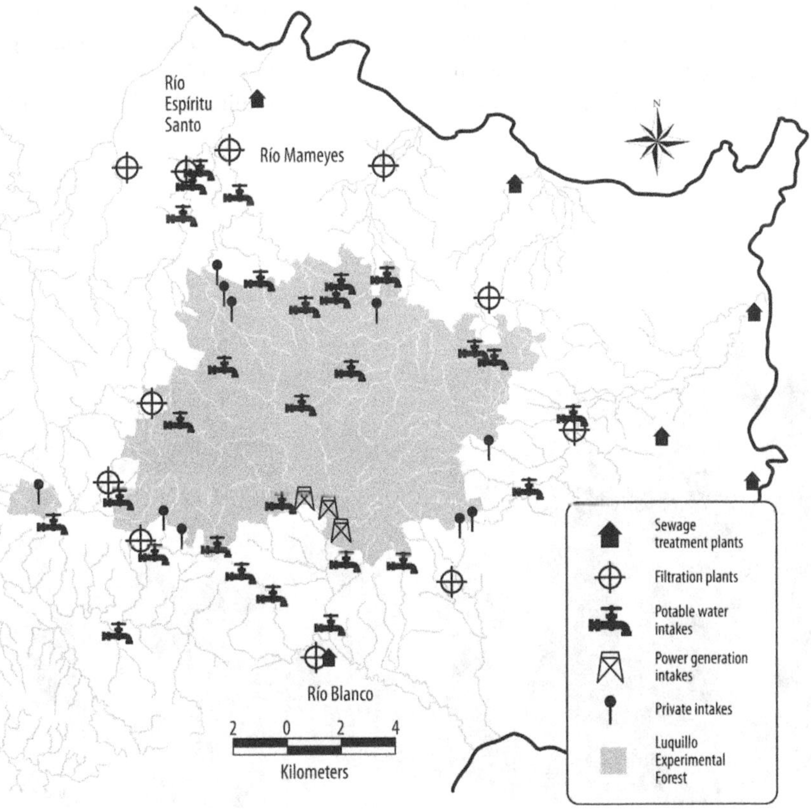

Figure 7.6 Sites of water withdrawals (intakes for potable water, power generation, and private), sewage treatment plants, and filtration plants in the Luquillo Experimental Forest. (Modified from Pringle 2000b.)

much of the tropics, the input of raw sewage into aquatic systems poses a severe water-quality problem (Hunter and Arbona 1995; Jobin 1998). This situation is exacerbated by water intakes, which reduce the capacity of river flow to dilute sewage by reducing freshwater runoff. The dire environmental situation of tropical rivers can prevent the use of rivers and coastal estuaries for tourism and recreation (Pringle and Scatena 1999b). With the singular exception of the Luquillo Mountains (Pringle and Scatena 1999a, 1999b), the scientific understanding needed in order to conserve water supplies and maintain quality in streams of the Caribbean is generally lacking. However, our research in the Luquillo Mountains is relevant to the management and conservation of streams and rivers not only in the Caribbean and Latin America but also in the temperate zone (Pringle 2000a, 2000b, 2001; Pringle et al. 2000a, 2000b; Postel and Richter 2003).

Earlier, we suggested that an understanding of ecological cycles and ecological life histories was important for ecosystem management. This is best illustrated by research on shrimp that showed that predictable temporal cycles characterize reproduction and migration. These predictable ecological cycles allow the development of pumping schedules that minimize the entrainment of biota. Larval shrimp drift during the night, with a nocturnal peak slightly after dusk (March et al. 1998). Water abstraction entrains larvae, juveniles, and even adults, significantly reducing population levels (Pringle 1997; Benstead et al. 1999). For example, in the lower Río Espíritu Santo, water abstraction caused 42 percent mortality of drifting first-stage

Figure 7.7 The Lago Guayo dam, a large structure that blocks shrimp and fish migration and causes the upstream decimation of native shrimp and fish populations in Puerto Rico. (Photograph by Effie A. Greathouse.)

shrimp larvae via entrainment during downstream migration (figure 7-8). All drifting larvae were killed during the dry season (Benstead et al. 1999). By stopping water abstraction for 5 hours during peak migration periods, larval mortality due to entrainment can be reduced to only 11 to 20 percent (Benstead et al. 1999).

The effects of dams can also be mitigated by fish ladders and the maintenance of a minimum flow over the dam (March et al. 2003). However, all dams alter riverine conditions. This problem can be averted through alternative water intake designs. A different design of the water intake pipes at the Río Mameyes eliminated the need to use dams, and thus allowed water abstraction without changing the free-flowing nature of the river. The new intake design was supplemented with administrative actions based on hydrological estimates, coupled with life-history information of organisms. Studies calculated the minimum flows required in order to maintain the ecological functioning of streams (Scatena and Johnson 2001). The enforcement of minimum flows would facilitate ecosystem functioning during drought periods and ensure a sustainable balance between human use and conservation of the biota, in addition to attendant ecosystem services. Recently, some new water storage reservoirs in Puerto Rico have been constructed off the river channel so as to avoid damming the river and exposing the reservoir to excessive sedimentation.

Roads and Landslides

Roads and landslides are land covers in which the soil is either covered by pavement (roads) or lost to such a degree that the saprolite or deep soil strata are exposed. The conditions resulting from both types of disturbance are inhospitable to plant growth (Walker 1999). Roads and landslides cover about 1 percent of the surface of

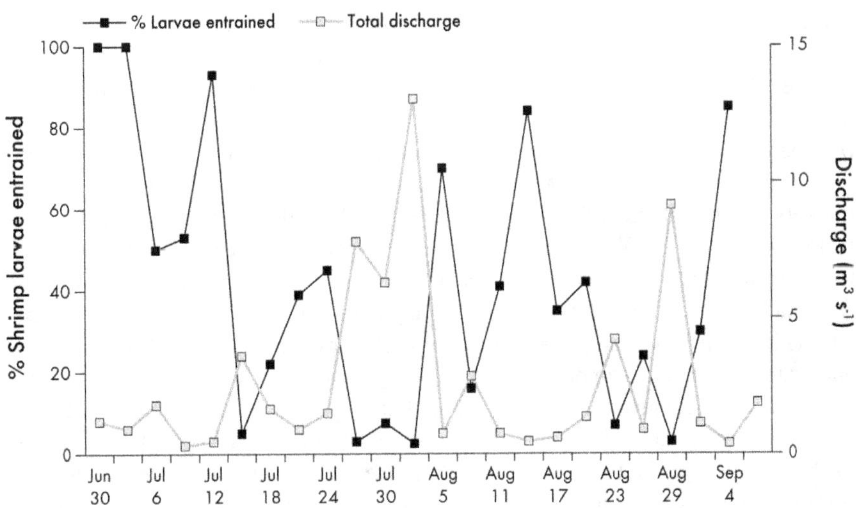

Figure 7.8 Percentage of larval entrainment by a major water intake on the Espíritu Santo River (left axis) and discharge over the dam (right axis) during the period of June 30 to September 4, 1995. (Modified from Benstead et al. 1999.)

the Luquillo Mountains (Larsen and Torres-Sánchez 1992) but are the most severe disturbances in the forest because of the burial or removal of all organic matter (Walker et al. 1996). Landslides potentially redistribute nutrients in the Luquillo Mountains, as phosphorus-rich mineral soil is exposed or added to streams, and carbon- and nitrogen-rich organic matter is buried to varying depths. However, such redistributions have yet to be quantified. Studying ecosystem recovery on roads and landslides provides valuable insights about primary succession, ecosystem assembly, and revegetation that are applicable to many severe disturbances, including in urban areas, construction zones, mined areas, or flooded areas (Walker and del Moral 2003; Walker et al. 2007). Contrasts between secondary succession on abandoned farmlands and in hurricane-affected forests in Puerto Rico and primary succession on roads and landslides clarify the role of soils and surviving vegetation in recovery following disturbance.

Road construction and poor maintenance enhance the likelihood of landslides, as >80 percent of landslides in the Luquillo Mountains occur along road corridors (Larsen and Parks 1997). Roads also act as corridors for the movement and establishment of introduced species (Walker and Boneta 1995), although little spread of introduced grasses into adjacent forests has been detected in the upper Luquillo Mountains (Olander et al. 1998). Successional changes on abandoned paved roads at lower elevations in the Luquillo Mountains occur quickly. Within 11 years of road abandonment, the litter mass, soil bulk density, soil moisture, soil organic matter, and total soil nitrogen reached adjacent forest levels (Heyne 2000). At higher elevations, changes were slower on road fill (Olander et al. 1998) and differed from those noted in a study conducted at a lower elevation by Heyne (2000). The species composition, however, did not resemble that of adjacent forests in any of the forests in the 60-year chronosequence studied (Heyne 2000).

Plant succession on landslides is governed by slope stability and nutrient availability (Guariguata 1990; Walker et al. 1996). Upper slip faces are often unstable and low in nutrients, so only climbing ferns that spread vegetatively can survive (Walker 1994). The middle chute zone of landslides is generally more nutrient rich but very unstable, so shrubs and trees might grow but often reslide. The deposition zone is the most stable and fertile zone, and succession to forest can occur there within 50 years (Zarin and Johnson 1995a, 1995b; Myster and Walker 1997). Fern thickets can inhibit tree colonization (Walker 1994), and seed dispersal can be slow from landslide edges (Walker and Neris 1993). Large landslides are generally slower to revegetate than narrow or small landslides that are affected by local slumping of residual forest soil and short-distance propagule dispersal. Frequent resliding, limitations in nutrient and propagule dispersal (Fetcher et al. 1996; Shiels et al. 2006), and growth inhibition can delay forest recovery for centuries (Walker et al. 1996).

Landslide management requires the application of knowledge from the ecological, engineering, and geological sciences, as each makes a significant contribution to the others. The first conservation alternative for dealing with landslides is to prevent them, because once they occur, the time tax of recovery can be long. Once the landslide occurs, the management options include stabilizing the landslide and either (1) allowing succession to proceed naturally or (2) accelerating natural succession or revegetating the slide through planting.

Because most landslides are associated with roads, their prevention is best achieved through better road design and maintenance, and especially improved drainage design. The management of landslides in roadless areas involves forecasting which conditions are likely to create them. Larsen and Simon (1993) suggested that landslides in the Luquillo Mountains are triggered by storms that exceed 100 to 200 mm of rain. Shallow landslides result from storms of short duration, whereas longer storms result in much deeper landslides.

Efforts to stabilize landslides in the LEF include minimal treatments with mulches, silt fences, and fertilizer to encourage plant growth. Also used are contouring, plantings, and jute cloth coverings. Greater interventions include the use of gabions and the redirection of water flow along lined channels (M. Ortíz and P. Ríos, USDA Forest Service, personal communication, 1999). The long-term success of such efforts depends largely on stochastic factors such as rainfall and the rate and direction of succession that result in a stabilizing vegetative cover.

Sometimes landslides involve massive land movements, such as the 300,000 m^3 landslide that bisected State Road 191, the main road that traverses the Luquillo Mountains. Given the importance of the road, government efforts to stabilize the landslide and restore the road were undertaken at a cost of millions of dollars. In this particular case, it was impossible to stabilize the slopes with the resources available, and the restoration efforts had to be abandoned after several years. Today, some 30 years later, the road remains closed, and natural succession has taken over the site of the landslide. The example illustrates the limits of human manipulation of natural phenomena.

The prediction and manipulation of succession on landslides is still problematic. Adding perches that attract birds onto landslides facilitates propagule and seed arrival (Shiels and Walker 2003). Fertilizers can increase plant growth (Fetcher et al. 1996) but might promote a dense cover of ferns or grasses, which hamper tree establishment and prevent the longer-term landslide stabilization provided by trees (Walker et al. 2010). Natural stabilization by thicket-forming ferns appears to be the best long-term path to forest recovery on landslides. Sloughing of nutrient-rich forest soil (Shiels et al. 2006) and the decomposition of pioneer tree ferns and *Cecropia* trees (Shiels 2006) eventually lead to forest development on landslides in the Luquillo Mountains. Tree planting can speed primary succession and is most successful when proper soil and symbionts such as mycorrhizae are provided (Lodge and Calderón 1991; Myster and Fernández 1995). Matching plant species to appropriate microsites and layering exposed surfaces with moss to provide better germination sites for seeds might aid landslide recovery (Myster and Sarmiento 1998). Also, the redirecting of roads can reduce the angle of the slope and thus the potential for landslides.

The experience in the Luquillo Mountains and elsewhere in Puerto Rico raises the issue of the inevitability of the association of roads and landslides. In the karst region, for example, chronic landslides raised the cost of constructing 1 km of road to over $30 million. Despite this expenditure of funds and decades of roadwork, PR 10 remains unstable (Lugo et al. 2001). Given these experiences, planners and road builders have two main options when dealing with wet, steep terrain. First, they need to select road alignments carefully and arrive at realistic cost-benefit analyses in order to avoid the costly surprises of PR 10. Second, they can opt to avoid building the road altogether, as was decided in the Luquillo Mountains with the repair of PR 191.

Regardless of the choices made, roads will continue to be ubiquitous components of landscapes and present problems to managers. Lugo and Gucinski (2000) proposed a unified approach to the management and analysis of the function and effects of roads on forested rural landscapes. The approach is based on considering roads as ecosystems (techno-ecosystems) and conducting analyses of road ecology prior to making policy or management decisions. An ecosystem approach to road issues has four advantages: (1) it allows for the analysis of all types of roads, irrespective of geographic location; (2) it provides a holistic framework for analyzing all aspects of roads, from their alignment to their operation and decommissioning, as well as all road functions, irrespective of value judgments; (3) it provides a holistic focus to road management; and (4) it supplements landscape management approaches based on spatial concepts. Lugo and Gucinski (2000) recommended five precautions when evaluating road ecosystems:

1. Identify the type of road under consideration.
2. Differentiate the effects and conditions of individual road segments from those of road networks.
3. Be explicit about to which phase of road development the argument applies, because different phases of development have different effects on the landscape.
4. Ascertain the age of the road and evaluate the degree of landscape adjustment to the road, and vice versa.
5. Do not prejudge human-induced changes in landscapes as automatically good or bad for the ecology or economy of a region.

Figure 1 in their work (Lugo and Gucinski 2000) is a useful model of a road ecosystem that applies to most tropical conditions.

Living with Environmental Change

The emergence of humans as the dominant agents of change on Earth was one of the most biologically significant consequences of the Industrial Revolution. As an example of the extent to which humans influence the biosphere, Sanderson et al. (2002) quantified the human footprint on the planet and found that humans directly influence 83 percent of the land surface and 98 percent of the area where it is possible to grow rice. In contrast, protected areas represent less than 10 percent of the land surface of Earth, and it is obvious that protected areas alone cannot resolve environmental problems facing the world. The most practical alternative for achieving a sustainable future for humans is to learn to cope with environmental change and apply conservation measures to all lands and waters of the planet. In this section we present examples of the consequences of environmental change in the Luquillo Mountains and assess the usefulness of these examples for informing management beyond the Luquillo Mountains.

Current and future agents of environmental change in the Luquillo Mountains include natural disturbances, urbanization, land cover change, and climate change.

These agents of change are interconnected and known to influence the ecological space and development of the ecosystems of the Luquillo Mountains. For example, after Hurricane Hugo, when much of the forest in the Luquillo Mountains was defoliated, the forest experienced a drought because of the change in the cloud level over the forest (Beard et al. 2005; Heartsill-Scalley et al. 2007). Cloud level and associated water inputs are also influenced by urbanization in the lowlands and on the periphery of the mountains (van der Molen 2002). Urbanization creates a heat island that influences the elevation to which air must rise in order to form clouds (Malkus and Stern 1953). Climate change has affected and will continue to affect the Luquillo Mountains (Scatena 1998), but we do not know with certainty the magnitude or direction of the change. We can formulate scenarios of likely outcomes of environmental change using current knowledge. Here, we provide two likely scenarios that are already in progress in the Luquillo Mountains.

The first likely scenario is a progressive change in the species composition of forests and aquatic ecosystems. Changes in the species composition of ecosystems owing to introduced species are a ubiquitous result of environmental change in the Luquillo Mountains. The recent natural invasion of the Luquillo Mountains by an ecotype of Africanized bee (*Apis mellifera scutellata* [Ruttner]) is one example. Africanized bees have competitively displaced the preexisting (as well as introduced) honeybee species (*Apis mellifera*) in places without any obvious change in environmental conditions. Other progressive changes in species composition include the loss of amphibian species (Joglar 1998) and the spread of the invasive tree *Syzygium jambos* (Brown et al. 2006).

Ecologists argue about the role and causes of species invasions (Vermeij 1996; Lodge and Shrader-Frechette 2003; Lugo and Brandeis 2005), but they agree on the fact that the presence of introduced species increases with increasing anthropogenic disturbances. High dams are an example for aquatic systems. The dams create new aquatic environments, made by people, where introduced aquatic plant species such as *Eichornia crassipes* dominate. In streams above large reservoirs, migration failures cause reductions of native aquatic biota, which allow the spread of introduced aquatic species (Holmquist et al. 1998).

These progressive changes in species composition can lead to the emergence of new ecosystems that perform desired ecological services and support economic development (Lugo 1996; Lugo and Helmer 2004). The process is notable in degraded sites, but it also occurs, at a slower rate, in less disturbed conditions. Tropical plant and animal species, both native and introduced, have the capacity to invade most ecosystems and, through self-organization, form terrestrial and aquatic ecosystems of species mixtures that are new to the island (Lugo and Brandeis 2005). This "creativity" and adaptability of the biota in the face of significant environmental change provides examples and experiences that will be useful to other tropical countries where landscapes have not yet reached the levels of modification seen in Puerto Rico.

The second likely scenario deals with ecosystem-level adjustments that are likely in an environment with an increased frequency of disturbances and change (Lugo 2000). For example, an increased level of disturbances such as hurricanes would lead to the following outcomes:

- A larger fraction of the natural landscape will be set back in successional stage (i.e., there will be younger ecosystems). For example, modeling different hurricane intensities and frequencies showed that a range of forest types are possible, from mature forests with large trees in areas of low hurricane frequency to areas in which forest trees are not allowed to mature when hurricane frequencies are high (O'Brien et al. 1992).
- Forest aboveground biomass and height will decrease because vegetation growth will be interrupted more frequently or will suffer greater impact.
- Combinations of familiar species will change as species capable of thriving under disturbance conditions increase in frequency at the expense of species that require long periods of disturbance-free conditions in order to mature.

The mitigation of environmental change requires global measures because of the magnitude of the forces that regulate climate and land use. At local scales, individual countries have control over the management of land cover (Lugo 2002), which influences climate at mesoscales. Land cover management should pursue those options that reduce changes in atmosphere–land interactions and are more amenable to the movement of species into desirable and appropriate habitats as change proceeds. Thus, landscape management is a regional approach to mitigating the consequences of global environmental change.

Future Directions for Tropical Ecology

The tropics contain most of the world's biodiversity (Wilson 1988) and ecosystem types (Lugo and Brown 1991). The moist tropics alone support half of the world's population (Gladwell and Bonell 1990). Four-fifths of the world's population increase will occur in the tropics (Pereira 1989). The resulting mosaic of social and natural ecosystems is one of enormous complexity and interdependence. For tropical ecology, this scenario presents a formidable challenge. It requires a new vision and era of conservation. We need new ways to evaluate the increasingly intimate relationship between humans and their environment. In response to this reality, scientific societies are proposing new approaches to ecological research (Bawa et al. 2004; Palmer et al. 2004). These approaches, coupled with changes taking place in civil society, are harbingers of a new era of conservation.

A New Era of Conservation

The new era of conservation taxes our knowledge, understanding, and imagination. Understanding how ecosystems function and are assembled is the key to conservation success, a success best ensured through LTER of the type presented in this book. Aldo Leopold (1953) wrote that the first rule of intelligently tinkering with nature is to save all the parts, a task that is made more difficult with increasing human pressure on the biota. In order to prevail, we must do our utmost to avoid species extinctions. One strategy is the maintenance of global and regional interconnected systems of reserves (Andelman and Willig 2003). This solution is compromised by climate change, which shifts environmental conditions and can strand reserves in the wrong climatic setting. Climate change, for example, might convert moist forests to dry forests, thus endangering moist-forest species that cannot adjust to the reduction in water availability. Global change also endangers small reserves. Reserves with a large perimeter-to-area ratio are more vulnerable to invasion and edge effects.

A reduction in the pressures occurring on reserves is possible if we concentrate human activity in small areas (Lugo 1991). Such a solution currently requires increased supplies of fossil fuels to power the intense level of human activity and transport the vital materials needed to sustain urban systems (Odum and Odum 2001). However, the sustainability of fossil fuel supplies for powering urban systems is uncertain (Hubbert 1968; Campbell 1997). At the same time, we must be ready to deal with the consequences of increased concentrations of atmospheric carbon dioxide (CO_2) produced by the combustion of fossil fuels. Regardless of how we elect to arrange humans on the landscape, all lands and waters require conservation attention in order to sustain human activity and protect ecosystems and species.

The conservation of the biota is leading to new areas of scientific activity such as ecological engineering (Mitsch and Jørgensen 1989), ecological economics (Maxwell and Costanza 1989; Hall et al. 2001), and restoration ecology (Jordan et al. 1987). In all these new fields of science, a common denominator is the use of designed ecosystems to obtain needed products and services. No longer do we deal with natural ecosystems in the search for ecological solutions to human problems; we now manipulate and create new ecosystems for specific purposes, such as with the use of microbes and wetlands for treating sewage and cleaning water. The biota serves as a reservoir of genetic information; each species contains genetic combinations that allow it to function under particular sets of environmental conditions. Therefore, the biota has functional capabilities that humans can use judiciously in the design of new ecosystems. We could use a green infrastructure in cities, such as a vegetation wall to absorb sound, rather than the current inanimate, gray infrastructure built of concrete or steel. For flood control, flood plains or wetlands function as well as, or better than, concrete canals and reservoirs. New ecosystems, often containing introduced species, have been designed to repair degraded lands such as abandoned mines (Parrotta et al. 1997). Green infrastructure provides aesthetic and pollution-control services to cities, lowers the heat island effect, is self-maintaining, provides open green spaces that provide connectivity between natural habitats for the movement of organisms, and reduces pressure on native ecosystems. In short, the new era of conservation should be an era of the protection of biodiversity and intelligent tinkering for products and services and for coping with constant global change. This new era of conservation will require novel ways of evaluating ecosystems and their services and functions.

Uncertainty and Surprise

Uncertainty and surprise are inevitable aspects of the future that sometimes help, but usually derail, conservation plans. Natural ecosystems and anthropogenic landscapes are too complex to permit accurate forecasting based on current knowledge, but this does not mean that all surprises will be negative. In Puerto Rico, we were surprised by several ecological events that no one had predicted (listed below). Usually, once revealed, the surprise is easily explained, but in other examples, such as the extinction of frogs (see below), the explanation is still elusive. The lesson for conservation is to work on managing uncertainty while expecting surprises, with the understanding that the changes they entail are not necessarily detrimental to conservation objectives. The strategy must be to first assess the nature of the change, before making judgments about its value, and then be ready to adapt to the change, incorporating new knowledge into the conservation activity.

The following examples of surprises are presented in rough chronological order.

- The linking of tabonuco trees by root grafts, as revealed by the failure of tree poisoning in tabonuco trees, was a surprise to foresters employing the standard technique of poisoning selected trees in order to thin stands. At the time, it was not known that root grafts connect tabonuco trees (Lugo and Scatena 1995), and that nonpoisoned trees in the tree union could keep the poisoned trees and logged stumps alive for decades. The poisoning of root-grafted trees is no longer a management option.
- The high resistance of the rain forest to high levels of ionizing gamma radiation (Odum et al. 1970b) was a surprise, as pine forests in the United States had proven to be sensitive to this treatment (Woodwell and Rebuck 1967). This was one of the first experimental indications of the resilience of tropical forests.
- The development of species-rich understories under plantations of introduced pines was surprising because it was previously thought that monocultures of introduced species would inhibit understory development (Lugo 1992). This led to the use of introduced-species tree plantations to restore native tree species to degraded sites.
- The sudden increase in reproductive effort by wild populations of Puerto Rican Parrot after Hurricane Hugo was a surprise. Long-term study of these birds had consistently shown a low reproductive output in the wild (Snyder et al. 1987), but somehow the hurricane reversed the trend and mitigated to some extent the losses of birds during the storm. This observation provided clues for managing parrot reproduction in the aviary.
- The endangerment and local extinction of amphibians in places where the habitat has not changed (Joglar 1998) is a surprise with negative consequences, because species might be lost for reasons we still do not understand. The outcome has been to increase the monitoring of amphibian populations in undisturbed sites.
- Researchers had not anticipated the importance of the legacy of tree species' spatial distribution in response to past land uses (Thompson et al. 2002). Before this study we had not fully understood that knowledge of past land use history is a requirement for the interpretation of species distribution data.
- Life history studies yield many surprises regarding the adaptations of organisms to ecological space—for example, the 40-plus-year-old and woody

seedlings of *Manilkara bidentata* that suddenly responded to canopy opening by a hurricane (You and Petty 1991), or the extremely slow upstream migrations and age-size distributions of snails that reflected old age groups in such small organisms (Blanco and Scatena 2005). These surprises underscore the importance of slow variables to ecosystem management and illustrate biological responses (seedling growth) tuned to low-frequency environmental signals (hurricanes).

- Predictions of land cover change did not anticipate the collapse of agricultural activity (Lugo 2002), and thus forest cover increased island-wide in spite of increased population density. This surprise underscores our inability to predict the direction of major land use/land cover processes.
- The dominance of introduced tree species in the secondary forests of Puerto Rico was unexpected. Although introduced species have always been part of the flora and were known to compose 28 percent of the tree flora, scientists were not aware that introduced species were forming and dominating new forest types until island-wide inventories starting in 1980 demonstrated this (Birdsey and Weaver 1982). This finding led to the realization that natural processes of self-organization are already integrating the forest composition with past land uses and current environmental changes.
- The similarity of the turnover rates of biomass in elfin and tabonuco forests in spite of their differences in structure and productivity (table 7-2) explained how low-productivity forests can survive in environments with high levels of natural disturbances.

Adaptive management is the best way to deal with uncertainty and a surprise (Bormann et al. 1999). The core idea behind the concept of adaptive management is that we need to learn from and adapt to changes in ecological space. Conservation activities should be conducted as if they were an unfinished experiment. For example, as part of management, a disturbance is applied to a system with a purpose and an expectation of a product. The response of the system is not always as expected, and thus it has to be monitored and evaluated against the expectations. Long-term study and monitoring, as well as the scientific process, are critical for understanding those human activities that occur on large scales and impact ecosystems for a long time. Such types of study require institutions and procedures, such as sound data management and record keeping, that provide continuity to the study regardless of the turnover of people.

What Is Next?

Tropical science will continue to be challenged by the complexity of tropical ecosystems. This complexity compounds most research problems that we attempt to solve. The future requires more and better science in support of management and conservation policy. Such science needs to monitor the effects of human activities on the planet's ecosystems. In order for science to be effective in future ecosystem management, scientists must increasingly use the scientific approach to focus on the synthesis of knowledge and communicate its relevance to society. Scientists and managers must produce materials that are easy for nonscientists, such as policymakers and the public, to understand. More important, scientists must work in closer collaboration with nonscientific citizens (figure 7-9). According to Ludwig (2001), the era of management is over. He believes that the problems of resource conservation facing society are so complex that they cannot be solved by scientists alone or by any one sector of society. Instead, "scientists must be prepared to share their advisory and decision-making roles with a variety of interested parties and participate with them on equal footing" (Ludwig 2001:758). This has been put in practice in the Luquillo Mountains in the development of the land management plan for the El Yunque National Forest, with positive outcomes. During this effort, scientists maintained their objective approach to understanding and developed a relationship of mutual respect with forest managers. The persistence of humans and the ecosystems on which they depend requires a new global coalition with long-term

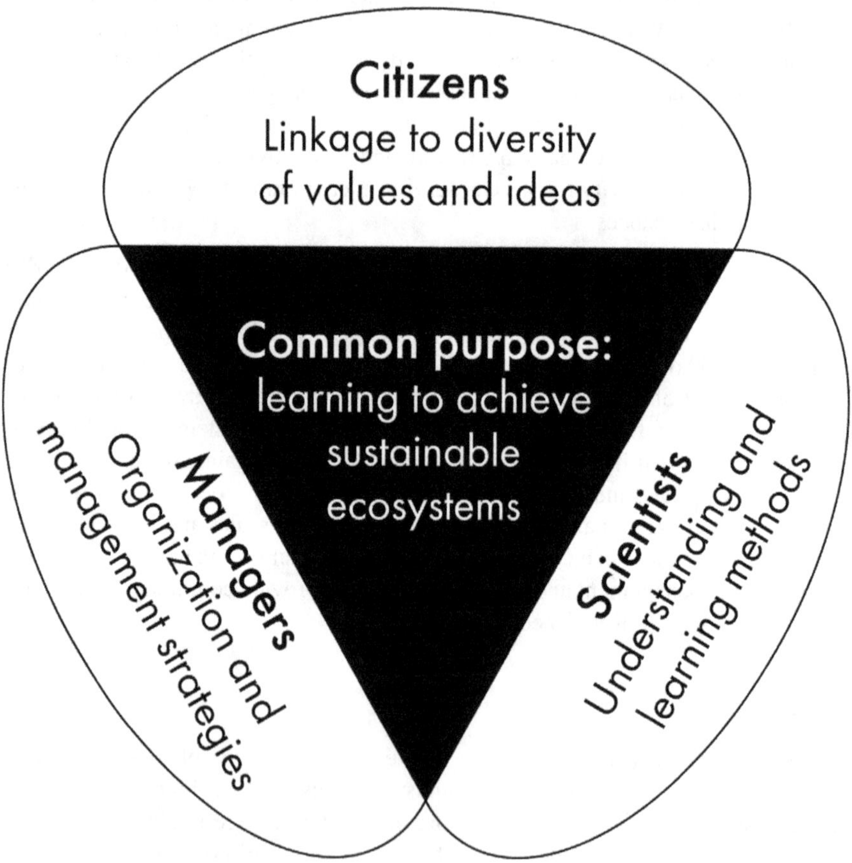

Figure 7.9 Learning as a common ground for building new, mutually beneficial relations among citizens, managers, and scientists in order to achieve sustainable ecosystems (Bormann et al. 1999).

understanding of contemporary systems and the mechanisms that support them. If this is achieved, a humane balance between use and preservation can be sustained for future generations in the face of risk and uncertainty.

Summary

Research in the Luquillo Experimental Forest has made significant contributions to the conservation and management of tropical forests and watersheds. The research has focused on species life histories and ecosystem processes at three levels of spatial organization over the long term. These spatial levels are the catena, the watershed, and the interfaces between different ecosystem components, such as the leaf–atmosphere, terrestrial–aquatic (riparian), or aerobic–anaerobic substrates. We discuss six examples of how research results have contributed to the addressing of specific management situations. These examples include the management of forests for ecosystem services, restoring degraded lands to productivity, restoring biodiversity to degraded lands, sustaining water supplies while conserving aquatic biodiversity, the management of roads and landslides, and living with environmental change. Moreover, the Luquillo LTER has contributed to changes in paradigms in the field of tropical ecology, particularly in terms of the importance of disturbances and the resilience of tropical forests as a result of both natural and anthropogenic disturbances. Future avenues of research activity will have to deal with novel environmental conditions, biotic surprises, and uncertainty. Managing under these conditions requires flexibility and support from cutting-edge research activity, which foresees a new era of conservation.

Literature Cited

Aide, M. T., J. K. Zimmerman, L. Herrera, M. Rosario, and M. Serrano. 1995. Forest recovery in abandoned tropical pastures in Puerto Rico. *Forest Ecology and Management* 77:77–86.

Aide, M. T., J. K. Zimmerman, M. Rosario, and H. Marcano. 1996. Forest recovery in abandoned cattle pastures along an elevational gradient in northeastern Puerto Rico. *Biotropica* 28:537–548.

Andelman, S. J., and M. R. Willig. 2003. Present patterns and future prospects for biodiversity in the Western Hemisphere. *Ecology Letters* 6:818–824.

Arendt, W. J. 2006. Adaptation of an avian supertramp: Distribution, ecology, and life history of the Pearly-Eyed Thrasher (*Margarops fuscatus*). General Technical Report IITF-GTR-27. Río Piedras, PR: USDA Forest Service, International Institute of Tropical Forestry.

Bawa, K. S., W. J. Kress, and N. M. Nadkarni. 2004. Beyond paradise—Meeting the challenges in tropical biology in the 21st century. *Biotropica* 36:276–284.

Beard, K., A. K. Eschtruth, K. A. Vogt, D. J. Vogt, and F. N. Scatena. 2003. The effects of the frog *Eleutherodactylus coqui* on invertebrates and ecosystem processes at two scales in the Luquillo Experimental Forest, Puerto Rico. *Journal of Tropical Ecology* 19:607–617.

Beard, K. H., K. A. Vogt, and A. Kulmatiski. 2002. Top-down effects of a terrestrial frog on forest nutrient dynamics. *Oecologia* 133:583–593.

Beard, K. H., K. A. Vogt, D. J. Vogt, F. N. Scatena, A. P. Covich, R. Sigurdardottir, T. C. Siccama, and T. A. Crowl. 2005. Structural and functional responses of a subtropical forest to 10 years of periodic hurricanes and droughts. *Ecological Monographs* 75:345–361.

Benstead, J. P., J. G. March, and C. M. Pringle. 2000. Estuarine stages and upstream migration of freshwater shrimps in two tropical rivers, Puerto Rico. *Biotropica* 32:545–547.

Benstead, J. P., J. G. March, C. M. Pringle, and F. N. Scatena. 1999. Effects of a low-head dam and water abstraction on migratory tropical stream biota. *Ecological Applications* 9:656–668.

Benzing, D. H. 1991. Aerial roots and their environments. Pages 867–886 in Y. Waisel, A. Eshel, and U. Kafkafi, editors, *Plant roots: The hidden half.* New York: Marcel Dekker, Inc.

Birdsey, R. A., and P. L. Weaver. 1982. The forest resources of Puerto Rico. Resource Bulletin SO-85. New Orleans, LA: USDA Forest Service, Southern Forest Experiment Station.

Blanco, J. F., and F. N. Scatena. 2005. Floods, habitat hydraulics and upstream migration of *Neritina virginea* (Gastropoda: Neritidae) in northeastern Puerto Rico. *Caribbean Journal of Science* 41(1):55–74.

Bormann, B. T., J. R. Martin, F. H. Wagner, G. W. Wood, J. Alegría, P. G. Cunningham, M. H. Brookes, P. Friesema, J. Berg, and J. R. Henshaw. 1999. Adaptive management. Pages 505–534 in W. T. Sexton, A. J. Malk, R. C. Szaro, and N. C. Johnson, editors, *Ecological stewardship. A common reference for ecosystem management.* Oxford, England: Elsevier.

Brokaw, N. V. L. 1998. *Cecropia schreberiana* in the Luquillo Mountains of Puerto Rico. *The Botanical Review* 64:91–120.

Brown, S., and A. E. Lugo. 1982. The storage and production of organic matter in tropical forests and their role in the global carbon cycle. *Biotropica* 14:161–187.

Brown, S., and A. E. Lugo. 1990. Tropical secondary forests. *Journal of Tropical Ecology* 6:1–32.

Brown, S., and A. E. Lugo. 1994. Rehabilitation of tropical lands: A key to sustaining development. *Restoration Ecology* 2:97–111.

Brown, K. A., F. N. Scatena, and J. Gurevitch. 2006. Effects of an invasive tree on community structure and diversity in a tropical forest in Puerto Rico. *Forest Ecology and Management* 226:145–152.

Burney, D. A., and L. P. Burney. 1994. Holocene charcoal stratigraphy from Laguna Tortugero, Puerto Rico and the timing of human arrival on the island. *Archaeological Science* 21:273–281.

Butterfield, R. P., and R. H. Fisher. 1994. Untapped potential: Native species for reforestation. *Journal of Forestry* 92:37–40.

Callaway, R. M., and L. R. Walker. 1997. Competition and facilitation: A synthetic approach to interactions in plant communities. *Ecology* 78:1958–1965.

Campbell, C. J. 1997. Depletion patterns show change due for production of conventional oil. *Oil and Gas Journal Special* 29(Dec):33–37.

Carpenter, S., B. Walker, J. M. Anderies, and N. Abel. 2001. From metaphor to measurement: Resilience of what to what? *Ecosystems* 4:765–781.

Chestnut, T. J., and W. H. McDowell. 2000. C and N dynamics in the riparian and hyporheic zones of a tropical stream, Luquillo Mountains, Puerto Rico. *Journal of the North American Benthological Society* 19:199–214.

Chinea, J. D. 1999. Changes in the herbaceous and vine communities at the Bisley Experimental Watersheds, Puerto Rico, following Hurricane Hugo. *Canadian Journal of Forest Research* 29:1433–1437.

Corujo Flores, I. N. 1980. A study of fish populations in the Espíritu Santo River estuary. M.S. thesis. University of Puerto Rico-Río Piedras, PR.

Costanza, R., and H. E. Daly. 1987. Toward an ecological economics. *Ecological Modeling* 38:1–7.

Covich, A. P., and W. H. McDowell. 1996. The stream community. Pages 433–460 in D. P. Reagan and R. B. Waide, editors, *The food web of a tropical rain forest.* Chicago: University of Chicago Press.

Crépin, A. S. 2007. Using fast and slow processes to manage resources with thresholds. *Environmental & Resources Economics* 36:191–213.

Cronon, W. 1996. The trouble with wilderness; or, getting back to the wrong nature. Pages 69–90 in W. Cronon, editor, *Uncommon ground: Rethinking the human place in nature.* New York: W.W. Norton.

Crook, K. E., F. N. Scatena, and C. M. Pringle. 2007. Water withdrawn from the Luquillo Experimental Forest, 2004. General Technical Report IITF-GTR-34. Rio Piedras, PR: USDA Forest Service, International Institute of Tropical Forestry.

Crosby, A. W. 1986. *Ecological imperialism: The biological expansion of Europe, 900–1900.* Cambridge, England: Cambridge University Press.

Cruz, A. 1987. Avian community organization in a mahogany plantation in a Neotropical island. *Caribbean Journal of Science* 23:286–296.

Cruz, A. 1988. Avian resource in a Caribbean pine plantation. *Journal of Wildlife Management* 52:274–279.

Daily, G. C., and K. Ellison. 2002. *The new economy of nature. The quest to make conservation profitable.* Washington, DC: Island Press.

Dale, V. H., L. A. Joyce, S. McNulty, R. P. Neilson, M. P. Ayres, M. D. Flannigan, P. J. Hanson, L. C. Irland, A. E. Lugo, C. J. Peterson, D. Simberloff, F. J. Swanson, B. J. Stocks, and B. M. Wotton. 2001. Climate change and forest disturbances. *BioScience* 59:723–734.

Dale, V. H., F. J. Swanson, and C. M. Crisafulli, editors. 2005. *Ecological responses to the 1980 eruption of Mount St. Helens.* New York: Springer.

Devoe, N. N. 1989. Differential seeding and regeneration in openings and beneath closed canopy in subtropical wet forests. Doctor of Forestry dissertation. Yale University, New Haven, CT.

Diamond, J. 2005. *Collapse. How societies choose to fail or succeed.* New York: Viking Penguin Group.

Dirzo, R. 2001. Tropical forests. Pages 251–276 in O. E. Sala, F. S. Chapin III, and E. Huber-Sannwald, editors, *Global biodiversity in a changing environment: Scenarios for the 21st century.* New York: Springer-Verlag.

Duryea, M. L., E. Kampf, R. C. Littell, and C. D. Rodríguez-Pedraza. 2006. Hurricanes and the urban forest: II. Effects on tropical and subtropical tree species. *Arboriculture and Urban Forestry* 33:98–112.

Erdman, D. S. 1984. Exotic fishes in Puerto Rico. Pages 162–176 in W. R. Courtenay, Jr., and J. R. Stauffer, Jr., editors, *Distribution, biology and management of exotic fishes.* Baltimore, MD: Johns Hopkins University Press.

Ewel, J. J. 1977. Differences between wet and dry successional tropical ecosystems. *Geo-Eco-Trop* 1:103–117.

Fetcher, N., B. L. Haines, R. Cordero, J. D. Lodge, L. R. Walker, R. Roecker, D. Fernández, and W. Lawrence. 1996. Responses of tropical plants to mineral nutrients and solar radiation on a landslide in Puerto Rico. *Journal of Ecology* 84:331–341.

Foster, D. R., M. Fluet, and E. R. Boose. 1999. Human or natural disturbance: Landscape-scale dynamics of the tropical forests of Puerto Rico. *Ecological Applications* 9:555–572.

Francis, J. K. 1995. Forest plantations in Puerto Rico. Pages 210–223 in A. E. Lugo and C. A. Lowe, editors, *Tropical forests: Management and ecology*. New York: Springer-Verlag.

Francis, J. K., and C. A. Lowe, editors. 2000. Silvics of native and exotic trees of Puerto Rico and the Caribbean islands. General Technical Report IITF-GTR-15. Río Piedras, PR: USDA Forest Service, International Institute of Tropical Forestry.

Freeman, M. C., C. M. Pringle, E. A. Greathouse, and B. J. Freeman. 2003. Ecosystem-level consequences of migratory faunal depletion caused by dams. Pages 255–266 in K. E. Limburg and J. R. Waldman, editors, *Biodiversity, status and conservation of the world's shads*. Bethesda, MD: American Fisheries Society.

García Montiel, D., and F. N. Scatena. 1994. The effect of human activity on the structure and composition of a tropical forest in Puerto Rico. *Forest Ecology and Management* 63:57–78.

Gladwell, J. S., and M. Bonell. 1990. An international programme for environmentally sound hydrological and water management strategies in the humid tropics. Proceedings of the International Symposium on Tropical Hydrology and Fourth Caribbean Islands Water Tesources Congress. Technical Publication Series TPS-90-2. Denver, CO: American Water Works Association.

Goldammer, J. G. 1992. Tropical forests in transition. Ecology of natural and anthropogenic disturbance processes—An introduction. Pages 1–16 in J. G. Goldammer, editor, *Tropical forests in transition*. Basel, Switzerland: Birkhäuser Verlag.

González, G., and T. R. Seastedt. 2000. Comparison of the abundance and composition of litter fauna in tropical and subalpine forests. *Pedobiologia* 44:545–555.

González, G., and T. R. Seastedt. 2001. Soil fauna and plant litter decomposition in tropical and subalpine forests. *Ecology* 82:955–964.

González, G., and X. Zou. 1999. Earthworm influence on N availability and the growth of *Cecropia schreberiana* in tropical pasture and forest soils. *Pedobiologia* 43:824–829.

González, G., X. Zou, and S. Borges. 1996. Earthworm abundance and species composition in abandoned tropical croplands: Comparison of tree plantations and secondary forests. *Pedobiologia* 40:385–391.

Greathouse, E. A., C. M. Pringle, and J. G. Holmquist. 2006a. Conservation and management of migratory fauna: Dams in tropical streams of Puerto Rico. *Aquatic Conservation-Marine and Freshwater Ecosystems* 16:695–712.

Greathouse, E. A., C. M. Pringle, W. H. McDowell, and J. G. Holmquist. 2006b. Indirect upstream effects of dams: Consequences of migratory consumer extirpation in Puerto Rico. *Ecological Applications* 16:339–352.

Groffman, P. M., N. J. Boulware, W. C. Zipperer, R. V. Pouyat, L. E. Band, and M. F. Colosimo. 2002. Soil nitrogen cycling processes in urban riparian zones. *Environmental Science and Technology* 36:4547–4552.

Guariguata, M. R. 1990. Landslide disturbance and forest regeneration in the Upper Luquillo Mountains. *Journal of Ecology* 78:814–832.

Gunderson, L. 2000. Ecological resilience in theory and application. *Annual Review of Ecology and Systematics* 31:425–439.

Hall, C. A. S., C. Leon, W. Ravenscroft, and H. Wang. 2000. Temporal and spatial overview of Costa Rican agricultural production. Pages 349–401 in C. A. S. Hall, G. Leclerc, and C. L. Perez, editors, *Quantifying sustainable development: The future of tropical economies*. San Diego, CA: Academic Press.

Hall, C., D. Lindenberger, R. Kümmel, T. Kroeger, and W. Eichhorn. 2001. The need to reintegrate the natural sciences with economics. *BioScience* 51:663–673.

Heartsill-Scalley, T., and T. M. Aide. 2003. Riparian vegetation and stream condition in a tropical agricultural-secondary forest mosaic. *Ecological Applications* 13:225–234.

Heartsill-Scalley, T., F. N. Scatena, C. Estrada, W. H. McDowell, and A. E. Lugo. 2007. Disturbance and long-term patterns of rainfall and throughfall nutrient fluxes in a subtropical wet forest in Puerto Rico. *Journal of Hydrology* 333:472–485.

Heyne, C. M. 2000. Soil and vegetation recovery on abandoned paved roads in a humid tropical rain forest, Puerto Rico. M.S. thesis. University of Nevada, Las Vegas, NV.

Holdridge, L. R. 1967. *Life zone ecology*. San José, Costa Rica: Tropical Science Center.

Holling, C. S. 1973. Resiliency and stability of ecological systems. *Annual Review of Ecology and Systematics* 4:1–23.

Holmquist, J. G., J. M. Schmidt Gengenbach, and B. B. Yoshioka. 1998. High dams and marine-freshwater linkages: Effects on native and introduced fauna in the Caribbean. *Conservation Biology* 12:621–630.

Hubbert, M. K. 1968. Energy resources. Pages 157–242 in *Resources and man*. San Francisco, CA: National Academy of Science and W.H. Freeman.

Hunter, J. M., and S. I. Arbona. 1995. Paradise lost: An introduction to the geography of water pollution in Puerto Rico. *Social Science & Medicine* 40:1331–1355.

Hunter, M. L. 1990. *Wildlife, forests and forestry: Principles of managing forests for biological diversity*. Upper Saddle River, NJ: Prentice Hall.

Jobin, W. 1998. *The Caribbean region*. Pages 201–214 in W. Jobin, editor, *Sustainable management for dams and waters*. Boca Raton, FL: Lewis Publishers.

Joglar, R. L. 1998. *Los coquíes de Puerto Rico. Su historia natural y conservación* [*The coquís of Puerto Rico. Their natural history and conservation*]. Río Piedras, PR: Universidad de Puerto Rico.

Johnson, S. L., and A. P. Covich. 2000. The importance of night-time observation for determining habitat preferences of stream biota. *Regulated Rivers: Research and Management* 16:91–99.

Jordan, W. R., M. E. Gilpin, and J. D. Aber. 1987. *Restoration ecology. A synthetic approach to ecological research*. Cambridge, England: Cambridge University Press.

Kartchner, S. C., and T. A. Crowl. 2002. Recreational use of montane streams of the Puerto Rico rainforest. Luquillo LTER REU Program Report. Logan: Utah State University.

Kent, R., H. T. Odum, and F. N. Scatena. 2000. Eutrophic overgrowth in the self-organization of tropical wetlands illustrated with a study of swine wastes in rainforest plots. *Ecological Engineering* 16:255–269.

Kosonen, M., A. Otsamo, and J. Kuusipalo. 1997. Financial, economic and environmental profitability of reforestation of *Imperata* grasslands in Indonesia. *Forest Ecology and Management* 99:247–259.

Larsen, M. C., and J. E. Parks. 1997. How wide is a road? The association of roads and mass-wasting in a forested montane environment. *Earth Surface Processes and Landforms* 22:835–848.

Larsen, M. C., and A. Simon. 1993. A rainfall intensity-duration threshold for landslides in a humid-tropical environment, Puerto Rico. *Geografiska Annaler* 75A:13–23.

Larsen, M. C., and A. J. Torres-Sánchez. 1992. Landslides triggered by Hurricane Hugo in eastern Puerto Rico, September 1989. *Caribbean Journal of Science* 28:113–120.

Laurance, W. F., S. G. Laurance, L. V. Ferreira, J. M. Rankin-de-Morena, C. Gaston, and T. E. Lovejoy. 1997. Biomass collapse in Amazonian forest fragments. *Science* 278:1117–1118.

Leopold, L. B., editor. 1953. *Round River: From the journals of Aldo Leopold*. Oxford, England: Oxford University Press.

Liegel, L. H. 1982. Growth, development and hurricane resistance of Honduras pine in Puerto Rico. Pages 28–48 in J. González-Liboy and J. Figueroa, editors, *Noveno simposio de recursos naturales*. San Juan, PR: Departamento de Recursos Naturales de Puerto Rico.

Liegel, L. H. 1984. Assessment of hurricane rain/wind damage in *Pinus caribaea* and *Pinus oocarpa* provenance trials in Puerto Rico. *Commonwealth Forestry Review* 63:47–53.

Likens, G. E., and F. H. Bormann. 1995. *Biogeochemistry of a forested ecosystem*. New York: Springer-Verlag.

Liu, Z. G., and X. M. Zou. 2002. Exotic earthworms accelerate plant litter decomposition in a Puerto Rican pasture and a wet forest. *Ecological Applications* 12:1406–1417.

Lodge, D. J., and F. J. Calderón. 1991. Distribution of plant species in landslides can be explained by their dependency on VA-mycorrhizal fungi and responsiveness to high phosphorus availability. *Bulletin of the Ecological Society of America* 72:178.

Lodge, D. M., and K. Shrader-Frechette. 2003. Nonindigenous species: Ecological explanation, environmental ethics, and public policy. *Conservation Biology* 17:31–37.

Lowrance, R., L. S. Altier, J. D. Newbold, R. R. Schnabel, P. M. Groffman, J. M. Denver, D. L. Correll, J. W. Gilliam, J. L. Robinson, R. B. Brinsfield, K. W. Staver, W. Lucas, and A. H. Todd. 1997. Water quality functions of riparian forest buffers in Chesapeake Bay watersheds. *Environmental Management* 21:687–712.

Ludwig, D. 2001. The era of management is over. *Ecosystems* 4:758–764.

Ludwig, D., B. H. Walker, and C. S. Holling. 2002. Models and metaphors of sustainability, stability, and resilience. Pages 21–48 in L. H. Gunderson and J. L. Pritchard, editors, *Resilience and the behavior of large-scale systems*. Washington, DC: Island Press.

Lugo, A. E. 1988. The future of the forest. Ecosystem rehabilitation in the tropics. *Environment* 30:16–20, 41–45.

Lugo, A. E. 1991. Cities in the sustainable development of tropical landscapes. *Nature and Resources* 27:27–35.

Lugo, A. E. 1992. Comparison of tropical tree plantations with secondary forests of similar age. *Ecological Monographs* 62:1–41.

Lugo, A. E. 1994. Preservation of primary forests in the Luquillo Mountains, Puerto Rico. *Conservation Biology* 8:1122–1131.

Lugo, A. E. 1996. Caribbean island landscapes: Indicators of the effects of economic growth on the region. *Environment and Development Economics* 1:128–136.

Lugo, A. E. 1997. The apparent paradox of re-establishing species richness on degraded lands with tree monocultures. *Forest Ecology and Management* 99:9–19.

Lugo, A. E. 2000. Effects and outcomes of Caribbean hurricanes in a climate change scenario. *The Science of the Total Environment* 262:243–251.

Lugo, A. E. 2002. Can we manage tropical landscapes? An answer from the Caribbean perspective. *Landscape Ecology* 17:601–615.

Lugo, A. E. 2004. The outcome of alien tree invasions in Puerto Rico. *Frontiers in Ecology and the Environment* 2:265–273.

Lugo, A. E., J. S. Baron, T. P. Frost, T. W. Cundy, and P. Dittberner. 1999a. Ecosystem processes and functioning. Pages 219–254 in R. C. Szaro, N. C. Johnson, W. T. Sexton, and A. J. Malk, editors. *Ecological stewardship: A common reference for ecosystem management*. Oxford, England: Elsevier Science.

Lugo, A. E., and T. J. Brandeis. 2005. A new mix of alien and native species coexist in Puerto Rico's landscapes. Pages 484–509 in D. F. R. P. Burslem, M. A. Pinard, and S. E. Hartley, editors, *Biotic interactions in the tropics: Their role in the maintenance of species diversity*. Cambridge, England: Cambridge University Press.

Lugo, A. E., and S. Brown. 1991. Comparing tropical and temperate forests. Pages 319–330 in J. Cole, G. Lovett, and S. Findlay, editors, *Comparative analysis of ecosystems: Patterns, mechanisms, and theories*. New York: Springer-Verlag.

Lugo, A. E., S. L. Brown, R. Dodson, T. S. Smith, and H. H. Shugart. 1999b. The Holdridge life zones of the conterminous United States in relation to ecosystem mapping. *Journal of Biogeography* 26:1025–1038.

Lugo, A. E., J. Figueroa Colón, and F. N. Scatena. 2000. The Caribbean. Pages 593–622 in M. G. Barbour and W. D. Billings, editors, *North American terrestrial vegetation*. Cambridge, England: Cambridge University Press.

Lugo, A. E., and J. L. Frangi. 1993. Fruit fall in the Luquillo Experimental Forest, Puerto Rico. *Biotropica* 25:73–84.

Lugo, A. E., and H. Gucinski. 2000. Function, effects, and management of forest roads. *Forest Ecology and Management* 133:249–262.

Lugo, A. E., and E. Helmer. 2004. Emerging forests on abandoned land: Puerto Rico's new forests. *Forest Ecology and Management* 190:145–161.

Lugo, A. E., T. d. M. López, O. Ramos, and L. L. Vélez. 2004. Urbanización de los terrenos en la periferia de El Yunque [Urbanization of lands in the periphery of El Yunque]. General Technical Report IITF-66. Washington, DC: USDA Forest Service, International Institute of Tropical Forestry.

Lugo, A. E., and L. Mastroantonio. 1999. Institute of Tropical Forestry: The first fifty years. General Technical Report IITF-GTR-7. Río Piedras, PR: USDA Forest Service, International Institute of Tropical Forestry.

Lugo, A. E., L. Miranda-Castro, A. Vale, T. d. M. López, E. Hernández-Prieto, A. García-Martinó, A. R. Puente-Rolón, A. G. Tossas, D. A. McFarlane, T. Miller, A. Rodríguez, J. Lundberg, J. Thomlinson, J. Colón, J. H. Schellekens, O. Ramos, and E. Helmer. 2001. Puerto Rican karst—A vital resource. General Technical Report WO-65. Washington, DC: USDA Forest Service.

Lugo, A. E., J. A. Parrotta, and S. Brown. 1993. Loss in species caused by tropical deforestation and their recovery through management. *Ambio* 22:106–209.

Lugo, A. E., and F. N. Scatena. 1992. Epiphytes and climate change research in the Caribbean: A proposal. *Selbyana* 13:123–130.

Lugo, A. E., and F. N. Scatena. 1995. Ecosystem-level properties of the Luquillo Experimental Forest with emphasis on the tabonuco forest. Pages 59–108 in A. E. Lugo and C. Lowe, editors, *Tropical forests: Management and ecology*. New York: Springer-Verlag.

Lugo, A. E., F. N. Scatena, W. Silver, S. Molina Colón, and P. G. Murphy. 2002. Resilience of tropical wet and dry forests in Puerto Rico. Pages 195–225 in L. H. Gunderson and L. Pritchard, Jr., editors, *Resilience and the behavior of large-scale systems*. Washington, DC: Island Press.

Lugo, A. E., and J. K. Zimmerman. 2002. Ecological life histories. Pages 191–213 in J. A. Vozzo, editor, *Tropical tree seed manual*. Agriculture Handbook 721. Washington, DC: USDA Forest Service.

Malkus, J. S., and M. E. Stern. 1953. The flow of a stable atmosphere over a heated island. *Journal of Meteorology* 10:30–41.

Mann, C. C. 2005. *1491: New revelations of the Americas before Columbus*. New York: Alfred A. Knopf.

March, J. G., J. P. Benstead, C. M. Pringle, and F. N. Scatena. 1998. Migratory drift of larval amphidromous shrimps in two tropical streams, Puerto Rico. *Freshwater Biology* 40:1–14.

March, J. G., J. P. Benstead, C. M. Pringle, and F. N. Scatena. 2003. Damming tropical island streams: Problems, solutions, and alternatives. *BioScience* 53:1069–1078.

March, J. G., and C. M. Pringle. 2003. Food web structure and basal resource utilization along a tropical island stream continuum, Puerto Rico. *Biotropica* 35:84–93.

March, J. G., C. Pringle, M. J. Townsend, and A. I. Wilson. 2002. Effects of freshwater shrimp assemblages on benthic communities along an altitudinal gradient of a tropical island stream. *Freshwater Biology* 47:377–390.

Marrero, J. 1950. La reforestación de tierras degradadas de Puerto Rico. *Caribbean Forester* 11:16–24.

Maxwell, J., and R. Costanza. 1989. An ecological economics for ecological engineering. Pages 57–77 in W. J. Mitsch and S. E. Jørgensen, editors, *Ecological engineering. An introduction to ecotechnology*. New York: John Wiley & Sons.

McClain, M. E., E. W. Boyer, C. L. Dent, S. E. Gergel, N. B. Grimm, P. M. Groffman, S. C. Hart, J. W. Harvey, C. A. Johnston, E. Mayorga, W. H. McDowell, and G. Pinay. 2003. Biogeochemical hot spots and hot moments at the interface of terrestrial and aquatic ecosystems. *Ecosystems* 6:301–312.

McClain, M. E., J. E. Richey, and T. Pena Pimental. 1994. Groundwater nitrogen dynamics at the terrestrial-lotic interface of a small catchment in the central Amazon basin. *Biogeochemistry* 27:113–127.

McDowell, W. H. 2001. Hurricanes, people, and riparian zones: Controls on nutrient losses from forested Caribbean watersheds. *Forest Ecology and Management* 154:443–451.

McDowell, W. H., W. B. Bowden, and C. E. Asbury. 1992. Riparian nitrogen dynamics in two geomorphologically distinct tropical rain forest watersheds—Subsurface solute patterns. *Biogeochemistry* 18:53–75.

McDowell, W. H., C. P. McSwiney, and W. B. Bowden. 1996. Effects of hurricane disturbance on groundwater chemistry and riparian function in a tropical rain forest. *Biotropica* 28:577–584.

Meffe, G. K. and C. R. Carroll, editors. 1997. *Principles of conservation biology*, 2nd ed. Sunderland, MA: Sinauer Associates, Inc.

Meine, C. 1987. The farmer as conservationist: Aldo Leopold on agriculture. *Journal of Soil and Water Conservation* 42:144–149.

Meybeck, M., D. Chapman, and R. Helmer. 1989. *Global freshwater quality. A first assessment*. Oxford, England: Basil Blackwell.

Millennium Ecosystem Assessment. 2005. *Ecosystems and human well-being: Synthesis*. Washington, DC: Island Press.

Mitsch, W. J., and S. E. Jørgensen, editors. 1989. *Ecological engineering: An introduction to ecotechnology*. New York: John Wiley & Sons.

Murphy, L. S. 1916. Forests of Puerto Rico: Past, present, and future and their physical and economic environment. Bulletin 354. Washington, DC: U.S. Department of Agriculture.

Myster, R. W., and D. S. Fernández. 1995. Spatial gradients and patch structure on two Puerto Rican landslides. *Biotropica* 27:149–159.

Myster, R. W., and F. O. Sarmiento. 1998. Seed inputs to microsite patch recovery on two tropandean landslides in Ecuador. *Restoration Ecology* 6:35–43.

Myster, R. W., and L. R. Walker. 1997. Plant successional pathways on Puerto Rican landslides. *Journal of Tropical Ecology* 13:165–173.

Naiman, R. J., H. Décamps, and M. E. McClain. 2005. *Riparia: Ecology, conservation, and management of streamside communities*. San Diego, CA: Elsevier.

Naumann, M. 1994. A water-use budget for the Caribbean National Forest of Puerto Rico. Special Report. Río Piedras, PR: USDA Forest Service, International Institute of Tropical Forestry.

Nieves, L. O. 1998. Ecological study of the freshwater stream fishes of the upper Mameyes River (Bisley) in the Luquillo Mountains of Puerto Rico. Ph.D. dissertation. George Mason University, Fairfax, VA.

O'Brien, S. T., B. P. Hayden, and H. H. Shugart. 1992. Global climate change, hurricanes, and a tropical forest. *Climate Change* 22:175–190.

Odum, H. T. 1971. *Environment power and society*. New York: Wiley.

Odum, H. T. 1988. Self-organization, transformity, and information. *Science* 242: 1132–1139.

Odum, H. T. 1989. Ecological engineering and self-organization. Pages 79–101 in W. J. Mitsch and S. E. Jørgensen, editors, *Ecological engineering*. New York: Wiley.

Odum, H. T. 1995. Tropical forest systems and the human economy. Pages 343–393 in A. E. Lugo and C. Lowe, editors, *Tropical forests: Management and ecology*. New York: Springer-Verlag.

Odum, H. T. 1996. *Environmental accounting: EMERGY and environmental decision making*. New York: John Wiley & Sons.

Odum, H. T., S. J. Doherty, F. N. Scatena, and P. A. Kharecha. 2000. EMERGY evaluation of reforestation alternatives in Puerto Rico. *Forest Science* 46:521–530.

Odum, H. T., G. Drewry, and E. A. McMahan. 1970a. Introduction to Section E. Pages E3–E15 in H. T. Odum and R. F. Pigeon, editors, *A tropical rain forest: A study of irradiation and ecology at El Verde, Puerto Rico*. Oak Ridge, TN: U.S. Atomic Energy Commission.

Odum, H. T., P. Murphy, G. Drewry, F. McCormick, C. Schinham, E. Morales, and J. McIntyre. 1970b. Effects of gamma radiation on the forest at El Verde. Pages D3–D75 in H. T. Odum and R. F. Pigeon, editors, *A tropical rain forest: A study of irradiation and ecology at El Verde, Puerto Rico*. Oak Ridge, TN: U.S. Atomic Energy Commission.

Odum, H. T., and E. C. Odum. 2001. *A prosperous way down: Principles and policies*. Boulder, CO: University Press of Colorado.

Olander, L. P., F. N. Scatena, and W. L. Silver. 1998. Impacts of disturbance initiated by road construction in a subtropical cloud forest in the Luquillo Experimental Forest, Puerto Rico. *Forest Ecology and Management* 109:33–49.

Ortiz Zayas, J. R., and F. N. Scatena. 2004. Integrated water resources management in the Luquillo Mountains, Puerto Rico: An evolving process. *Journal of Water Resources Development* 20:387–398.

Palmer, M. A., E. Bernhardt, E. Chornesky, S. Collins, A. Dobson, C. Duke, B. Gold, R. Jacobson, S. Kingsland, R. Kranz, M. Mappin, M. L. Martinez, F. Micheli, J. Morse, M. Pace, M. Pascual, S. Palumbi, O. J. Reichman, A. Simons, A. Townsend, and M. Turner. 2004. Ecology for a crowded planet. *Science* 304:1251–1252.

Parker, K. R., and J. A. Wiens. 2005. Assessing recovery following environmental accidents: Environmental variation, ecological assumptions, and strategies. *Ecological Applications* 15:2037–2051.

Parrotta, J. A. 1995. Influence of overstory composition on understory colonization by native species in plantations on a degraded tropical site. *Journal of Vegetation Science* 6:627–636.

Parrotta, J. A. 1999. Productivity, nutrient cycling, and succession in single- and mixed-species plantations of *Casuarina equisetifolia, Eucalyptus robusta*, and *Leucaena leucocephala* in Puerto Rico. *Forest Ecology and Management* 124:45–77.

Parrotta, J. A., O. H. Knowles, and J. M. Wunderle. 1997. Development of floristic diversity in 10-year-old restoration forests on a bauxite mined site in Amazonia. *Forest Ecology and Management* 99:21–42.

Pereira, H. C. 1989. *Policy practice in the management of tropical watersheds*. San Francisco, CA: Westview Press.

Perlin, J. 1989. *A forest journey: The role of wood in the development of civilization*. Cambridge, MA: Harvard University Press.

Peterjohn, W. T., and D. L. Correll. 1984. Nutrient dynamics in an agricultural watershed: Observations on the role of a riparian forest. *Ecology* 65:1466–1475.

Postel, S., and B. Richter. 2003. *Rivers for life: Managing water for people and nature*. Washington, DC: Island Press.

Pringle, C. M. 1997. Exploring how disturbance is transmitted upstream: Going against the flow. *Journal of the North American Benthological Society* 16:425–438.

Pringle, C. M. 2000a. River conservation in tropical versus temperate regions. Pages 367–379 in P. J. Boon, B. Davies, and G. E. Petts, editors, *Global perspectives on river conservation: Science, policy and practice*. New York: John Wiley & Sons.

Pringle, C. M. 2000b. Threats to U.S. public lands from cumulative hydrologic alterations outside of their boundaries. *Ecological Applications* 10:971–989.

Pringle, C. M. 2001. Hydrologic connectivity and the management of biological reserves: A global perspective. *Ecological Applications* 11:981–998.

Pringle, C. M., and G. A. Blake. 1994. Quantitative effects of atyid shrimp (Decapoda: Atyidae) on the depositional environment in a tropical stream: Use of electricity for experimental exclusion. *Canadian Journal of Fisheries and Aquatic Sciences* 51:1443–1450.

Pringle, C. M., M. C. Freeman, and B. J. Freeman. 2000a. Regional effects of hydrologic alterations on riverine microbiota in the New World: Tropical-temperate comparisons. *BioScience* 50:807–823.

Pringle, C. M., N. Hemphill, W. H. McDowell, A. Bednarek, and J. G. March. 1999. Linking species and ecosystems: Different biotic assemblages cause interstream differences in organic matter. *Ecology* 80:1860–1872.

Pringle, C. M., and F. N. Scatena. 1999a. Factors affecting aquatic ecosystem deterioration in Latin America and the Caribbean. Pages 104–113 in U. Hatch and M. E. Swisher, editors, *Tropical managed ecosystems: The Mesoamerican experience*. New York: Oxford University Press.

Pringle, C. M. and F. N. Scatena. 1999b. Freshwater resource development: Case studies from Puerto Rico and Costa Rica. Pages 114–121 in U. Hatch and M. E. Swisher, editors, *Tropical managed ecosystems: The Mesoamerican experience*. New York: Oxford University Press.

Pringle, C. M., F. Scatena, P. Paaby, and N. Nuñez. 2000b. River conservation in Latin America and the Caribbean. Pages 39–73 in P. J. Boon, B. Davies, and G. E. Petts, editors, *Global perspectives on river conservation: Science, policy and practice*. New York: John Wiley & Sons.

Roberts, R. C. 1942. Soil survey of Puerto Rico. USDA Series 1936, No. 8. Washington, DC: U.S. Department of Agriculture.

Ruiz, B. I., editor. 2002. Manual de reforestación para América tropical [Reforestation manual for tropical America]. General Technical Report IITF-GTR-18. Río Piedras, PR: USDA Forest Service, International Institute of Tropical Forestry.

Sala, O. E., F. S. Chapin, J. J. Armesto, E. Berlow, J. Bloomfield, R. Dirzo, E. Huber-Sanwald, L. F. Huenneke, R. B. Jackson, A. Kinzig, R. Leemans, D. M. Lodge, H. A. Mooney, M. Oesterheld, N. L. Poff, M. T. Sykes, B. H. Walker, M. Walker, and D. H. Wall. 2000. Global biodiversity scenarios for the year 2100. *Science* 287:1770–1774.

Salwasser, H. 1997. Conservation biology and the management of natural resources. Pages 9–10 in G. K. Meffe and C. R. Carroll, editors, *Principles of conservation biology*. Sunderland, MA: Sinauer Associates, Inc.

Sánchez de León, Y., X. Zou, S. Borges, and A. Honghua Ruan. 2003. Recovery of native earthworms in abandoned tropical pastures. *Conservation Biology* 17:999–1006.

Sanderson, E. W., M. Jaiteh, M. A. Levy, K. A. Redford, A. V. Wannebo, and G. Woolmer. 2002. The human footprint and the last of the wild. *BioScience* 52:891–904.

Saunders, D. A., R. J. Hobbs, and C. R. Margules. 1991. Biological consequences of ecosystem fragmentation: A review. *Conservation Biology* 5:18–32.

Scatena, F. N. 1989. An introduction to the physiography and history of the Bisley Experimental Watersheds in the Luquillo Mountains of Puerto Rico. General Technical Report 72. New Orleans, LA: USDA Forest Service, Southern Forest Experiment Station.

Scatena, F. N. 1990. Selection of riparian buffer zones in humid tropical steeplands. Pages 328–337 in R. R. Ziemer, C. L. O'Loughlin, and L. S. Hamilton, editors, *Research needs and applications to reduce erosion and sedimentation in tropical steeplands.* AISH Pub. No. 192. Louvain, Belgium: International Association of Hydrological Sciences.

Scatena, F. N. 1995. Relative scales of time and effectiveness of watershed processes in a tropical montane forest of Puerto Rico. Pages 103–111 in J. E. Costa, A. J. Miller, K. W. Porter, and P. Wilcock, editors, *Natural and anthropogenic influences in fluvial geomorphology*. Monograph 89. Washington, DC: American Geophysical Union Press.

Scatena, F. N. 1998. An assessment of climate change in the Luquillo Mountains of Puerto Rico. Pages 193–198 in *Third international symposium on water resources. Fifth Caribbean islands water resources congress*. San Juan, PR: American Water Resources Association.

Scatena, F. N. 2001. Ecological rhythms and the management of humid tropical forests: Examples from the Caribbean National Forest, Puerto Rico. *Forest Ecology and Management* 154:453–464.

Scatena F. N., S. J. Doherty, H. T. Odum, and P. Kharecha P. 2002. An EMERGY evaluation of Puerto Rico and the Luquillo Experimental Forest. General Technical Report IITF-GTR-9. Río Piedras, PR: USDA Forest Service, International Institute of Tropical Forestry.

Scatena, F. N., and S. L. Johnson. 2001. Instream-flow analysis for the Luquillo Experimental Forest, Puerto Rico: Methods and analysis. General Technical Report GTR-11. Río Piedras, PR: USDA Forest Service, International Institute of Tropical Forestry.

Scatena, F. N., and M. C. Larsen. 1991. Physical aspects of Hurricane Hugo in Puerto Rico. *Biotropica* 23:317–323.

Scatena, F. N., and A. E. Lugo. 1995. Geomorphology, disturbance, and the soil and vegetation of two subtropical steepland watersheds of Puerto Rico. *Geomorphology* 13:199–213.

Scatena, F. N., S. Moya, C. Estrada, and J. D. Chinea. 1996. The first five years in the reorganization of aboveground biomass and nutrient use following Hurricane Hugo in the Bisley Experimental Watersheds, Luquillo Experimental Forest, Puerto Rico. *Biotropica* 28:424–440.

Scherr, S., A. White, and A. Khare. 2004. For services rendered. The current status and future potential of markets for the ecosystem services provided by tropical forests. ITTO Technical Series No. 21. Yokohama, Japan: International Tropical Timber Organization.

Schubert, T. H. 1979. *Trees for urban use in Puerto Rico and the Virgin Islands.* General Technical Report SO-27. New Orleans, LA: USDA Forest Service, Southern Forest Experiment Station, Institute of Tropical Forestry.

Shiels, A. B. 2006. Leaf litter decomposition and substrate chemistry of early succession species on landslides in Puerto Rico. *Biotropica* 38:348–353.

Shiels, A. B., and L. R. Walker. 2003. Bird perches increase forest seeds on Puerto Rican landslides. *Restoration Ecology* 11:1–9.

Shiels, A. B., L. R. Walker, and D. B. Thompson. 2006. Organic matter inputs create variable resource patches on Puerto Rican landslides. *Plant Ecology* 184:223–236.

Silver, W. L. 1992. Effects of small-scale and catastrophic disturbance on carbon and nutrient cycling in a lower montane subtropical wet forest in Puerto Rico. Ph.D. dissertation. Yale University, New Haven, CT.

Silver, W. L., S. Brown, and A. E. Lugo. 1996a. Biodiversity and biogeochemical cycles. Pages 49–67 in G. Orians, R. Dirzo, and J. H. Cushman, editors, *Biodiversity and ecosystem process in tropical forests*. Heidelberg, Germany: Springer-Verlag.

Silver, W. L., S. Brown, and A. E. Lugo. 1996b. Effects of changes in biodiversity on ecosystem function in tropical forests. *Conservation Biology* 10:17–24.

Silver, W. L., D. J. Herman, and M. K. Firestone. 2001. Dissimilatory nitrate reduction to amonium in upland tropical forest soils. *Ecology* 82:2410–2416.

Silver, W. L., Kueppers, L. M., Lugo, A. E., Ostertag, R., and Matzek, V. 2004. Carbon sequestration and plant community dynamics following reforestation of tropical pasture. *Ecological Applications* 14:1115–1127.

Silver, W. L., A. E. Lugo, and M. Keller. 1999. Soil oxygen availability and biogeochemistry along rainfall and topographic gradients in upland wet tropical forest soils. *Biogeochemistry* 44:301–328.

Silver, W. L., R. Ostertag, and A. E. Lugo. 2000. The potential for carbon sequestration through reforestation of abandoned tropical agricultural and pasture lands. *Restoration Ecology* 8:394–407.

Simmons, R. C., A. J. Gold, and P. M. Groffman. 1992. Nitrate dynamics in riparian forests—Groundwater studies. *Journal of Environmental Quality* 21:659–665.

Smith, K. L. 2008. Managing freshwater inflow to estuaries in northeastern Puerto Rico: Ecological and institutional considerations. Ph.D. dissertation. Department of Ecology, University of Georgia, Athens.

Snyder, N. F. R., J. W. Wiley, and C. B. Kepler. 1987. *The parrots of Luquillo: Natural history and conservation of the Puerto Rican Parrot*. Los Angeles, CA: Western Foundation of Vertebrate Zoology.

Stewart, M. M., and L. L. Woolbright. 1996. Amphibians. Pages 273–320 in D. P. Reagan and R. B. Waide, editors, *The food web of a tropical rain forest*. Chicago: University of Chicago Press.

Thompson, J., N. Brokaw, J. K. Zimmerman, R. B. Waide, E. M. Everham III, D. J. Lodge, C. M. Taylor, D. García Montiel, and M. Fluet. 2002. Land use history, environment, and tree composition in a tropical forest. *Ecological Applications* 12:1344–1363.

Tilley, R. D., and W. T. Swank. 2003. EMERGY-based environmental systems assessment of a multi-purpose temperate mixed-forest watershed of the southern Appalachian Mountains, USA. *Journal of Environmental Management* 69:213–227.

Triska, F. J., J. H. Duff, and R. J. Avanzino. 1993. Patterns of hydrological exchange and nutrient transformation in the hyporheic zone of a gravel-bottom stream: Examining terrestrial-aquatic linkages. *Freshwater Biology* 29:259–274.

U.S. Department of Agriculture, Forest Service. 1997. Revised land and resource management plan: Caribbean National Forest, Luquillo Experimental Forest, Puerto Rico. Management Bulletin R8-MB 80G. Atlanta, GA: USDA Forest Service Southern Region.

van der Molen, M. K. 2002. *Meteorological impacts of land use change in the maritime tropics*. Amsterdam, The Netherlands: Vrije Universiteit.

Vermeij, G. 1996. An agenda for invasion biology. *Conservation Biology* 78:3–9.

Vogt, K. A., K. H. Beard, S. Himmann, J. O. H. Palmiotto, D. J. Vogt, F. N. Scatena, and B. P. Hecht. 2002a. Indigenous knowledge informing management of tropical forests: The link between rhythms in plant secondary chemistry and lunar cycles. *Ambio* 31:485–490.

Vogt, K. A., J. Gordon, J. Wargo, D. Vogt, H. Asbjornsen, P. A. Palmiotto, H. Clark, J. O'Hara, W. S. Keeton, T. Patel-Weynand, and E. Witten, editors. 1997. *Ecosystems: Balancing science with management*. New York: Springer-Verlag.

Vogt, K. A., M. Grove, H. Asbjornsen, K. Maxwell, D. J. Vogt, R. Sigurdardottir, B. C. Larson, L. Schibli, and M. Dove. 2002b. Linking social and natural science spatial scales. Pages 143–175 in L. Jianguo and W. M. Taylor, editors, *Integrating landscape ecology into natural resource management*. Cambridge, England: Cambridge University Press.

Vogt, K. A., O. Schmitz, K. H. Beard, J. O'Hara, and M. Booth. 2000. Conservation biology—Contemporary issues. Pages 865–881 in S. Levin, editor, *Encyclopedia of biodiversity*. New York: Academic Press.

Vogt, K. A., D. J. Vogt, P. Boon, A. Covich, F. N. Scatena, H. Asjbornsen, J. L. O'Hara, J. Perez, T. Siccama, and J. Bloomfield. 1996. Litter dynamics along stream, riparian and upslope areas following Hurricane Hugo, Luquillo Experimental Forest, Puerto Rico. *Biotropica* 28:458–470.

Wadsworth, F. H. 1949. The development of the forest land resources of the Luquillo Mountains of Puerto Rico. Ph.D. dissertation. University of Michigan, Ann Arbor, MI.

Wadsworth, F. H. 1970. A review of past research in the Luquillo Mountains. Pages B33–B46 in H. T. Odum and R. F. Pigeon, editors, *A tropical rain forest: A study of irradiation and ecology at El Verde, Puerto Rico*. Oak Ridge, TN: U.S. Atomic Energy Commission.

Wadsworth, F. H. 1995. A forest research institution in the West Indies: The first 50 years. Pages 33–56 in A. E. Lugo and C. Lowe, editors, *Tropical forests: Management and ecology*. New York: Springer-Verlag.

Wadsworth, F. H. 1997. Forest production for tropical America. Agriculture Handbook 710. Washington, DC: USDA Forest Service.

Wadsworth, F. H., and R. A. Birdsey. 1983. Un nuevo enfoque de los bosques de Puerto Rico [A new focus to the forests of Puerto Rico]. Pages 12–27 in *Puerto Rico Department of Natural Resources ninth symposium on natural resources*. San Juan, PR: Puerto Rico Department of Natural Resources.

Waide, R. B., J. K. Zimmerman, and F. N. Scatena. 1998. Controls of primary productivity: Lessons from the Luquillo Mountains of Puerto Rico. *Ecology* 79:31–37.

Walker, L. R. 1994. Effects of fern thickets on woodland development on landslides in Puerto Rico. *Journal of Vegetation Science* 5:525–532.

Walker, L. R., editor. 1999. *Ecosystems of disturbed ground*. Amsterdam, The Netherlands: Elsevier.

Walker, L. R., and W. Boneta. 1995. Plant and soil responses to fire on a fern-covered landslide in Puerto Rico. *Journal of Tropical Ecology* 11:473–479.

Walker, L. R., and R. del Moral. 2003. *Primary succession and ecosystem rehabilitation*. Cambridge, England: Cambridge University Press.

Walker, L. R., F. H. Landau, E. Velazquez, A. B. Shiels, and A. D. Sparrow. 2010. Early successional woody plants facilitate and ferns inhibit forest development on Puerto Rican landslides. *Journal of Ecology* 98:625–635.

Walker, L. R., and L. Neris. 1993. Posthurricane seed rain dynamics in Puerto Rico. *Biotropica* 25:408–418.

Walker, L. R., J. Walker, and R. J. Hobbs, editors. 2007. *Linking restoration and ecological succession*. New York: Springer.

Walker, L. R., D. Zarin, N. Fetcher, R. Myster, and A. Johnson. 1996. Ecosystem development and plant succession on landslides in the Caribbean. *Biotropica* 28:566–576.

Wang, H. H., and F. N. Scatena. 2003. Regeneration after hurricane disturbance of big-leaf and hybrid mahogany plantations in Puerto Rico. Pages 237–257 in A. E. Lugo, J. C. Figueroa Colón, and M. Alayón, editors, *Big-leaf mahogany: Genetics, ecology, and management*. New York: Springer-Verlag.

Weaver, P. L. 1987. Structure and dynamics in the colorado forests of the Luquillo Mountains of Puerto Rico. Ph.D. dissertation. Michigan State, East Lansing, MI.

Weaver, P. L. 1989. Taungya plantings in Puerto Rico. *Journal of Forestry* 87:37–41.

Weaver, P. L., and G. P. Bauer. 1986. Growth, survival and shoot borer damage in mahogany plantings in the Luquillo Forest, Puerto Rico. *Turrialba* 36:509–522.

Weaver, P. L., and P. G. Murphy. 1990. Forest structure and productivity in the Luquillo Mountains of Puerto Rico. *Biotropica* 22:69–82.

Weinbren, M. P., B. M. Weinbren, W. B. Jackson, and J. B. Villella. 1970. Studies of the roof rat (*Rattus rattus*) in the El Verde forest. Pages E169–E181 in H. T. Odum and R. F. Pigeon, editors, *A tropical rain forest: A study of irradiation and ecology at El Verde, Puerto Rico*. Oak Ridge, TN: U.S. Atomic Energy Commission.

Wells, M., and K. Brandon. 1992. *People and parks: Linking protected area management with local communities*. Report to The International Bank for Reconstruction and Development (World Bank), Washington, DC.

Wiens, J. A. 1976. Population responses to patchy environments. *Annual Review of Ecology and Systematics* 7:81–120.

Wilshusen, P., S. Brechin, C. Fortwangler, and P. West. 2002. Reinventing a square wheel: Critique of a resurgent "protection paradigm" in international biodiversity conservation. *Society and Natural Resources* 15:17–40.

Wilson, E. O. 1988. The current state of biological diversity. Pages 3–18 in E. O. Wilson and F. M. Peters, editors, *Biodiversity*. Washington, DC: National Academy of Sciences.

Woodwell, G. M., and A. L. Rebuck. 1967. Effects of chronic gamma radiation on the structure and diversity of an oak-pine forest. *Ecological Monographs* 37:53–69.

You, C., and W. H. Petty. 1991. Effects of Hurricane Hugo on *Manilkara bidentata*, a primary tree species in the Luquillo Experimental Forest of Puerto Rico. *Biotropica* 23:400–406.

Zarin, D. J., and A. H. Johnson. 1995a. Base saturation, nutrient cation, and organic matter increases during early pedogenesis on landslide scars in the Luquillo Experimental Forest, Puerto Rico. *Geoderma* 65:317–330.

Zarin, D. J., and A. H. Johnson. 1995b. Nutrient accumulation during primary succession in a montane tropical forest, Puerto Rico. *Soil Science Society of America Journal* 59:1444–1452.

Zimmerman, J. K., E. M. Everham III, R. B. Waide, D. J. Lodge, C. M. Taylor, and N. V. L. Brokaw. 1994. Responses of tree species to hurricane winds in subtropical wet forest in Puerto Rico: Implications for tropical tree life histories. *Journal of Ecology* 82:911–922.

Zimmerman, J. K., T. Mitchell Aide, M. Rosario, and L. Herrera. 1995. Effects of land management and a recent hurricane on forest structure and composition in the Luquillo Experimental Forest, Puerto Rico. *Forest Ecology and Management* 77:65–76.

Zimmerman, J. K., M. R. Willig, L. R. Walker, and W. L. Silver. 1996. Introduction: Disturbance and Caribbean ecosystems. *Biotropica* 28:414–423.

Zon, R., and W. N. Sparhawk. 1923. *Forest resources of the world*. New York: McGraw-Hill.

Zou, X., and G. González. 1997. Changes in earthworm density and community structure during secondary succession in abandoned tropical pastures. *Soil Biology and Biochemistry* 29:627–629.

8

Long-Term Research in the Luquillo Mountains

Synthesis and Foundations for the Future

Michael R. Willig, Christopher P. Bloch, Alan P. Covich, Charles A. S. Hall, D. Jean Lodge, Ariel E. Lugo, Whendee L. Silver, Robert B. Waide, Lawrence R. Walker, and Jess K. Zimmerman

Key Points

- The biota responds to disturbance and, equally important, influences the frequency, magnitude, and intensity of disturbances.
- Environmental gradients provide a context for contrasting the roles of particular species with respect to resilience and resistance during the interplay between disturbance and succession.
- Disturbance increases the complexity of interactions (i.e., macro- and microclimatic, biogeochemical, biotic) that control the flow of energy and cycling of materials through ecosystems.
- Soil microorganisms, as well as the timing, quantity, and quality of litter deposition, play a critical role in affecting the dynamics of carbon and nutrient cycling over short and long temporal scales.
- Disturbance affects the life history and demographic parameters of species at fine spatial scales and creates a mosaic of patches at large spatial scales, which together influence the dispersal of individuals among patches (i.e., the degree of connectedness) in a species-specific fashion. Such a cross-scale perspective provides a spatially explicit metacommunity framework for understanding the assembly of species in disturbance-mediated environments.
- Differences in biodiversity affect ecosystem processes through species complementarity, organismal traits, and trophic interactions. These effects are mediated by scale and ultimately determine the resistance and resilience of ecosystems to disturbance.

- The effects of multiple disturbances on riparian and stream communities have complex spatial and temporal linkages. Life histories of species that connect freshwater and marine communities with those in headwater tributaries and riparian forests provide pathways for pulsed flows of energy and materials.
- Anthropogenic disturbance facilitates invasions by introduced tree species, sometimes culminating in the emergence of new forest communities dominated by introduced taxa. The development of these new emerging forests does not necessarily result in the loss of native species or a reduction in species richness.
- In mature forests not subject to intense anthropogenic degradation, introduced species can occur sporadically as rare species in hurricane-induced gaps, but these populations rapidly decrease in numbers after canopy closure.
- The recognition and study of emerging new forests is important for developing an ecological understanding of how organisms respond to anthropogenic disturbances, including global climate change.
- Forecasting environmental change requires the integration of biophysical and social science perspectives. We outline an approach for developing an integrated social-ecological system for the Luquillo Experimental Forest of Puerto Rico.

Introduction

A leitmotif of research in the Luquillo Mountains is that a deep understanding of the spatial and temporal dynamics of a tropical system is predicated on consideration of the effects of disturbance and associated succession at the levels of populations, communities, and biogeochemical processes. This focus is useful because patterns of biotic change represent integrators of a disturbance regime. In addition, successional theory provides a framework within which to interpret the temporal dynamics of ecosystems in disturbance-mediated environments. Moreover, biotic responses and disturbances of the physical environment have reciprocal influences with both positive and negative feedback. Consequently, the rate and pattern of biotic change will influence and to some degree control future disturbances. Studying disturbance without understanding biotic feedbacks is useful only when the disturbance is independent of biotic influences (e.g., some earthquakes or volcanic events).

The effects of hurricanes, landslides, agricultural clearings, river flooding, and most other disturbances in the Luquillo Mountains, as well as in most other biotic systems, are influenced by human activities, the landscape configuration of biotic and abiotic characteristics, and the stage of successional development. For example, the effects from any hurricane are related to the stage of forest recovery from previous hurricanes, as well as to the hurricane's wind speed, directionality, and rate of movement across the terrain. Landslides are most frequent on slopes destabilized by roads and in early or late stages of recovery from previous landslides. Intermediate stages of vegetative succession on landslides have not developed tall trees and presumably unstable aboveground biomass, and instead they stabilize slopes through root growth and the interception of rain. Forest regrowth on abandoned pastures is a function of interactions between agricultural weeds and native woody species, as well as between nonnative and native seed dispersers. The natural flooding of streams and rivers can be exacerbated by landslides or other erosional processes or reduced by stable, well-vegetated banks. Thus, disturbance initiates succession, which in turn mediates changes associated with future disturbances. Similar feedback and complexity characterize drought effects on riparian forests. These trees decrease water loss by dropping leaves that provide energy subsidies to stream detritivores. Such leaf loss results in pulses of inceased sunlight and nutrients (leaching from leaves) that stimulate the growth of periphyton and populations of aquatic grazers.

Despite considerable effort over more than a century, we cannot predict how particular communities will change over time (Walker and del Moral 2003). This inability is not surprising, considering the diverse responses of many interacting organisms to the frequency, intensity, and extent of multiple interacting disturbance types. However, both intellectual and societal motivations compel us to forecast rates and trajectories of successional change.

Paradigm Shifts

Until the 1950s, the dominant view of succession was that species change was largely predictable, with convergence to a stable climax condition after any of a number of initial disturbances (Odum 1969). An opposing view that gained credence in the following decades espoused an unpredictable outcome based on individualistic interactions of species subjected to a constantly changing disturbance regime (Drury and Nisbet 1973). An uneasy combination of these two approaches currently dominates successional theory (Glenn-Lewin et al. 1992). Although functional attributes of some successional sequences are predictable, forecasts of complete successional trajectories remain elusive (Walker and del Moral 2003), in part because of conditions generated by global change.

Ecological research in the Luquillo Mountains is poised to develop a comprehensive understanding of the spatial and temporal dynamics of populations, communities, and nutrients via an integration of succession and disturbance from both theoretical and empirical perspectives, including applications to critical issues in management. Diverse personnel with multiple perspectives using complementary approaches, a long history of collaboration, and a varied physical setting with a well-documented history of land uses are particular strengths of the Long-Term Ecological Research Program in the Luquillo Mountains of Puerto Rico. For example, the responses of soil nutrients, decomposers, autotrophs, herbivores, and carnivores all have been documented at 1- and 5-year intervals following Hurricane Hugo (Walker et al. 1991; Walker et al. 1996a). The unique, decades-long research history in the Luquillo Mountains provides long-term data sets, chronosequences for successional studies, and a legacy of comprehensive experiments in ecology.

The remarkable physical setting of the Luquillo Mountains provides excellent opportunities for research on the spatial and temporal dynamics that are affected by disturbance and succession. First, the presence of many ecological gradients (e.g., urban to rural, high to low elevation, wet to dry forests, high to low disturbance intensity, terrestrial to aquatic) encourages experiments across distinctly different environments and facilitates robust generalizations. Second, the Luquillo Mountains mark the high-precipitation and high-temperature endpoints for the 26 sites included in the U.S. Long-Term Ecological Research (LTER) Network, thereby providing opportunities for cross-site comparisons with respect to even more expansive ecological gradients. Third, the numerous mountains within Puerto Rico and the location of Puerto Rico within the island-rich Caribbean present the potential to replicate mountains or even islands, thereby providing strong and expansive inference for ecological conclusions. Fourth, multiple disturbance types frequently interact and have a significant effect on ecological patterns and processes in the Luquillo Mountains. Fifth, the wide range of disturbance severities that initiate both primary and secondary succession represents a point of departure for testing different successional hypotheses and paradigms.

Succession—Looking to the Future

Research in the Luquillo Mountains is well suited to address several unresolved issues about succession within a broader context of disturbance. Some of these involve large spatial (landscape or larger) or temporal (decadal data sets) scales. Measurements across climatic, elevational, disturbance, or fertility gradients provide strong tests of ecological theory. Interactions between disturbance and succession, especially the effects of multiple disturbances, provide opportunities for understanding the mechanistic bases of complex biotic systems. The role of particular taxa during succession and the influence of taxa on ecosystem processes are species-based issues that continue to be explored in the Luquillo Mountains, particularly the roles of microbes, plants, and animals. Comparing effects on nutrients, as well as responses of populations and communities, offers a powerful way to address the relative importance of each with respect to resistance and resilience, providing a science-based understanding of restoration that can link science to management and policy.

Extensive Environmental Gradients

Many ecologists generally focus research at a small spatial scale (i.e., plots), with environmental measurements recorded over the course of a typical grant period (3 to 6 years). However, the LTER Program in the Luquillo Mountains was designed to address larger spatial scales and longer temporal dynamics. Nonetheless, more emphasis in the future on even larger spatial scales and the incorporation of additional types of long-term measurements that are synoptic in nature will enhance our understanding of landscape-level interactions between disturbance and succession.

Gradient analysis is an underused tool that can clarify the mechanistic bases of the distribution and abundance of species, their activities, and their succession. Climatic gradients of the Luquillo Mountains (e.g., temperature and precipitation change from low to high elevation) have been used recently in terms of static community descriptions and some soil processes. Extending manipulative experiments along such gradients (and incorporating fertility and disturbance) will clarify the successional responses to disturbance at landscape levels. For example, much is known about landslide succession in the Luquillo Mountains, but the overall importance of landslides compared to that of background treefall gaps along an elevational gradient with respect to carbon cycling or forest succession is unknown. Manipulations of fertility along steep fertility and elevational gradients might illustrate the successional responses of vegetation to variations in climate, disturbance, and soil fertility (Walker et al. 1996b). Response variables could include the performance of existing plant species and transplanted individuals of the same or different species. The one relevant experiment in the Luquillo Mountains (Fetcher et al. 1999) assessed the effects of wind on seedlings at multiple elevations. Eventually, comparisons of replicated gradients will offer even more comprehensive insights (e.g., how species responses to a hurricane vary across elevation among several watersheds or mountain ranges).

Successional Trajectories and Multiple Disturbances

The complex disturbance regime of the Luquillo Experimental Forest (LEF) provides opportunities for transformative research. Specifically, successional trajectories after a disturbance are modified by subsequent episodic disturbances. The effects of these repeated disturbances on succession, soil stability, canopy birds, soil microbiota, and many other forest and associated aquatic components are unknown. For example, are effects of multiple disturbances additive, and if not, are the synergisms positive or negative? Do population-level or community-level responses cascade in a particular order that represents successional replacements? In this regard, landslide succession is altered by selective damage to recovering vegetation by hurricanes. During Hurricane Hugo, many large fallen trees and branches accumulated in stream channels. This woody material formed debris dams that retained leaf litter and sediment washed into the streams during the storm. These debris dams formed because many older trees had grown along the riparian zone during the more than 50 years since the previous hurricane. At the time of Hurricane Georges, the same stream channels were not filled with fallen trees and sediments, and leaf litter was not retained in headwater channels, because few trees had regrown in the riparian zone during the 9-year interval between Hurricanes Hugo and Georges. Consequently, wood was not present to slow stream discharge, and floods had greater effects in removing bottom-dwelling-organisms (A. P. Covich, personal observation). Furthermore, unlike the situation after Hurricane Georges, streams that flooded during Hurricane Hugo experienced a drought in subsequent months. Comparisons of various interactions between disturbances of the same (e.g., hurricane to hurricane) or different (e.g., road construction to landslide) types will deepen the understanding of forest dynamics and the interplay of disturbance history and succession (see Willig and Walker 1999).

The Role of Species

The roles of particular species or functional groups of species provide potential avenues for developing useful insights. Experimental manipulations of keystone species might reveal mechanisms whereby the biota influences successional trajectories. Palms, *Cecropia*, tree ferns, ground ferns, snails, walking sticks, crabs, and rats are candidates for more extensive research regarding their responses to disturbance and roles during succession. Do these species inhibit or facilitate species change, and is it done directly or through their interactions with other species? How do the effects of these species on nutrient turnover and biomass accumulation affect the mode, tempo, or direction of succession? Additions or removals of key species might provide experimental corroboration of their importance to succession. Alternatively, key functions might not be associated uniquely with any one species. Rather, ecosystem attributes such as soil stability, soil aeration, or rates of nutrient turnover might be the direct drivers of regeneration. For example, differences in soil oxygen levels might be the most critical constraint on responses to disturbance and successional dynamics. If assembly rules exist in the Luquillo Mountains, they likely revolve around the establishment of processes that are critical to succession. The LTER Program in the Luquillo Mountains is well prepared to provide comprehensive answers to such questions. For example, Zimmerman et al. (1996) identified six types of response to hurricanes exhibited by nutrients, plants, and animals (see figure 2-3). This exciting discovery—that a limited set of curves summarizes responses to a disturbance—provides a fruitful avenue for further investigation, particularly when placed in a successional context. Do organisms with similar functional roles respond in similar ways to similar disturbances, thereby constituting response groups during succession? How do the various characteristics of disturbance (e.g., intensity, extent) modify responses, particularly aspects of resistance and resilience? Finally, what are the evolutionary forces that select for adaptive traits of these "response groups" in rain forest ecosystems?

Disturbance, Succession, and Society

Research in the Luquillo Mountains must continue to address societal needs. Predictive modeling of the responses of populations, communities, and biogeochemical fluxes and pools to increased water extraction, increased roadway construction, the erosion of riparian zones, and changes in the distribution of precipitation or increased hurricane frequency depend on knowledge of the successional responses to disturbance. Forest restoration and management directly benefit from realistic goals cast in a successional context. For example, can any successional stages be skipped and time saved in restoring forests derived from abandoned agricultural land? Are the processes the same for roadside or riparian restoration at low and high elevations? Are endpoint goals realistic, given the constraints of soils or vegetation?

Ecological Vignettes

We present here a number of ecological vignettes to characterize the depth and breadth of integrated or emerging ecological research in the Luquillo Mountains.

These vignettes were selected to represent a range of products from recent syntheses of the LTER Program in Puerto Rico. In all cases, they arise from the study of disturbance and succession, and they promise to advance ecological theory from basic and applied perspectives. Although speculative in places, these vignettes point to promising areas of future research that will advance the site-specific understanding of ecological patterns and processes and catalyze the development of theory associated with them.

First, we describe insights from a long-term experiment to decouple the influence of the immediate effects of a hurricane (increased temperature and decreased litter moisture versus the simultaneous addition of biomass from branch and leaf fall) on the subsequent structure and functioning of tabonuco forest. Second, we consider the advantages of employing a metacommunity perspective for understanding responses of the biota to disturbance and to environmental gradients in general. Third, we explore ways in which research in the Luquillo Mountains can provide a deeper understanding of the linkage between biotic structure and ecosystem functioning. Fourth, we reveal how food webs connect terrestrial and aquatic compartments of tropical forest ecosystems in dynamic ways following hurricane disturbances, with implications for integrated ecosystem management. Fifth, we discuss the consequences of invasions of introduced species in the structure and functioning of Puerto Rican ecosystems. Sixth, we advance the view that distinctively new forests with uncompromised functionality can arise in tropical landscapes after extensive historical deforestation. Seventh, we explore the benefits of integrating social and natural science perspectives in environmental models that forecast spatial and temporal dynamics of tropical ecosystems.

Disentangling Mechanisms of Ecosystem Response to Disturbance

Tropical forests are exposed to a wide range of disturbances that differ in spatial extent, severity, intensity, and frequency (Boose et al. 1994; Foster and Boose 1995). Considerable research in the Luquillo Mountains has documented the effects of different disturbances on plants, animals, and ecosystem processes over short (Frangi and Lugo 1991; Walker et al. 1991; Scatena et al. 1993; Silver and Vogt 1993; Wunderle 1995; Zimmerman et al. 1995a; Everham and Brokaw 1996; Ostertag et al. 2005) and intermediate (Crow 1980; Weaver 1986, 2002; Gregory and Sabat 1996; Walker et al. 1996a, 1996b) time scales. The documentation of changes in organismal, population, community, and ecosystem characteristics following disturbance has provided a rich context in which to explore fundamental mechanisms controlling the dynamics during the recovery of strongly interconnected biotic systems. Nonetheless, the many simultaneous and interacting attributes of disturbance make it difficult to distinguish proximate and ultimate causes.

The mechanisms responsible for ecosystem responses to disturbance are complex because they differ according to the spatial and temporal scales of the analysis. Large-scale disturbances (e.g., hurricanes) create a spatial mosaic of effects across a landscape, with the initial effect and subsequent response to disturbance being affected by the topography, aspect, and initial vegetation structure (Wunderle et al. 1992; Boose et al. 1994; Bellingham et al. 1995, 1996; Everham and Brokaw 1996; Lundquist et al. 2011). This complexity is particularly evident in highly dissected environments with steep slopes, such as the Luquillo Mountains. Secondary succession is characterized by changes in resource availability, which in turn are linked to population and community dynamics, creating a complexity of interacting causes and effects.

In order to disentangle the complex and interacting factors associated with responses to disturbance, we present a conceptual model that links direct and indirect effects of disturbance with key physical, biological, and chemical characteristics of the environment. We emphasize the rapid mechanistic response to disturbance by the microbial community because it is responsible for most of the processing of organic matter in ecosystems—primarily through decomposition—and for controlling the supply of key resources (nutrients) to primary producers (Lodge 1996). Microbes also play important long-term roles in nutrient retention and loss from ecosystems (Lodge and McDowell 1991; Lodge 1993; Lodge et al. 1994; Zimmerman et al. 1995b; Miller and Lodge 1997; Silver et al. 2001). They contribute to the composition of the soil atmosphere shared with roots and other organisms and are dominantly responsible for the production of three globally important greenhouse gases (carbon dioxide [CO_2], nitrous oxide [N_2O], and methane [CH_4]) that are produced in great quantity in tropical forests (Keller et al. 1986; Silver et al. 1999, 2005a, 2005b; Teh et al. 2005). Although we chose microbes as a concise and illustrative example, other organisms or processes could be substituted within this framework because many other types of organismal processes are interconnected in strong and complex ways.

The framework starts with three key effects of disturbance: the reduction of structural complexity, increased plant mortality, and transfers of organic material to the forest floor. Although these are not the only changes associated with large-scale disturbances in a tropical forest, they are arguably the most important in terms of rapid and sustained environmental changes to the ecosystem. These three attributes of disturbance alter physical (e.g., moisture and temperature) and chemical (e.g., availability of carbon, nutrients, and toxic elements and compounds) conditions, as well as biological processes (e.g., nutrient uptake, photosynthesis, and herbivory) that eventually feed back to the microbial community composition and metabolic activity.

Reduction of Structural Complexity

One of the primary effects of large-scale disturbances such as hurricanes is the reduction of structural complexity. Severe hurricanes relocate the canopy to the soil surface, or at least to within a few meters of the soil surface (Brokaw and Grear 1991; Lodge et al. 1991; Wunderle et al. 1992; Wunderle 1995). Several associated physical changes occur as a consequence of this canopy disturbance. Decreased canopy cover results in increased light levels at or near the soil surface (Fernández and Fetcher 1991; Bellingham et al. 1996) and a corresponding increase in temperature. In terrestrial habitats, greater temperatures and increased air circulation at the ground surface lead to increased evaporation from litter and soils, and this can decrease litter and soil moisture (Lodge 1996; Richardson et al. 2010), although reduced transpiration can offset soil moisture losses (Silver and Vogt 1993; Richardson et al. 2010) (figure 8-1). Several of these physical changes can affect microbial activity, such as decomposition (figure 8-1). All else being equal, higher temperatures generally stimulate biochemical and physiological activity of microbes, resulting in higher rates of decomposition and nutrient mineralization (Lloyd and Taylor 1994; Wang et al. 2002). Decreased moisture could provide either a positive or negative feedback to microbial processes (Lodge et al. 1994).

In dry microsites, a decrease in litter and soil moisture is likely to result in plant and microbial moisture stress, slowing the rate of decomposition and associated nutrient mineralization (Miller and Lodge 1997). Basidiomycete fungi, the microbes

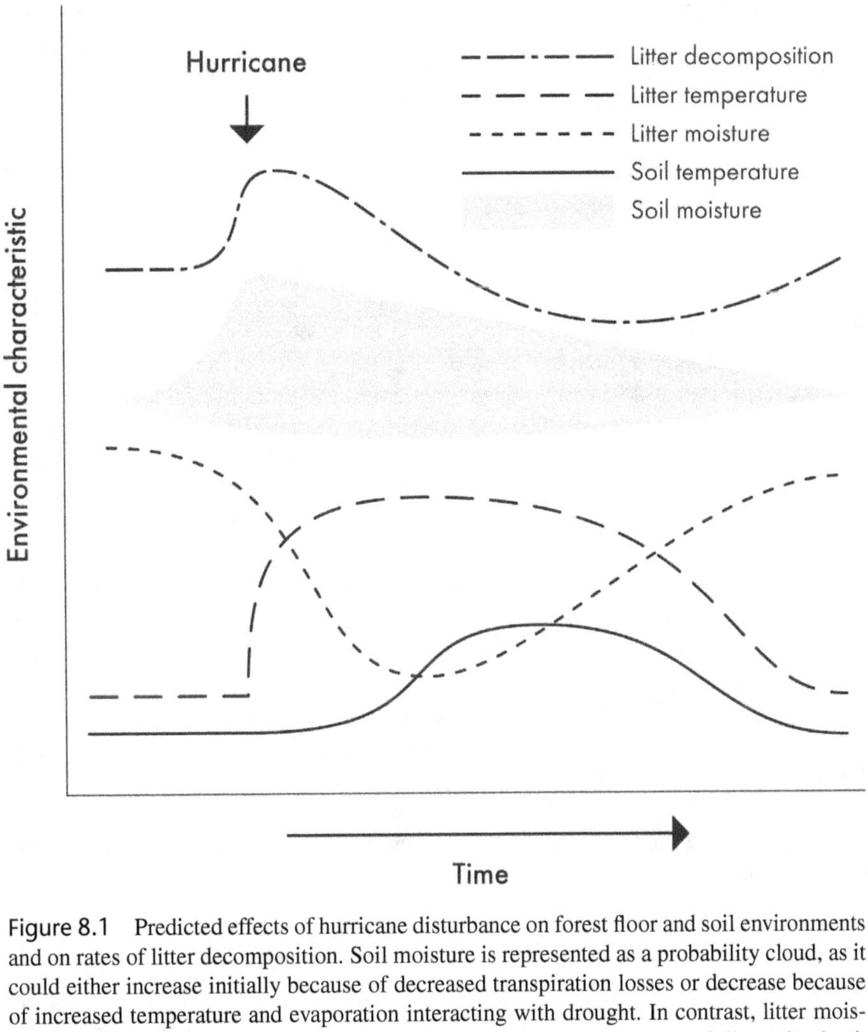

Figure 8.1 Predicted effects of hurricane disturbance on forest floor and soil environments and on rates of litter decomposition. Soil moisture is represented as a probability cloud, as it could either increase initially because of decreased transpiration losses or decrease because of increased temperature and evaporation interacting with drought. In contrast, litter moisture should decrease because of higher temperature and evaporation, especially as the depth of storm debris decreases over time because of higher initial rates of decomposition and reduced litterfall inputs.

primarily responsible for delignification through white-rot, are especially susceptible to drying in the litter layer following disturbance (Lodge and Cantrell 1995; Lodge 1996; Miller and Lodge 1997). Basidiomycete litter decomposer fungi use rootlike structures (rhizomorphs, cords, and strands) to translocate nutrients among resource bases (Lodge 1996). The abilities of basidiomycetes to translocate nutrients and degrade lignin increase the overall rate of litter decomposition (Lodge et al. 2008). Furthermore, the products of degradation by basidiomycetes differ from those of other microbial groups, which affects soil fertility (Hintikka 1970). Thus, although increased temperature can stimulate decomposition, if higher temperatures are coupled with lower moisture in the litter layer, decomposition rates might decline and decomposition products might be altered via the differential inhibition of basidiomycete fungi. Drying following Hurricane Hugo was associated with increased root mortality and the generation of large nitrate (NO_3) pools in the tabonuco forest, probably owing to increased nitrification and reduced uptake by plants (Silver and Vogt 1993; Zimmerman et al. 1995b; Miller and Lodge 1997). Larger nitrogen (N) pools could also diminish the activity of basidiomycete fungi (Carriero et al. 2000; Berg et al. 2001; Lodge 2001; Schröter et al. 2003; Mack et al. 2004; Lodge et al. 2008). Moreover, drying can reduce the capacity of microbial biomass to immobilize N (Lodge et al. 1994; Miller and Lodge 1997).

Drying associated with increased soil temperatures or drought could have a positive feedback on microbial activity in very wet forest microsites and in lower topographic zones where soils are saturated. In saturated soils, a reduction in soil moisture might relieve microbes from oxygen (O_2) limitation and increase rates of aerobic respiration associated with decomposition and nutrient cycling (McGroddy and Silver 2000; Silver et al. 2001). Increased soil O_2 could decrease rates of denitrification and increase rates of CH_4 oxidation (Teh et al. 2005), leading to a general decline in the emissions of greenhouse gases.

Water temperatures and concentrations of dissolved oxygen are relatively stable in headwater streams even after the forest canopy is opened to full sun by disturbances (hurricane and landslides) if groundwater inflows are significant components of the discharge and turbulent flow occurs along steep, rocky channels. In stream channels receiving primarily surface runoff, diurnal water temperatures will also increase, especially during periods of low flow such as droughts. In isolated pools during low-flow periods of drought, microbial respiration can lower levels of dissolved oxygen following pulses of leaf-litter inputs and accumulations of litter following windstorms. In general, microbial conditioning in streams has an important role in litter processing, especially because bacteria and fungi grow rapidly when pulses of nutrients are available after disturbances of riparian zones (e.g., Wright and Covich 2005b).

Increased Plant Mortality

Hurricanes generally increase plant mortality through both direct and indirect effects (Walker et al. 1991; Bellingham et al. 1995; Zimmerman et al. 1995a; Everham and Brokaw 1996). In addition to causing substantial structural changes, plant mortality affects soil moisture and nutrient dynamics. The reduction of live plant biomass in turn decreases the rate of water loss via transpiration and, if high rainfall continues (which was not the case after Hurricane Hugo), increases soil moisture (Silver and Vogt 1993; Richardson et al. 2010) and runoff to streams. Decreased plant activity also reduces nutrient uptake from soils, thereby increasing the standing stocks of nutrients adsorbed on soil surfaces or dissolved in soil solution (Steudler et al. 1991). As with drying, increased soil moisture could either favor or inhibit aerobic respiration, depending on the redox potential of the soil. Increased soil moisture could lead to the soil O_2 limitation of plant and microbial processes, slowing decomposition and increasing rates of N_2O and CH_4 emissions (Silver et al. 1999, 2001; Teh et al. 2005). Greater nutrient availability is likely to enhance microbial activity, and it can augment the growth rates of surviving plants (Scatena et al. 1996) or increase nutrient losses via denitrification or leaching (Steudler et al. 1991; McDowell et al. 1996).

Transfers of Organic Matter to the Forest Floor

Hurricane winds defoliate trees, break fine branches, sever roots, and snap or topple stems, resulting in large inputs of fresh organic material to the forest floor (Lodge et al. 1991; Silver et al. 1996; Ostertag et al. 2003) and to stream channels (Crowl et al. 2001). These inputs of organic matter initially insulate soils from evaporative water loss, thereby increasing litter and soil moisture and decreasing soil temperature (figure 8-1; Richardson et al. 2010). However, litter mass decreases during subsequent decomposition, and its insulating effect is thus reduced. In combination with the higher temperatures that result from an open canopy, the remaining litter (and sometimes soil) layer can become dry in these gaps (figure 8-1). We conducted a factorial experiment in order to disentangle the confounded effects of canopy opening and debris deposition on litter and soil processes (figure 8-2). Litter moisture decreased in plots that were trimmed to simulate hurricane damage, unlike in control plots that were neither cut nor subject to debris addition (figure 8-3), resulting in slower rates of decomposition in the upper, green leaf layer (Lodge, unpublished data). Increased soil temperature and moisture affected microbial communities and, together with additions of carbon [C] and nutrients in organic matter, increased soil microbial activity significantly (Silver, unpublished data).

Control of Carbon and Nitrogen Fluxes following Disturbances

Canopy disturbance results in a large pulse of carbon and nitrogen to the forest floor and soil and, with time, to streams. The pulse of carbon and nitrogen from fine debris associated with Hurricane Hugo in tabonuco forest (Lodge et al. 1991) was more than the cumulative total for a year without hurricane disturbance, but the pulse was slightly less than yearly litterfall totals in various forest types after Hurricane Georges (Ostertag et al. 2003). Although the decomposition of this litter pulse was rapid, with the forest floor returning to prehurricane standing stocks of litter within 2 to 10 months after Hurricane Georges, the rate of mass loss might not have been elevated above normal background levels (Ostertag et al. 2003). The addition of green leaves having higher nutrient concentrations than

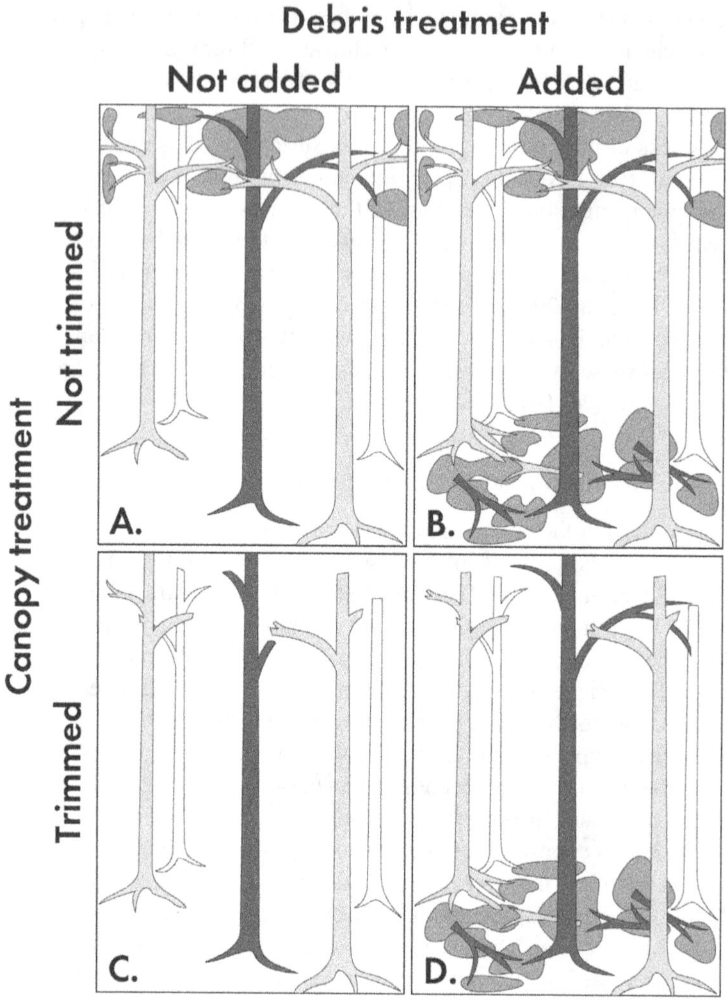

Figure 8.2 The canopy-trimming experiment employed a factorial design (canopy vs. debris treatments) to disentangle the interacting effects of increased inputs of organic matter associated with hurricane windfall (i.e., debris addition) and the effects of solar insolation and warming associated with canopy removal (canopy trimming). (A) Control (i.e., not trimmed and no debris addition). (B) Debris addition without canopy trimming. (C) Canopy trimming without debris addition, (D) Canopy trimming with debris addition.

normally senesced leaves (Lodge et al. 1991), together with the large mass of fallen litter, which helped retain moisture on the forest floor, counteracted greater evaporative losses resulting from canopy opening, thus maintaining high rates of decomposition. Consequently, rates of decomposition were high following both Hurricanes Hugo and Georges, despite the possible negative effects of high concentrations of secondary compounds (e.g., tannins, phenolics, or alkaloids) in fresh leaves of some species in certain forest types that are toxic to microbes and

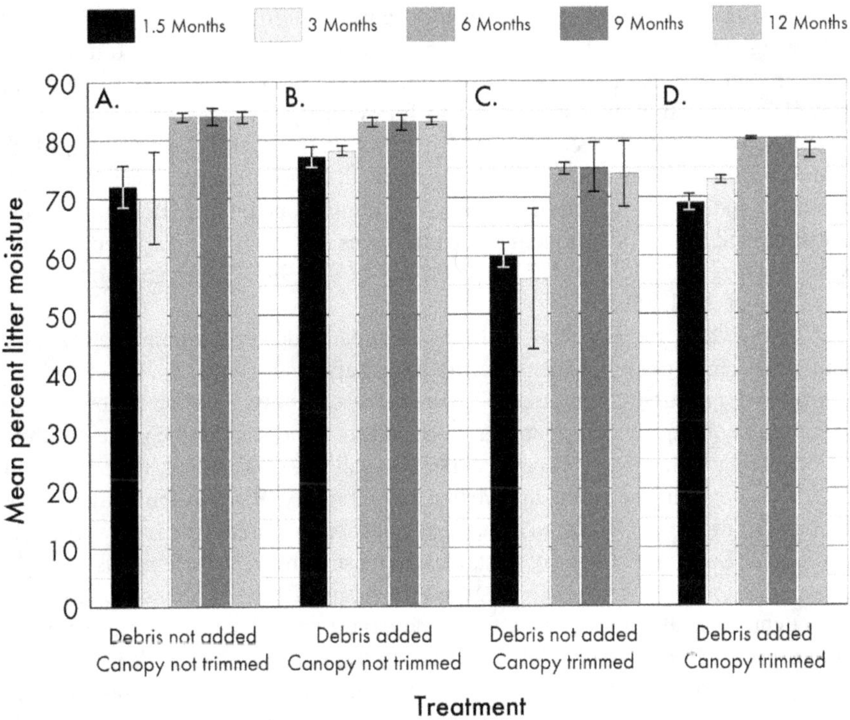

Figure 8.3 Changes in the percentage of litter moisture over 1 year in a cohort of senesced leaves from the canopy-trimming experiment (see figure 8-2) for each of four combinations of levels of treatment: (A) control (i.e., not trimmed and no debris addition), (B) debris addition without canopy trimming, (C) canopy trimming without debris addition, and (D) canopy trimming with debris addition. Data are from a weighed layer of freshly fallen leaves sandwiched between 1-mm-mesh screens in litter baskets and placed directly on the forest floor. Treatments with debris addition were capped with fresh green leaves (mass equivalent to the mean input of fine litter during Hurricane Hugo). Debris addition increased the litter moisture of the cohort below and buffered the drying effect of canopy trimming.

can inhibit decomposition (Silver, unpublished data). This maintenance of high posthurricane decomposition rates held true despite different characteristics of the storms and poststorm environments, and it prevented any substantial accumulation of soil organic carbon when averaged over the landscape (Silver et al. 1996; Ostertag et al. 2005). Although those results conflict with model predictions by Sanford et al. (1991) that coarse woody debris would increase tree productivity in tabonuco forest by increasing phosphorus availability associated with increased soil carbon, Lodge (unpublished data) and Zalamea et al. (2007) found higher total and extractable soil carbon, respectively, under decaying logs than in carefully paired samples taken near logs. Zalamea et al. (2007) found that sodium hydroxide (NaOH)-extractable C and water-extractable organic matter were higher in the soil influenced by 15-year-old logs. Such pulses can create hot spots that are readily exploited by plants (Lodge et al. 1994). The fate of nitrogen was less certain, but it might be coupled to weather patterns preceding, during, and following storms. For example, Hurricane Georges was preceded and followed by rains that maintained high soil moisture in the warm, open canopy environment, and Hurricane Hugo was preceded and followed by drier conditions that were associated with a high necromass of fine roots (Parrotta and Lodge 1991; Silver and Vogt 1993). Regardless, hurricane rains, decreased plant uptake, and high substrate and energy availability generally stimulate short-term rates of N mineralization and lead to subsequent pulses in losses via leaching and gaseous emissions (Lodge and McDowell 1991; Steudler et al. 1991; Silver and Vogt 1993; Zimmerman et al. 1995b).

Consequently, the export of nitrogen in streams decreased immediately after Hurricane Hugo owing to microbial immobilization stimulated by labile carbon. Nitrogen concentrations in streams then increased above baseline following the storm as a consequence of microbial mineralization induced by the pulse of labile C interacting with drought (Lodge and McDowell 1991). The droughts before and after Hurricane Hugo caused massive fine root mortality (Parrotta and Lodge 1991), thereby limiting plant uptake of mineralized nutrients (Lodge and McDowell 1991; Miller and Lodge 1997). In contrast, exports in stream water were not as strongly elevated following Hurricane Georges, which was neither preceded nor followed by a drought. Globally, tropical forests are the largest natural source of nitrous oxide (Prather et al. 1995), a radiatively important N trace gas. Nitrous oxide is produced during nitrification and denitrification, although denitrification is thought to predominate under humid conditions in which soils experience periodic anaerobiosis and have sufficient labile C to fuel the process (Groffman and Tiedje 1989). Following Hurricane Hugo, nitrate accumulated in soils owing to high nitrification and low assimilation rates (Silver and Vogt 1993). The high nitrification probably resulted in nitrous oxide emissions; short-term anaerobic events stimulated by high biological activity during infrequent storms also could have resulted in considerable denitrification to nitrous oxide (Steudler et al. 1991). Research in the LEF and other tropical forests (Silver et al. 2001, 2005b) suggests that nitrogen also can be retained following disturbances via dissimilatory nitrate reduction to ammonium (DNRA). An anaerobic microbial process, DNRA rapidly reduces nitrate to ammonium, which can be assimilated easily by roots and soil microbes. DNRA can be limited, primarily by nitrate, in humid tropical forest soils, effectively competing with denitrification and contributing to nitrogen retention following disturbance events. The fate of nitrogen derived from pulsed hurricane inputs thus depends greatly on the soil moisture and soil oxygen regime following the disturbance.

In addition to controlling DNRA, soil microbes can play an important role in retaining nitrogen in the ecosystem via assimilatory processes. Nitrogen immobilization by soil microbial biomass increased 3 months after Hurricane Hugo and continued for about 5 years in tabonuco forest (Zimmerman et al. 1995b). Nutrient immobilization from the decay of woody debris from Hurricane Hugo might have contributed to slow canopy recovery, as plots in which hurricane debris was removed within a month of the disturbance recovered canopies more quickly than did those without such treatment (Zimmerman et al. 1995b). However, bole diameters increased more rapidly in the control than with the debris-removal treatment after 4 years (Walker et al. 1996b). This is consistent with predictions by Sanford et al. (1991) based on expected increases in the availability of nutrients from the decomposed debris. Less structural modification of the forest characterized Hurricane Georges, partly because it had slower windspeeds, but also probably because trees had previously been pruned by Hurricane Hugo (chapter 5). As such, it is unclear whether the amount of N immobilized by soil microbes depends on the ratio of fine to woody debris, resulting from differences in hurricane windspeeds, or is a consequence of the conditions that characterized the response of the forest to earlier disturbance.

The rate of decay of surface litter and nutrient mineralization in fine litter is controlled largely by fungi and litter arthropods. Nonetheless, the manner in which a large pulse of storm debris affects fungal processes on the forest floor is unclear. The quality of storm-produced litter is different from that of litterfall that comprises senescent leaves, and storm-produced litter has greater concentrations of nutrients that normally would be translocated from the leaves before abscission (Lodge et al. 1991). In some cases, the mixing of green and senescent litter can inhibit litter decomposition, perhaps through the effects of phenolic compounds contained in the green leaves (but see Xu et al. 2004). Basidiomycete fungi that form rootlike structures (rhizomorphs, cords and hyphal strands) that allow them to translocate N, phosphorus (P), and other nutrients between old and new substrata are favored by greater litter depths but are often diminished by the dry conditions associated with litter in canopy openings (Lodge 1993; Miller and Lodge 1997; Lodge et al. 2008) (figure 8-3) and by higher nitrogen availability in such disturbed areas (Carriero et al. 2000; Berg et al. 2001; Lodge 2001; Schröter et al. 2003; Mack et al. 2004; Sjöbe et al. 2004).

Rates of tree growth were higher in tabonuco forest in the decade after a major hurricane than in subsequent decades (Briscoe and Wadsworth 1970; Weaver 2001; Uriarte et al. 2004), consistent with ecosystem model predictions (Sanford et al. 1991; Wang and Hall 2004). The extent to which this stimulatory effect is the result of decreased competition (thinning effect), increased nitrogen mineralization (Wang and Hall 2004), or increased P availability associated with organic and inorganic inputs (Sanford et al. 1991) and lower redox potential (Chacón et al. 2005) is unclear. Rapid recruitment of fast-growing, N-demanding species such as *Cecropia schreberiana* occurred after Hurricane Hugo (Guzmán-Grajales and Walker 1991; Scatena et al. 1996; Brokaw 1998) and might have played an important role in stemming the loss of N (Silver 1992; Walker 2000). However, recruitment of *C. schreberiana* was low following Hurricane Georges. If repeated disturbance depletes the soil seed bank of common secondary successional species, the role of these species in N retention will decline.

Mechanisms of Response: Directions for the Future

A long history of monitoring and observational studies in the Luquillo Mountains has provided a wealth of data detailing the severity of disturbance and documenting biotic and abiotic responses in a comprehensive manner. Nonetheless, the mechanistic basis of these responses is difficult to uncover because the major effects of large-scale disturbances (reduced structural complexity, increased plant mortality, translocation of organic matter, and consequent changes in abiotic conditions) are confounded as "treatment effects" in an uncontrolled experiment or observational study. Thus, understanding the mechanistic basis of the observed severity of a hurricane and subsequent population, community, and ecosystem responses requires approaches capable of isolating or decoupling these effects.

Two such approaches that facilitate mechanistic understanding are experimental manipulations and mathematical modeling. Modeling has a long history in the Luquillo Mountains, whereas long-term experiments have been implemented only recently. In an attempt to disentangle the confounded effects of hurricanes, we initiated a long-term manipulative experiment that will guide further modeling of successional dynamics. The experiment involves two factors: trimming of the canopy and the transfer of litter and branches to the forest floor in quantities similar to those associated with Hurricane Hugo. These treatments were applied using a factorial design (figure 8-2), resulting in four types of experimental plots: (1) unmanipulated control (i.e., intact canopy with no litter added), (2) intact canopy with litter deposition, (3) trimmed canopy with no litter addition, and (4) trimmed canopy with litter deposition. A suite of microenvironmental, biotic, and edaphic factors and processes were measured prior to the treatment application and will be continued throughout the experimental period. In order to simulate the increased frequency of intense hurricanes that is predicted for the Caribbean region (Webster et al. 2005; Hopkinson et al. 2008), the treatments will be reapplied every 6 years. Changes in litter moisture in response to the four treatment combinations (figure 8-3) suggest that the addition of debris to the forest floor shields it from the drying effects of canopy opening. In relation to the processes discussed here, factorial and repeated-measures analyses of data from the canopy-trimming experiment will help to disentangle the confounded effects of disturbance on the fungal processes on the forest floor. They will allow us to carefully track rates of decomposition and nutrient immobilization in different litter cohorts and quantify the connectedness between litter cohorts via fungal organs for nutrient translocation with and without the associated canopy opening. We will also be able to determine whether increased rates of nutrient translocation are associated with increased rates of fine litter decomposition. The repetition of canopy trimming at regular intervals should help elucidate the latter point. By repeatedly disturbing the system at relatively close intervals, the canopy-trimming experiment will also illuminate seedling dynamics and the capacity of seedling recruitment to absorb excess nitrogen released as a consequence of disturbances.

Synopsis

Our framework of responses to disturbance highlights the role of a few key physical changes associated with hurricane disturbances as potential controls on microbial activity. Changes in temperature clearly play an important role, even in a tropical forest that experiences little day-to-day variation in temperature. Temperature likely is responsible, either directly or indirectly, for the large decrease in litterfall production from low to high elevations in the LEF (Weaver and Murphy 1990), although the mechanisms are unclear (Silver 1998). Temperature changes with elevation also might contribute to patterns in soil respiration along the elevational gradient (McGroddy and Silver 2000). Patterns of soil moisture likely have strong influences on microbial processes in tropical forests. Soil water stress (e.g., rapid wetting and drying cycles) leads to the lysing of microbial cells and the release of nutrients and organic compounds to the soil (Lodge et al. 1994). High soil moisture leads to conditions in which O_2 consumption exceeds diffusive transport (Silver et al. 1999). The direct effects of soil moisture and the indirect effects of soil temperature on soil moisture are difficult to predict at an ecosystem scale because of the many disturbance-related changes that feed back on microclimatic conditions. The canopy-trimming experiment is designed to explore how single and repeated disturbances affect these microenvironmental, biotic, and edaphic dynamics. The data on moisture and temperature, combined with detailed studies of litter quantity and quality, decomposition, and microbial identity and activity, will allow us to disentangle the wide range of factors that affect detrital processing and C and N dynamics following hurricanes.

Metacommunity Perspectives on Environmental Gradients and Disturbance

Spatial variation and heterogeneity in the Luquillo Mountains, with attendant variation in ecological characteristics (chapter 3), provides a rich template by which to address contemporary questions concerning the organization of biotic communities. In particular, hurricane-induced heterogeneity in environmental characteristics at the local scale, along with elevation-induced variation in environmental characteristics at the landscape scale, can determine key aspects of biodiversity and species composition. A consideration of metacommunity theory in this context promises significant insights regarding answers to the recurrent question of what determines the nature of ecological communities.

A community comprises a suite of species at a particular site, whereas the set of all communities that occur in a particular area represents a metacommunity within the landscape (Leibold and Mikkelson 2002). Effectively, the distinction between community and metacommunity reflects a change in focal scale and extent (Scheiner et al. 2000). Within a metacommunity, constituent species populations variously link the site-specific communities to each other through different degrees of dispersal. An incidence matrix that details the presence or absence of species, or their abundances, for all communities that collectively occur in an area is a convenient visualization of a metacommunity. Importantly, the distinction between community and metacommunity frequently is guided by operational expedience, methodological constraints, or insights based on considerations of natural history. Nonetheless, the concept has considerable heuristic value. It facilitates an assessment of the manner in which underlying variation in species composition within a landscape changes over time or in response to disturbance. From a landscape perspective, the metacommunity concept is useful in understanding why different species inhabit different sites and why different sites contain different suites of species. Via two examples—one at a relatively large extent, defined by the elevational gradient in the

Luquillo Mountains (tabonuco, palo colorado, and elfin forests), and one at a relatively small extent, defined by the Luquillo Forest Dynamics Plot (tabonuco forest)—we begin to explore the utility of the metacommunity approach in clarifying spatial and temporal dynamics.

Elevational Gradients of Biodiversity

In the Luquillo Mountains, foresters and ecologists have long recognized geographic patterns of species composition, but the degree to which they are related to environmental gradients and the extent to which these patterns change as a result of disturbance are less understood (see Weaver 1991; Lugo 2005). At higher elevations in the Luquillo Mountains, the climate becomes cloudier, wetter (2,500 to 4,500 mm rainfall per year), and cooler (25°C to 18.5°C), and forests become shorter, denser, and less productive (Waide et al. 1998). Four forest types have been historically recognized within this elevational gradient: tabonuco forest, palo colorado forest, elfin woodland, and sierra palm forest (Wadsworth 1951; Weaver 1994). Tabonuco forest, palo colorado forest, and elfin woodland forest are thought to form a geographic sequence from low to high elevations, with sierra palm forest occurring throughout all elevations, particularly on poorly drained slopes. Recently, these forests were quantitatively described and classified and related to environmental controls such as climate, substrate, and topographic position (Gould et al. 2006), providing a useful classification of vegetation and the mechanisms underlying its spatial distribution. However, sampling sites were selected based on a priori conceptions of compositional differences among forest types, and locations considered to occupy ecotones and forest edges were avoided (contrast with Barone et al. 2008). Consequently, the data and subsequent analyses do not provide incontrovertible evidence that each of these four forest types is a distinct entity that reflects geographic discontinuities in species composition.

A metacommunity perspective (Holyoak et al. 2005) can shed light on this topic, as it inherently considers the ways in which variation among sites in species composition can result in particular geographic patterns. Metacommunities can be characterized by three metrics: coherence, species turnover, and boundary clumping. *Coherence* is the degree to which variation among sites (i.e., communities) can be represented by a single axis of variation, and potentially by a single underlying environmental gradient to which the biota responds. *Species turnover* reflects the extent of species replacement along this continuum, and *boundary clumping* reflects the degree to which the edges of species ranges are distributed in a nonrandom fashion, thereby suggesting the possibility of zonation. Quantifying all three measures (figure 8-4) allows the organization of the biota within a landscape to be classified as a (1) random, (2) checkerboard, (3) nested, (4) Clementsian, (5) Gleasonian, or (6) evenly spaced metacommunity (Leibold and Mikkelson 2002; Presley et al. 2010). Such analyses should provide useful insights into changes along geographic gradients and complement recent expositions of the structure and composition of forests in Puerto Rico (Gould et al. 2006).

A recent study addressed the metacommunity structure of trees along two transects in the Luquillo Mountains (paralleling the Río Mameyes and Quebrada

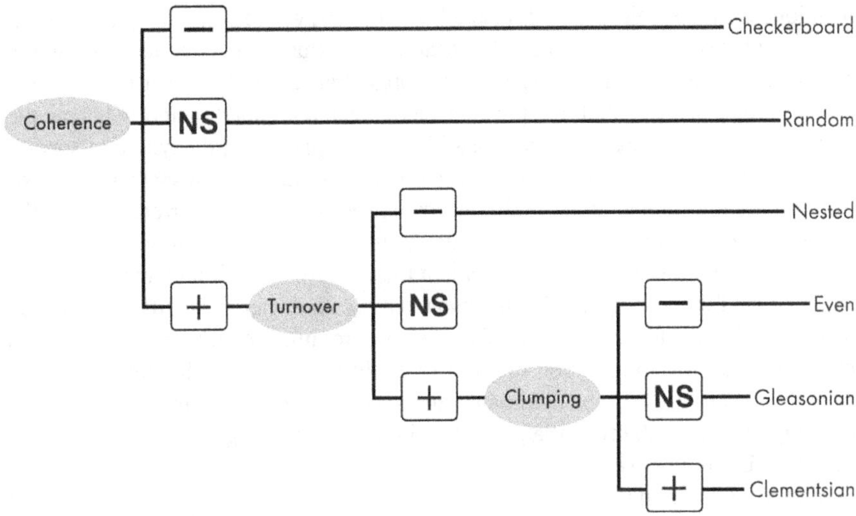

Figure 8.4 Hierarchical assessment of patterns in the distribution of species (i.e., random, checkerboards, nested, Clementsian, Gleasonian, or even gradients) based on an ordinated (reciprocal averaging) incidence matrix of sites and a statistical consideration of coherence, turnover, and boundaries. A significant positive test is indicated by a plus (+), a significant negative test is indicated by a minus (–), and a nonsignificant test is indicated by NS. For metacommunities with a nonrandom distribution of range boundaries, Morisita's index > 1 indicates clumped boundaries and Clementsian structure, whereas Morisita's index < 1 indicates hyperdispersed range boundaries and evenly spaced structure. Details of this approach for understanding the spatial organization of metacommunities appear in Leibold and Mikkelson (2002).

Sonadora; see chapter 3). This study, based on the approaches of Hoagland and Collins (1997) and Leibold and Mikkelson (2002), quantified complex patterns of nestedness and clumping of species boundaries (Barone et al. 2008) that differed between transects. Although the upper boundaries of species distributions were clumped on both transects, only one cluster of upper boundaries was detected on the Mameyes transect (850 m), whereas three clusters were visible on the Sonadora transect (500 m, 700 m, and 900 m). Neither lower boundaries of species distributions nor modal abundances of species were clumped on either transect. Finally, one test suggested that species distributions were nested, whereas another suggested that they were antinested. Taken together, these analyses do not support the contention that distinctive Clementsian community types exist in the Luquillo Mountains corresponding to tabonuco, palo colorado, or elfin forest. Indeed, Barone et al. (2008) never refer to tabonuco, palo colorado, or elfin forest.

Using a comprehensive framework (Leibold and Mikkelson 2002; Presley et al. 2011), the metacommunity structure of terrestrial gastropods has been explored extensively in the Luquillo Mountains (Presley et al. 2010, Willing et al. 2011) along extensive elevational transects. One transect passed through traditionally recognized tabonuco, palo colorado, and elfin forest (mixed-forest transect), reflecting variation in abiotic characteristics and in forested habitat types, whereas another parallel transect included only palm forest patches (palm-forest transect), reflecting variation in abiotic characteristics alone. The mixed-forest transect exhibited a Clementsian structure (species distributions forming recognizable compartments), whereas the palm-forest transect exhibited a quasi-Gleasonian structure (idiosyncratic species distributions that are independent of each other). The variation in species composition among elevational strata within each transect was highly correlated with the elevation along each transect, even though patterns of composition were not correlated between transects. This suggests that the identity of environmental characteristics, or the form of response by the fauna to those characteristics, differed between mixed- and palm-forest transects. Despite the proximity of the transects to each other, the patchy configuration of palm forest in the Luquillo Mountains, and the pervasive distribution of the dominant palm species (*Prestoea acuminata*), the relative importance of abiotic variables and habitat in structuring gastropod metacommunities differed between transects.

Future research in the Luquillo Mountains should take advantage of elevational transects established within each of three replicate watersheds: Sonadora, Icacos, and Mameyes Rivers. This infrastructural network provides a critical resource for measuring in tandem suites of environmental measurements concerning mesometeorological characteristics (e.g., precipitation, temperature), population and community characteristics (e.g., species abundances, species richness), and biogeochemical characteristics (i.e., fluxes and pools of nutrients). As such, it enables the application of hierarchical models of metacommunity organization to a variety of taxa, and it provides an opportunity to understand the way in which environmental correlates of elevation mold the spatial organization of the biota and attendant ecosystem processes.

Cross-Scale Interactions and Disturbance

Linking patterns of metacommunity structure to underlying ecological mechanisms is challenging. The spatial and temporal dynamics of biotic systems might arise as a consequence of cross-scale interactions—processes at one spatial or temporal scale that interact with processes at another scale to create nonlinear dynamics or thresholds (Peters and Havstad 2006). These interactions alter the association between pattern and process across scales because broad-scale drivers, especially those associated with disturbance, change local conditions and alter the configuration of patches in landscapes, thereby molding system dynamics (Peters et al. 2007a, 2007b). In general, disturbances affect the life-history and demographic parameters of species at fine spatial scales by altering the local abiotic, biotic, or structural environment. Environmental differences among local patches, as well as their spatial configuration, affect the interpatch dispersal of individuals (i.e., a transfer process). This alters the effective degree of connectedness among patches in a species-specific manner, coincident with the niche characteristics of species.

In the course of secondary succession, the biotic, abiotic, and structural characteristics of local sites change because of interactions between fine-scale processes and transfer processes among sites. Importantly, alterations of environmental characteristics at a local scale alter the nature and configuration of patches at the landscape scale. Such cross-scale interactions (figure 8-5) can significantly influence spatial patterns of biodiversity in complex ways, especially during posthurricane secondary succession. This view parallels conceptions that a strictly hierarchical view of systems might be insufficient to capture dynamic aspects of complex and evolving systems (Gunderson and Holling 2002).

A landscape perspective that explicitly considers cross-scale interactions (Peters and Havstad 2006; Peters et al. 2007a, 2007b) as well as metacommunity dynamics (Holyoak et al. 2005) represents an emerging view of biotic responses to disturbance (Willig et al. 2007). The metacommunity dynamics of gastropods, a taxonomic subset of the community best referred to as an assemblage (Fauth et al. 1996), in the Luquillo Forest Dynamics Plot (chapter 5) illustrate the potential of this quantitative approach to have broad applicability for furthering the understanding of community succession in the Luquillo Mountains and elsewhere. This example focuses on nestedness (Patterson and Atmar 1986; Atmar and Patterson 1993), the propensity of species-poor sites to be proper subsets of more species-rich sites. It is one of the most frequently studied patterns of assemblage structure. It has been incorporated into studies involving both large and small extents of time and

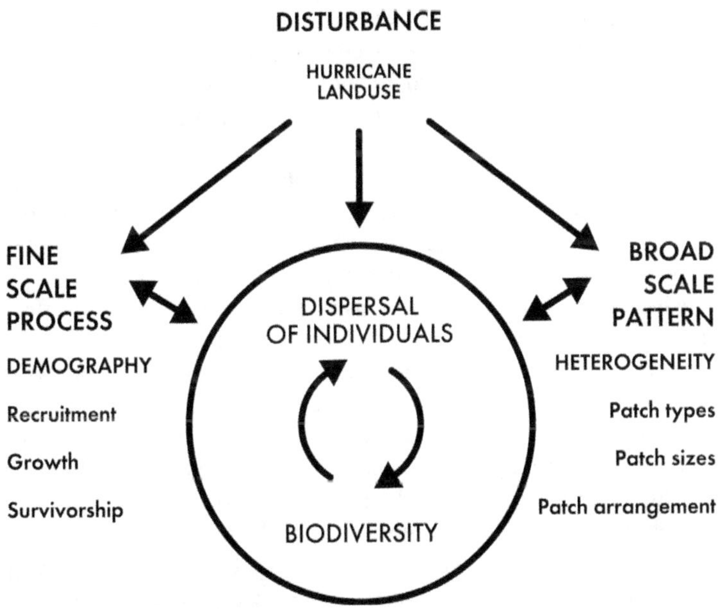

Figure 8.5 Disturbances such as hurricanes affect broad-scale patterns of habitat heterogeneity (i.e., characteristics of patchiness) in addition to fine-scale demographic processes of species associated with environmental characteristics of particular patches. The dispersal of individuals of different species and patterns of biodiversity are critically dependent on cross-scale linkages between growth, recruitment, and survivorship at the scale of particular patches, and on landscape heterogeneity regarding the types, sizes, and arrangement of patches in a landscape comprising multiple patches. (Modified from Willig et al. 2007.)

space (Kaufman et al. 2000; Leibold and Mikkelson 2002; Norton et al. 2004; Azeria et al. 2006), and it responds to disturbance (Bloch et al. 2007).

In the Luquillo Mountains, the immediate consequence of disturbance associated with hurricanes was a decrease in the nestedness of terrestrial gastropods (Bloch et al. 2007; Willig et al. 2007). Thereafter, the nestedness of gastropods increased over time; this was the case after both Hurricane Hugo and Hurricane Georges (figure 8-6). However, the rate of increase in nestedness was greater after Hurricane Hugo than after Hurricane Georges. The magnitude of nestedness differed among areas that differed in historical land use (Thompson et al. 2002). These differences in nestedness, on average, persisted after both hurricanes. Moreover, the rates of increase in nestedness did not differ as a consequence of differences in historical land use. In concert, these observations suggest that the reassembly of metacommunities (for a discussion of assembly rules, see Diamond 1975; Weiher and Keddy 2001) adheres to general patterns during posthurricane succession. Moreover, the application of neutral theory (Hubbell 2001) to metacommunity dynamics holds great promise for research in the Luquillo Mountains. This is especially true if the spatial distribution of intensities or severities of disturbance can be used to predict the patch-to-patch (or site-to-site) dispersal of individuals of different species under assumptions of equivalence (dispersal-based assembly) or nonequivalence (niche-based assembly). Indeed, the balance between these two mechanisms in terms of affecting metacommunity dynamics might itself differ during secondary succession in disturbance-prone landscapes.

Prior to hurricane disturbance, when canopy cover is extensive and environmental conditions at the level of the understory are more homogeneous, the gastropod assemblage in tabonuco forest might represent a metacommunity with a high degree of nestedness and connectivity among sites. If high connectivity enhances the likelihood that individuals will be randomly distributed among sites, a nested pattern of species occurrence manifests, regardless of the species abundance distribution in the landscape (Higgins et al. 2006). Such interconnectedness among sites probably was extensive before Hurricane Hugo, as no major hurricanes had affected the forest in decades. Each hurricane altered abundances of species at some sites more than others and caused local extirpations, thereby decreasing the degree of nestedness. Hurricanes also modify habitat by causing treefalls and moving branches and leaves from the canopy to the forest floor (chapter 5). The resultant environmental conditions (i.e., increased light and temperature, decreased humidity) generally are more stressful for terrestrial gastropods, especially in severely disturbed sites. Such abiotic changes differentially affected local populations via fine-scale processes throughout secondary succession. The reconfigured landscape of patches created by hurricanes modifies transfer processes among sites (i.e., the dispersal of particular species). Gastropod activity is reduced by unfavorable microclimatic conditions (see Cook 2001) because movement is costly in terms of water balance. Consequently, the dispersal of gastropods across severely disturbed patches is likely limited, and species that are driven to local extinction might not be rescued via immigration from subpopulations in other patches.

As forest cover and vegetation structure regenerated during succession, microclimatological conditions improved from the perspective of gastropods. These

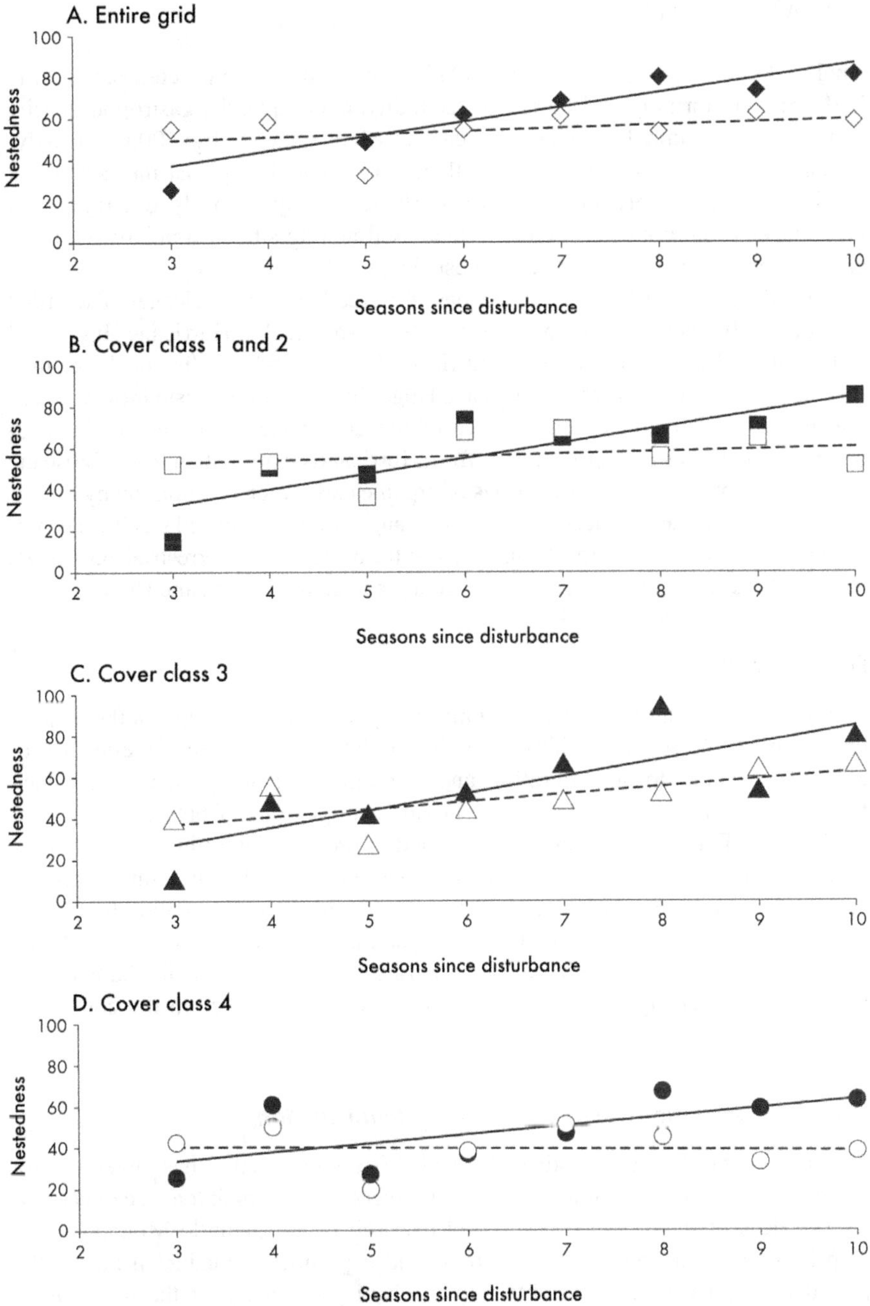

Figure 8.6 Long-term variation in the nestedness of terrestrial gastropods during wet (even numbers on abscissa) and dry (odd numbers on abscissa) seasons on the Luquillo Forest Dynamics Plot, with a focus on trajectories of change after Hurricane Hugo (solid symbols and solid lines) and after Hurricane Georges (open symbols and dashed lines). Spatial extents for analyses include (a) the entire Luquillo Forest Dynamics Plot, (b) cover classes 1 and 2, (c) cover class 3, and (d) cover class 4 (see chapter 3). Nestedness is presented using the N_c metric of Wright and Reeves (1992), standardized to eliminate the effects of matrix size (Wright et al. 1998). (Modified from Willig et al. 2007.)

ameliorated conditions at local sites likely altered fine-scale processes, resulting in decreased mortality and increased fecundity. Consequently, gastropod density increased over time following hurricanes (Bloch and Willig 2006). Density-dependent selective pressures should then have favored dispersal into areas that had been more severely modified by disturbance. Simultaneously, environmental homogeneity and connectivity likely increased among sites. Cumulatively, these effects would enhance the degree of nestedness in the assemblage.

The effect of disturbance on nestedness differed between hurricanes. The initial nestedness after Hurricane Georges was higher than that after Hurricane Hugo in all historical land use areas. Because Hurricane Georges did not disrupt the canopy structure to the same degree as Hurricane Hugo did (chapter 5), nestedness was less affected as an immediate consequence of the disturbance. The smaller input of branches and leaves to the forest floor from Hurricane Georges than from Hurricane Hugo also provided smaller quantities of organic carbon for assimilation by microbial food sources and induced only minor changes in understory plants during subsequent secondary succession. Consequently, the nestedness of terrestrial gastropods increased more slowly after Hurricane Georges than after Hurricane Hugo.

Future Directions

Future research from the metacommunity perspective should focus on the characterization of fine-scale, spatially explicit population- and assemblage-level processes, especially local emigration and immigration among sites in tabonuco forest. Concomitant quantification of broad-scale patterns of heterogeneity over the Luquillo Forest Dynamics Plot is needed in order to determine patch connectivity. To this end, it is vital to implement a synoptic network of measurements and sensors that simultaneously and syntopically assess features of the abiotic and biotic environment to which snails respond, so that the mechanistic bases of transfer processes at a small scale can be integrated with broader scale characterizations of heterogeneity to understand the effects of disturbance in the context of succession.

Disturbance, Biotic Structure, and Ecosystem Function

The relationship between biodiversity and ecosystem function has engendered a contentious debate on the importance of biodiversity to the maintenance of ecosystem services of interest to humans, and research in the Luquillo Mountains can help illuminate this debate. Observations and experiments conducted during the past decade provide new insights into mechanisms underlying the relationship between biodiversity (particularly species richness) and ecosystem processes (particularly productivity). However, there is substantial and often acrimonious disagreement about the relative importance of different mechanisms by which species richness influences productivity, with the complementarity of species and the traits of particular species proposed as the key factors (Mooney 2002). Synthesis of the ideas emerging from this debate has helped scientists to develop a more focused research agenda (Loreau et al. 2002). Recommendations arising from this synthesis correspond with research opportunities available in Puerto Rico and the research focus of the Luquillo LTER program (Hooper et al. 2005).

Specifically, the state of knowledge of ecosystems in the Luquillo Mountains provides a solid base for the advancement of four key areas of investigation: (1) the mechanism (taxonomic and functional diversity, community structure) by which changes in biodiversity affect ecosystem processes (understanding diversity–function relationships), (2) the importance of the trophic level at which changes in biodiversity take place, (3) the assessment of the temporal stability of ecosystem properties and its relationship to disturbance, and (4) the scale at which changes in biodiversity operate to affect ecosystem processes. Uncertainty about the mechanism by which biodiversity affects ecosystem function arises in part because of the relationships among taxonomic diversity, functional diversity, and community structure (Stevens et al. 2003; Willig 2003; Hooper et al. 2005; Stevens et al. 2006; chapter 6). These relationships can be addressed in the Luquillo Mountains through studies of the variation in these characteristics in time and space, which might be caused by either abiotic gradients or disturbance. The historical focus on trophic structure and dynamics in the Luquillo Mountains (Reagan and Waide 1996) establishes the foundation for comparative studies in which biodiversity manipulations take place at different trophic levels. Natural and experimental disturbances in the Luquillo Mountains provide the opportunity to evaluate the long-term stability of many elements of biotic structure and ecosystem processes. Multiple gradients in the biotic structure at different scales (e.g., local, landscape, within Puerto Rico, within the topics, tropic-temperate) provide additional opportunities to compare the effects of biotic structure on functionality at different scales. Because of these research opportunities, the Luquillo LTER program is in a position to advance a mechanistic understanding of the relationship between biodiversity and a number of critical ecosystem functions.

Taxonomic Diversity, Functional Diversity, and Community Structure

A mechanistic understanding of how biodiversity relates to ecosystem processes depends on knowledge of the functional traits of species (Hooper et al. 2005). Species traits provide a critical link between biodiversity and ecosystem processes by affecting energy and nutrient fluxes or by modifying abiotic conditions that indirectly affect these factors (Chapin et al. 2000; Covich et al. 2004a). Moreover, the expression of these traits depends on spatial and temporal variation in elements of biodiversity such as species richness, evenness, composition, and interactions (Chapin et al. 2000). In the Luquillo Mountains, studies of the traits of stream organisms have contributed to a clear understanding of the relationship between diversity and ecosystem processes in relatively simple communities (see below). However, a similar understanding of the traits of terrestrial organisms and the factors affecting the expression of these traits is lacking, and this is an impediment to the advancement of a comprehensive understanding of the relationship between biodiversity and functionality in the Luquillo Mountains.

The Luquillo Mountains have 830 plant species, including at least 250 tree species (chapter 3), thereby providing ample opportunities for biodiversity research.

The taxonomy of most organisms in the Luquillo Mountains is well understood, and this critically informs the comparative study of species traits. Disturbances themselves, as well as the environmental variability produced by disturbance and succession, encourage frequent, significant shifts in the local components of biodiversity (e.g., species composition, diversity, evenness), providing an observational framework for the study of diversity–function relationships. The trophic structure of one of the major Luquillo ecosystem types, tabonuco forest, has been articulated in a comprehensive fashion (Reagan and Waide 1996), detailing important ecological information concerning predator–prey and competitive relationships. Long-term measurements of flows and storages are available for many biogeochemical processes. Lacking are coordinated efforts to determine the biological characteristics of organisms and to understand how these properties interact to mold ecosystem function. Studies of such interactions are critical for an understanding of the relationship between biodiversity and ecosystem function.

A key gap in knowledge concerning biodiversity–function relationships relates to the scarcity of research on species-rich natural communities (Chapin et al. 2000). Moreover, much of the research on the relationship between diversity and function has emphasized small, artificial communities in which functional responses (e.g., biomass, cover, net primary productivity [NPP]) approach an asymptote at a low number of species (Tilman et al. 1996; Hector et al. 1999). The diversity and variability of forest communities in the Luquillo Mountains, coupled with frequent disturbance, provide an opportunity to evaluate the contribution of high species richness to ecosystem function. Moreover, life-history studies of numerous tree species (McCormick 1994; Zimmerman et al. 1995a; Lugo and Zimmerman 2002), as well as comprehensive investigations concerning the productivity of forest trees under a variety of conditions (Wadsworth 1947; Crow and Weaver 1977; Weaver 1979, 1983; Wadsworth et al. 1989), provide rich data from which to assess the importance of species traits to ecosystem productivity. For example, long-term studies of the response of tree assemblages to disturbance in the Luquillo Forest Dynamics Plot can be coupled to ecosystem-level responses (e.g., successional change, NPP, decomposition) through knowledge of species traits such as the response to light levels, growth and mortality rates, and leaf chemistry (see chapter 6 for examples).

Biodiversity has been identified as an important factor in determining rates of decomposition, an essential process in all ecosystems (Salonius 1981). In the Luquillo Mountains, research has established a detailed understanding of the factors that contribute to rates of decomposition (see above). Microbes and their attendant invertebrate associates are keystone taxa that control decomposition, and many of the microbial organisms important in decomposition in the Luquillo Mountains have been studied in detail.

Willig et al. (1996) examined the effect of disturbance on the functional diversity of microbes in the Luquillo Mountains. They found that the functional diversity of microbes was related positively to the degree of damage from a hurricane, but they were not able to detect an effect of historical land use. However, the particular species having catabolic profiles that make them critical to the process of decomposition (e.g., the ability to metabolize lignin versus starch) require identification and study, focusing on the possible existence of mutualisms among different microorganisms (Paerl and

Pinckney 1996). Specifically, the degree to which decomposing organisms form highly integrated functional communities (consortia), with different microorganisms contributing unique enzymatic functions (Chapin et al. 2000), is unknown, especially in tropical systems. Moreover, the effect of disturbance on these functional consortia and the means by which they are reestablished require further study if the successional dynamics of decomposition are to be understood from a mechanistic perspective.

Specialization by decomposer microfungi for particular leaf types within naturally occurring mixtures of litter on the rainforest floor strongly contributes to the high diversity of these fungi and results in more efficient processing of debris (i.e., faster rates of decomposition) than if such specialization were absent. Indeed, Polishook et al. (1996) found that two-thirds of the microfungal taxa were present in only one of two contrasting litter types from natural mixtures on the forest floor in tabonuco forest. Moreover, Santana et al. (2005) found that dominant microfungi from a particular leaf species caused greater mass loss of leaves than did microfungi that were dominant decomposers in other leaf species. Matching the microfungi to leaf substrates based on physical or chemical characteristics of the source and substrate plants or the source-substrate phylogenetic relatedness of the plants significantly increased rates of mass loss relative to plant mass loss when leaves were inoculated with microfungi from mismatched source-substrates. Thus, the high diversity of microfungi associated with decomposing leaves, and the strong association of these microfungi with particular leaf types, likely influences rates of decomposition. Although basidiomycete fungi that degrade lignin but are not host-specific had a stronger effect on mass loss than did microfungi, Santana et al. (2005) found that microfungi and basidiomycetes acted synergistically during decomposition.

Similarly, the traits of terrestrial invertebrates that determine their contribution to decomposition have been established only in general terms (Pfeiffer 1996; González 2002). For example, the fragmentation of litter by invertebrates increases the surface-to-volume ratio of fragments and therefore increases the rates of leaching and microbial decomposition. However, pellets of unassimilated material that pass through the guts of some millipedes decompose at a rate similar to that of the parent leaf material (Nicholson et al. 1966; Webb 1977). A better understanding of the relationship between biodiversity and the rate of ecosystem processes such as decomposition and nutrient cycling requires more detailed knowledge of the collective functional traits of decomposer assemblages (Balser et al. 2002; Mikola et al. 2002). The correspondence between taxonomic and functional diversity in macroinvertebrates has yet to be examined, but the moderate diversity of these groups in the Luquillo Mountains (Garrison and Willig 1996) makes the goal of understanding this correspondence attainable. Comprehensive research on terrestrial decomposer communities would advance understanding of the trait-based linkages between biodiversity and ecosystem function during recurrent cycles of distribution and succession.

Multitrophic Manipulations to Understand Diversity–Function Relationships

An improved understanding of the relationship between biodiversity and ecosystem function requires more sophisticated theoretical and experimental treatments than have been attempted to date (Hooper et al. 2005). The relative paucity of theoretical studies involving multitrophic interactions limits understanding of the relationship between diversity and function. Theoretical studies that do manipulate species richness in multitrophic systems suggest variable responses that depend on elements of trophic structure such as connectivity. Hooper et al. (2005) suggested that a theory that includes multitrophic interactions will lead to more complex responses than will models based on a single trophic level.

Experimental work on the relationship between biodiversity and ecosystem function has also focused principally on manipulations of a single trophic level, generally primary producers. Observations and experiments that manipulate consumers are less common but critical for determining how effects that cascade across multiple trophic levels affect ecosystem functionality (Hooper et al. 2005). Studies that manipulate heterotrophs generally have more variable and idiosyncratic results than those that manipulate primary producers. Additional experiments aimed at elucidating the effects of changes in diversity at multiple trophic levels are necessary in order to address the subtleties of the diversity–function relationship (Hooper et al. 2005).

In the Luquillo Mountains, the relative simplicity of the food web and the dominance of small vertebrates and large invertebrates as top predators (Reagan et al. 1996) facilitate manipulative studies of terrestrial consumers. The few experiments that have been conducted demonstrate significant effects of consumers on terrestrial ecosystem processes (e.g., earthworms on nutrient cycling [González and Zou 1999; Liu and Zou 2002], litter invertebrates on decomposition rates [González and Seastedt 2001], canopy invertebrates on leaf consumption [Schowalter 1995], and understory vertebrates on herbivory [Beard et al. 2003] and nutrient cycling [Beard et al. 2002; chapter 6]). In one experiment, Dial and Roughgarden (1995) manipulated anolis lizards in the forest canopy, which caused cascading effects on populations of invertebrate herbivores and leaf consumption. Experiments that manipulate the diversity or composition of consumers instead of primary producers are likely to yield new insights into the relationship between biodiversity and ecosystem function (Hooper et al. 2005). Moreover, experiments that manipulate diversity or composition at multiple trophic levels often yield complex and revealing results because of interactions within and across trophic levels (Hulot et al. 2000; Bradford et al. 2002; Holt and Loreau 2002).

Observational approaches to the examination of the effects of trophic structure on ecosystem function rely on comparisons of community composition (and, thus, trophic structure) across time or space. Disturbance and succession provide natural experiments in which community composition changes along a temporal trajectory, with concomitant effects on trophic structure and ecosystem processes. Although much is known about community changes after disturbance in the Luquillo Mountains (chapter 5), this information has not been integrated into models of food web structure that predict functional changes (e.g., decomposition rates) over time. The development and implementation of models linking trophic structure and ecosystem function are critical challenges for the future, especially as a means of developing a mechanistic understanding of the effects of an altered trophic structure on ecosystem processes.

Elevational variation in community composition and trophic structure provides a spatial approach for linking biodiversity and ecosystem function. Because species richness in the Luquillo Mountains declines with elevation across a broad range of taxa (Waide et al. 1998), leading to a simpler trophic structure, elevational comparisons can shed light on the diversity–function relationship, provided that other environmental characteristics (e.g., slope, aspect, soil type, forest structure) can be controlled in analyses to disentangle the effects of correlated factors. Ongoing surveys of variation in community compositions with elevation (e.g., trees, gastropods, insects) provide information with which to examine elevational changes in trophic structure. For example, Richardson et al. (2005) compared invertebrate communities in the litter of palm (*Prestoea montana*) and matched dicot forest stands that changed in plant species composition along a gradient of elevation. They found that the composition of the invertebrate community was affected more by forest type than by changes in temperature or rainfall.

A productive new research approach for the Luquillo Mountains would link new theoretical approaches to the diversity–function relationship with multitrophic experimental manipulations informed by observational studies of trophic webs along the elevational gradient. Previous, detailed work on the food web of tabonuco forest (Reagan and Waide 1996) provides a point of departure for this new research focus. In addition, such a focus is compatible with the research objectives of the LTER Decadal Plan (U.S. Long Term Ecological Research Network 2007), which provides a mechanism for expanding research on trophic webs to other sites in the LTER Network.

Stability, Disturbance, and the Relationship between Structure and Function

The relationship between stability (the extent to which the parameters that characterize an ecosystem remain unchanged in response to perturbation; chapter 2) and diversity has long been a popular theme in ecology (MacArthur 1955; May 1974; Pimm 1984). Theoretical examinations of this relationship are plentiful and provide a wealth of hypotheses for examination. However, Ives and Carpenter (2007) suggested that the focus of research properly belongs on stability and the multiple factors that influence it, including biodiversity. Because stability is measured in many ways and is influenced by a complex of interacting factors, a critical need exists to understand the mechanisms responsible for stability. An enhanced mechanistic understanding of stability requires long-term measurements to assess temporal stability, as well as experimental manipulations to examine the factors affecting recovery from disturbance (Hooper et al. 2005). Long-term studies of natural (Waide and Lugo 1992; Lugo and Waide 1993) and experimental disturbances are central elements of research in the Luquillo Mountains and provide opportunities to examine the mechanisms underlying the long-term stability of biotic structure (e.g., vertebrates, invertebrates, microbes) and ecosystem processes (e.g., decomposition, nutrient cycling) affected by repeated experimental perturbations.

Since its inception, the Luquillo LTER program has used disturbance as a unifying theme for integrating studies of population, community, ecosystem, and landscape ecology (figure 8-7). Periodic disturbances decouple structural and functional characteristics of ecosystems (Willig and Walker 1999) and provide an opportunity to observe how such relationships change during secondary succession. Within the Luquillo Mountains, the rate of change after disturbance can be rapid or slow depending on the particular structural (e.g., biomass) or functional (e.g., nutrient retention) element under consideration (Zimmerman et al. 1996; chapter 2). For example, forest structure might change for decades after a hurricane, whereas nutrient retention might rapidly return to predisturbance levels. Mechanistic models of stability must provide explanations for the full range of responses to disturbance in order to link theory and empirical results successfully (Ives and Carpenter 2007). Tests of theory about the effects of species and functional diversity on stability require an understanding of the mechanistic basis of the diversity–stability relationship. One approach to developing such a mechanistic understanding involves long-term measurements of systems in which differences in diversity are not confounded by other characteristics (such as climate or disturbance) (Hooper et al. 2005).

The Greater Caribbean Basin provides the necessary conditions for examining ecosystem stability across a gradient of biodiversity. In the Caribbean, biogeographic factors create an east-west gradient of biodiversity from Puerto Rico to the Yucatan Peninsula. Across this biodiversity gradient, the major drivers of climate and disturbance are similar, and thus the gradient provides an opportunity to examine the relationship between diversity and stability for several forested ecosystems, including subtropical wet, dry, and mangrove forests. One ongoing effort to make such comparisons involves the Luquillo LTER program (wet Caribbean forest), the Florida Coastal Everglades LTER site (mangroves), the Atlantic Neotropical Domain of the National Ecological Observatory Network (dry Caribbean forest at Guánica), and three sites from the Mexican LTER Network: Celestún (mangrove

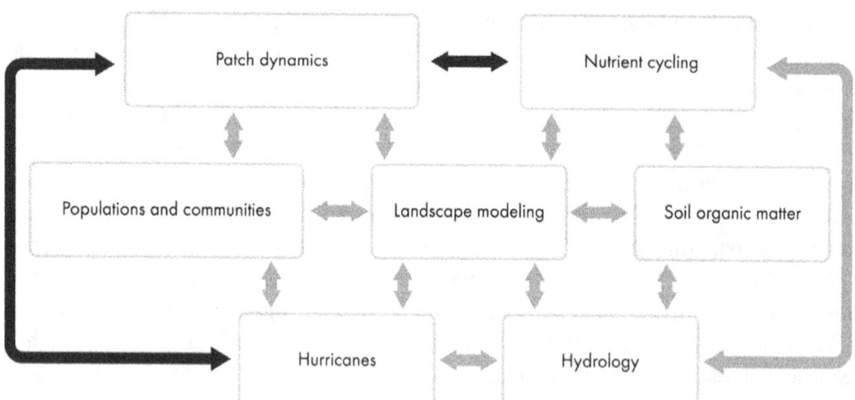

Figure 8.7 Relationships (gray arrows) among research components (boxes) of the Luquillo Mountains LTER proposal that were originally funded by the National Science Foundation. Since that time, additional foci on the connections between hurricanes and patch dynamics, and between patch dynamics and nutrient cycling (black arrows), have been undertaken as well.

forest sites in the Yucatan peninsula), Los Tuxtlas (wet forest), and Chamela (dry forest).

The Effects of Scale on the Relationship between Biotic Structure and Function

Biodiversity in general and biotic structure in particular, as well as the distribution of biomass among species and ecosystem compartments, are determined by biogeographic, biogeochemical, and climatic factors, which often are distributed along spatial gradients of latitude or elevation. On the broadest scale, the Luquillo Mountains anchor gradients of climate and biotic structure within the LTER Network. The warm and wet climate, relatively high species richness and landscape heterogeneity, and complex forest structure in the Luquillo Mountains provide important opportunities for comparisons with temperate mainland sites. For example, the Long-Term Intersite Decomposition Experiment showed that the diverse, warm Luquillo site had higher decomposition rates than less diverse, cooler sites, but it had similar rates of nitrogen mineralization (Parton et al. 2007). The Lotic Intersite Nitrogen Experiment demonstrated that ammonium turnover in streams is more rapid in the N-rich Luquillo site than in most temperate sites, owing largely to high nitrification rates (e.g., Peterson et al. 2001). The second phase of this multisite experiment showed that the total biotic uptake and the denitrification of nitrate increase with stream nitrate concentrations across 72 streams in eight biomes, but the efficiency of these processes declines with concentration (Mulholland et al. 2008). The Luquillo LTER program operates one of 15 sites in the Center for Tropical Forest Science (Smithsonian Institution) network of large, long-term forest plots. Cross-site comparisons among these plots have produced much new understanding of tropical forests and biodiversity maintenance (Brokaw et al. 2004; Losos et al. 2004; Condit et al. 2005).

At a less extensive scale, comparisons with other tropical island and mainland sites at latitudes similar to that of Puerto Rico allow examination of the relationship between structure and function over a biodiversity gradient. For example, sites along the Gulf coast of Mexico are similar to those of Puerto Rico in terms of climate and disturbance regime, but they generally have a higher taxonomic richness (chapter 3). These factors provide the necessary conditions for comparing the effects of variation in biodiversity on ecosystem processes (e.g., nutrient cycling) and properties (e.g., stability).

Within the Luquillo Mountains, the interaction of landform- and landscape-scale gradients affects the rate of change of biotic structure and ecosystem processes and can lead to strong differences over relatively small spatial extents. The pattern of plant species occurrence is related strongly to position along the catena within tens of meters of forest streams (Scatena and Lugo 1995). Plant species richness, community composition, and physiognomy change substantially over 700 m of vertical elevation from the forest boundary to the mountain tops (Weaver and Murphy 1990; Barone et al. 2008). Consequently, the biotic structure is quite distinct between the extremes of this elevational gradient (Waide et al. 1998) (figure 8-8). At mid-elevations, landform and landscape gradients interact to produce an interdigitation of forest communities, with

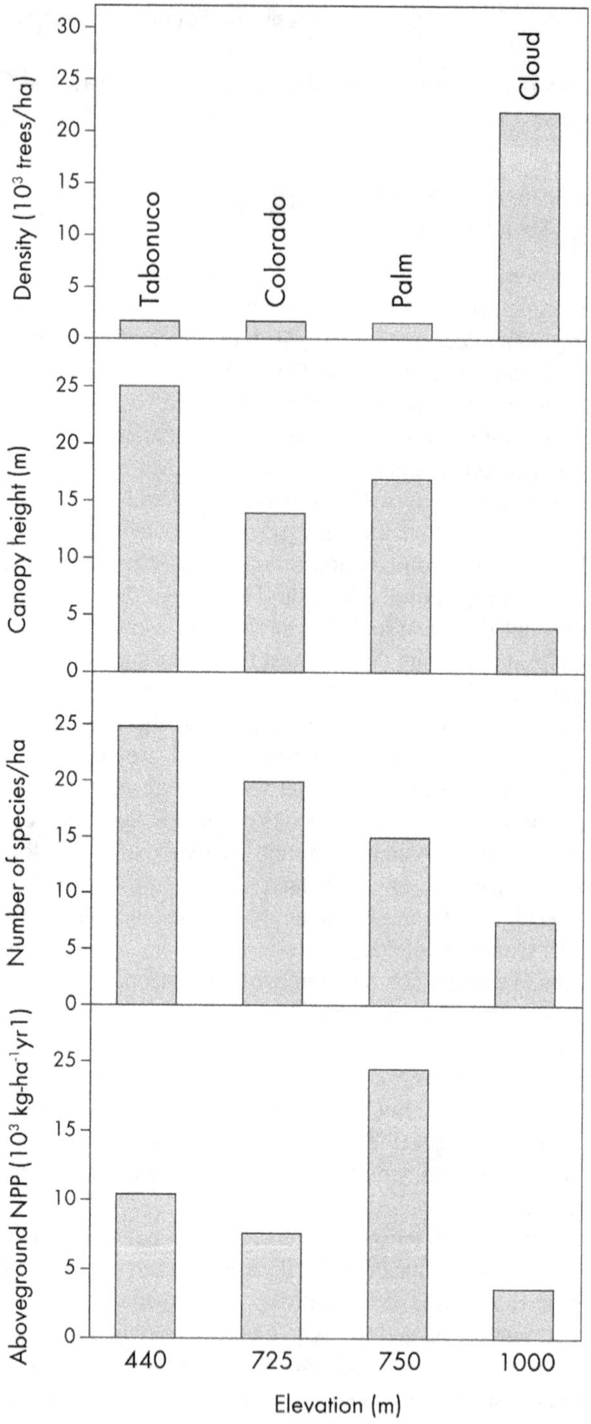

Figure 8.8 Changes in forest structure and net primary productivity (NPP) with elevation in the Luquillo Mountains, Puerto Rico. Data are from Weaver and Murphy (1990) and references therein. Elevations (in meters above sea level) are those at which particular studies were conducted; full elevational ranges of forest types are indicated in the text.

lower elevation species sometimes occupying ridgetops and higher elevation species occurring in the valleys (Odum 1970). We know less about changes in ecosystem processes (e.g., decomposition, primary production) across scales, and even less about the traits of organisms that might modulate these processes (but see González and Zou 1999; González and Seastedt 2001; González 2002). Therefore, the study of the correspondence of changes in biotic structure and ecosystem function across scales is a research priority.

Chapin (2003) presents a framework with which to predict which plant traits (and therefore which species) have the strongest effects on ecosystem processes. This framework incorporates five state factors that determine the characteristics of ecosystems (climate, parent material, potential biota, topography, and time) (Jenny 1941) and five interactive controls (functional types of organisms, resources, modulators [e.g., temperature, pH], disturbance, and human activities) (figure 8-9) that mediate the effects of state factors. Interactive controls respond to changes in state factors, and both affect and are affected by other interactive controls. As a result, state factors have both direct and indirect effects that must be measured (or controlled) in order to assess the effect of species traits on ecosystem processes. For example, Grace et al. (2007) employed a multivariate statistical model to assess the relationship between species diversity and biomass production while controlling for the effects of other integrative factors. The interplay among fluctuating interactive

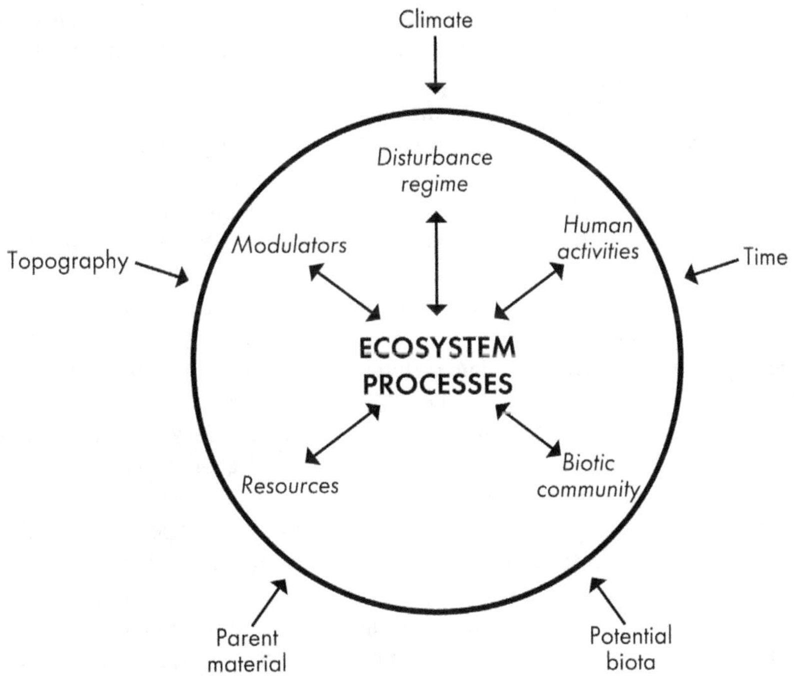

Figure 8.9 The relationship between state factors (listed outside the circle), interactive controls (listed inside the circle), and ecosystem processes. The circle represents the boundary of the ecosystem. (Modified from Chapin et al. 2002 and Chapin 2003.)

controls regulates ecosystem dynamics in the context of broader-scale changes in state factors (Chapin 2003). Species traits are most likely to have strong effects on ecosystem processes when they alter the interactive controls on those processes (Chapin et al. 2002). For example, differences in litter quality among plant species can alter resource supply rates by influencing rates of nutrient cycling (Chapin 2003).

Factors that affect the interactive controls of ecosystem processes (species identity, community composition, and physiognomy) (Chapin 2003) all vary with elevation in the Luquillo Mountains (Brown et al. 1983). However, a lack of critical information impedes our ability to use Chapin's (2003) framework to predict the effect of changing species, community characteristics, and physiognomic factors with elevation. Although functional traits and species identities are related closely (Chapin 2003), we do not know whether the distribution of species and functional traits are congruent in the Luquillo Mountains, or whether species replacements result in the continuity of functional traits with elevation. Moreover, we lack the knowledge to determine a priori which traits have the strongest effects on processes in Luquillo ecosystems or how these effects are exerted on interactive controls. The response of ecosystem processes to changes in functional traits might be linked to diversity at particular trophic levels. For example, primary productivity depends on the number and functional diversity of producers, whereas decomposition might be more closely linked to the functional diversity of microbial consumers (Chapin 2003). Our knowledge of the functional diversity of producers in the Luquillo Mountains is substantially stronger than our knowledge of the functional diversity of microbial consumers. The spatial and temporal dynamics of ecosystems along the elevational gradient might be subject to cross-scale interactions that create nonlinear patterns (see above). Moreover, anthropogenic changes in state factors are likely to increase the likelihood of nonlinear responses (Chapin et al. 2000; Ives and Carpenter 2007). Further research is needed to address these issues in the Luquillo Mountains, and particularly to determine whether the responses of ecosystem processes to changes in biotic structure at different scales are nonlinear.

Because the many factors that alter biotic structure often change in tandem across space and time, an integrated understanding of the nature and pace of ecological change requires a conceptual framework that includes the interactions among the elements of biotic structure (Willig and Walker 1999). This conceptual framework must also accommodate differences in spatial and temporal scales at which biotic structure changes (Peters et al. 2007a, 2007b; Willig et al. 2007). Substantial evidence exists that the state factors affecting ecosystems are changing at a global scale (Chapin 2003) and that these global changes have potentially important effects on biomes (Millennium Ecosystem Assessment 2005). However, changes in species composition at landform or landscape scales might be more important for ecosystem functioning than are global changes in the atmospheric composition and climate (Chapin et al. 2002). Moreover, rapid changes in land use, acting through regional climate shifts or alterations in the species pool, might have more immediate landform or landscape effects on ecosystem function than do slower global changes. The hurricane-dominated disturbance regime of the Luquillo Mountains interacts with global and regional climate changes operating as press disturbances. The development of a conceptual framework that incorporates changes at multiple spatial and temporal scales is one of the most important challenges facing researchers in the Luquillo Mountains.

Food Webs Link Forests and Streams following Disturbances

Research in the Luquillo Mountains can make valuable contributions toward an understanding of the food web connecting terrestrial and aquatic portions of tropical ecosystems. Our research uses an integrated analysis of stream and riparian food webs to allow us to understand responses to different types of disturbance. These disturbances remove biomass and alter dominance relationships of species, with significant consequences to the food web structure. In order to understand the spatial and temporal dynamics of stream food webs and their responses to disturbances, it is important to examine (1) the underlying physical structure of the drainage basin (e.g., steepness of slopes, locations and sizes of waterfalls, and depths of pools that together affect distributions of species); (2) the disturbance history of the watershed (e.g., effects of recent and past landslides, hurricanes, extreme floods and droughts, and land uses); (3) the life history characteristics of the freshwater species and their adaptations to extremely variable flows; and (4) the distributions and diversity of riparian trees, including the phenology of leaf fall, leaf chemistry, and patterns of wood inputs and shading that influence sources of energy in different locations within the drainage basin. These factors interact in a hydrological and ecological network in which horizontal and vertical flow paths of various strengths determine food web structure and function. Landslides and treefalls from riparian forests and steep hillslopes have significant long-term effects on stream environments by altering sources of nutrients and energy, as well as by changing rates of deposition of sediments and modifying pool depths and channel configurations. All these interactions can affect food web dynamics through food limitations (e.g., sediments that cover leaf litter and algae), increased vulnerability to predators (e.g., shallow pools have few refuges), and the physical removal of stream organisms to downstream or overbank locations. Moreover, inputs ("subsidies") to streams of terrestrially derived nutrients and organic matter link riparian and aquatic food webs (Covich 1988b, 2006b; Crowl et al. 2006). The movement of stream-produced nutrients and organic matter (e.g., emergent aquatic insects, shrimps, and amphibious crabs) into the terrestrial food web (e.g., spiders, bats, anolis lizards, and wading birds) is another important connection between food webs. Rates of runoff and groundwater inputs from the surrounding hillslopes greatly affect the concentrations of dissolved nutrients and sediments that, in turn, influence species distributions and abundance in the rivers of the LEF (Covich and McDowell 1996).

Physical Habitat and Succession Affect Food Web Composition

Stream food webs on islands such as Puerto Rico are characterized by relatively low species richness as a result of the combined effects of biogeography and disturbance history (chapter 3). Distances from mainland to insular rivers, as well as the ages and geological origins of tropical islands, greatly influence the geographic distributions of freshwater species and food web structure (Craig 2003; Smith et al. 2003; Covich 2006a; Boulton et al. 2008; Kikkert et al. 2009; Hein et al. 2011). Insular food webs are characterized by a high degree of connectivity among species because of the presence of dominant omnivorous species. This complex structure results in multiple energy pathways and the potential for rapid rates of biotic responses to disturbance that appear to be characteristic of many insular headwater streams. In addition to modifying the light and temperature environment, riparian forest communities along headwater streams provide important inputs of leaf litter and wood (Pyron et al. 1999), as well as other energy subsidies (e.g., dead insects) that sustain complex food webs (Crowl et al. 2001; Covich et al. 2006; Crowl et al. 2006).

Tropical insular streams generally differ from mainland streams by having smaller and more linear drainage basins, as well as fewer freshwater and riparian species. These relatively narrow basins result in the rapid rise and fall of high discharges that influence the downstream transport of food resources (leaf litter and algae) and which constitute important pulsed disturbances (Wohl and Covich, unpublished data). These events further limit the number of species that are adapted to these extremely high flows. Some species are able to migrate upstream and recolonize headwaters following disturbances (Covich and McDowell 1996; Fievet et al. 2001; Pyron and Covich 2003; Blanco and Scatena 2006). Many insular species evolved from marine species and adapted to low salinities. However, even those species that are adapted to migrate upstream to feed and reproduce must return to higher salinities during some phase of their life history. In general, the steep terrain and extremes in rainfall and runoff act as filters for the colonization of headwater streams, so that only a subset of potential colonists reaches the highest elevations. The resulting distribution of species along elevational gradients constitutes a hierarchical series of nested food webs composed of subsets of the riverine community (Covich 1988b; Greathouse and Pringle 2006; Kikkert et al. 2009; Hein et al. 2011).

Although montane streams on tropical islands are subject to many of the same types of disturbances and geologic processes as their mainland counterparts, the biotic responses to changes are often more evident on islands because they can occur rapidly over a large proportion of the watershed and are observed readily in steep terrain with high numbers of streams per unit area. For example, primary stream succession in relatively small insular watersheds is initiated when new channels are created on steep slopes by erosional processes over long periods of time (decades), as well as by the rapid (hours, days) removal of materials during major landslides. Some older channels and deep pools are filled with sediments and disappear during large landslides. Such rapid changes in the physical structure (depth, volume, flow velocity, turbidity, turbulence, and sizes of substrata) of pool habitats cause frequent turnover in the species composition and food web structure. Consequently, over decadal scales, the Luquillo Mountains provide numerous opportunities for examining how riparian and aquatic food webs are linked to successional dynamics throughout the watershed.

The steep terrain of the LEF contains numerous small streams. Importantly, these streams constitute a hierarchical network that ultimately comprises a few larger channels. Each location in the drainage network has distinct types and sizes of pools, runs, and waterfalls (Pyke 2008). As in all forested montane watersheds, headwater tributaries contain small, shallow pools and fast-flowing riffles that retain sediments and organic matter (Church 2002; Benda et al. 2004; Richardson and Danehy 2007). Larger rivers have deeper pools and longer segments of rapids in downslope locations, often with longer retention times than are typical of headwater streams (Thorp and Delong 1994; Townsend 1996; Fisher et al. 1998; Rice et al. 2001).

Pulsed Flows in Drainage Networks Affect Food Web Resilience

Pulsed additions of nutrients and organic matter from hurricanes, landslides, and land use practices affect stream nutrient cycling and retention in riparian zones (McDowell 2001). These additions result from extreme flow events, from treefalls in the riparian zone, and from bank erosion, as well as from larger scale, often repetitive landslides on steep slopes. Frequent natural, small-scale disturbances (e.g., treefalls, local bank erosion, and landslides) produce scattered inputs of nutrients that are relatively transient. However, infrequent, large disturbances contribute significant pulsed inputs of fine and coarse sediments that affect the stream channel substrata for long periods. A dynamic series of disturbances with short-term and long-term effects influences the community composition over time because species respond to changes in substrata-related resources (e.g., retention of leaf litter by roots, stable rock surfaces for algal growth, and crevices for protective cover from predators).

Intense tropical rainstorms result in extreme flows (Wohl and Covich, unpublished data). Such variability affects the distribution of species and the retention of leaf litter in montane streams. High-flow events displace some individuals to considerable distances downstream. A series of these events homogenizes distributions of detritivores, as well as leaf litter and other food resources in headwater tributaries. The downstream transport of highly turbid water to wide, well-illuminated channels decreases light penetration and algal growth, an important source of food for herbivores and omnivores.

In years with only brief, low-intensity rainfall events, stream flows do not produce pulses of organic matter and nutrients. Under these low-flow or no-flow conditions, spatial distributions of different riparian trees within the drainage network can be important in determining the quality of local habitats and species abundances. These differences in litter inputs are especially significant during periods of prolonged drought, when leaf fall increases and is retained locally in pools and riffles (Covich et al. 2003, 2006). If the periods between large storm flows are sufficiently long (several months), then local conditions in pools and riparian influences can dominate food web composition and dynamics.

Continuous inputs of leaf litter throughout the year, as well as pulsed inputs after storm events, provide important sources of energy in forested headwater streams. Daily inputs of leaf litter differ among riparian tree species. Some (e.g., *Prestoea montana*, *Casearia arborea*, and *Dacryodes excelsa*) produce a relatively continuous supply of dead leaves to the stream throughout the year. Others (e.g., *Buchenavia tetraphylla*) produce a pulse of leaf input during February and March (Thompson et al. 2002). The input of wood and the accumulations of leaf litter (especially large palm leaves) in debris dams result in the retention of organic matter until the debris dams are disrupted by larger floods. Such retention increases the reliability of energy supplies to detritivores and provides structural habitats for many species during periods of low and high stream flows.

Intense winds and rainfall associated with hurricanes and other tropical storms affect the composition and structure of riparian forests (and constituent tributaries) over timescales as short as hours or as long as several decades. Intense tropical hurricanes remove the riparian canopy, resulting in increased sunlight and nutrient availability in headwater streams. These physical and chemical changes enhance rapid growth by algae and aquatic grazers. At the same time, aquatic detritivores benefit from a large pulsed input of leaf litter, so that these riparian connections are especially strong. Thus, posthurricane conditions lead to several months of high biological productivity within the headwater tributaries.

Downstream connections within the entire drainage network are important because these channels distribute sediments, nutrients, and other materials. These connections provide corridors for upstream and downstream movements of migratory species. Debris dams and other large accumulations of organic matter associated with bank-side roots and in-channel rocky substrata reduce peak flows of water and the downstream transport of suspended sediments and dissolved nutrients. However, extreme peak flows of water transport significant amounts of nutrients and suspended sediments into larger rivers along the coast and, ultimately, into marine ecosystems.

Biotic Responses to Disturbances

Stream species respond to disturbances in different ways depending on their mobility and life history characteristics. Many are well adapted to avoid extreme floods by seeking small side channels, slow-moving waters along streambanks, or undercut banks and burrows. Often, they rapidly recolonize upstream habitats from these refuges after extremely high flows. Some species (highly mobile fishes and larger species of shrimp) are well adapted to move rapidly upstream after being displaced downstream or onto flood plains and into lateral pools along the main channels at lower elevations. Because of their mobility (e.g., swimming and crawling by fishes, many decapods, and gastropods, or flying by most adult aquatic insects), stream animals are resilient to natural disturbances and often return rapidly to their predisturbance food web structure (e.g., species composition, relative abundance).

Increasingly, stream ecologists examine the responses of food webs to drought or other disturbances (Covich et al. 1991, 1996, 2006; Power and Dietrich 2002; Romanuk et al. 2006). Responses are often rapid but seasonal in temperate-zone ecosystems (Wallace and Hutchens 2000; Nakano and Murakami 2001; Power and Dietrich 2002; Power 2006). In tropical ecosystems, flow-mediated disturbance events can occur frequently and at any time of year. Some disturbances produce long-lasting legacies by changing the species composition and dominance of riparian forest trees (Heartsill-Scalley 2005; Lecerf et al. 2005). Moreover, disturbances that lead to the establishment of nonnative riparian tree species can have long-lasting effects on the quality, quantity, and temporal distributions of leaf litter, as well as on shading. For example, since its introduction several decades ago, nonnative bamboo has come to dominate many riparian zones in the Luquillo Mountains, affecting leaf-litter production and breakdown by microbes and benthic invertebrates (O'Connor et al. 2000). Similar effects occur in other insular tropical streams where nonnative trees are introduced, such as *Hibiscus tiliaceus* L. (Larned et al. 2003).

Tropical Stream Food Webs in a Landscape Perspective

Long-term studies of food web structures are critical for understanding the direction and magnitude of net flows of energy or nutrients between terrestrial and aquatic ecosystems and among different elevational zones. Particular species in these food webs determine the directions of movement for energy and nutrients. For example, spiders and bats in the riparian corridor increase connectivity between freshwater and terrestrial food webs when they consume emerging aquatic insects from headwater streams. Adult freshwater crabs move from the stream channel to the forest floor as they forage in the leaf litter and thus link terrestrial and freshwater components of food webs (Zimmerman and Covich 2003). These amphibious adult crabs return to the stream to reproduce, and thus transfer nutrients into headwater pools from the forest floor (up to 50 m from the stream). In addition, the upstream movement of numerous postlarval shrimps, snails, and fishes transports nutrients and energy from coastal lagoons to headwaters (Covich and McDowell 1996; March et al. 2001; March and Pringle 2003; Pyron and Covich 2003; Blanco and Scatena 2006). Despite movements of matter and energy associated with animal activity, the large amount of water flowing downslope from higher elevations produces a net downstream transport of dissolved and suspended materials. At lower elevations, the movement of water from river channels onto the flood plains transfers large amounts of organic and inorganic materials into these habitats (Ballinger and Lake 2006).

Detrital Processing in Stream Food Webs

The breakdown rates of leaf litter are affected by changes in riparian tree species richness and composition, in the composition of microbial and macroinvertebrate communities, and, especially, in species that shred leaves (Crowl et al. 2001, 2006; Lecerf et al. 2005; Wright and Covich 2005a, 2005b). However, relatively little is known regarding the distributions of freshwater detritivores, changes in the rates of litter processing, and species-specific relationships in tropical stream ecosystems that differ greatly in the seasonality of rainfall and phenology of riparian leaffall (Boulton et al. 2008). Factors controlling rates of leaf-litter processing in tropical streams might differ regionally. For example, the relative importance of physical, microbial, and invertebrate-based processing of leaf litter seems to differ between insular streams and those on the mainland, where more invertebrate species occur and shred leaf litter. Some of the fastest rates of leaf breakdown have been reported for shredders in the Luquillo Mountains, where freshwater shrimp (*Xiphocaris elongata*) rapidly shred leaf litter (Crowl et al. 2006). The importance of rapid leaf shredding by invertebrates is demonstrated in streams in some locations but not others (Cheshire et al. 2005; Gonçalves et al. 2006; Rincón and Martínez 2006; Rueda-Delgado et al. 2006; Wantzen and Wagner 2006).

Most stream research has been done in northern temperate regions, where many programs focus research on food webs and ecosystem processes. The value of concepts developed in these temperate-zone ecosystems for understanding tropical ecosystems remains uncertain (Graça et al. 2001; Mathuriac and Chauvet 2002; Iwata et al. 2003; Greathouse and Pringle 2006; Boulton et al. 2008). Research in Puerto Rico has highlighted some of the most fundamental differences (e.g., continuous inputs of leaf litter and wood, high diversity of riparian forests, high frequency of extremely high-flow events) and similarities (e.g., importance of detrital inputs and shading from closed canopies, functions of extreme high and low flows, and significance of cumulative effects) among ecosystems at different latitudes. Such geographic comparisons are important for understanding how types, intensities, and frequencies of disturbances affect food webs in tropical streams relative to those in other regions (Covich et al. 2006; Crowl et al. 2006; Wantzen and Wagner 2006; Boulton et al. 2008). Recent research on riparian and stream connections in temperate-zone ecosystems has focused on the diversity of connections among terrestrial and stream communities at one or more elevations (Nakano and Murakami 2001; Power and Deitrich 2002; Sabo and Power 2002; Allan et al. 2003; Decamps et al. 2004; Baxter et al. 2005; Naiman et al. 2005; Ballinger and Lake 2006; Paetzold et al. 2006). Studies in Puerto Rico have emphasized that gradients of biodiversity and food web structure (figure 8-10) in stream communities arise because of the locations of differently sized waterfalls and the steepness of the channels. Patterns of species distribution result from (1) the types and timing of disturbance events, (2) different distances of upstream migrations of freshwater invertebrates (*Macrobrachium* and *Atya* [shrimps] and *Neritina* [snails]) into headwaters, and (3) limited upstream migrations of predatory mountain mullet (*Agonostomus monticola)* and eels (*Anguilla rostrata*) related to geomorphic barriers. The limited upstream migrations of predatory eels affect prey populations of shrimp differently from those of mountain mullet, because eels are more common at lower elevations (Lamson et al. 2006; Covich et al. 2009). Studies of invertebrate migrations and their effects on food webs have not been emphasized for temperate streams. However, studies of river shrimp (*Macrobrachium ohione*) in the lower Mississippi River are underway (Bauer 2004). The results of these studies will provide a basis for comparative analyses of migratory pathways with several species of *Macrobrachium* from Puerto Rico, as well as with those of other low-latitude locations.

At smaller scales, vertical gradients within sediment-filled river channels characterize subsurface flows through groundwater and porous sediments within channels (the hyporheic zone). Inflows of groundwater to channels enhance the persistence of those stream segments during prolonged drought, when runoff is not available. Several locations become important refuges for species that lack adaptations for living in intermittent streams. In many streams, these vertical gradients of upwelling and lateral inflows create highly complex subsurface flow paths (Boulton et al. 1998; Poole 2002; Fisher et al. 2004). These subsurface waters provide dissolved nutrients, organic matter, and refuge for benthic invertebrates and microbes.

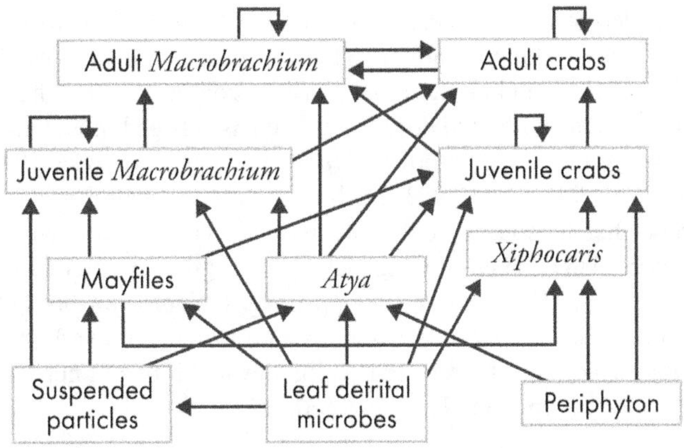

Figure 8.10 Headwater food webs in the Luquillo Mountains are dominated by omnivorous decapods (shrimps [*Atya*, *Macrobrachium*, and *Xiphocaris*] and crabs) that occupy habitats that lack fish predators. Riparian trees provide leaf litter (energy subsidies) that represents important food resources for detritivores (microbes, invertebrate shredders, and filter feeders) and determine light regimes that influence instream algal production (periphyton) for grazers. Arrows denote the flow of energy between food web compartments. Recursive arrows identify cannibalistic, intraspecific predation (i.e., food loops).

Up-welling waters from the subsurface zone typically contain dissolved nutrients that increase algal production at the channel surface. Down-welling zones generally transport dissolved oxygen and organic matter to depths where microbial and invertebrate communities process materials. These complex gradients are well studied in temperate-zone rivers (Fisher et al. 1998, 2004; Poole 2002; Lowe et al. 2006) but are just beginning to be considered in the tropics.

Gradients and Organization of Food Webs

Studies on linear profiles and gradients are used widely for forecasting the distributions of functional feeding groups of invertebrates (primarily aquatic insects as detritivores, grazers, and predators) along rivers (Vannote et al. 1980). In many forested watersheds, inputs of leaf litter from riparian trees represent energy sources for detritivores in small, narrow tributaries of first- and second-order streams (Cummins 1974; Henderson and Walker 1986). Grazers typically dominate communities in wider channels where sunlight is the main source of energy. Predators occur at lower elevations where herbivorous prey are abundant. The pattern of organic-matter processing in headwaters results from species (shredders) that break down coarse leaf material into fine suspended particulates that are consumed by downstream filter-feeding species. These processing chains in forested headwaters depend on adequate flow and turbulent transport. The biotic linkages (from shredders to filter feeders) represent flow-mediated ecosystem processing that is drought sensitive in both temperate-zone (Heard and Richardson 1995; Whiles and Dodds 2002) and tropical (Crowl et al. 2001; Covich et al. 2003; Wright and Covich 2005a, 2005b) streams. For example, filter-feeding shrimp (e.g., *Atya lanipes*) and leaf-litter shredders (e.g., *Xiphocaris elongata*) constitute detrital processing chains when both co-occur in flowing habitats. These species alter their rates of consumption of food (leaf litter and algae) in response to combinations of flow velocities and biotic interactions with predatory fishes such as eels and mountain mullet and larger, predatory shrimps such as *Macrobrachium carcinus* (Covich 1988a, 1988b; Crowl and Covich 1994; Crowl et al. 2001; Covich et al. 2009). Atyid shrimp function as grazers when flow rates are too low to suspend organic particulates for filter feeding (Covich 1988a). Their grazing increases the productivity of algae growing on submerged rocks. These shrimp use appendages adapted for both filtering and grazing (cheliped fans composed of many setae) to remove overlying sediments during feeding, which results in increased light availability and nutrient recycling (Pringle et al. 1993; March and Pringle 2003).

Geomorphic Network Approaches to Food Webs

Important insights for understanding entire drainage basins have emerged from observational and experimental analyses of species responses to disturbances within riparian and stream communities of Puerto Rico. These studies illuminate how species-specific interactions, such as processing chains of detritivores that function as shredders and filter feeders, respond to different riparian inputs (Covich and McDowell 1996; Crowl et al. 2001, 2006) and to disturbance events such as floods and droughts (Covich et al. 1991, 1996, 2003). Previous agricultural land uses modified the physical terrain and soils, consequently altering the plant species composition. This legacy has persistent and significant effects on riparian tree species and the composition of stream food webs (Beard et al. 2005; Zimmerman and Covich 2007). Extreme floods, landslides, and erosion of stream banks continue to alter the sediment composition and habitat quality for many riparian tree species and stream invertebrates. These physically driven disturbances create patchiness that alters species distributions in different locations of drainage networks.

Early studies of temperate stream ecosystems established that terrestrial inputs of water and nutrients affect the diversity of detritivores and the productivity of aquatic food webs along stream profiles (Cummins 1974; Hynes 1975; Vannote et al. 1980). More recent studies established that drainage network connections (Gomi et al. 2002; Benda et al. 2004; Fisher et al. 2004) and the geomorphic template are critical for understanding how changes in land use or climate alter the structure and functioning of drainage basins (Church 2002; Likens 2004). Recent studies in the Luquillo Mountains provide tropical comparisons in which the degree of patchiness is high as a consequence of 2 decades of natural disturbances (e.g., Hurricanes Hugo and Georges). Agricultural clearing and other human land uses resulted in legacies that produced a complex mosaic of habitats and riparian tree distributions that still influences stream food webs. Flows of water and materials connect forest and stream food webs in ways that accelerate responses to frequent disturbances and create a high degree of food web resiliency.

Recognition that aquatic and terrestrial habitats are highly interconnected by hydrologic processes is critical for understanding ecosystem dynamics and managing riparian areas. Because drainage networks link land-based nutrients to stream food webs, effective land use policies require an ecosystem-level approach for controlling nutrient availability (Meyer and Wallace 2001; Power and Dietrich 2002; Pringle 2003; Thoms and Parsons 2003; Freeman et al. 2007; Meyer et al. 2007). Hydrologic connections link stream food webs across a wide range of elevations within a drainage basin (Lewis et al. 2001; Junk and Wantzen 2004). These hydrologic and ecological connections are especially important during prolonged droughts, because habitats with sustainable stream inflows become refuges for species adapted to persistent stream flows (Covich et al. 2003, 2006). Ecologists continue to explore linkages among aquatic and terrestrial ecosystems (Kling et al. 2000; Hershey et al. 2006; Richter et al. 2006) that integrate biotic distributions relative to the physical terrain and long-term variability in flow regimes. Research on tropical stream ecosystems is needed in order to document the importance of anticipated extremes in flow that arise from climatic changes and the increased diversion of water for human needs (Covich et al. 2004a, 2004b; Giller et al. 2004; Malmqvist et al. 2008).

Introduced Species in Ecological Perspective

The effect of introduced species invasions on native species composition and on ecosystem functioning and services is of critical concern to society (Ewel et al. 1999). Unfortunately, the ecological consequences of introduced species invasions are unclear. Some connect introduced species invasions to native species extinctions (Allendorf and Lundquist 2003; Lodge and Shrader-Frechette 2003), homogenized landscapes (McKinney and Lockwood 2001), or even the genetic alteration of native species through hybridization (Lockwood and McKinney 2001). Others consider the level of native species extinctions resulting from invasions to be exaggerated (Case 1996; Vermeij 1996) and perceive the landscape to be diversified as a result of enrichment with introduced species (Davis 2003). Because importance values or rank-abundances of species (*sensu* Whittaker 1970) change after invasions, and because an altered species composition affects the rates of ecosystem processes and the magnitudes of state variables, extreme caution must be exercised before advocating or introducing species into new environments (Ewel et al. 1999).

Research in the Luquillo Mountains and elsewhere in Puerto Rico has documented the presence and some of the ecological roles of introduced plant and animal species. This research provides a basic understanding upon which to build an ecological perspective of introduced species invasions in tropical environments ranging from primary to urban forests. We briefly present 10 observations from studies involving introduced species and conclude with a discussion of implications for biodiversity science and management. The first observation focuses on primary and mature native forests and excludes anthropogenically disturbed sites (e.g., roadsides, recreation areas, tree plantations). The second observation includes experimental disturbances of mature native forests. The remaining eight involve sites subjected to anthropogenic disturbances.

First, many introduced animal species and one introduced tree species have been reported in the otherwise mature or primary forests of the Luquillo Mountains

Table 8.1 Examples of naturalized introduced species within mature or primary forests of the Luquillo Mountains

Scientific name	Common name	Source
Animals		
Mus musculus	House mouse	Odum et al. 1970a
Rattus rattus	Roof rat	Odum et al. 1970a; Weinbren et al. 1970; Snyder et al. 1987
Rattus norvegicus	Wharf rat	Weinbren et al. 1970; Snyder et al. 1987
Herpestes auropunctatus	Mongoose	Weinbren et al. 1970; Snyder et al. 1987
Felis catus	Cat	Weinbren et al. 1970; Snyder et al. 1987
Bufo marinus	Common toad	Odum et al. 1970b
Apis mellifera	Honeybee	Snyder et al. 1987
Aedes aegypti	Mosquito	Weinbren and Weinbren 1970
Trees		
Syzygium jambos	Pomarosa	Smith 1970
Calophyllum calaba	María	Thompson et al. 2007
Genipa americana		Thompson et al. 2007
Simarouba glauca		Thompson et al. 2007
Swietenia macrophylla	Caoba	Thompson et al. 2007

(table 8-1). Some of these species have been present in the forest for centuries and maintain stable, but low, population densities. For example, active honeybee (*Apis mellifera*) hives have an average density of 1 per 3.4 ha at upper elevations (Snyder et al. 1987). Rat (*Rattus rattus*) density ranges from 32 to 40 ha^{-1} (Odum et al. 1970b) and fluctuates annually (Weinbren et al. 1970). Introduced plant species such as bamboo (*Bambusa vulgaris*) or African tulip tree (*Spathodea campanulata*) are common in some areas and provide food (leaves and seeds) to native shrimp in reaches of streams for which there are no records of historical deforestation by humans.

Second, experimental disturbance (cutting and gamma irradiation) of mature tabonuco forest resulted in the establishment of introduced plant taxa such as *Swietenia* (Duke 1970) and eight species from roadsides (Smith 1970) that do not survive conditions of canopy closure. Twenty-three years after irradiation, regeneration in the experimental area was only by native species (Taylor et al. 1995), a finding recently corroborated by Thompson et al. (2007).

Third, introduced tree species occur in mature forest sites that were logged selectively over 60 years ago (e.g., the Luquillo Forest Dynamics Plot [LFDP]), but densities are low, and local ranges are not expanding (Thompson et al. 2007). After a hurricane, the invasive introduced tree species *Spathodea campanulata* germinated but failed to establish beyond the sapling stage as a consequence of canopy closure. Most introduced species on the LFDP occur on a sector that had been farmed and logged, and fewer occur on the sector that was not farmed or where the canopy has been closed at least since 1936 (Thompson et al. 2007).

Fourth, in some areas of the Luquillo Mountains, pastures dominate after deforestation, agricultural use, and abandonment. In instances where introduced species were planted as monocultures in these pastures, forest cover was restored, but

Table 8.2 Examples of Puerto Rican taxa with increased numbers of species resulting from the introduction and naturalization of species.

Taxon	Number of species			Source
	Native	*Introduced*	*Total*	
Plants			3,126	Liogier and Martorell 2000
Trees	547	203	773	Little et al. 1974
Orchids	137	6	143	Ackerman 1992
Ferns			408	Proctor 1989
Birds	240	87	327	Biaggi 1997
Earthworms	18	11	29	Borges 1996
Ants	*a		*a	Torres and Snelling 1997

**a* Surveyed Puerto Rico and 44 adjacent islands over an 18-year period and found 31 extinctions and 146 new colonizing species. In all but two cases, the number of ant species increased.

native species quickly invaded plantation understories (Lugo 1992) and eventually dominated the canopy (Silver et al. 2004). Introduced species remain, but with low dominance after 60 years of forest succession (Silver et al. 2004). Native species reinvade the site at a rate of one species per year (Lugo et al. 1993).

Fifth, hurricanes can destroy plantations of introduced species and accelerate the establishment of native tree species (Wadsworth and Englerth 1959; Liegel 1984). In other instances, mechanical injury by hurricanes to forests with mixed species composition (introduced and native) was more a function of tree growth rate (faster growing trees experienced greater effects than did slower growing ones) or location relative to wind direction than of the biogeographic origin of the species (Ostertag et al. 2005).

Sixth, the conversion of forest to pastures results in the invasion of introduced earthworms and the local extinction of native earthworms (González et al. 1996). The abandonment of pastures and their subsequent recovery results in a community comprising both introduced and native earthworm species, in which introduced species dominate in terms of numbers but not biomass (Sánchez de León et al. 2003).

Seventh, as a result of the introduction of species into Puerto Rico, the species richness of many areas has increased (table 8-2). These increases were not associated with native species extinctions, which are low in Puerto Rico, at least for plants (Lugo 1988; Figueroa Colón 1996) and birds (Brash 1984; Biaggi 1997). The large increase in bird species richness as a result of introductions led Biaggi (1997) to state that any future compendium on birds requires attention to the introduced species, because they increase in number on a daily basis (p. 327). He listed 87 introduced bird species from Puerto Rico, compared to 116 native resident bird species and 92 migratory ones.

Eighth, island-wide forest inventories document that introduced species of trees dominate most of the Puerto Rican landscape, particularly in regions with high anthropogenic disturbance (Lugo and Brandeis 2005). One introduced species, *Spathodea campanulata*, is the most common tree on the island. The abundance of introduced species was greatest in moist life zones that had been deforested, farmed, and abandoned, and it was lowest in dry and wet life zones and in regions with

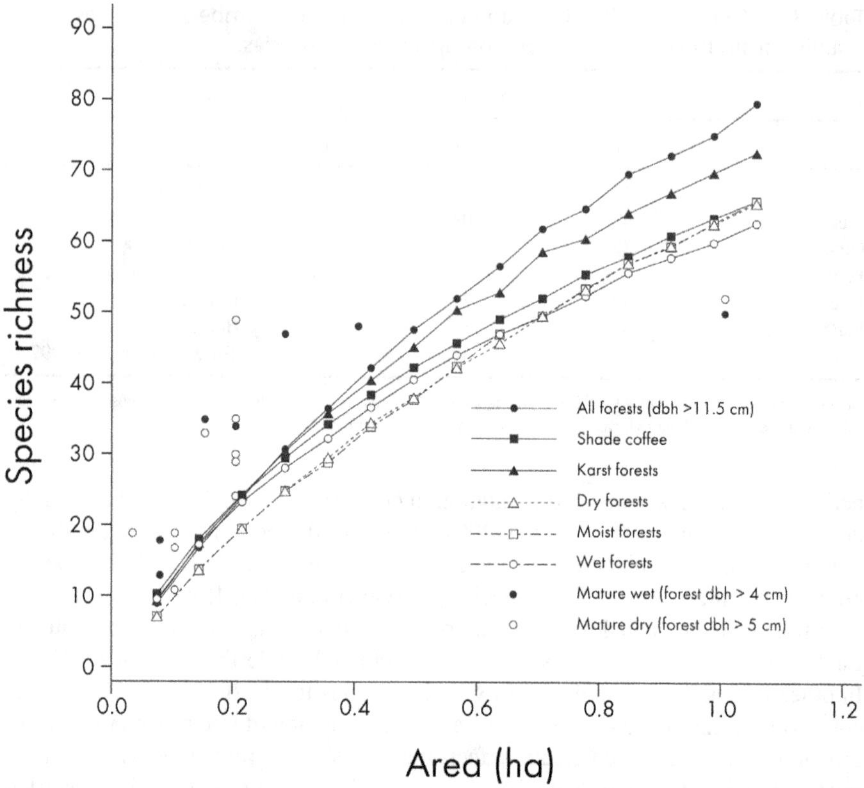

Figure 8.11 Species-area curves for undisturbed mature native dry and wet forests (point data) and new forest types dominated by introduced species (data connected by lines). The data for new forest types are from an island-wide forest inventory and cover different types of forests; there also is a combined curve for all forests (Lugo and Brandeis 2005). Note the differences in the minimum diameter at breast height (dbh), which result in different amounts of underestimation in the number of species in the emerging new forests. Native forests in Puerto Rico saturate at about 60 species ha^{-1} (Lugo 2005), but the emerging forests have higher species densities.

mature or undisturbed native forests. Rather than causing forests to become depauperate, the presence of introduced species augmented species density (figure 8-11).

Ninth, urban forests had the highest proportion of introduced species in Puerto Rico (Lugo and Brandeis 2005). Sixty-six percent of the importance value (an index that reflects abundance, frequency of occurrence, and biomass) in urban forests was attributed to introduced species.

Tenth, the dominance of introduced species in Puerto Rican forests declines over time. Native species grow under the canopies of nonnative species and regain dominance in more mature forests that support combinations of introduced and native species (Wadsworth and Birdsey 1983; Lugo 2004b; Lugo and Helmer 2004).

Ecological research on introduced species contributes to the emerging fields of conservation biology and conservation biogeography (Lomolino 2004). The Puerto Rico case study offers empirical evidence about a situation in which intense and chronic human activity created environmental havoc through deforestation, fragmentation, conversion to agriculture, and abandonment after centuries of soil-degrading activity. From this scenario emerged a natural recolonization of abandoned lands by trees, mostly introduced species. Some 60 to 80 years after abandonment, forests that do not resemble the original ones in terms of species composition cover almost half of the island (Lugo and Helmer 2004). They contain a considerable proportion of introduced species, as well as native and endemic elements of the flora.

At the same time, species extinctions in Puerto Rico have been negligible relative to predictions of island-biogeographic theory or ideas about fragmentation of forest cover (Lugo 1988). These methods for estimating species extinctions do not consider recovery mechanisms after deforestation and abandonment and assume linear or exponential relationships between the percentage of area deforested and the percentage of the species pool that becomes extinct. Island-wide deforestation of > 90 percent should have resulted in large losses of species in Puerto Rico and a depauperate biota. However, extensive extinctions did not occur, and current forests are diverse and functional (Lugo 2004a). By "functional," we mean that ecological processes in these emerging forests (e.g., rates of primary productivity, nutrient cycling, and decomposition rates) compare favorably with those in native, undisturbed ones (Lugo 1992).

Documenting and understanding the consequences of massive deforestation and subsequent invasion by introduced species advances ecological understanding and provides insights to guide the restoration and conservation of tropical forests. Introduced species invasions and establishment in Puerto Rico occur naturally in response to anthropogenic disturbances. In the absence of anthropogenic disturbances, only a few introduced species become established in native forests, where they function as rare specialists.

The consideration of species invasion from the perspective of ecological space is consistent with the deconstructive approach to species richness favored by Marquet et al. (2004). They argued that understanding patterns of species richness requires the consideration of evolutionary (extinction/speciation dynamics), environmental (external properties and states), and physiological or life history characteristics (internal properties and states). This approach is particularly important in a disturbance-mediated system because ecological characteristics continually change across geographic space and along time sequences at the same geographic location. Indeed, disturbance and subsequent succession alter the environmental characteristics associated with geographic space, providing dynamic opportunities for many species to establish or go locally extinct, potentially influencing the species composition and richness at multiple scales. Knowledge of the physiological and life history characteristics of species is thus essential if one is to understand the acclimation, adaptation, and even evolution of species to new ecological conditions, including the novel ones that emerge because of human activity.

Understanding the dynamics of species assembly in tropical forests, including mechanisms that favor or retard invasion, is critical for the development of policies concerning the conservation of biodiversity. This is particularly true because human activity is accelerating changes in the composition of species, either through the introduction of new species or by altering site conditions and redirecting succession. Regardless of the identity of the factors that cause changes in the species composition, conservation scientists need to understand how communities reassemble, self-organize, and form new ecosystems after disturbance or invasion by introduced species (chapter 7).

In summary, tropical environs can be quite resilient to anthropogenically induced habitat loss and fragmentation. Moreover, invasion by introduced species frequently does not lead to the loss of native species or to significantly altered ecosystem services. The challenge for the future is to understand the circumstances that lead to resilience and to inform restoration and reclamation efforts by articulating a mechanistic approach to guide successional trajectories to endpoints consistent with societal goals of sustainability.

Emerging New Forest Types on the Tropical Landscape

Research in the Luquillo Mountains has illuminated the nature of emerging new forest types. The paleoecological and biogeographical literature is replete with examples of changes in forest types as a consequence of environmental change (Behrensmeyer et al. 1992; Colinvaux 1996; Jackson 2004). Modifications to the biota that result in different types of ecosystems are a matter of historical record (Graham 2003a, 2003b) and characterize the natural spatiotemporal dynamics of ecosystems. Alterations in species composition and ecosystem processes that result from global change are a growing concern, especially as they become connected more clearly to human activities (Mack et al. 2000). At least four legitimate issues arise concerning future changes in species composition in contemporary ecosystems: (1) the potential loss of endemic species, (2) spatial homogenization of the biota as a result of the spread of introduced species and the extinction of native species, (3) the loss of ecosystem services, and (4) uncertainty regarding long-term consequences (Mack et al. 2000; Lockwood and McKinney 2001). Such issues are best considered in the context of emerging new forest ecosystems.

An emerging new forest ecosystem is one with a species composition (e.g., distribution of importance values among species) that is novel for the landscape on which it occurs (Lugo and Helmer 2004). This process is a natural one; the emergence of new forest ecosystems occurs because species invade, establish, and interact, even in the absence of human intervention. This differentiates emerging new forest ecosystems from those purposefully established by humans (e.g., plantations). An emerging new forest is different from a forest with no known history of anthropogenic disturbance in that it results from human activities that were *not* executed in order to achieve biotic change per se. Rather, new forests emerge at severely modified sites at which the succession of native species failed to reestablish a native forest. Emerging new forest ecosystems are characterized by three key elements. First, they occur mostly on sites that were modified severely by humans and which are incapable of sustaining many native tree species. Second, they become established through natural processes of dispersal, establishment, and species interaction. Third, they involve novelty; that is, they encompass a combination of species that is different from what is typical for a particular landscape.

In emerging new forests in Puerto Rico, introduced species are dominant components of the flora (Lugo and Helmer 2004). Over the long term (~80 years), introduced species share dominance with native and endemic species that reinvade subsequent to initial colonization by introduced species (Lugo 2004b). Generally, emerging new forests in Puerto Rico occur on sites that support dry, moist, or wet forests and on calcareous, volcanic, or alluvial substrates. They are ubiquitous wherever humans have modified the ecological characteristics of geographic space significantly. Compared to native forests, the notable characteristics (Lugo and Helmer 2004) of emerging new forests in Puerto Rico are that they (1) have a small complement of endemic species, (2) are young (<100 y), (3) originate in highly fragmented landscapes, (4) are structurally simple (low basal area and low species richness), (5) contain few large trees (≥30 cm diameter at breast height), (6) arise on soil with low organic matter and high soil bulk density, (7) exhibit high temporal turnover of species, and (8) are similar in canopy structure and physiognomy to each other as well as to native forests.

The degree of difference in the species composition between new emerging forests and native forests with no known historical record of anthropogenic activity is a function of the type and intensity of disturbance. Disturbances of all types (natural or anthropogenic) offer opportunities for invasions and the reassembly of species (chapter 5). However, undisturbed tropical forests remain resistant to invasion even when subjected to natural disturbances (Denslow and DeWalt 2008), and emerging forests must cope with natural disturbances. The synergy between natural and anthropogenic disturbances, many of which produce novel combinations of ecological characteristics, predisposes sites to support emerging forests. The type and intensity of disturbance are thus responsible for changes in species composition, especially the distribution of importance values among species in forests. Forest responses to combined natural and anthropogenic disturbance can be characterized into four states:

1. When a natural disturbance (e.g., a hurricane) affects a mature or primary forest, the change in species composition is minimal, particularly in the long term (Crow 1980). In the short term, secondary forest species can become abundant, but primary forest species remain dominant. Introduced species are rare or absent.
2. When a natural disturbance (e.g., a hurricane) affects a mature forest that was deforested previously and used for agriculture (Scatena et al. 1996; Thompson et al. 2002), changes in the species composition and the distribution of importance values among species persist. Introduced species are rare or absent.
3. The abandonment of intensively used agricultural fields (Lugo and Helmer 2004) gives rise to species compositions that are different from those of native forests. Introduced species dominate early in succession and remain present in forests at maturity (Lugo 2004b).

4. The abandonment of highly degraded land leads to arrested succession and an herbaceous system in which trees fail to grow. Tree planting is required in order to restore forest conditions, and the early species composition is determined by the identities of planted species (Parrotta 1995; Silver et al. 2004). Over time, and without further human intervention, a new combination of species emerges, including native and introduced species.

Research in Puerto Rico (e.g., Lugo and Brandeis 2005) suggests that ecologists need to recognize variant ecosystem types (from pristine to artificial) associated with a gradient of human activity. Within this gradient, successional processes lead to particular combinations of species with familiar physiognomy and structure. These floras generally function in a manner comparable to that of native forests. However, the evaluation of various ecosystem functions is considerably less advanced than the description of the structure and species composition of these new forests. Once ecologists recognize the reality of emerging new forest ecosystems, the level of research concerning their functional characteristics will increase, thereby helping to resolve the current debate about the role of introduced species in a human-dominated environment.

Comprehensive study of the emergence of new forest ecosystems will provide new perspectives and avenues of investigation regarding four ecological issues. First, populations of some endemic species have an opportunity to flourish in emerging new forests, as has occurred in Puerto Rico (Lugo and Brandeis 2005). This happens as new forests mature and conditions for increased species diversity develop through restored soil fertility and microclimate.

Second, homogenization of the biota by the spread of introduced species and the extinction of native species (McKinney and Lockwood 1999) is not consistent with observations from Puerto Rico, as relatively few extinctions have been caused by introduced species, and when forests recover after deforestation, both native and introduced species generally persist (Lugo 2004a). However, during the early stages of establishment of emerging forests, species diversity is low, weedy species predominate, and rare species are absent. These trends reverse over time. Consequently, homogenization has a temporal trajectory that requires additional study. Moreover, the potential hybridization of introduced and native species (Ellstrand and Schierenbeck 2000) is a possibility that requires new research.

Third, little research has focused on the functionality of emerging ecosystems and the ecosystem services that they provide to society. This is an area in urgent need of comprehensive research, especially in light of the contention that ecosystem services are not compromised during the establishment of emerging new forest systems (Lugo and Helmer 2004).

Finally, uncertainty and surprise are fundamental characteristics of the behavior of complex systems, especially those containing an appreciable number of species that did not evolve in syntopy. Because novelty is a fundamental aspect of emerging new forests, uncertainty and surprise regarding successional trajectories, functionality, and ecosystem services are unavoidable. The only avenue by which to advance the understanding of this issue is long-term research on the properties of tropical ecosystems that span the gradient of anthropogenic disturbance.

The Integration of Social and Natural Sciences in Understanding and Forecasting Ecological Change

In the Luquillo Mountains and elsewhere, the integration of social and natural sciences is critical to a comprehensive understanding of pressing environmental issues that face society in the 21st century. After decades of focusing research on pristine areas of the tropics, ecologists now recognize that the principal influence on many processes in tropical ecosystems, including the pristine ones, is the activity of a growing human population (Turner et al. 1990; Clark 1996; Foster et al. 1998; Hall 2000; Watson et al. 2001; Lambin et al. 2003; LeClerc and Hall 2007). Initially, studies of human effects in the tropics focused on the effects of deforestation on biodiversity and global biogeochemical cycles (Gómez-Pompa et al. 1972; Detwiler and Hall 1988; Turner et al. 1990). In addition, the economic development of the tropics has been of interest to social scientists since the 1950s. However, few linked that development with environmental effects, even though a clear association exists between economic development and land use change (Hall and Ko 2005; Hall 2006). The outright conversion of tropical forests is only one of many aspects of human-induced ecological change in the tropics; less intensive development that leaves large proportions of forests standing still affects such environmental attributes as biodiversity, hydrology, and meteorology (Bonnell and Bruijnzeel 2004). As has been documented widely in the temperate zone as well (e.g., Foster et al. 1998), the general pattern of initial colonization, deforestation, and land degradation in the tropics often is followed by generalized economic development and industrialization, associated in turn with rural-to-urban migration and the abandonment of agriculture in economically marginal areas (Hall 2000; Rudel et al. 2002; United Nations 2002). In other words, as the basis for the economy shifts from solar to fossil energy, pressure on the land is reduced. In a few tropical regions, this has led to the establishment of large areas of secondary forest; Puerto Rico is the best documented case of this (Grau et al. 2003). As urbanization continues, these secondary forests might themselves be deforested as suburban areas expand away from urban centers (e.g., Thomlinson and Rivera 2000). All of these processes are underway on a large scale in Puerto Rico (Grau et al. 2003).

Future long-term research in the Greater Luquillo Ecosystem (the Luquillo Mountains and environs) should consider two important questions. First, to what degree are changes in forest cover, and the socioeconomic factors driving these changes, a general feature of other regions in the tropics (i.e., to what degree might Puerto Rico be a model for the rest of the tropics)? If the situation in Puerto Rico is unique, then perhaps there is little value in studying this aspect of the Greater Luquillo Ecosystem. We argue, however, that it is not. Instead, the situation in Puerto Rico, where the net effect is reforestation, illustrates a general relation between humans and tropical forests that is driven by the changing characteristics of tropical economies, especially the degree to which they become industrialized at the expense of land-intensive labor. Second, to what degree is it necessary to understand and account for human ecology in explaining ecosystem change? In other words, how complex are human interactions with the environment? We contend that such interactions are best understood in an integrated socioecological context that explicitly defines the connections between humans and other components of nature (Hall 2000; Hall et al. 2001; Pickett et al. 2001; Hall and Klitgaard 2006; LeClerc and Hall 2007). Because Puerto Rico is small with a steep topography, its social and ecological systems are tightly coupled, presenting an excellent opportunity to address complex human-environmental dynamics. We address generality and complexity, and we outline a strategy for integrating human dynamics into the long-term study of the Greater Luquillo Ecosystem, as has been done for similar programs in the mainland United States (Gragson and Grove 2006).

The traditional view of human dynamics in the tropics is that of an unavoidable pattern of an expanding human population, growing out along newly created roads into tropical wilderness areas, leading to deforestation and destruction (Gómez-Pompa et al. 1972; Barbier 2005). In large part, this perspective has been correct. Many tropical countries have lost, and continue to lose, much of their forest cover (Turner et al. 1990; Hall 2000; Watson et al. 2001; Lambin et al. 2003). However, a few exceptions to this pattern currently exist in the tropics in places where economic growth, principally fueled by fossil fuels, has resulted in the growth of secondary forest, paralleling a pattern first seen in the United States and Western Europe (e.g., Andre 1998; Foster et al. 1998). Clear examples include some Caribbean islands (including Puerto Rico), northwestern Costa Rica, (Janzen 2000, 2002), Taiwan, peninsular Malaysia, and portions of the Andes (Rudel et al. 2002; Grau et al. 2003). One controversial viewpoint argues that these patterns apply to the entire tropics, which would guarantee that human development will not lead to a catastrophic loss of tropical biodiversity (Wright and Mueller-Landau 2006).

Grau et al. (2003) presented a detailed assessment of such an ecodemographic transition in Puerto Rico. During the first half of the 20th century, the population of Puerto Rico increased from 1 million to just over 2 million inhabitants, and forest cover was reduced to approximately 5 percent of the island (much of this in shade coffee). However, Puerto Rico was one of the earliest parts of the tropics to be developed explicitly for industrial manufacturing, as a result of close ties with the United States; the desire of many in the United States to take advantage of the relatively well-educated, inexpensive, and compliant labor force; and the success of the program "operation bootstrap," initiated in 1948 (Dietz 1987). Although the population almost doubled again during the second half of the century, forest cover increased from 5 to 35 percent, such that much of the island now supports secondary forest (figure 8-12). These changes were initiated by the abandonment of marginal agricultural lands in mountainous regions and a concentration of the human population in urban areas (Rudel et al. 2000) concomitant with a dramatic increase in the portion of the economy devoted to manufacturing and an increase in oil consumption from a very low level to about 70 million barrels per year in 2000. The rural population in Puerto Rico actually decreased during the second half of the 20th century, even though the overall population was growing quite rapidly (Grau et al. 2003).

Grau et al. (2003) were careful not to extrapolate the results from Puerto Rico too broadly, because much of the economic dynamics that drove the change in land use might result from the special political relationship between Puerto Rico and the United States. Some recent data suggest, however, that increased free trade and

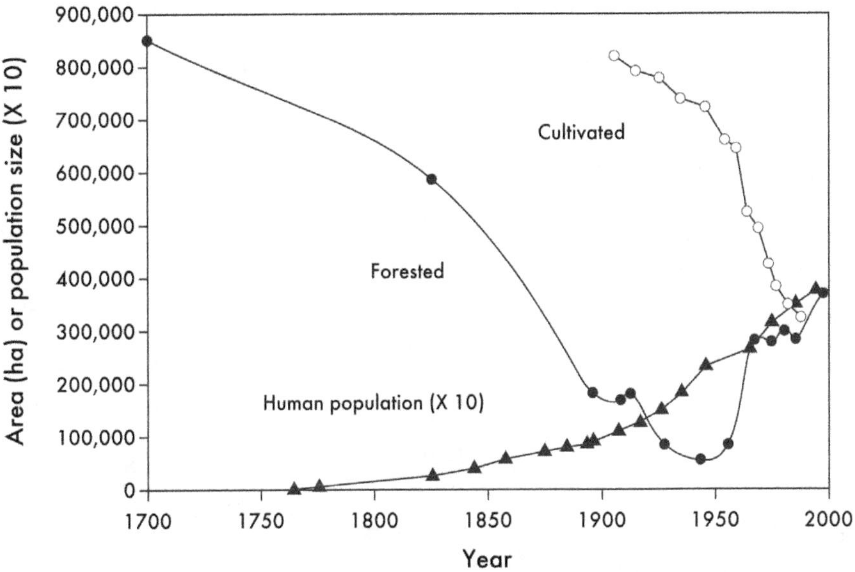

Figure 8.12 Temporal trends in land cover and human population size in Puerto Rico over 3 centuries.

other aspects of globalization, including the continued expansion of fossil-fuel-based economic activities, allow Puerto Rico to serve as a model for other parts of the tropics. An example can be drawn from patterns of urbanization. As revealed by a recent United Nations Population Division study (United Nations 2002), the current proportion of the human population in urban areas in Puerto Rico, and projected increases in coming decades (figure 8-13), are similar to such patterns throughout the tropical countries of the Americas (Wright and Mueller-Landau 2006). Levels of urbanization (i.e., the proportion of the entire population living in urban areas) in the American tropics range from 60 to 85 percent, bracketing the current value in Puerto Rico (72 percent). These values are very different from those of central Africa and Southeast Asia, where current levels of urbanization are about 35 percent and are not expected to exceed 60 percent by the year 2030. Guyana is the only tropical American country that currently has such a low level of urbanization (33 percent). The high levels of urbanization in most countries of tropical America offer hope that future effects of the growth of human populations, which will occur predominantly in urban areas, will minimize the anthropogenic modification of forests and other native habitats in rural areas and lead to an increase of secondary forest, as occurred in Puerto Rico (Grau et al. 2003). The important point arising from this crude assessment is that Puerto Rico, in the context of the Neotropics, has followed a pattern similar to those of many other countries.

Levels of urbanization, of course, are only one dimension of socioeconomic change in developing economies. For example, the level of urbanization in Brazil is higher than that in the United States (figure 8-13), but no one would claim that the standard of living or the effect of the Brazilian citizens on their forests and environment are the

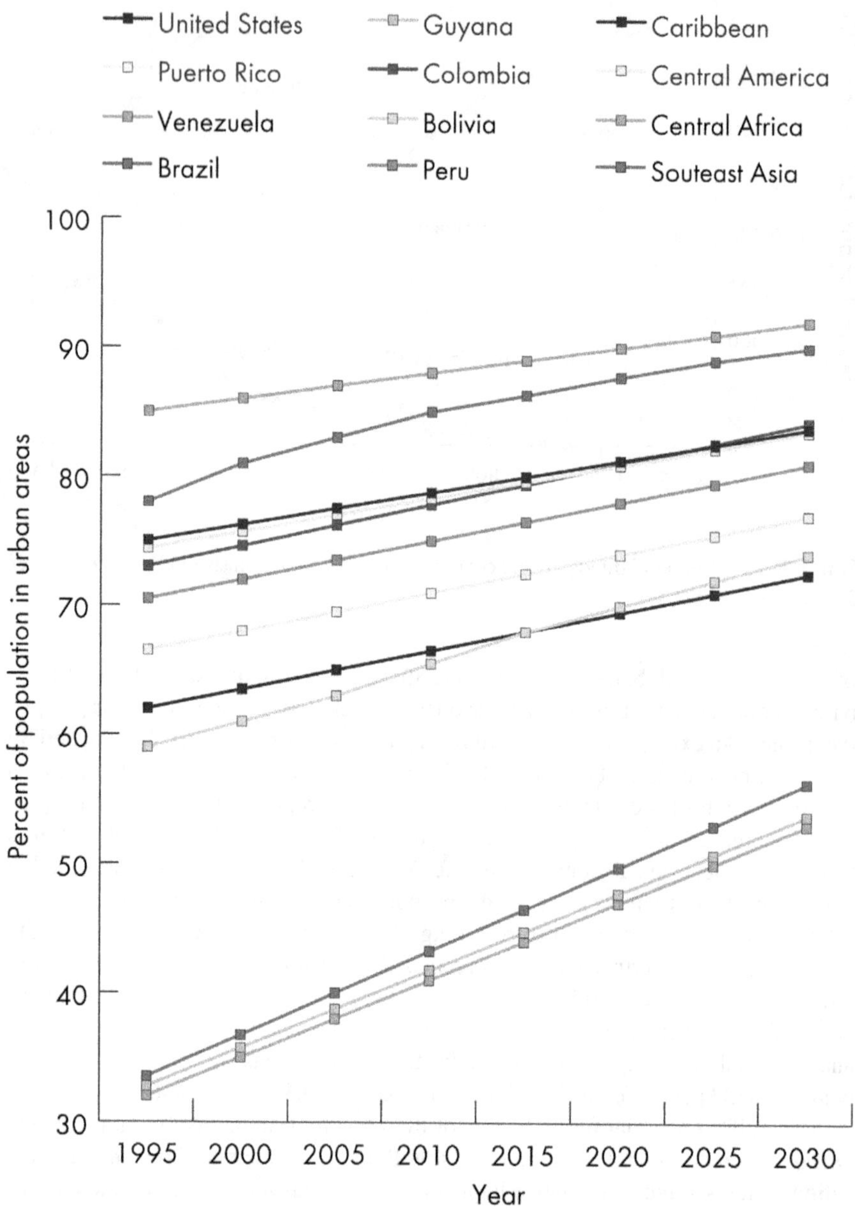

Figure 8.13 Predicted percentage of human population in urbanized areas for various tropical countries or regions. Puerto Rico (included both individually and pooled with all Caribbean countries) and the United States serve as references. The order of population projections for 2030 (highest to lowest) corresponds to that of countries (associated symbols) in the legend.

same as those in the United States. To understand such differences, one must delve deeper into the relationship between economic development and human influences on the environment. The reversal in forest losses with development, as observed in Puerto Rico (Rudel et al. 2000) and elsewhere, is of particular interest to those seeking to understand the relationship between human activities and land use change, and it might reflect a more general phenomenon of human ecology. Forest cover does not necessarily decrease monotonically with increasing human population in the developing tropics. Rather, human effects on the environment are nonlinear and complex. In particular, it is critical to understand and assess the relationship between population growth and the degree of industrialization, because the real driver of deforestation probably is the degree to which the human economy depends directly on the quantity of solar energy intercepted and used by agriculture, pastures, and other solar, non-forest-based economic activities. Because there has been a trend in parts of the tropics to displace these solar-based economies with fossil-fuel-based economic activities such as manufacturing, tourism, and even ecotourism, there exists, for some regions, the possibility of continued growth of the human population and the economy with a concomitant decrease in the use of land-based resources. Whether the environmental effects stemming from the use of fossil fuels are greater than the effects of deforestation is for others to ponder. Moreover, human perceptions of the value and meaning of nature and the manner in which government policies and regulations drive patterns of development deserve additional study and consideration from the perspective of sustainability (Pickett et al. 2001).

The initial loss of forest cover but subsequent gain with increasing economic development has been called the "forest transition" (Rudel 1998; Rudel et al. 2002; Perz and Skole 2003). If real and, more important, general, this forest transition could be an example of an environmental Kuznets curve (EKC) (Kuznets 1955; Dinda 2004), an empirical observation that the degradation of the environment increases with economic development until, at some point, further development leads to a decline in human effects. A common example is the pollution of air and water, which generally increases initially with development but then often decreases (at least locally) with continued development. The actuality, degree, mechanisms, and value of EKCs have been widely debated, with biophysical (e.g., declining soil quality), demographic (e.g., rural-urban migration), economic (e.g., capital scarcity, transitions from agrarian to industrial to service-based industries), political (strong democratic traditions), and sociological (e.g., demand for a cleaner environment) factors prominent among the many explanations proffered (Rudel 1998; Ehrhardt-Martínez et al. 2002; Perz and Skole 2003; Dinda 2004; Khanna and Plassman 2004). It is clear that different environmental factors have different Kuznet curves (Khanna and Plassman 2004), with the change in forest cover being one of the earliest and, therefore, key transitions (Rudel 1998; Ehrhardt-Martínez et al. 2002). However, recently there have been claims that the manner in which development now proceeds in tropical countries might be so different from the way in which their temperate counterparts developed that the forest transition concept (Klooster 2003) and EKCs in general might not apply (Stern 2004). Most critical, the reliance of most economies on fossil fuels, which generate CO_2 pollution and lead to global warming (IPCC 2007), is the most important EKC, and one for which modern economies are nowhere near the turning point, as energy use and release of CO_2 show no sign of decreasing for most countries. More generally, and contrary to the viewpoint of Wright and Mueller-Landau (2006), increased urbanization and economic development in tropical countries must lead to a larger ecological footprint for these cities, with concomitant demands for increased resources and increasing local and global pollution (e.g., Hall 2000).

A separate issue is the degree to which present models, including economic models, are adequate to represent in a comprehensive and integrative way the changes that are taking place in Puerto Rico and that are expected to take place in the future. In particular, contemporary economic models might not be adequate to represent or guide the future (Hall 2000; Hall and Klitgaard 2006; LeClerc and Hall 2007). Instead of, or in addition to, neoclassical models, a more biophysical approach might be necessary in order to understand economics and the effect of humans on the environment. An example is the identity of the real drivers that allowed the reforestation of Puerto Rico. The standard explanation is that Puerto Rico has changed from a principally agriculturally based economy to an increasingly urban manufacturing- and service-based economy. Another way to view this, which is more aligned with a biophysical perspective, is that the energy basis of the Puerto Rican economy has shifted from a solar basis early in the 20th century to a fossil-fuel-based one today. Thus it is (only) the availability of cheap oil that has allowed the development of today's economy, which in turn has allowed the forests to regenerate. Given increasing evidence for "the end of cheap oil" (e.g., Campbell and Leharrère 1998) and the very large dependence of Puerto Rico on oil as a basis for its manufacturing and service economy, it is not clear that this approach (increased use of fossil fuels) will remain feasible for a great deal longer. If not, a resurgence of intensive deforestation in Puerto Rico and other such areas might be forthcoming.

How would a detailed understanding of human ecology in Puerto Rico improve our understanding of the Greater Luquillo Ecosystem? A conceptual diagram illustrates the key factors for the development of an integrated perspective on human and natural systems in the Luquillo area (figure 8-14). This model distinguishes human, geomorphic, and ecological processes and proposes that the most significant interactions among these processes are regulated by the disturbance regime. External drivers, such as long-term climate change, are important in regulating the disturbance regime (e.g., increased sea surface temperatures might cause an increased frequency of severe hurricanes [Emanuel 2005; Webster et al. 2005]). The key integrating feature of this model is that ecosystem services feed back from geomorphic and ecological processes to the human component.

There are a number of ways that this conceptual approach can lead to an integrative understanding of human and natural systems. These include several areas of current interest, focusing on how increasing urbanization (1) feeds back on the climate and (2) is regulated by geomorphology, and (3) how this in turn feeds back on agricultural production and the provision of ecosystem services.

The feedback of human processes (i.e., the expansion of the urban zone) on local climate is illustrated by an "urban heat island" associated with San Juan, where city temperatures are higher by about 2°C (usually) to as much as 10°C (occasionally in

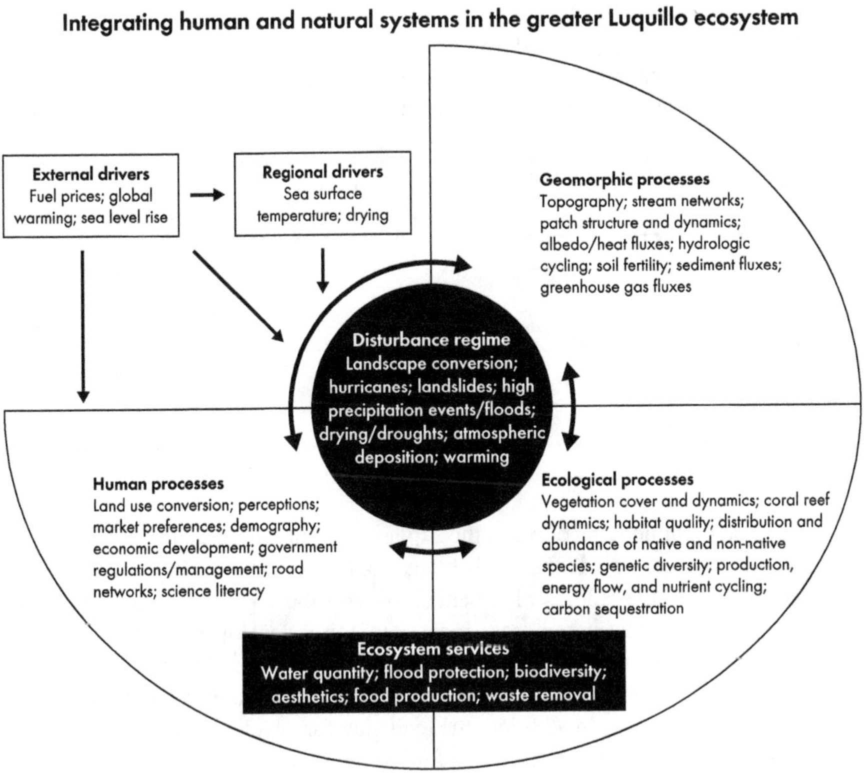

Figure 8.14 A conceptual diagram for the integration of human and natural systems in understanding and forecasting long-term change in the greater Luquillo ecosystem (the Luquillo Experimental Forest and adjacent urban areas, including San Juan). The entire ecosystem is divided into three interacting components: human, geomorphic, and ecological processes. "Disturbance regime" is placed at the center of the three interacting components to emphasize the degree to which disturbance regulates these interactions. Similarly, ecosystem services are a critical feedback between ecological/geomorphic processes and human processes. External drivers emphasize prices (particularly liquid fuel) and local and regional climate.

the dry season) because of the existence of the urban construction itself (e.g., Velázquez-Lozada et al. 2006; Murphy 2007). The transition from the forested to the suburban environment has as great an effect on meteorological characteristics as that between the urban and the suburban environment (A. Chen and D. Murphy, personal observation). Thus, development to the east of San Juan is extending the urban heat island toward the Luquillo Mountains, a major source of municipal water. This is an important concern because global climate models predict drying of the Caribbean region (Neelin et al. 2006). Urban heat islands such as that associated with San Juan might be expected to exacerbate the effect of global climate change, both locally and in adjacent forest (Velázquez-Lozada et al. 2006). As possible evidence of this interaction, Wu et al. (2007) recently found that the proportion of rainfall leaving the Fajardo watershed as streamflow is decreasing.

Table 8.3 Influence of landscape attributes on changes in land cover in Puerto Rico. A positive relationship is indicated by "+," a negative relationship is indicated by "–," and a landscape attribute with effects that were not tested with respect to a particular landscape conversion is indicated by "nt." (Modified from Grau et al. 2003.)

Landscape attributes	Agricultural land lost to urban development	Agricultural land lost to secondary forest
Distance to existing urban areas	+	nt
Population	nt	–
Elevation	–	+
Distance to roads	–	+
Percentage of slope	–	+
Farm size	nt	–
Distance to reserved area	–	+

In addition to influencing climate, the expansion of urban areas negatively affects secondary forests, as well as the biodiversity contained in them. Grau et al. (2003) summarized landscape studies of urbanization and the cover of secondary forest (table 8-3), showing that urban areas most commonly develop near existing urban areas and on flat topography. In contrast, secondary forests tend to develop initially at high elevations on steep topography near existing reserve areas and then move progressively downslope. In addition, the tendency for development to occur on the coastal plain has significant implications for the future of agriculture in Puerto Rico. Much of the development takes place on what were once prime agricultural lands, implying difficulties for feeding people should the current, petroleum-based economy become less viable. As a point of concern, Puerto Rico lost 6 percent of its prime agricultural lands to urbanization between 1977 and 1994 (López et al. 2001).

Human actions also feed back on the ecological connectivity of streams. Dams and water withdrawals strongly influence the biotic structure of the island's streams, strongly affecting migratory species (e.g., Greathouse et al. 2006a, 2006b). As such, there might be significant economic benefits to maintaining free-flowing streams (González-Cabán and Loomis 1997). This has led to an ongoing effort to develop an integrated understanding of stream and road networks (NSF Biocomplexity Project 2009) that should provide a firm basis for developing a wider understanding of the socioecological system of the Greater Luquillo Ecosystem.

Each of these examples emphasizes the influence of land cover change on climate, rainfall, and biodiversity, as modified by the geomorphic setting. What is clearly lacking is a detailed understanding of the dynamics of the human system and how ecological systems feed back to them. For the immediate future, an integrated research strategy should focus on the human population (including institutions and perceptions), biodiversity, and water as key elements in tropical ecosystems. The goal of the strategy should be to develop a series of validated models, the integration of which would facilitate the prediction of human population density and distribution, forest cover, and the biodiversity of key taxa, as well as water quality, quantity, and biodiversity in streams flowing through the landscape. Other factors deemed to be of importance (e.g., NPP, soil fertility) should be measured, modeled, and integrated into this strategy. The models should incorporate three characteristics:

1. A spatially explicit land use model would allow us to predict land cover and effect on aquatic resources under a variety of development scenarios. A number of good candidates for modeling approaches currently exist (Hall et al. 1995; Acevedo et al. 2001; Engelen and Gutiérrez 2003; Veldkamp and Verburg 2004). The model should combine biophysical factors (e.g., using the watershed approach [Pickett et al. 2001]) with key sociological, economic, and political factors. With respect to political considerations, an important issue in Puerto Rico would be the degree to which the promulgation and enforcement of public planning could guide future urban development. That is, the model should be relevant not only to biophysical and social scientists, but to regional planners as well (e.g., Engelen and Gutiérrez 2003).
2. A model should link land use cover to spatial patterns of and changes in local temperature and precipitation, the latter being a key driver of aquatic systems. Wu et al. (2006, 2007) have developed a series of models for the Luquillo Mountains that predict cloud cover, evapotranspiration, and stream flow as a function of land cover. Such models could be adapted readily to a larger scale. A model has been developed for the entire island of Puerto Rico that indicates that deforestation will reduce annual precipitation (van der Molen 2002). The development of a model integrating regional and local climate changes and their effects at the scale of the Greater Luquillo Ecosystem might be a challenge. However, if this were possible, then, in combination with global climate models for the northern Atlantic region (Neelin et al. 2006), it would facilitate the prediction of stream water quantity through time. The integration of a land use model with a precipitation and hydrology model would be used to predict water quality, which in turn would be used to predict the biodiversity of key taxa.
3. The diversity of secondary forests and some key terrestrial taxa (e.g., terrestrial arthropods) in Puerto Rico is well described (summarized in Grau et al. 2003), and studies of other key taxa are underway. Land use cover could be combined with these data in a model that incorporates life zone (determined predominantly by temperature and precipitation [Ewel and Whitmore 1973]) or environmental gradients (Hall et al. 1992) straightforwardly in order to predict terrestrial biodiversity. A forest model also would have to incorporate factors controlling the distribution of trees and other key taxa along the elevational gradient in the Luquillo Mountains (Abbott-Wood 2002; Wang et al. 2002). In this way, one could predict changes in forest zonation caused by long-term drying of the climate in Puerto Rico owing to reduced forest cover or global climate change.

We now accept that humans are the principal agents of change in most tropical ecosystems. Accordingly, the Luquillo LTER is beginning to integrate this perspective explicitly into a number of facets of its research program. We propose that this can be achieved by developing models that incorporate human effects on the environment, as long as effects from social factors in the human population are given sufficient weight. A wild card in our analysis would be the longer term effects associated with "the peaking of oil." This peak, and its presumed accompaniment, "the end of cheap oil," is likely to have enormous effects on land use in Puerto Rico as, perhaps, the tourism industry contracts, the cost of commuting increases, and the need for locally grown food increases. It might be likely that these processes will be dominant forcing functions for future land use in Puerto Rico (e.g., Hall 2000; Hall and Klitgaard 2006). A long-term strategy for understanding ecosystem dynamics in the Greater Luquillo Ecosystem (specifically, land use, biodiversity, and aquatic resources) must consider both direct and indirect effects of human populations. Once constructed, such a model could be used to guide local decision making (e.g., land use planning) and serve as a rubric for understanding the potential for human effects on ecosystems of other Neotropical countries.

Concluding Remarks

Long-term associations among scientists with multiple disciplinary backgrounds, as well as the inclusion of young scientists with fresh perspectives, contribute to the vigor and productivity of the Luquillo LTER Program. As such, the program acts as an incubator for new research ideas that emerge from the in-depth understanding of complex ecological systems within the context of evolving theory about populations, communities, and biogeochemical processes. Our divergent perspectives challenge dogmatic assumptions and expand the frontiers of ecological understanding in an integrated, innovative, and synthetic fashion.

Following in this tradition, it is clear that research in the program will continue to expand our understanding of geographic and ecological gradients of Puerto Rico via synoptic approaches, more intimately incorporate social and natural science perspectives, explore larger scale manipulative and observational experiments to produce a mechanistic understanding of responses to disturbance and successional change, and assume greater relevance to society by addressing important issues that are central to management, conservation, and policy. At the same time, our future likely will continue to leverage our strengths as (1) the hot and wet environmental anchor of the U.S. LTER Network, (2) a portal for network research to engage tropical issues, and (3) a complex, disturbance-mediated ecological system that is sensitive to modifications arising from global change.

Summary

Research in the Luquillo Mountains has documented the variety of ways in which the biota responds to disturbance and the way in which the biota influences the frequency, magnitude, and intensity of disturbances. Disturbance increases the complexity of interactions (i.e., macro- and microclimatic, biogeochemical, biotic) that control the flow of energy and the cycling of materials through ecosystems. It affects the life history and demographic parameters of species at fine spatial scales and creates a mosaic of patches at large spatial scales that, together, influence the dispersal of individuals among patches (i.e., the degree of connectedness) in a species-specific fashion. Such a cross-scale perspective provides a spatially explicit framework for understanding the assembly of species in disturbance-mediated environments. Moreover, differences in biodiversity affect ecosystem processes through species complementarity, organismal traits, and trophic interactions. Soil microorganisms, as well as the timing, quantity, and quality of litter deposition, play a critical role in affecting the dynamics of carbon and nutrient cycling over short and long temporal scales. These effects are mediated by scale, ultimately determining the resistance and resilience of ecosystems to disturbance.

Within this context, environmental gradients provide a platform for contrasting the role of particular species with respect to resilience and resistance during the interplay between disturbance and succession. In addition, multiple or sequential disturbances have complex spatial and temporal linkages, especially in riparian and stream communities, where species that connect freshwater and marine communities with those in headwater tributaries and riparian forests provide pathways for pulsed flows of energy and materials. From a terrestrial perspective, anthropogenic disturbance facilitates invasions by introduced tree species, sometimes culminating in the emergence of new forest communities dominated by introduced taxa. In mature forests not subject to intense anthropogenic degradation, introduced species might occur sporadically as rare species in hurricane-induced gaps, but these populations rapidly decrease in number after canopy closure. Thus, the development of new emerging forests does not necessarily result in the loss of native species or a reduction in species richness. The recognition and study of emerging new forests are important for understanding how organisms respond to anthropogenic disturbances, including global climate change. Finally, forecasting change requires the integration of biophysical and social science perspectives, an approach we have developed for studying interconnected ecosystems of the greater Luquillo region of Puerto Rico, extending from ridgetop to coastal environments.

Literature Cited

Abbott-Wood, C. 2002. Landscape forest modeling of the Luquillo Experimental Forest. M.S. thesis. University of North Texas, Denton, TX.

Acevedo M. F., S. Pamarti, M. Ablan, D. L. Urban, and A. Mikler. 2001. Modeling forest landscapes: Parameter estimation from gap models over heterogeneous terrain. *Simulation* 77:53–68.

Ackerman, J. D. 1992. *The orchids of Puerto Rico and the Virgin Islands*. Río Piedras, PR: University of Puerto Rico Press.

Allan, J. D., M. S. Wipfli, J. P. Caouette, A. Prussian, and J. Rodgers. 2003. Influence of streamside vegetation on inputs of terrestrial invertebrates to salmonid food webs. *Canadian Journal of Fisheries and Aquatic Sciences* 60:309–320.

Allendorf, F. W., and L. L. Lundquist. 2003. Introduction: Population biology, evolution, and control of invasive species. *Conservation Biology* 17:24–30.

Andre, M. F. 1998. Depopulation, land-use change and transformation in the French Massif Central. *Ambio* 4:351–353.

Atmar, W., and B. D. Patterson. 1993. The measure of order and disorder in the distributions of species in fragmented habitat. *Oecologia* 96:373–382.

Azeria, E. T., A. Carlson, T. Pärt, and C. G. Wiklund. 2006. Temporal dynamics and nestedness of an oceanic island bird fauna. *Global Ecology and Biogeography* 15:328–338.

Ballinger, A., and P. S. Lake. 2006. Energy and nutrient fluxes from rivers and streams into terrestrial food webs. *Marine and Freshwater Research* 57:15–28.

Balser, T. C., A. Kinzig, and M. K. Firestone. 2002. Linking soil microbial communities and ecosystem functioning. Pages 265–293 in A. Kinzig, S. W. Pacala, and D. Tilman, editors, *The functional consequences of biodiversity: Empirical progress and theoretical extensions*. Princeton, NJ: Princeton University Press.

Barbier, E. B. 2005. *Natural resources and economic development.* Cambridge, England: Cambridge University Press.

Barone, J. A., J. Thomlinson, P. A. Cordero, and J. K. Zimmerman. 2008. Metacommunity structure of tropical forest along an elevation gradient in Puerto Rico. *Journal of Tropical Ecology* 24:525–534.

Bauer, R. T. 2004. *Remarkable shrimps: Natural history and adaptations of the carideans.* Norman, OK: University of Oklahoma Press.

Baxter, C. V., K. D. Fausch, and W. C. Sanders. 2005. Tangled webs: Reciprocal flows of invertebrate prey link streams and riparian zones. *Freshwater Biology* 50:201–220.

Beard, K. H., A. Eschtruth, K. A. Vogt, D. J. Vogt, and F. N. Scatena. 2003. Amphibian effects on its prey and ecosystem: Evidence from two spatial scales. *Journal of Tropical Ecology* 19:607–617.

Beard, K. H., K. A. Vogt, and A. Kulmatiski. 2002. Top-down effects of a terrestrial frog on nutrient concentrations in a subtropical forest. *Oecologia* 133:583–593.

Beard, K. H., K. A. Vogt, D. J. Vogt, F. N. Scatena, A. P. Covich, R. Sigurdardottir, T. C. Siccama, and T. A. Crowl. 2005. The response of structural and functional characteristics of a tropical forested ecosystem to multiple, natural disturbances in legacy environments. *Ecological Monographs* 75:345–361.

Behrensmeyer, A. K., J. D. Damuth, W. A. DiMichele, R. Potts, H. D. Sues, and S. L. Wing, editors. 1992. *Terrestrial ecosystems through time: Evolutionary paleoecology of terrestrial plants and animals*. Chicago: University of Chicago Press.

Bellingham, P. J., E. V. J. Tanner, and J. R. Healey. 1995. Damage and responsiveness of Jamaican montane tree species after disturbance by a hurricane. *Ecology* 76:2562–2580.

Bellingham, P. J., E. V. J. Tanner, P. M. Rich, and T. C. R. Goodland. 1996. Changes in light below the canopy of a Jamaican montane rainforest after a hurricane. *Journal of Tropical Ecology* 12:699–722.

Benda, L., N. L. Poff, D. Miller, T. Dunne, G. Reeves, G. Pess, and M. Pollock. 2004. The network dynamics hypothesis: How channel networks structure riverine habitats. *BioScience* 54:413–427.

Berg, M., P. C. De Ruiter, W. Didden, M. Janssen, T. Schouten, and H. Verhoef. 2001. Community food web, decomposition, and nitrogen mineralization in a stratified Scots pine forest soil. *Oikos* 94:130–142.

Biaggi, V. 1997. *Las aves de Puerto Rico* [*The birds of Puerto Rico*], 4th ed. Río Piedras, PR: Editorial de la Universidad de Puerto Rico.

Blanco, J. E., and F. N. Scatena. 2006. Hierarchical contributions of river-basin connectivity, water chemistry, hydraulics, and substrate to the distribution of diadromous snails in Puerto Rican streams. *Journal of the North American Benthological Society* 25:82–98.

Bloch, C. P., C. L. Higgins, and M. R. Willig. 2007. Effects of large-scale disturbance on metacommunity structure of terrestrial gastropods: Temporal trends in nestedness. *Oikos* 116:395–406.

Bloch, C. P., and M. R. Willig. 2006. Context-dependence of long-term responses of terrestrial gastropod populations to large-scale disturbance. *Journal of Tropical Ecology* 22:111–122.

Bonnell, M., and L. A. Bruijnzeel, editors. 2004. *Forests, water and people in the humid tropics: Past, present and future hydrological research for integrated land and water management*. UNESCO. Cambridge, England: Cambridge University Press.

Boose, E. R., D. R. Foster, and M. Fluet. 1994. Hurricane impacts to tropical and temperate forest landscapes. *Ecological Monographs* 64:369–400.

Borges, S. 1996. The terrestrial oligochaetes of Puerto Rico. *Annals of the New York Academy of Sciences* 776:239–248.

Boulton, A. J., L. Boyero, A. P. Covich, M. Dobson, S. Lake, and R. Pearson. 2008. Are tropical streams ecologically different from temperate streams? Pages 257–284 in D. Dudgeon, editor, *Tropical stream ecology*. San Diego, CA: Academic Press.

Boulton, A. J., S. Findlay, P. Marmonier, E. H. Stanley, and H. M. Valett. 1998. The functional significance of the hyporheic zone in streams and rivers. *Annual Review of Ecology and Systematics* 29:59–81.

Bradford, M. A., T. H. Jones, R. D. Bardgett, H. I. J. Black, B. Boag, M. Bonkowski, R. Cook, T. Eggers, A. C. Gange, S. J. Grayston, E. Kandeler, A. E. McCaig, J. E. Newington, J. I. Prosser, H. Setälä, P. L. Staddon, G. M. Tordoff, D. Tscherko, and J. H. Lawton. 2002. Impacts of soil faunal community composition on model grassland ecosystems. *Science* 298:615–618.

Brash, A. R. 1984. *Avifauna reflections of historical landscape ecology in Puerto Rico*. New Haven, CT: Tropical Resources Institute, Yale University.

Briscoe, C. B., and F. H. Wadsworth. 1970. Stand structure and yield in the tabonuco forest of Puerto Rico. Pages B-79–B-89 in H. T. Odum and R. F. Pigeon, editors, *A tropical rain forest: A study of irradiation and ecology at El Verde, Puerto Rico*. Oakridge, TN: U.S. Atomic Energy Commission.

Brokaw, N. V. L. 1998. *Cecropia schreberiana* in the Luquillo Mountains of Puerto Rico. *The Botanical Review* 64:91–120.

Brokaw, N., S. Fraver, J. Grear, J. Thompson, J. Zimmerman, R. Waide, E. Everham, S. Hubbell, and R. Foster. 2004. Disturbance and canopy structure in two tropical forests. Pages 177–194 in E. Losos and E. G. Leigh, editors, *Forest diversity and dynamism: Findings from a large-scale plot network*. Chicago: Chicago University Press.

Brokaw, N. V. L., and J. S. Grear. 1991. Forest structure before and after Hurricane Hugo at three elevations in the Luquillo Mountains, Puerto Rico. *Biotropica* 23:386–392.

Brown, S., A. E. Lugo, S. Silander, and L. Liegel. 1983. Research history and opportunities in the Luquillo Experimental Forest. General Technical Report Number SO-44. New Orleans, LA: USDA Forest Service, Southern Experiment Station.

Campbell, C. J., and J. H. Leharrère. 1998. The end of cheap oil. *Scientific American* March:78–83.

Carriero, M. M., R. L. Sinsabaugh, D. A. Repert, and D. F. Parkhurst. 2000. Microbial enzyme shifts explain litter decay responses to simulated nitrogen deposition. *Ecology* 81:2359–2365.

Case, T. J. 1996. Global patterns in the establishment and distribution of exotic birds. *Conservation Biology* 78:69–90.

Chacón, N., N. Dezzeo, and S. Flores. 2005. Effect of particle-size distribution, soil organic carbon content and organo-mineral aluminium complexes on acid phosphatases of seasonally flooded forest soils. *Biology and Fertility of Soils* 41:69–72.

Chapin, F. S., III. 2003. Effects of plant traits on ecosystem and regional processes: A conceptual framework for predicting the consequences of global change. *Annals of Botany* 91:455–463.

Chapin, F. S., III, P. A. Matson, and H. A. Mooney. 2002. *Principles of terrestrial ecosystem ecology*. New York: Springer-Verlag.

Chapin, F. S., III, E. S. Zavaleta, V. T. Eviner, R. L. Naylor, P. M. Vitousek, H. L. Reynolds, D. U. Hooper, S. Lavorel, O. E. Sala, S. E. Hobbie, M. C. Mack, and S. Diaz. 2000. Consequences of changing biodiversity. *Nature* 405:234–242.

Cheshire, K., L. Boyero, and R. C. Pearson. 2005. Food webs in tropical Australian streams: Shredders are not scarce. *Freshwater Biology* 50:748–769.

Church, M. 2002. Geomorphic thresholds in riverine landscapes. *Freshwater Biology* 47:541–557.

Clark, D. B. 1996. Abolishing virginity. *Journal of Tropical Ecology* 12:735–739.

Colinvaux, P. A. 1996. Quaternary environmental history and forest diversity in the neotropics. Pages 359–405 in J. B. C. Jackson, A. F. Budd, and A. G. Coates, editors, *Evolution and environment in tropical America*. Chicago: University of Chicago Press.

Condit, R., P. Ashton, H. Balslev, N. Brokaw, S. Bunyavejchewin, G. Chuyong, L. Co, H. S. Dattaraja, S. Davies, S. Esufali, C. E. N. Ewango, R. Foster, N. Gunatilleke, S. Gunatilleke, C. Hernandez, S. Hubbell, R. John, D. Kenfack, S. Kiratiprayoon, P. Hall, T. Hart, A. Itoh, J. V. LaFrankie, I. Liengola, D. Lagunzad, S. Loo de Lao, E. Losos, E. Magård, J.-R. Makana, N. Manokaran, H. Navarrete, S. Mohammed Nur, T. Okhubo, R. Pérez, C. Samper, L. Hua Seng, R. Sukumar, J.-C. Svenning, S. Tan, D. Thomas, J. Thompson, M. I. Vallejo, G. Villa Muñoz, R. Valencia, T. Yamakura, and J. K. Zimmerman. 2005. Tropical tree α-diversity: Results from a worldwide network of large plots. *Biologiske Skrifter* 55:565–582.

Cook, A. 2001. Behavioural ecology: On doing the right thing, in the right place, at the right time. Pages 447–487 in G. M. Barker, editor, *The biology of terrestrial molluscs*. Wallingford, UK: CAB International.

Covich, A. P. 1988a. Atyid shrimp in the headwaters of the Luquillo Mountains, Puerto Rico: Filter feeding in natural and artificial streams. *Verhandlungen der Internationale Vereinigung für Theorestische und Angewandte Limnologie* 23:2108–2113.

Covich, A. P. 1988b. Geographical and historical comparisons of neotropical streams: Biotic diversity and detrital processing in highly variable habitats. *Journal of the North American Benthological Society* 7:361–386.

Covich, A. P. 2006a. Dispersal-limited biodiversity of tropical insular streams. *Polish Journal of Ecology* 54:523–547.

Covich, A. P. 2006b. Protecting benthic biodiversity to insure detrital processing and ecosystem services: Importance of invertebrate shredders in stream networks. *Ecotropicos* 19:109–127.

Covich, A. P., M. C. Austen, F. Barlocher, E. Chauvet, B. J. Cardinale, C. L. Biles, P. Inchausti, O. Dangles, M. Solan, M. O. Gessner, B. Statzner, and B. Moss. 2004a. The role of biodiversity in the functioning of freshwater and marine benthic ecosystems. *BioScience* 54:767–775.

Covich, A. P., T. A. Crowl, and T. Heartsill-Scalley. 2006. Effects of drought and hurricane disturbances on headwater distributions of palaemonid river shrimp (*Macrobrachium* spp.) in the Luquillo Mountains, Puerto Rico. *Journal of the North American Benthological Society* 25:99–107.

Covich, A. P., T. A. Crowl, C. L. Hein, M. J. Townsend, and W. H. McDowell. 2009. Importance of geomorphic barriers to predator-prey interactions in river networks. *Freshwater Biology* 54:450–465.

Covich, A. P., T. A. Crowl, S. L. Johnson, and M. Pyron. 1996. Distribution and abundance of tropical freshwater shrimp along a stream corridor: Response to disturbance. *Biotropica* 28:484–492.

Covich, A. P., T. A. Crowl, S. L. Johnson, D. Varza, and D. L. Certain. 1991. Post-hurricane Hugo increases in atyid shrimp abundance in a Puerto Rican montane stream. *Biotropica* 23:448–454.

Covich, A. P., T. A. Crowl, and F. N. Scatena. 2003. Effects of extreme low flows on freshwater shrimps in a perennial tropical stream. *Freshwater Biology* 48:1199–1206.

Covich, A. P., K. C. Ewel, R. O. Hall, P. G. Giller, D. Merritt, and W. Goedkoop. 2004b. Ecosystem services provided by freshwater benthos. Pages 137–159 in D. Wall, editor, *Sustaining biodiversity and ecosystem services in soils and sediments*. Washington, DC: Island Press.

Covich, A. P., and W. H. McDowell. 1996. The stream community. Pages 433–459 in D. P. Reagan and R. B. Waide, editors, *The food web of a tropical rain forest*. Chicago: University of Chicago Press.

Craig, D. A. 2003. Geomorphology, development of running water habitats, and evolution of blackflies on Polynesian islands. *BioScience* 53:1079–1093.

Crow, T. R. 1980. A rainforest chronicle: A 30-year record of change in structure and composition at El Verde, Puerto Rico. *Biotropica* 12:42–55.

Crow, T. R., and P. L. Weaver. 1977. Tree growth in a moist tropical forest of Puerto Rico. Forest Service Research Paper ITF-22. Río Piedras, PR: Institute of Tropical Forestry, USDA Forest Service.

Crowl, T. A., and A. P. Covich. 1994. Responses of a freshwater shrimp to chemical and tactile stimuli from a large decapod predator. *Journal of the North American Benthological Society* 13:291–298.

Crowl, T. A., W. H. McDowell, A. P. Covich, and S. L. Johnson. 2001. Species-specific responses in leaf litter processing in a tropical headwater stream (Puerto Rico). *Ecology* 82:775–783.

Crowl, T. A., V. Welsh, T. Heartsill-Scalley, and A. P. Covich. 2006. Effects of different types of conditioning on leaf-litter shredding by *Xiphocaris elongata*, a neotropical freshwater shrimp. *Journal of the North American Benthological Society* 25:196–206.

Cummins, K. W. 1974. Structure and function of stream ecosystems. *BioScience* 24: 631–640.

Davis, M. A. 2003. Biotic globalization: Does competitition from introduced species threaten biodiversity? *BioScience* 53:481–489.

Decamps, H., G. Pinay, R. J. Naiman, G. E. Petts, M. E. McClain, A. Hilbricht-Ilkowska, T. A., Hanley, R. M. Holmes, J. Quinn, J. Gibert, A. M. P. Tabacchi, F. Schiemer, E. Tabacchi, and M. Zaleski. 2004. Riparian zones: Where biogeochemistry meets biodiversity in management practice. *Polish Journal of Ecology* 52:3–18.

Denslow, J. S., and S. J. DeWalt. 2008. Plant invasions in tropical forests: Patterns and mech anisms. Pages 409–426 in W. P. Carson and S. A. Schnitzer, editors, *Tropical forest community ecology*. Chicago: University of Chicago Press.

Detwiler, R. P., and C. A. S. Hall. 1988. Tropical forests and the global carbon cycle. *Science* 239:42–47.

Dial, R., and J. Roughgarden. 1995. Experimental removal of insectivores from rain forest canopy: Direct and indirect effects. *Ecology* 76:1821–1834.

Diamond, J. M. 1975. Assembly of species communities. Pages 342–444 in M. L. Cody and J. M. Diamond, editors, *Ecology and evolution of communities*. Cambridge, MA: Harvard University Press.

Dietz, J. L. 1987. *The economic history of Puerto Rico*. Princeton, NJ: Princeton University Press.

Dinda, S. 2004. Environmental Kuznets curve hypothesis: A survey. *Ecological Economics* 49:431–455.

Drury, W. H., and I. C. T. Nisbet. 1973. Succession. *Journal of the Arnold Arboretum* 54:331–368.

Duke, J. A. 1970. Postirradiation woody seedlings. Pages D189–D192 in H. T. Odum and R. F. Pigeon, editors, *A tropical rain forest: A study of irradiation and ecology at El Verde, Puerto Rico*. Oakridge, TN: U.S. Atomic Energy Commission.

Ehrhardt-Martinez, K., E. M. Crenshaw, and J. C. Jenkins. 2002. Deforestation and the environmental Kuznets curve: A cross-national investigation of intervening mechanisms. *Social Science Quarterly* 83:226–243.

Ellstrand, N. C., and K. A. Schierenbeck. 2000. Hybridization as a stimulus for the evolution of invasiveness in plants? *Proceedings of the National Academy of Sciences* 97: 7043–7050.

Emanuel, K. 2005. Increasing destructiveness of tropical cyclones over the past 30 years. *Nature* 436:686–688.

Engelen, G., and E. R. Gutiérrez. 2003. *Xplorah: Support for integrated spatial planning in Puerto Rico*. Río Piedras, PR: Research Institute for Knowledge Systems and Graduate School for Planning, University of Puerto Rico.

Everham, E. M., III, and N. V. L. Brokaw. 1996. Forest damage and recovery from catastrophic wind. *The Botanical Review* 62:113–185.

Ewel, J. J., D. J. O'Dowd, J. Bergelson, C. C. Daehler, C. M. D'Antonio, L. D. Gómez, D. R. Gordon, R. J. Hobbs, A. Holt, K. R. Hooper, C. E. Hughes, M. LaHart, R. R. B. Leakey, W. G. Lee, L. L. Loope, D. H. Lorence, S. M. Louda, A. E. Lugo, P. B. McEvoy, D. M. Richardson, and P. M. Vitousek. 1999. Deliberate introductions of species: Research needs. *BioScience* 49:619–630.

Ewel, J. J., and J. L.Whitmore. 1973. The ecological life zones of Puerto Rico and the U.S. Virgin Islands. Research Paper ITF-18. New Orleans, LA: USDA Forest Service.

Fauth, J. E., J. Bernardo, M. Camara, W. J. Resetarits, Jr., J. Van Buskirk, and S. A. McCollum. 1996. Simplifying the jargon of community ecology: A conceptual approach. *The American Naturalist* 147:282–286.

Fernández, D. S., and N. Fetcher. 1991. Changes in light availability following Hurricane Hugo in a subtropical montane forest in Puerto Rico. *Biotropica* 23:393–399.

Fetcher, N., R. A. Cordero, and J. Voltzow. 1999. Lack of differentiation: Plant response to elevation, population origin, and wind in the Luquillo Mountains, Puerto Rico. *Biotropica* 32:225–234.

Fievet, E., S. Doledec, and P. Lim. 2001. Distribution of migratory fishes and shrimps along multivariate gradients in tropical island streams. *Journal of Fish Biology* 59:390–402.

Figueroa Colón, J. 1996. Phytogeographic trends, centers of high species richness and endemism, and the question of extinctions in the native flora of Puerto Rico. *Annals of the New York Academy of Sciences* 776:89–102.

Fisher, S. G., N. B. Grimm, E. Marti, and R. Gomez. 1998. Hierarchy, spatial configuration, and nutrient cycling in streams. *Australian Journal of Ecology* 23:1–52.

Fisher, S. G., R. A. Sponseller, and J. B. Heffernan. 2004. Horizons in stream biogeochemistry: Flowpaths to progress. *Ecology* 85:2369–2379.

Foster, D. R., and E. R. Boose. 1995. Hurricane disturbance regimes in temperate and tropical forest ecosystems. Pages 305–339 in M. P. Coutts and J. Grace, editors, *Wind and trees*. Cambridge, England: Cambridge University Press.

Foster, D. R., G. Motzkin, and B. Slater. 1998. Land use history as long-term broad scale disturbance: Regional forest dynamics in central New England. *Ecosystems* 1: 96–119.

Frangi, J. L., and A. E. Lugo. 1991. Hurricane damage to a flood plain forest in the Luquillo Mountains of Puerto Rico. *Biotropica* 23:324–335.

Freeman, M. C., C. M. Pringle, and C. R. Jackson. 2007. Hydrologic connectivity and the contribution of stream headwaters to ecological integrity at regional scales. *Journal of the American Water Resources Association* 43:5–14.

Garrison, R. W., and M. R. Willig. 1996. Arboreal invertebrates. Pages 183–271 in D. P. Reagan and R. B. Waide, editors. *The food web of a tropical rain forest*. Chicago: University of Chicago Press.

Giller, P. S., A. P. Covich, K. C. Ewel, R. O. Hall, Jr., and D. M. Merritt. 2004. Vulnerability and management of ecological services in freshwater systems. Pages 137–159 in D. Wall, editor, *Sustaining biodiversity and ecosystem services in soils and sediments*. Washington, DC: Island Press.

Glenn-Lewin, D. C., R. K. Peet, and T. T. Veblen, editors. 1992. *Plant succession: Theory and prediction*. London: Chapman & Hall.

Gómez-Pompa, A., C. Vázquez-Yanes, and S. Guevara. 1972. The tropical rain forests: A nonrenewable resource. *Science* 177:762–765.

Gomi, T., R. C. Sidle, and J. S. Richardson. 2002. Understanding processes and downstream linkages in headwater streams. *BioScience* 52:905–916.

Gonçalves, J. F., M. A. S. Graça, and M. Callisto. 2006. Leaf-litter breakdown in 3 streams in temperate, Mediterranean, and tropical Cerrado climates. *Journal of the North American Benthological Society* 25:344–355.

González, G. 2002. Soil organisms and litter decomposition. Pages 315–329 in R. S. Ambasht and N. K. Ambasht, editors, *Modern trends in applied terrestrial ecology*. London: Kluwer Academic/Plenum Publishers.

González, G., and T. R. Seastedt. 2001. Soil fauna and plant litter decomposition in tropical and subalpine forests. *Ecology* 82:955–964.

González, G., and X. Zou. 1999. Earthworms influence on N availability and the growth of *Cecropia schreberiana* in tropical pasture and forest soils. *Pedobiologia* 43:1–6.

González, G., X. Zou, and S. Borges. 1996. Earthworm abundance and species composition in abandoned tropical croplands: Comparison of tree plantations and secondary forests. *Pedobiologia* 40:385–391.

González-Cabán, A., and J. Loomis. 1997. Economic benefits of maintaining ecological integrity of Río Mamayes in Puerto Rico. *Ecological Economics* 21:63–75.

Gould, W. A., G. González, and G. Carrero Rivera. 2006. Structure and composition of vegetation along an elevational gradient in Puerto Rico. *Journal of Vegetation Science* 17:653–664.

Graça, M. A. S., C. Cresssa, M. O. Gessner, M. J. Felo, K. A. Callies, and C. Barrios. 2001. Food quality, feeding preferences, survival and growth of shredders from temperate and tropical streams. *Freshwater Biology* 46:947–957.

Grace, J. B., T. M. Anderson, M. D. Smith, E. Seabloom, S. J. Andelman, G. Meche, E. Weiher, L. K. Allain, H. Jutila, M. Sankaran, J. Knops, M. Ritchie, and M. R. Willig. 2007. Integrating abiotic factors and biotic interactions into our understanding of diversity and primary production in natural grassland communities. *Ecology Letters* 10:680–689.

Gragson, T. L., and M. Grove. 2006. Social science in the context of the Long-Term Ecological Research Program. *Society and Natural Resources* 19:93–100.

Graham, A. 2003a. Geohistory models and Cenozoic paleoenvironments of the Caribbean region. *Systematic Botany* 28:378–386.

Graham, A. 2003b. Historical phytogeography of the Greater Antilles. *Brittonia* 55:357–383.

Grau, H. R., T. M. Aide, J. K. Zimmerman, J. R. Thomlinson, E. Helmer, and X. Zou. 2003. The ecological consequences of socioeconomic and land use changes in post agriculture Puerto Rico. *BioScience* 53:1159–1168.

Greathouse E. A., and C. M. Pringle. 2006. Does the river continuum concept apply on a tropical island? Longitudinal variation in a Puerto Rican stream. *Canadian Journal of Fisheries and Aquatic Sciences* 63:134–152.

Greathouse, E. A., C. M. Pringle, and J. G. Holmquist. 2006a. Conservation and management of migratory fauna: Dams in tropical streams of Puerto Rico. *Aquatic Conservation: Marine and Freshwater Ecosystems* 16:695–712.

Greathouse, E. A., C. M. Pringle, W. H. McDowell, and J. G. Holmquist. 2006b. Indirect upstream effects of dams: Consequences of migratory consumer extirpation in Puerto Rico. *Ecological Applications* 16:134–152.

Gregory, F. A., and A. M. Sabat. 1996. The effect of hurricane disturbance on the fecundity of sierra palms (*Prestoea montana*). *Bios* 67:135–139.

Groffman, P. M., and J. M. Tiedje. 1989. Denitrification in north temperate forest soils: Relationships between denitrification and environmental factors at the landscape scale. *Soil Biology and Biochemistry* 21:621–626.

Gunderson, L., and C. S. Holling. 2002. *Panarchy: Understanding transformations in human and natural systems*. Washington, DC: Island Press.

Guzmán-Grajales, S. M., and L. R. Walker. 1991. Differential seedling responses to litter after Hurricane Hugo in the Luquillo Experimental Forest, Puerto Rico. *Biotropica* 23:407–413.

Hall, C. A. S., editor. 2000. *Quantifying sustainable development: The future of tropical Economies*. San Diego, CA: Academic Press.

Hall, C. A. S. 2006. Integrating concepts and models from development economics with land use change models in the tropics. *Environment, Development and Sustainability* 8: 19–53.

Hall, C. A. S., and K. Klitgaard. 2006. The need for a new, biophysical-based paradigm in economics for the second half of the age of oil. *Journal of Transdisciplinary Research* 1:4–22.

Hall, C. A. S., and J.-Y. Ko. 2005. Energy and international development: A systems approach to international development. Pages 73–90 in E. Ortega and S. Ulgiati, editors, *Proceedings of IV biennial international workshop "Advances in energy studies."* Sao Paulo, Brazil: Grafica da Unicamp, Campinas.

Hall, C. A. S., D. Lindenberger, R. Kummel, T. Kroeger, and W. Eichhorn. 2001. The need to reintegrate the natural sciences with economics. *BioScience* 51:663–673.

Hall, C. A. S., J. A. Stanford, and R. Hauer. 1992. The distribution and abundance of organisms as a consequence of energy balances along multiple environmental gradients. *Oikos* 65:377–390.

Hall, C. A. S., H. Tain, Y. Qi, G. Pontius, and J. Cornell. 1995. Modelling spatial and temporal patterns of tropical land use change. *Journal of Biogeography* 22:753–757.

Heard, S. B., and J. S. Richardson. 1995. Shredder-collector facilitation in stream detrital food webs: Is there enough evidence? *Oikos* 72:359–366.

Heartsill-Scalley, T. 2005. Characterization of riparian zone vegetation and litter production in tropical, montane rainforest along an environmental gradient. Ph.D. dissertation. Utah State University, Logan.

Hector, A., B. Schmid, C. Beierkuhnlein, M. C. Caldeira, M. Diemer, P. G. Dimitrakopoulos, J. A. Finn, H. Freitas, P. S. Giller, J. Good, R. Harris, P. Högberg, K. Huss-Danell, J. Joshi, A. Jumpponen, C. Körner, P. W. Leadley, M. Loreau, A. Minns, C. P. H. Mulder, G. O'Donovan, S. J. Otway, J. S. Pereira, A. Prinz, D. J. Read, M. Scherer-Lorenzen, E.-D. Schulze, A.-S. D. Siamantziouras, E. M. Spehn, A. C. Terry, A. Y. Troumbis, F. I. Woodward, S. Yachi, and J. H. Lawton. 1999. Plant diversity and productivity experiments in European grasslands. *Science* 286:1123–1127.

Hein, C. L., A. S. Pike, J. F. Blanco, A. P. Covich, F. N. Scatena, C. P. Hawkins, and T. A. Crowl. 2011. Effects of coupled natural and anthropogenic factors on the community structure of diadronmus fish and shrimp species in tropical island streams. *Freshwater Biology* 56:1002–1015.

Henderson, P. A., and I. Walker. 1986. On the leaf litter community of the Amazonian blackwater stream Tarumazinho. *Journal of Tropical Ecology* 2:1–17.

Hershey, A. E., S. Beaty, K. Fortino, M. Keyse, P. P. Mou, W. J. O'Brien, L. Ulseth, G. A. Gettel, P. W. Lienesch, C. Lueche, M. E. McDonald, C. H. Mayer, M. C. Miller, C. Richards, J. A. Schuldt, and S. C. Whalen. 2006. Effect of landscape factors on fish distributions in arctic Alaskan lakes. *Freshwater Biology* 51:39–55.

Higgins, C. L., M. R. Willig, and R. E. Strauss. 2006. The role of stochastic processes in producing nested patterns of species distributions. *Oikos* 114:159–167.

Hintikka, V. 1970. Studies on white-rot humus formed by higher fungi in forest soils. *Communicationes Instituti Forestalis Fenniae* 69:1–68.

Hoagland, B. W., and S. L. Collins. 1997. Gradient models, gradient analysis, and hierarchical structure in plant communities. *Oikos* 78:23–30.

Holt, R. D., and M. Loreau. 2002. Biodiversity and ecosystem functioning: The role of trophic interactions and the importance of system openness. Pages 246–262 in A. P. Kinzig, S. W. Pacala, and D. Tilman, editors, *The functional consequences of biodiversity: Empirical progress and theoretical extensions*. Princeton, NJ: Princeton University Press.

Holyoak, M., M. A. Leibold, and R. D. Holt, editors. 2005. *Metacommunities: Spatial dynamics and ecological communities*. Chicago: University of Chicago Press.

Hooper, D. U., F. S. Chapin III, J. J. Ewel, A. Hector, P. Inchausti, S. Lavorel, J. H. Lawton, D. M. Lodge, M. Loreau, S. Naeem, B. Schmid, H. Setälä, A. J. Symstad, J. Vandermeer, and D. A. Wardle. 2005. Effects of biodiversity on ecosystem functioning: A consensus of current knowledge. *Ecological Monographs* 75:3–35.

Hopkinson, C., A. E. Lugo, M. Alber, A. P. Covich, and S. J. Van Bloem. 2008. Forecasting effects of sea-level rise and windstorms on coastal and inland ecosystems. *Frontiers in Ecology and Environment* 6:255–263.

Hubbell, S. P. 2001. *The unified neutral theory of biodiversity and biogeography*. Princeton, NJ: Princeton University Press.

Hulot, F. D., G. Lacroix, F. Lescher-Moutoue, and M. Loreau. 2000. Functional diversity governs ecosystem response to nutrient enrichment. *Nature* 405:340–344.

Hynes, H. B. N. 1975. The stream and its valley. *Verhandlungen der Internationalen Vereinigung für Theoretische und Angerwandte Limnologie* 19:1–15.

IPCC. 2007. *Climate change 2007: The physical basis*. Geneva, Switzerland: IPCC Secretariat.

Ives, A. R., and S. R. Carpenter. 2007. Stability and diversity of ecosystems. *Science* 317:58–62.

Iwata, T., M. Inoue, S. Nakano, H. Miyasaka, A. Doi, and A. P. Covich. 2003. Shrimp abundance and habitat relationships in tropical rain-forest streams, Sarawak, Borneo. *Journal of Tropical Ecology* 19:387–395.

Jackson, S. T. 2004. Quaternary biogeography: Linking biotic response to environmental variability across timescales. Pages 47–65 in M. V. Lomolino and L. R. Heaney, editors, *Frontiers of biogeography: New directions in the geography of nature*. Sunderland, MA: Sinauer Associates, Inc.

Janzen, D. H. 2000. Costa Rica's Area de Conservación Guanacaste: A long march to survival through non-damaging biodevelopment. *Biodiversity* 1:7–20.

Janzen, D. H. 2002. Tropical dry forest: Area de Conservación Guanacaste, northwestern Costa Rica. Pages 559–583 in M. R. Perrow and A. J. Davy, editors, *Handbook of ecological restoration. Volume 2. Restoration in practice*. Cambridge, England: Cambridge University Press.

Jenny, H. 1941. *Factors of soil formation.* New York: McGraw-Hill.

Junk, W. J., and K. M. Wantzen. 2004. The flood pulse concept: New aspects, approaches, and applications—An update. Pages 117–149 in R. L. Welcomme and T. Petr, editors, *Proceedings of the Second International Symposium on the Management of Large Rivers for Fisheries*. Vol. 2. RAP Publication 2004/16. Food and Agriculture Organization and Mekong River Commission. Bangkok, Thailand: FAO Regional Office for Asia and the Pacific.

Kaufman, D. M., G. A. Kaufman, and D. W. Kaufman. 2000. Faunal structure of small mammals in tallgrass prairie: An evaluation of richness and spatiotemporal nestedness. *Fort Hays Studies Special Issue* 1:47–70.

Keller, M., W. A. Kaplan, and S. C. Wofsy. 1986. Emissions of N_2O, CH_4, and CO_2 from tropical forest soils. *Journal of Geophysical Research* 91:11791–11802.

Khanna, N., and F. Plassman. 2004. The demand for environmental quality and the environmental Kuznets Curve hypothesis. *Ecological Economics* 51:225–236.

Kikkert, D. A., T. A. Crowl, and A. P. Covich. 2009. Upstream migration of amphidromous shrimp in the Luquillo Experimental Forest, Puerto Rico: Temporal patterns and environmental cues. *Journal of the North American Benthological Society* 28:233–246.

Kling, G. W., G. W. Kipphut, M. M. Miller, and W. J. O'Brien. 2000. Integration of lakes and streams in a landscape perspective: The importance of material processing on spatial patterns and temporal coherence. *Freshwater Biology* 43:477–497.

Klooster, D. 2003. Forest transitions in Mexico: Institutions and forests in a globalized countryside. *Professional Geographer* 55:227–237.

Kuznets, S. 1955. Economic growth and income inequality. *American Economic Review* 49:1–28.

Lambin, E. F., H. J. Geist, and E. Lepers. 2003. Dynamics of land-use and land-cover change in tropical regions. *Annual Review of Environment and Resources* 28:205–241.

Lamson, H. M., J.-C. Shiao, Y. Iizuka, W.-N. Tzeng, and D. K. Cairns. 2006. Movement patterns of American eels (*Anguilla rostrata*) in a coastal watershed, based on otolith microchemistry. *Marine Biology* 149:1567–1576.

Larned, S. T., R. A. Kinzie III, A. P. Covich, and C. T. Chong. 2003. Detritus processing by endemic and non-native Hawaiian stream invertebrates: A microcosm study of species-specific effects. *Archiv für Hydrobiologie* 156:241–254.

Lecerf, A., M. Dobson, C. K. Dang, and E. Chauvet. 2005. Riparian plant species loss alters trophic dynamics in detritus-based stream ecosystems. *Oecologia* 146:432–442.

LeClerc, G., and C. A. S. Hall, editors. 2007. *Making global development work: Scientific alternatives to neoclassical economic models.* Albuquerque, NM: University of New Mexico Press.

Leibold, M. A., and G. M. Mikkelson. 2002. Coherence, species turnover, and boundary clumping: Elements of meta-community structure. *Oikos* 97:237–250.

Lewis, W. M., Jr., S. K. Hamilton, M. A. Rodríquez, J. D. Sanders III, and D. H. Lasi. 2001. Ecological determinism on the Orinoco floodplain. *BioScience* 50:681–692.

Liegel, L. H. 1984. Assessment of hurricane rain/wind damage in *Pinus caribaea* and *Pinus oocarpa* provenance trials in Puerto Rico. *Commonwealth Forestry Review* 63:47–53.

Likens, G. E. 2004. Some perspectives on long-term biogeochemical research from the Hubbard Brook ecosystem. *Ecology* 85:2355–2362.

Liogier, A. H., and L. F. Martorell. 2000. *Flora of Puerto Rico and adjacent islands: A systematic synopsis.* 2nd ed. (revised). San Juan, PR: Editorial de la Universidad de Puerto Rico.

Little, E. L., R. O. Woodbury, and F. H. Wadsworth. 1974. *Trees of Puerto Rico and the Virgin Islands.* Vol. 2. Agriculture Handbook 449. Washington, DC: USDA Forest Service.

Liu, Z. G., and X. M. Zou. 2002. Exotic earthworms accelerate plant litter decomposition in a Puerto Rican pasture and a wet forest. *Ecological Applications* 12:1406–1417.

Lloyd, J., and J. A. Taylor. 1994. On the temperature-dependence of soil respiration. *Functional Ecology* 8:315–323.

Lockwood, J. L., and M. L. McKinney, editors. 2001. *Biotic homogenization.* New York: Kluer Academic/Plenum Publishers.

Lodge, D. J. 1993. Nutrient cycling by fungi in wet tropical forests. Pages 37–57 in S. Isaac, J. C. Frankland, R. Watling, and A. J. S. Whalley, editors, *Aspects of tropical mycology.* BMS Symposium Series, Vol. 19. Cambridge, England: Cambridge University Press.

Lodge, D. J. 1996. Microorganisms. Pages 53–108 in D. P. Reagan and R. B. Waide, editors, *The food web of a tropical rain forest.* Chicago: University of Chicago Press.

Lodge, D. J. 2001. Implications for nitrogen additions from air pollutants on litter decay fungi and ecosystem processes. *Mycological Research* 105:898–899.

Lodge, D. J., and S. Cantrell. 1995. Fungal communities in wet tropical forests: Variation in time and space. *Canadian Journal of Botany* Suppl. 1:S1391–S1398.

Lodge, D. J., and W. H. McDowell. 1991. Summary of ecosystem-level effects of Atlantic hurricanes. *Biotropica* 23:373–378.

Lodge, D. J., W. H. McDowell, J. Macy, S. K. Ward, R. Leisso, K. C. Campos, and K. Kuhnert. 2008. Distribution and role of mat-forming basidiomycetes in a tropical forest. Pages 195–208 in L. Boddy, J. C. Frankland, and P. van West, editors, *Ecology of saprophytic basidiomycetes.* Amsterdam, The Netherlands: Academic Press.

Lodge, D. J., W. H. McDowell, and C. P. McSwiney. 1994. The importance of nutrient pulses in tropical forests. *Trends in Ecology and Evolution* 9:384–387.

Lodge, D. J., F. N. Scatena, C. E. Asbury, and M. J. Sánchez. 1991. Fine litterfall and related nutrient inputs resulting from Hurricane Hugo in subtropical wet and lower montane rain forests of Puerto Rico. *Biotropica* 23:336–342.

Lodge, D. M., and K. Shrader-Frechette. 2003. Nonindigenous species: Ecological explanation, environmental ethics, and public policy. *Conservation Biology* 17:31–37.

Lomolino, M. V. 2004. Conservation biogeography. Pages 293–296 in M. V. Lomolino and L. R. Heaney, editors, *Frontiers of biogeography: New directions in the geography of nature.* Sunderland, MA: Sinauer Associates, Inc.

López, T. del M., T. M. Aide, and J. R. Thomlinson. 2001. Urban expansion and the loss of prime agricultural lands in Puerto Rico. *Ambio* 30:49–54.

Loreau, M., A. Downing, M. C. Emmerson, A. González, J. Hughes, P. Inchausti, J. Joshi, J. Norberg, and O. Sala. 2002. A new look at the relationship between diversity and stability. Pages 79–91 in M. Loreau, S. Naeem, and P. Inchausti, editors, *Biodiversity and ecosystem functioning: Synthesis and perspectives.* Oxford, England: Oxford University Press.

Losos, E. C., M. S. Ashton, N. Brokaw, R. Bunyavejchwin, R. Condit, G. B. Chuyong, L. Co, H. S. Dattaraja, S. J. Davies, S. Esufali, C. E. N. Ewango, R. B. Foster, N. Gunatilleke, S. Gunatilleke, T. H. Hart, C. Hernández, S. P. Hubbell, A. Itoh, R. John, M. Kanzaki, D. S. K. Kenfack, J. V. LaFrankie, H.-S. Lee, I. Liengola, S. Lao, J.-R. Makana, N. Manokaran, M. Navarette Hernández, T. Ohkugo, R. Perez, N. Pongpattananurak, C. Samper, K. Sringernyuang, R. Sukumar, I.-F. Fun, H. S. Sureh, S. Tan, D. W. Thomas, J. Thompson, M. I. Vallejo, G. Villa Muñoz, R. Valencia, T. Yamakura, and J. K. Zimmerman. 2004. The structure of tropical forests. Pages 69–78 in E. C. Losos and E. G. Leigh, Jr., editors, *Tropical forest diversity and dynamism: Findings from a large-scale plot network.* Chicago: University of Chicago Press.

Lowe, W. H., G. E. Likens, and M. E. Power. 2006. Linking scales in stream ecology. *Bioscience* 56:591–597.

Lugo, A. E. 1988. Estimating reductions in the diversity of tropical forest species. Pages 58–70 in E. O. Wilson and F. M. Peter, editors, *Biodiversity*. Washington, DC: National Academies Press.

Lugo, A. E. 1992. Comparison of tropical tree plantations with secondary forests of similar age. *Ecological Monographs* 62:1–41.

Lugo, A. E. 2004a. The homogeocene in Puerto Rico. Pages 366–375 in D. J. Zarin, J. R. R. Alavalapati, F. E. Putz, and M. Schmink, editors, *Working forests in the neotropics: Conservation through sustainable management?* New York: Columbia University Press.

Lugo, A. E. 2004b. The outcome of alien tree invasions in Puerto Rico. *Frontiers in Ecology and the Environment* 2:265–273.

Lugo, A. E. 2005. Los bosques [The forests]. Pages 395–548 in R. L. Joglar, editor, *Biodiversidad de Puerto Rico. Vertebrados terrestres y ecosistemas* [*Biodiversity of Puerto Rico. Terrestrial vertebrates and ecosystems*]. San Juan, PR: Editorial del Instituto de Cultura Puertorriqueña.

Lugo, A. E., and T. J. Brandeis. 2005. A new mix of alien and native species coexist in Puerto Rico's landscapes. Pages 484–509 in D. F. R. P. Burslem, M. A. Pinard, and S. E. Hartley, editors, *Biotic interactions in the tropics: Their role in the maintenance of species diversity*. Cambridge, England: Cambridge University Press.

Lugo, A. E., and E. Helmer. 2004. Emerging forests on abandoned land: Puerto Rico's new forests. *Forest Ecology and Management* 190:145–161.

Lugo, A. E., J. A. Parrotta, and S. Brown. 1993. Loss in species caused by tropical deforestation and their recovery through management. *Ambio* 22:106–209.

Lugo, A. E., and R. B. Waide. 1993. Catastrophic and background disturbance of tropical ecosystems at the Luquillo Experimental Forest. *Journal of Biosciences* 18: 475–481.

Lugo, A. E., and J. Zimmerman. 2002. Ecological life histories. Pages 191–213 in J. Vozzo, editor, *Tropical tree seed manual*. Agriculture Handbook 721. Washington, DC: USDA Forest Service.

Lundquist, J. E., A. E. Camp, M. L. Tyrrell, S. J. Seybold, P. Cannon, and D. J. Lodge. 2011. Earth, wind, and fire: Abiotic factors and the impacts of global environmental change on forest health. Pages 195–244 in J. D. Castello and S.A. Teale, editors, *Forest Health: An Integrated Perspective*. Cambridge, England: Cambridge University Press.

MacArthur, R. H. 1955. Fluctuations of animal populations and a measure of community stability. *Ecology* 36:533–536.

Mack, M., E. A. G. Schuur, M. S. Bret-Harte, G. Shaver, and F. S. Chapin III. 2004. Ecosystem carbon storage in arctic tundra reduced by long-term nutrient fertilization. *Nature* 431:440–443.

Mack, R. N., D. Simberloff, W. M. Lonsdale, H. Evans, M. Clout, and F. A. Bazzaz. 2000. Biotic invasions: Causes, epidemiology, global consequences, and control. *Ecological Applications* 10:689–710.

Malmqvist, B., S. D. Rundle, A. P. Covich, A. Hildrew, A. G. Robinson, and C. R. Townsend. 2008. Prospects for streams and rivers: An ecological perspective. Pages 19–29 in N. V. C. Polunin, editor, *Aquatic ecosystems—Trends and global prospects*. Cambridge, England: Cambridge University Press.

March, J. G., J. P. Benstead, C. M. Pringle, and M. R. Ruebel. 2001. Linking shrimp assemblages with rates of detrital processing along an elevational gradient in a tropical stream. *Canadian Journal of Fisheries and Aquatic Sciences* 58:470–478.

March, J. G., and C. M. Pringle. 2003. Foodweb structure and basal resource utilization along a tropical island stream continuum, Puerto Rico. *Biotropica* 35:84–93.

Marquet, P. A., M. Fernández, S. A. Navarrete, and C. Valdovinos. 2004. Diversity emerging: Towards a deconstruction of biodiversity patterns. Pages 191–209 in M. V. Lomolino and L. R. Heaney, editors, *Frontiers of biogeography: New directions in the geography of nature*. Sunderland, MA: Sinauer Associates, Inc.

Mathuriac, C., and E. Chauvet. 2002. Breakdown of leaf litter in a Neotropical stream. *Journal of the North American Benthological Society* 21:384–396.

May, R. M. 1974. *Stability and complexity in model ecosystems*. Princeton, NJ: Princeton University Press.

McCormick, J. F. 1994. A review of the population dynamics of selected tree species in the Luquillo Experimental Forest, Puerto Rico. Pages 224–257 in A. E. Lugo and C. Lowe, editors, *Tropical forests: Management and ecology*. New York: Springer-Verlag.

McDowell, W. H. 2001. Hurricanes, people, and riparian zones: Controls on nutrient losses from forested Caribbean watersheds. *Forest Ecology and Management* 154:443–451.

McDowell, W. H., C. P. McSwiney, and W. B. Bowden. 1996. Effects of hurricane disturbance on groundwater chemistry and riparian function in a tropical rain forest. *Biotropica* 28:577–584.

McGroddy, M., and W. L. Silver. 2000. Variations in belowground carbon storage and soil CO_2 flux rates along a wet tropical climate gradient. *Biotropica* 32:614–624.

McKinney, M. L., and J. L. Lockwood. 1999. Biotic homogenization: A few winners replacing many losers in the next mass extinction. *Trends in Ecology and Evolution* 14:450–452.

McKinney, M. L., and J. L. Lockwood. 2001. Biotic homogenization: A sequential and selective process. Pages 1–17 in J. L. Lockwood and M. L. McKinney, editors, *Biotic homogenization*. New York: Kluer Academic/Plenum Publishers.

Meyer, J. L., D. L. Strayer, J. B. Wallace, S. L. Eggert, G. S. Helfman, and N. E. Leonard. 2007. The contribution of headwater streams to biodiversity in river networks. *Journal of the American Water Resources Association* 43:86–103.

Meyer, J. L., and J. B. Wallace. 2001. Lost linkages and lotic ecology: Rediscovering small streams. Pages 295–317 in M. Press, N. Huntly, and S. Levin, editors, *Ecology: Achievement and challenge*. Oxford, England: Blackwell Science.

Mikola, J., R. D. Bardgett, and K. Hedlund. 2002. Biodiversity, ecosystem functioning, and soil decomposer food webs. Pages 169–180 in M. Loreau, S. Naeem, and P. Inchausti, editors, *Biodiversity and ecosystem functioning: Synthesis and perspectives*. Oxford, England: Oxford University Press.

Millennium Ecosystem Assessment. 2005. *Ecosystems and human well-being: Biodiversity synthesis*. Washington, DC: World Resources Institute.

Miller, R. M., and D. J. Lodge. 1997. Fungal responses to disturbance—Agriculture and forestry. Pages 65–84 in K. Esser, P. A. Lemke, and D. T. Wicklow, editors, *Environmental and microbial relationships*. The Mycota, Vol. V. Berlin: Springer-Verlag.

Mooney, H. A. 2002. The debate on the role of biodiversity in ecosystem functioning. Pages 12–17 in M. Loreau, S. Naeem, and P. Inchausti, editors, *Biodiversity and ecosystem functioning: Synthesis and perspectives*. Oxford, England: Oxford University Press.

Mulholland, P. J., A. M. Helton, G. C. Poole, R. O. Hall, Jr., S. K. Hamilton, B. J. Peterson, J. L. Tank, L. R. Ashkenas, L. W. Cooper, C. N. Dahm, W. K. Dodds, S. E. G. Findlay, S. V. Gregory, N. B. Grimm, S. L. Johnson, W. H. McDowell, J. L. Meyer, H. M. Valett, J. R. Webster, C. P. Arango, J. J. Beaulieu, M. J. Bernot, A. J. Burgin, C. L. Crenshaw, L. T. Johnson, B. R. Niederlehner, J. M. O'Brien, J. D. Potter, R. W. Sheibley, D. J. Sobota, and S. M. Thomas. 2008. Stream denitrification across biomes and its response to anthropogenic nitrate loading. *Nature* 452:202–206.

Murphy, D. 2007. The relation between land-cover and the urban heat island in northeastern Puerto Rico. M.S. thesis. State University of New York, College of Environmental Science and Forestry, Syracuse, NY.

Naiman, R. J., H. Decamps, and M. E. McClain. 2005. *Riparia: Ecology, conservation, and management of streamside communities*. Amsterdam, The Netherlands: Elsevier.

Nakano, S., and M. Murakami. 2001. Reciprocal subsidies: Dynamic interdependence between terrestrial and aquatic food webs. *Proceedings of the National Academy of Sciences* 98:166–170.

Neelin, J. D., M. Münnich, H. Su, J. E. Meyerson, and C. E. Holloway. 2006. Tropical drying trends in global warming models and observations. *Proceedings of the National Academy of Sciences* 103:6110–6115.

Nicholson, P. B., K. L. Bocock, and O. W. Neal. 1966. Studies on the decomposition of the faecal pellets of a millipede [*Glomeris marginata* (Villers)]. *Journal of Ecology* 54:755–766.

Norton, J., J. W. Lewis, and D. Rollinson. 2004. Temporal and spatial patterns of nestedness in eel macroparasite communities. *Journal of Parasitology* 129:203–211.

NSF Biocomplexity Project. nd. Biocomplexity project summary: Modeling complex interactions of overlapping river and road networks in a changing landscape. http://biocomplexity.warnercnr.colostate.edu. Accessed March 30, 2009.

O'Connor, P. J., A. P. Covich, F. N. Scatena, and L. L. Loope. 2000. Non-indigenous bamboo along headwater streams of the Luquillo Mountains, Puerto Rico: Leaf fall, aquatic leaf decay and patterns of invasion. *Journal of Tropical Ecology* 16:499–516.

Odum, E. P. 1969. The strategy of ecosystem development. *Science* 164:262–270.

Odum, H. T. 1970. Rain forest structure and mineral-cycling homeostasis. Pages H3–H52 in H. T. Odum and R. F. Pigeon, editors, *A tropical rain forest: A study of irradiation and ecology at El Verde, Puerto Rico*. Oak Ridge, TN: U.S. Atomic Energy Commission, Division of Technical Information.

Odum, H. T., W. Abbott, R. K. Selander, F. B. Golley, and R. F. Wilson. 1970a. Estimates of chlorophyll and biomass of the tabonuco forest of Puerto Rico. Pages I3–I19 in H. T. Odum and R. F. Pigeon, editors, *A tropical rain forest: A study of irradiation and ecology at El Verde, Puerto Rico*. Oak Ridge, TN: U.S. Atomic Energy Commission, Division of Technical Information.

Odum, H. T., G. Drewry, and E. A. McMahan. 1970b. Introduction to Section E. Pages E3–E15 in H. T. Odum and R. F. Pigeon, editors, *A tropical rain forest: A study of irradiation and ecology at El Verde, Puerto Rico*. Oak Ridge, TN: U.S. Atomic Energy Commission, Division of Technical Information.

Ostertag, R., F. N. Scatena, and W. L. Silver. 2003. Forest floor decomposition following hurricane litter inputs in several Puerto Rican forests. *Ecosystems* 6:261–273.

Ostertag, R., W. L. Silver, and A. E. Lugo. 2005. Factors affecting mortality and resistance to damage following hurricanes in a rehabilitated subtropical moist forest. *Biotropica* 37:16–24.

Paerl, H. W., and J. L. Pinckney. 1996. A mini-review of microbial consortia: Their roles in aquatic production and biogeochemical cycling. *Microbial Ecology* 31:225–247.

Paetzold, A., J. F. Bernet, and K. Tockner. 2006. Consumer-specific responses to riverine subsidy pulses in a riparian arthropod assemblage. *Freshwater Biology* 51:1103–1115.

Parrotta, J. A. 1995. Influence of overstory composition on understory colonization by native species in plantations on a degraded tropical site. *Journal of Vegetation Science* 6:627–636.

Parrotta, J. A., and D. J. Lodge. 1991. Fine root dynamics in a subtropical wet forest following hurricane disturbance. *Biotropica* 23:343–347.

Parton, W. J., W. L. Silver, I. C. Burke, L. Grassens, M. E. Harmon, W. S. Currie, J. Y. King, E. C. Adair, L. A. Brandt, S. C. Hart, and B. Fasth. 2007. Global-scale similarities in nitrogen release patterns during long-term decomposition. *Science* 315:361–364.

Patterson, B. D., and W. Atmar. 1986. Nested subsets and the structure of insular mammalian faunas and archipelagos. *Biological Journal of the Linnean Society* 28:65–82.

Perz, S. G., and D. L. Skole. 2003. Secondary forest expansion in the Brazilian Amazon and the refinement of forest transition theory. *Society & Natural Resources* 16:277–294.

Peters, D. P. C., B. T. Bestelmeyer, and M. G. Turner. 2007a. Cross-scale interactions and changing pattern-process relationships: Consequences for system dynamics. *Ecosystems* 10:790–796.

Peters, D. P. C., and K. M. Havstad. 2006. Nonlinear dynamics in arid and semi-arid systems: Interactions among drivers and processes across scales. *Journal of Arid Environments* 65:196–206.

Peters, D. P. C., O. E. Sala, C. D. Allen, A. P. Covich, and M. Brunson. 2007b. Cascading events in linked ecological and social-economic systems: Predicting change in an uncertain world. *Frontiers in Ecology and Environment* 5:221–224.

Peterson, B. J., W. M. Wollheim, P. J. Mulholland, J. R. Webster, J. L. Meyer, J. L. Tank, E. Marti, W. B. Bowden, H. M. Valett, A. E. Hershey, W. H. McDowell, W. K. Dodds, S. K. Hamilton, S. Gregory, and D. D. Morrall. 2001. Control of nitrogen export from watersheds by headwater streams. *Science* 292:86–90.

Pfeiffer, W. J. 1996. Litter invertebrates. Pages 137–181 in D. P. Reagan and R. B. Waide, editors, *The food web of a tropical rain forest*. Chicago: University of Chicago Press.

Pickett, S. T. A., M. L. Cadenasso, J. M. Grove, C. H. Nilon, R. V. Pouyat, W. C. Zipperer, and R. Costanza. 2001. Urban ecological systems: Linking terrestrial ecological, physical, and socioeconomic components of metropolitan areas. *Annual Review of Ecology and Systematics* 32:127–157.

Pimm, S. L. 1984. The complexity and stability of ecosystems. *Nature* 307:321–326.

Polishook, J. D., G. F. Bills, and D. J. Lodge. 1996. Microfungi from decaying leaves of two rain forest trees in Puerto Rico. *Journal of Industrial Microbiology* 17:284–294.

Poole, G. C. 2002. Fluvial landscape ecology: Addressing uniqueness within the river discontinuum. *Freshwater Biology* 47:641–660.

Power, M. E. 2006. Environmental controls on food web regimes: A fluvial perspective. *Progress in Oceanography* 68:125–133.

Power, M. E., and W. E. Dietrich. 2002. Food webs in river networks. *Ecological Research* 17:451–471.

Prather, M., R. Derwent, D. Ehhalt, P. Fraser, E. Sahueza, and X. Zhou. 1995. Other trace gases and atmospheric chemistry. Pages 77–126 in J. T. Houghton, L. G. Meira Filho, J. Bruce, H. Lee, B. A. Callander, E. Haites, N. Harris, and K. Maskell, editors, *Climate change 1994: Radiative forcing of climate change*. Cambridge, England: Cambridge University Press.

Presley, S. J., M. R. Willig, C. P. Bloch, I. Castro-Arellano, C. L. Higgins, and B. T. Klingbeil. 2011. A complex metacommunity structure for gastropods along an elevational gradient. Biotropica 43:480–488.

Presley, S. J., C. L. Higgins, and M. R. Willig. 2010. A comprehensive framework for the evaluation of metacommunity structure. *Oikos* 119:908–917.

Pringle, C. M. 2003. The need for a more predictive understanding of hydrologic connectivity. *Aquatic Conservation: Marine and Freshwater Ecosystems* 13:467–471.

Pringle, C. M., G. A. Blake, A. P. Covich, K. M. Buzby, and A. Finley. 1993. Effects of omnivorous shrimp in a montane tropical stream—Sediment removal, disturbance of sessile invertebrates and enhancement of understory algal biomass. *Oecologia* 93: 1–11.

Proctor, G. R. 1989. *Ferns of Puerto Rico and the Virgin Islands*. New York: The New York Botanical Garden.

Pyke, A. S. 2008. Longitudinal patterns in stream channel geomorphology and aquatic habitat in the Luquillo Mountains of Puerto Rico. Ph.D. dissertation. University of Pennsylvania, Pittsburgh.

Pyron, M., and A. P. Covich. 2003. Migration patterns, densities, and growth of *Neritina punctulata* snails in Río Espíritu Santo and Río Mameyes, northeastern Puerto Rico. *Caribbean Journal of Science* 39:338–347.

Pyron, M., A. P. Covich, and R. W. Black. 1999. On the relative importance of pool morphology and woody debris to distributions of shrimp in a Puerto Rican headwater stream. *Hydrobiologia* 405:207–215.

Reagan, D. P., G. Camilo, and R. B. Waide. 1996. The community food web: Major properties and patterns of organization. Pages 461–488 in D. P. Reagan and R. B. Waide, editors, *The food web of a tropical rain forest.* Chicago: University of Chicago Press.

Reagan, D. P., and R. B. Waide, editors. 1996. *The food web of a tropical rain forest.* Chicago: University of Chicago Press.

Rice, S. P., M. T. Greenwood, and C. B. Joyce. 2001. Tributaries, sediment sources, and the longitudinal organisation of macroinvertebrate fauna along river systems. *Canadian Journal of Fisheries and Aquatic Sciences* 58:824–840.

Richardson, B. A., M. J. Richardson, G. González, A. B. Shiels, and D. S. Srivastava. 2010. A canopy trimming experiment in Puerto Rico: The response of litter invertebrate communities to canopy loss and debris deposition in a tropical forest subject to hurricanes. *Ecosystems* 11:286–301.

Richardson, B. A., M. J. Richardson, and F. N. Soto-Adames. 2005. Separating the effects of forest type and elevation on the diversity of litter invertebrate communities in a humid tropical forest in Puerto Rico. *Journal of Animal Ecology* 74:926–936.

Richardson, J. S., and R. J. Danehy. 2007. A synthesis of the ecology of headwater streams and their riparian zones in temperate forests. *Forest Science* 53:131–147.

Richter, B. D., A. T. Warner, J. L. Meyer, and K. Lutz. 2006. A collaborative and adaptive process for developing environmental flow recommendations. *River Research and Applications* 22:297–318.

Rincón, J., and I. Martínez. 2006. Food quality and feeding preferences of *Phylloicus* sp. (Trichoptera: Calamoceratidae). *Journal of the North American Benthological Society* 25:207–213.

Romanuk, T. N., L. J. Jackson, J. R. Post, E. McCauley, and N. D. Martinez. 2006. The structure of food webs along river networks. *Ecography* 29:3–10.

Rudel, T. K. 1998. Is there a forest transition? Deforestation, reforestation, and development. *Rural Sociology* 63:533–552.

Rudel, T. K., D. Bates, and R. Machinguiashi. 2002. A tropical forest transition? Agricultural change, out-migration, and secondary forests in the Ecuadorian Amazon. *Annals of the Association of American Geographers* 92:87–102.

Rudel, T. K., M. Perez-Lugo, and H. Zichal. 2000. When fields revert to forest: Development and spontaneous reforestation in post-war Puerto Rico. *Professional Geographer* 52:386–397.

Rueda-Delgado, G., K. M. Wantzen, and M. B. Tolosa. 2006. Leaf-litter decomposition in an Amazonian floodplain stream: Effects of seasonal hydrological changes. *Journal of the North American Benthological Society* 25:233–249.

Sabo, J. L., and M. E. Power. 2002. River-watershed exchange: Effects of riverine subsidies on riparian lizards and their terrestrial prey. *Ecology* 83:1860–1869.

Salonius, P. O. 1981. Metabolic capabilities of forest soil microbial populations with reduced species diversity. *Soil Biology and Biochemistry* 13:1–10.

Sánchez de León, Y., X. Zou, S. Borges, and H. Ruan. 2003. Recovery of native earthworms in abandoned tropical pastures. *Conservation Biology* 17:999–1006.

Sanford, R. L., Jr., W. J. Parton, D. S. Ojima, and D. J. Lodge. 1991. Hurricane effects on soil organic matter dynamics and forest production in the Luquillo Experimental Forest, Puerto Rico: Results of simulation modelling. *Biotropica* 23:364–372.

Santana, M., D. J. Lodge, and P. Lebow. 2005. Relationship of host recurrence in fungi to rates of tropical leaf decomposition. *Pedobiologia* 49:549–564.

Scatena, F. N., and A. E. Lugo. 1995. Geomorphology, disturbance, and the vegetation and soils of two subtropical wet steepland watersheds in Puerto Rico. *Geomorphology* 13:199–213.

Scatena, F. N., S. Moya, C. Estrada, and J. D. Chinea. 1996. The first five years in the reorganization of aboveground biomass and nutrient use following Hurricane Hugo in the Bisley Experimental Watersheds, Luquillo Experimental Forest, Puerto Rico. *Biotropica* 28:424–440.

Scatena, F. N., W. L. Silver, T. G. Siccama, A. H. Johnson, and M. J. Sánchez. 1993. Biomass and nutrient content of the Bisley Experimental Watersheds, Luquillo Experimental Forest, Puerto Rico, before and after Hurricane Hugo, 1989. *Biotropica* 25:15–27.

Scheiner, S. M., S. B. Cox, M. Willig, G. G. Mittelbach, C. Osenberg, and M. Kaspari. 2000. Species richness, species-area curves and Simpson's paradox. *Evolutionary Ecology Research* 2:791–802.

Schowalter, T. D. 1995. Canopy invertebrate community response to disturbance and consequences of herbivory in temperate and tropical forests. *Selbyana* 16:41–48.

Schröter, D., V. Wolters, and P. C. De Ruiter. 2003. C and N mineralisation in the decomposer food webs of a European forest transect. *Oikos* 102:294–308.

Silver, W. L. 1998. The potential effects of elevated CO_2 and climate change on tropical forest soils and biogeochemical cycling. *Climatic Change* 39:337–361.

Silver, W. L. 1992. The effects of small-scale and catastrophic disturbances on carbon and nutrient cycling in a lower montane subtropical wet forest in Puerto Rico. Ph.D. dissertation. Yale University, New Haven, CT

Silver, W. L., D. J. Herman, and M. K. Firestone. 2001. Dissimilatory nitrate reduction to ammonium in tropical forest soils. *Ecology* 82:2410–2416.

Silver, W. L., L. M. Kueppers, A. E. Lugo, R. Ostertag, and V. Matzek. 2004. Carbon sequestration and plant community dynamics following reforestation of tropical pasture. *Ecological Applications* 14:1115–1127.

Silver, W. L., A. E. Lugo, and M. Keller. 1999. Soil oxygen availability and biogeochemical cycling along elevation and topographic gradients in Puerto Rico. *Biogeochemistry* 44:301–328.

Silver, W. L., F. N. Scatena, A. H. Johnson, T. G. Siccama, and F. Watt. 1996. At what temporal scales does disturbance affect belowground nutrient pools? *Biotropica* 28:441–457.

Silver, W. L., A. W. Thompson, M. E. McGroddy, R. K. Varner, J. R. Robertson, J. D. Dias, H. Silva, P. Crill, and M. Keller. 2005a. Fine root dynamics and trace gas fluxes in two lowland tropical forest soils. *Global Change Biology* 11:290–306.

Silver, W. L., A. W. Thompson, A. Reich, J. J. Ewel, and M. K. Firestone. 2005b. Nitrogen cycling in tropical plantation forests: Potential controls on nitrogen retention. *Ecological Applications* 15:1604–1614.

Silver, W. L., and K. A. Vogt. 1993. Fine root dynamics following single and multiple disturbances in a subtropical wet forest ecosystem. *Journal of Ecology* 81:729–738.

Sjöbe, G., S. I. Nilsson, T. Pesson, and P. Karlsson. 2004. Degradation of hemicellulose, cellulose and lignin in decomposing spruce needle litter in relation to N. *Soil Biology and Biochemistry* 36:1761–1768.

Smith, G. C., A. P. Covich, and A. M. D. Brasher. 2003. An ecological perspective on the biodiversity of tropical island streams. *BioScience* 53:1048–1051.

Smith, R. F. 1970. The vegetation structure of a Puerto Rican rain forest before and after short-term gamma irradiation. Pages D103–D140 in H. T. Odum and R. F. Pigeon, editors, *A tropical rain forest: A study of irradiation and ecology at El Verde, Puerto*

Rico. Oak Ridge, TN: U.S. Atomic Energy Commission, Division of Technical Information.

Snyder, N. F. R., J. W. Wiley, and C. B. Kepler. 1987. *The parrots of Luquillo: Natural history and conservation of the Puerto Rican Parrot*. Los Angeles: Western Foundation of Vertebrate Zoology.

Stern, D. I. 2004. The rise and fall of the environmental Kuznets curve. *World Development* 32:1419–1439.

Steudler, P. A., J. M. Melillo, R. D. Bowden, M. S. Castro, and A. E. Lugo. 1991. The effects of natural and human disturbances on soil nitrogen dynamics and trace gas fluxes in a Puerto Rican wet forest. *Biotropica* 23:356–363.

Stevens, R. D., S. B. Cox, R. E. Strauss, and M. R. Willig. 2003. Patterns of functional diversity across an extensive environmental gradient: Vertebrate consumers, hidden treatments and latitudinal trends. *Ecology Letters* 6:1099–1108.

Stevens, R. D., M. R. Willig, and R. E. Strauss. 2006. Latitudinal gradients in the phenetic diversity of New World bat communities. *Oikos* 112:41–50.

Taylor, C. M., S. Silander, R. B. Waide, and W. J. Pfeiffer. 1995. Recovery of a tropical forest after gamma irradiation: A 23-year chronicle. Pages 258–285 in A. E. Lugo and C. Lowe, editors, *Tropical forests: Management and ecology*. New York: Springer-Verlag.

Teh, Y. A., W. L. Silver, and M. E. Conrad. 2005. Oxygen effects on methane production and oxidation in humid tropical forest soils. *Global Change Biology* 11:1283–1297.

Thomlinson, J. R., and L. Y. Rivera. 2000. Suburban growth in Luquillo, Puerto Rico: Some consequences of development on natural and semi-natural systems. *Landscape and Urban Planning* 49:15–23.

Thompson, J., N. Brokaw, J. K. Zimmerman, R. B. Waide, E. M. Everham III, D. J. Lodge, C. M. Taylor, D. García Montiel, and M. Fluet. 2002. Land use history, environment, and tree composition in a tropical forest. *Ecological Applications* 12:1344–1363.

Thompson, J., A. E. Lugo, and J. Thomlinson. 2007. Land use history, hurricane disturbance, and the fate of introduced species in a subtropical wet forest in Puerto Rico. *Plant Ecology* 192:289–301.

Thoms, M. C., and M. Parsons. 2003. Eco-geomorphology: An interdisciplinary approach to river science. *International Association of Hydrological Sciences* 276:113–120.

Thorp, J. H., and M. D. Delong. 1994. The riverine productivity model: An heuristic view of carbon sources and organic processing in large river ecosystems. *Oikos* 70:305–308.

Tilman, D., D. Wedin, and J. Knops. 1996. Productivity and sustainability influenced by biodiversity in grassland ecosystems. *Nature* 379:718–720.

Torres, J. A., and R. R. Snelling. 1997. Biogeography of Puerto Rican ants: A non-equilibrium case? *Biodiversity and Conservation* 6:1103–1121.

Townsend, C. R. 1996. Concepts in river ecology: Pattern and process in the catchment hierarchy. *Archiv für Hydrobiologie* 113(Suppl.):3–21.

Turner, B. L., B. C. Turner, R. W. Clark, J. F. Kates, J. T. Richards, and W. B. Meyer, editors. 1990. *The Earth as transformed by human action*. Cambridge, England: Cambridge University Press.

United Nations Secretariat Population Division of the Department of Economic and Social Affairs. 2002. *World urbanization prospects: The 2001 revision*. New York: United Nations.

Uriarte, M., C. D. Canham, J. Thompson, and J. K. Zimmerman. 2004. A maximum-likelihood, spatially explicit analysis of tree growth and survival in a tropical forest. *Ecological Monographs* 74:591–614.

U.S. Long Term Ecological Research Network [LTER]. 2007. *The decadal plan for LTER: Integrative science for society and the environment*. LTER Network Office Publication Series No. 24. Albuquerque, NM: LTER.

van Der Molen, M. K. 2002. *Meteorological impacts of land use change in the Maritime tropics*. Ph.D. dissertation. Vrije University, Amsterdam.

Vannote, R., G. W. Minshall, K. W. Cummins, J. R. Sedell, and C. E. Cushing. 1980. The river continuum concept. *Canadian Journal of Fisheries and Aquatic Sciences* 37:130–137.

Velázquez-Lozada, A., J. E. Gonzalez, and A. Winter. 2006. Urban heat island effect analysis for San Juan, Puerto Rico. *Atmospheric Environment* 40:1731–1741.

Veldkamp, A., and P. H. Verburg. 2004. Modeling land use change and environmental impact. *Journal of Environmental Management* 72:1–3.

Vermeij, G. 1996. An agenda for invasion biology. *Conservation Biology* 78:3–9.

Wadsworth, F. H. 1947. Growth in the lower montane rain forest of Puerto Rico. *Caribbean Forester* 8:27–44.

Wadsworth, F. H. 1951. Forest management in the Luquillo Mountains. I. The setting. *Caribbean Forestry* 12:1344–1363.

Wadsworth, F. H., and R. A. Birdsey. 1983. Un nuevo enfoque de los bosques de Puerto Rico [A new approach to the forests of Puerto Rico]. Pages 12–27 in *Puerto Rico Department of Natural Resources ninth symposium on natural resources*. San Juan, PR: Puerto Rico Department of Natural Resources.

Wadsworth, F. H., and G. H. Englerth. 1959. Effects of the 1956 hurricane on forests in Puerto Rico. *Caribbean Forester* 20:38–51.

Wadsworth, F. H., B. R. Parresol, and J. C. Figueroa Colón. 1989. Tree increment indicators in a subtropical wet forest. Pages 205–212 in W. M. Wan Razali, H. T. Chan, and S. Appanah, editors, *Proceedings of seminar on growth and yield mixed/moist forests*. Kuala Lumpur, Malaysia: Forest Research Institute.

Waide, R. B., and A. E. Lugo. 1992. A research perspective on disturbance and recovery of a tropical montane forest. Pages 173–190 in J. G. Goldammer, editor, *Tropical forests in transition: Ecology of natural and anthropogenic disturbance processes*. Basel, Switzerland: Berkhauser-Verlag.

Waide, R. B., J. K. Zimmerman, and F. N. Scatena. 1998. Controls of primary productivity in a montane tropical forest: Lessons from the Luquillo Mountains in Puerto Rico. *Ecology* 79:31–37.

Walker, L. R. 2000. Seedling and sapling dynamics in treefall pits in Puerto Rico. *Biotropica* 32:262–275.

Walker, L. R., N. V. L. Brokaw, D. J. Lodge, and R. B. Waide, editors. 1991. Special issue: Ecosystem, plant and animal responses to hurricanes in the Caribbean. *Biotropica* 23:313–521.

Walker, L. R., and R. del Moral. 2003. *Primary succession and ecosystem rehabilitation*. Cambridge, England: Cambridge University Press.

Walker, L. R., W. L. Silver, M. R. Willig, and J. K. Zimmerman, editors. 1996a. Special issue: Long term responses of Caribbean ecosystems to disturbance. *Biotropica* 28:414–614.

Walker, L. R., J. K. Zimmerman, D. J. Lodge, and S. Guzmán-Grajales. 1996b. An elevational comparison of growth and species composition in hurricane-damaged forests in Puerto Rico. *Journal of Ecology* 84:877–889.

Wallace, J. B., and J. J. Hutchens. 2000. Effects of invertebrates in lotic ecosystem process. Pages 73–96 in D. C. Coleman and P. F. Hendrix, editors, *Invertebrates as webmasters in ecosystems*. Wallingford, England: CABI Publishing.

Wang, H., and C. A. S. Hall. 2004. Modeling the effects of hurricane Hugo on spatial and temporal variation in primary productivity and soil carbon and nitrogen in the Luquillo Experimental Forest, Puerto Rico. *Plant and Soil* 263:69–84.

Wang, H. Q., C. A. S. Hall, F. Scatena, N. Fetcher, and W. Wu. 2002. Modeling the spatial and temporal variability in climate and primary productivity across the Luquillo Mountains, Puerto Rico. *Forest Ecology and Management* 179:69–94.

Wantzen, K. M., and R. Wagner. 2006. Detritus processing by invertebrate shredders: A Neotropical–temperate comparison. *Journal of the North American Benthological Society* 25:216–233.

Watson, R. T., I. R. Noble, B. Bolin, N. H. Ravindranath, D. J. Verardo, and D. J. Dokken, editors. 2001. *Land use, land use change, and forestry*. Cambridge, England: Cambridge University Press.

Weaver, P. L. 1979. Tree growth in several tropical forests of Puerto Rico. Research Paper SO-152. New Orleans, LA: USDA Forest Service, Southern Forest Experiment Station.

Weaver, P. L. 1983. Tree growth and stand changes in the subtropical life zones of the Luquillo Mountains of Puerto Rico. Research Paper SO-190. New Orleans, LA: USDA Forest Service, Southern Forest Experiment Station.

Weaver, P. L. 1986. Growth and age of *Cyrilla racemiflora* L. in montane forests of Puerto Rico. *Interciencia* 11:221–228.

Weaver, P. L. 1991. Environmental gradients affect forest composition in the Luquillo Mountains of Puerto Rico. *Interciencia* 16:141–151.

Weaver, P. L. 1994. Baño de Oro Natural Area, Luquillo Mountains, Puerto Rico. General Technical Report SO-111. New Orleans, LA: USDA Forest Service, Southern Forest Experimental Station.

Weaver, P. L. 2001. Thinning and regeneration in Puerto Rico's colorado forest, with comments about their effect on the Puerto Rican parrot. *Caribbean Journal of Science* 37:252–258.

Weaver, P. L. 2002. A chronology of hurricane induced changes in Puerto Rico's lower montane rain forest. *Interciencia* 27:252–258.

Weaver, P. L., and P. G. Murphy. 1990. Forest structure and productivity in Puerto Rico's Luquillo Mountains. *Biotropica* 22:69–82.

Webb, D. P. 1977. Regulation of deciduous forest litter decomposition by soil arthropod feces. Pages 57–69 in W. J. Mattson, editor, *The role of soil arthropods in forest ecosystems*. New York: Springer-Verlag.

Webster, P. J., G. J. Holland, J. A. Curry, and H.-R. Chang. 2005. Changes in tropical cyclone number, duration, and intensity in a warming environment. *Science* 309:1844–1846.

Weiher, E., and P. Keddy, editors. 2001. *Ecological assembly rules: Perspectives, advances, retreats*. Cambridge, England: University of Cambridge Press.

Weinbren, M. P., and B. M. Weinbren. 1970. Observations on the mosquito population in the irradiated forest at El Verde. Pages E159–E167 in H. T. Odum and R. F. Pigeon, editors, *A tropical rain forest: A study of irradiation and ecology at El Verde, Puerto Rico*. Oak Ridge, TN: U.S. Atomic Energy Commission, Division of Technical Information.

Weinbren, M. P., B. M. Weinbren, W. B. Jackson, and J. B. Villella. 1970. Studies of the roof rat (*Rattus rattus*) in the El Verde forest. Pages E169–E181 in H. T. Odum and R. F. Pigeon, editors, *A tropical rain forest: A study of irradiation and ecology at El Verde, Puerto Rico*. Oak Ridge, TN: U.S. Atomic Energy Commission, Division of Technical Information.

Whiles, M. R., and W. K. Dodds. 2002. Relationships between stream size, suspended particles, and filter-feeding macroinvertebrates in a Great Plains drainage network. *Journal of Environmental Quality* 31:1589–1600.

Whittaker, R. H. 1970. *Communities and ecosystems*. Toronto: The Macmillan Company.

Willig, M. R. 2003. Challenges to understanding dynamics of biodiversity in time and space. *Paleobiology* 29:30–33.

Willig, M. R., C. P. Bloch, N. Brokaw, C. Higgins, J. Thompson, and C. R. Zimmermann. 2007. Cross-scale responses of biodiversity to hurricane and anthropogenic disturbance in a tropical forest. *Ecosystems* 10:824–838.

Willig, M. R., D. L. Moorehead, S. B. Cox, and J. C. Zak. 1996. Functional diversity of soil bacteria communities in the tabonuco forest: The interaction of anthropogenic and natural disturbance. *Biotropica* 28:471–483.

Willig, M. R., and L. R. Walker. 1999. Disturbance in terrestrial ecosystems: Salient themes, synthesis, and future directions. Pages 747–767 in L. R. Walker, editor, *Ecosystems of disturbed ground*. Amsterdam, The Netherlands: Elsevier.

Willig, M. R., S. J. Presley, C. P. Bloch, I. Castro-Arellano, L. M. Cisneros, C. L. Higgins, and B. T. Klingbeil. 2011. Tropical metacommunities and elevational gradients: disentangling effects of forest type from other elevational factors. Oikos 120:1497–1508.

Wright, D. H., B. D. Patterson, G. M. Mikkelson, A. Cutler, and W. Atmar. 1998. A comparative analysis of nested subset patterns of species composition. *Oecologia* 113:1–20.

Wright, D. H., and J. H. Reeves. 1992. On the meaning and measurement of nestedness of species assemblages. *Oecologia* 92:416–428.

Wright, M. S., and A. P. Covich. 2005a. Relative importance of bacteria and fungi in a tropical headwater stream: Leaf decomposition and invertebrate feeding preference. *Microbial Ecology* 20:1–11.

Wright, M. S., and A. P. Covich. 2005b. The effect of macroinvertebrate exclusion on leaf breakdown rates in a tropical headwater stream. *Biotropica* 37:403–408.

Wright, S. J., and H. C. Mueller-Landau. 2006. The future of tropical forest species. *Biotropica* 38:287–301.

Wu, W., C. A. S. Hall, and F. N. Scatena. 2007. Modeling the impact of recent land cover changes on the stream flows in northeastern Puerto Rico. *Hydrological Processes* 21:2944–2956.

Wu, W., C. A. S. Hall, F. N. Scatena, and L. Quackenbush. 2006. Spatial modeling of evapotranspiration in the Luquillo experimental forest of Puerto Rico using remotely-sensed data. *Journal of Hydrology* 328:733–752.

Wunderle, J. M., Jr. 1995. Responses of bird populations in a Puerto Rican forest to Hurricane Hugo: The first 18 months. *Condor* 97:879–896.

Wunderle, J. M., Jr., D. J. Lodge, and R. B.Waide. 1992. Short-term effects of Hurricane Gilbert on terrestrial bird populations on Jamaica. *Auk* 109:148–166.

Xu, X., E. Hirata, T. Enoki, and Y. Tokashiki. 2004. Leaf litter decomposition and nutrient dynamics in a subtropical forest after typhoon disturbance. *Plant Ecology* 173:161–170.

Zalamea, M., G. González, C. L. Ping, and G. Michaelson. 2007. Soil organic matter dynamics under decaying wood in a subtropical wet forest: Effect of tree species and decay stage. *Plant Soil* 296:173–185.

Zimmerman, J. K. H., and A. P. Covich. 2003. Distribution of juvenile crabs (*Epilobocera sinuatifrons*) in two Puerto Rican headwater streams: Effects of pool morphology and past land-use legacies. *Archiv für Hydrobiologie* 158:343–357.

Zimmerman, J. K. H., and A. P. Covich. 2007. Damage and recovery of riparian sierra palms (*Prestoea acuminata* var. *montana*) after Hurricane Georges: Influence of topography, land use, and biotic characteristics. *Biotropica* 39:43–49.

Zimmerman, J. K., E. M. Everham III, R. B. Waide, D. J. Lodge, C. M. Taylor, and N. V. L. Brokaw. 1995a. Responses of tree species to hurricane winds in subtropical wet forest in Puerto Rico: Implications for tropical tree life histories. *Journal of Tropical Ecology* 82:911–922.

Zimmerman, J. K., W. M. Pulliam, D. J. Lodge, V. Quiñones-Orfila, N. Fetcher, S. Guzmán-Grajales, J. A. Parrotta, C. E. Asbury, L. R. Walker, and R. B. Waide. 1995b. Nitrogen immobilization by decomposing woody debris and the recovery of tropical wet forest from hurricane damage. *Oikos* 72:314–322.

Zimmerman, J. K., M. R. Willig, L. R. Walker, and W. L. Silver. 1996. Introduction: Disturbance and Caribbean ecosystems. *Biotropica* 28:414–423.

Index

Entries in **boldface** refer to illustrations.

Printed in the USA
CPSIA information can be obtained
at www.ICGtesting.com
LVHW060925280723
753705LV00005BA/22

9 780195 334692